EPA-600-R-16-236Fb
December 2016
www.epa.gov/hfstudy

Hydraulic Fracturing for Oil and Gas: Impacts from the Hydraulic Fracturing Water Cycle on Drinking Water Resources in the United States

Appendices

Office of Research and Development
U.S. Environmental Protection Agency
Washington, DC 20460

This page is intentionally left blank.

Disclaimer

This document has been reviewed in accordance with U.S. Environmental Protection Agency policy and approved for publication. Mention of trade names or commercial products does not constitute endorsement or recommendation for use.

Preferred citation: U.S. EPA (U.S. Environmental Protection Agency). 2016. Hydraulic Fracturing for Oil and Gas: Impacts from the Hydraulic Fracturing Water Cycle on Drinking Water Resources in the United States - Appendices. Office of Research and Development, Washington, DC. EPA/600/R-16/236Fb.

Contents

List of Tables

List of Figures

List of Text Boxes

This page is intentionally left blank.

Appendix A. The EPA's Study of the Potential Impacts of Hydraulic Fracturing for Oil and Gas on Drinking Water Resources

This page is intentionally left blank.

Appendix A. The EPA's Study of Hydraulic Fracturing for Oil and Gas and Its Potential Impact on Drinking Water Resources

In 2009, at the urging of the U.S. Congress, the EPA initiated a study of hydraulic fracturing for oil and gas and its relationship to drinking water resources (hereafter the EPA's hydraulic fracturing study). The national study culminates with this report, the *Hydraulic Fracturing for Oil and Gas: Impacts from the Hydraulic Fracturing Water Cycle on Drinking Water Resources in the United States*.

The EPA's hydraulic fracturing study consisted of many elements. It included independent research projects conducted by EPA scientists and contractors, and involved the analysis of existing data, scenario and modeling evaluations, laboratory studies, toxicological assessments, and case studies. A list of the ensuing EPA publications is presented in Table A-1. The EPA's hydraulic fracturing study also included the development of this report, which is a state-of-the-science synthesis of available data and information, as well as the EPA's own research.

Throughout, the EPA consulted with the Agency's independent Science Advisory Board (SAB) on the scope of its hydraulic fracturing study and the progress made on each of the research projects. The timeline of this work is presented in Figure A-1. The SAB also conducted a peer review of both the EPA's *Plan to Study the Potential Impacts of Hydraulic Fracturing on Drinking Water Resources* (U.S. EPA, 2011a, hereafter Study Plan) and the *Hydraulic Fracturing for Oil and Gas: Impacts from the Hydraulic Fracturing Water Cycle on Drinking Water Resources in the United States*, as described in Chapter 1.

Stakeholder engagement also played an important role in the development and implementation of the EPA's hydraulic fracturing study. The EPA held public meetings across the United States to hear feedback from stakeholders on the proposed study design and scope. In addition, while conducting the hydraulic fracturing study, the EPA engaged with technical, subject-matter experts on relevant topics in a series of technical workshops and roundtables (Figure A-1).

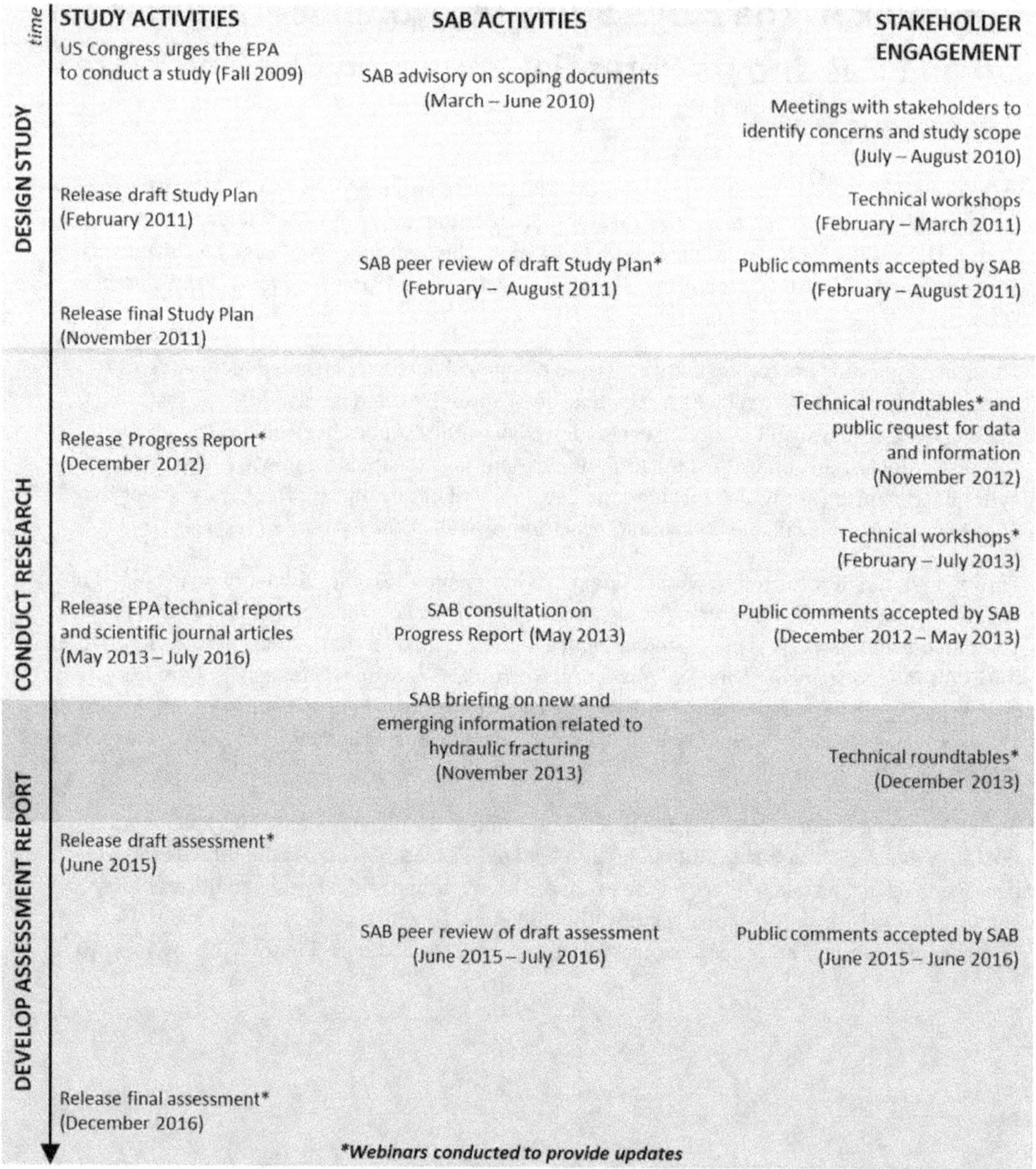

time	STUDY ACTIVITIES	SAB ACTIVITIES	STAKEHOLDER ENGAGEMENT
DESIGN STUDY	US Congress urges the EPA to conduct a study (Fall 2009)	SAB advisory on scoping documents (March – June 2010)	
			Meetings with stakeholders to identify concerns and study scope (July – August 2010)
	Release draft Study Plan (February 2011)		Technical workshops (February – March 2011)
		SAB peer review of draft Study Plan* (February – August 2011)	Public comments accepted by SAB (February – August 2011)
	Release final Study Plan (November 2011)		
CONDUCT RESEARCH	Release Progress Report* (December 2012)		Technical roundtables* and public request for data and information (November 2012)
			Technical workshops* (February – July 2013)
	Release EPA technical reports and scientific journal articles (May 2013 – July 2016)	SAB consultation on Progress Report (May 2013)	Public comments accepted by SAB (December 2012 – May 2013)
DEVELOP ASSESSMENT REPORT		SAB briefing on new and emerging information related to hydraulic fracturing (November 2013)	Technical roundtables* (December 2013)
	Release draft assessment* (June 2015)		
		SAB peer review of draft assessment (June 2015 – July 2016)	Public comments accepted by SAB (June 2015 – June 2016)
	Release final assessment* (December 2016)		

Webinars conducted to provide updates

Figure A-1. Timeline of activities in the EPA's hydraulic fracturing study.
On the left are activities related to the development of products from the EPA's hydraulic fracturing study, in the center are interactions between the EPA and the SAB, and on the right are stakeholder engagement activities.

A.1.   The EPA's Hydraulic Fracturing Study Publications Cited in This Assessment

In this section, we provide a table of publications that were completed as part of the EPA's hydraulic fracturing study and cited in this assessment. We also indicate projects that were originally part of the Study Plan but that did not result in a publication. The full list of publications under the EPA's hydraulic fracturing study is updated and available at https://www.epa.gov/hfstudy.

Table A-1. Titles, descriptions, and citations for the EPA's hydraulic fracturing study publications cited in this assessment.

Research project	Description	Citations/Notes
Analysis of existing data		
Literature Review	Review and assessment of existing papers and reports, focusing on peer-reviewed literature	Literature review is incorporated into this assessment.
Spills Database Analysis	Characterization of hydraulic fracturing-related spills using information obtained from selected state and industry data sources	U.S. EPA (2015j)
Service Company Analysis	Analysis of information provided by nine hydraulic fracturing service companies in response to a September 2010 information request on hydraulic fracturing operations	Analysis of data received is incorporated into this assessment.[a]
Well File Review	Analysis of information provided by nine oil and gas operators in response to an August 2011 information request for 350 well files	U.S. EPA (2015k) U.S. EPA (2016a) Analysis of data received is also incorporated into this assessment.[b]
FracFocus Analysis	Analysis of water and chemical use data for hydraulic fracturing wells compiled from FracFocus 1.0, the national hydraulic fracturing chemical registry operated by the Ground Water Protection Council and the Interstate Oil and Gas Compact Commission	U.S. EPA (2015a) U.S. EPA (2015b) U.S. EPA (2015c)
Scenario evaluations		
Subsurface Migration Modeling	Numerical modeling of subsurface fluid migration scenarios that explore the potential for fluids, including liquids and gases, to move from the fractured zone to drinking water aquifers	Kim and Moridis (2013) Kim et al. (2014) Kim and Moridis (2015) Kim et al. (2016) Reagan et al. (2015) Rutqvist et al. (2013) Rutqvist et al. (2015)

Research project	Description	Citations/Notes
Surface Water Modeling	Modeling of concentrations of selected chemicals at public water supplies downstream from wastewater treatment facilities that discharge treated hydraulic fracturing wastewater to surface waters	Weaver et al. (2016)
Water Availability Modeling	Assessment and modeling of current and future scenarios exploring the impact of water usage for hydraulic fracturing on drinking water availability in the Upper Colorado River Basin and the Susquehanna River Basin	U.S. EPA (2015d)
Laboratory studies		
Source Apportionment Studies	Identification and quantification of the source(s) of high bromide and chloride concentrations at public water supply intakes downstream from wastewater treatment plants discharging treated hydraulic fracturing wastewater to surface waters	U.S. EPA (2015l)
Wastewater Treatability Studies	Assessment of the efficiency of common wastewater treatment processes on removing selected chemicals found in hydraulic fracturing wastewater	None
Br-DBP Precursor Studies	Assessment of the ability of bromide and brominated compounds present in hydraulic fracturing wastewater to form brominated disinfection byproducts (Br-DBPs) during drinking water treatment processes	None
Analytical Method Development	Development of analytical methods for selected chemicals found in hydraulic fracturing fluids or wastewater	DeArmond and DiGoregorio (2013a) DeArmond and DiGoregorio (2013b) U.S. EPA (2014b) U.S. EPA (2014f)
Toxicity assessment		
Toxicity Assessment	Toxicity assessment of chemicals reportedly used in hydraulic fracturing fluids or found in hydraulic fracturing wastewater	Yost et al. (2016a) Yost et al. (2016b) Yost et al. (In Press)
Case studies		
Retrospective case studies: _Investigations of whether reported drinking water impacts may be associated with or caused by hydraulic fracturing activities_		
Las Animas and Huerfano Counties, Colorado	Investigation of potential drinking water impacts from coalbed methane extraction in the Raton Basin	U.S. EPA (2015h)

Research project	Description	Citations/Notes
Dunn County, North Dakota	Investigation of potential drinking water impacts from a well blowout during hydraulic fracturing for oil in the Bakken Shale	U.S. EPA (2015f)
Bradford County, Pennsylvania	Investigation of potential drinking water impacts from shale gas development in the Marcellus Shale	U.S. EPA (2014d)
Washington County, Pennsylvania	Investigation of potential drinking water impacts from shale gas development in the Marcellus Shale	U.S. EPA (2015g)
Wise County, Texas	Investigation of potential drinking water impacts from shale gas development in the Barnett Shale	U.S. EPA (2015i)

Prospective case studies *Investigation of potential impacts of hydraulic fracturing through collection of samples from a site before, during, and after well pad construction and hydraulic fracturing*

The EPA was unable to find suitable locations that met both the scientific criteria of a rigorous prospective study and the business needs of potential partners.

[a] Data received and incorporated into this document is cited as: U.S. EPA (U.S. Environmental Protection Agency). (2013). Data received from oil and gas exploration and production companies, including hydraulic fracturing service companies 2011 to 2013. Non-confidential business information source documents are located in Federal Docket ID: EPA-HQ-ORD2010-0674. Available at http://www.regulations.gov.

[b] Data received and incorporated into this document is cited as: U.S. EPA (U.S. Environmental Protection Agency). (2011). Sampling data for flowback and produced water provided to EPA by nine oil and gas well operators (non-confidential business information). US Environmental Protection Agency. http://www.regulations.gov/#!docketDetail;rpp=100;so=DESC;sb=docId;po=0;D=EPA-HQ-ORD-2010-0674.

A.2. Answers to the Secondary Research Questions

The EPA's Study Plan (U.S. EPA, 2011a) was organized around the five stages of the hydraulic fracturing water cycle. Each stage of the hydraulic fracturing water cycle was associated with a primary research question (Figure A-2). Nested within each primary research question was a set of secondary research questions. The primary and secondary research questions provided a framework for exploring how hydraulic fracturing water cycle activities could potentially impact drinking water resources. Research projects, undertaken using different types of research approaches (i.e., analysis of existing data, scenario evaluations, laboratory studies, toxicity assessment, and case studies), were designed to provide information relevant to answering the secondary research questions.

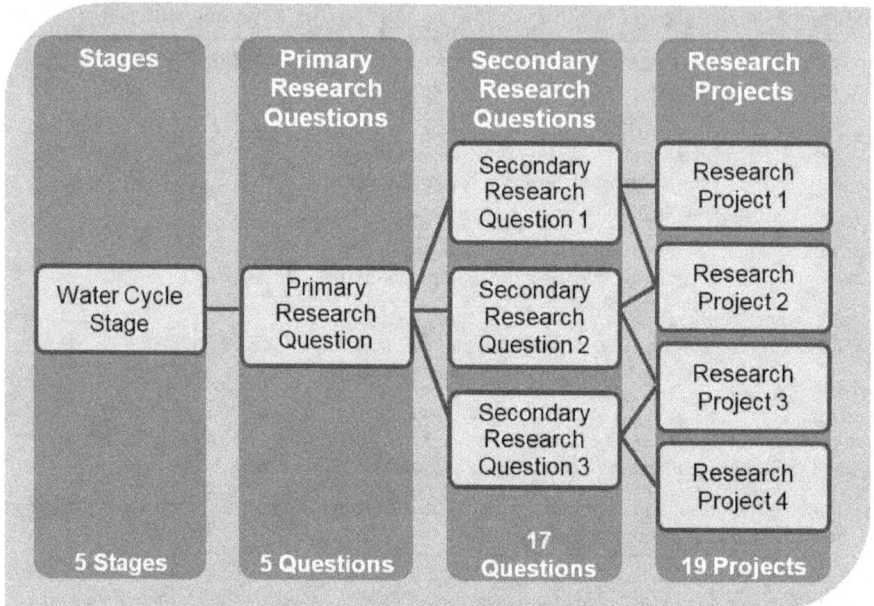

Figure A-2. Structure of the EPA's hydraulic fracturing study.
This diagram shows the generalized elements of the study and how they relate to one another.

The primary research questions included:

- Water acquisition: What are the potential impacts of large volume water withdrawals from groundwater and surface water on drinking water resources?

- Chemical mixing: What are the possible impacts of hydraulic fracturing fluid surface spills on or near well pads on drinking water resources?

- Well injection: What are the possible impacts of the injection and fracturing process on drinking water resources?

- Produced water handling: What are the possible impacts of flowback and produced water (collectively referred to as "hydraulic fracturing wastewater") surface spills on or near well pads on drinking water resources?

- Wastewater disposal and reuse: What are the possible impacts of inadequate treatment of hydraulic fracturing wastewater on drinking water resources?

In this section we present answers to the secondary research questions posed in the Study Plan as a way of providing continuity between the Study Plan and this assessment. Answers were informed by the knowledge accumulated and synthesized from the EPA's hydraulic fracturing study, including the scientific literature reviewed for this assessment.

A.2.1. Water Acquisition

- *What are the types of water used for hydraulic fracturing?*

The three major types of water used for hydraulic fracturing are surface water, groundwater, and reused hydraulic fracturing wastewater. Because trucking can be a major expense, operators tend to use water sources as close to the well pad as possible. Operators usually self-supply surface water or groundwater directly, but may also obtain water through public water systems or other suppliers. Hydraulic fracturing operations in the eastern United States rely predominantly on surface water, whereas operations in more semi-arid to arid western states use either surface water or groundwater. In some areas of the country, operators rely entirely on groundwater supplies (e.g., western Texas).

Fresh water (from both surface water and groundwater sources) currently supplies the vast majority of water used for hydraulic fracturing. However, the reuse of hydraulic fracturing wastewater for injection reduces the demand on fresh water sources. Nationally, the proportion of water used in hydraulic fracturing that comes from reused hydraulic fracturing wastewater is generally low; in a survey of literature values from 10 states, basins, or plays, we found a median value of 5%, with this percentage varying by location (Table 4-2). Available data on reuse trends indicate increasing reuse of wastewater over time in both Pennsylvania and West Virginia, likely due to the lack of nearby disposal options. Reuse as a percentage of water injected is typically lower in other areas, in part because of the availability of disposal wells (Chapter 8).

- *How much water is used per well?*

The median amount of water used nationally per hydraulically fractured well was approximately 1.5 million gal (5.7 million L) in 2011 and through early 2013 based on disclosures to FracFocus (U.S. EPA, 2015a, c). This increased to approximately 2.7 million gal (10.2 million L) in 2014, driven by a proportional increase in horizontal wells that, on average, use more water per well (estimated from data reported in Gallegos et al., 2015) (Figure 4-1). These national estimates represent a variety of fractured well types, including types requiring much less water per well than horizontal shale gas wells. Thus, published estimates for horizontal shale gas wells are typically higher (e.g., approximately 4 million gal (15 million L) per well (Vengosh et al., 2014), and should not be applied to all fractured wells to derive national estimates.

There was also wide variation within and among states and basins in the median per well water volumes reported in 2011 and 2012, from more than 5 million gal (19 million L) in Arkansas and Louisiana to less than 1 million gal (3.8 million L) in Colorado, Wyoming, Utah, New Mexico, and California (U.S. EPA, 2015a). This variation results from several factors, including geology, well length, and fracturing fluid formulation.

- *How might cumulative water withdrawals for hydraulic fracturing affect drinking water quantity?*

Hydraulic fracturing uses billions of gallons of water every year at the national and state scales, and even in some counties. When expressed relative to total water use or consumption at these scales, however, hydraulic fracturing generally accounts for only a small percentage, usually less than 1%. These percentages are higher in specific counties. Annual hydraulic fracturing water use was 10% or more compared to 2010 total water use in 6.5% of counties with FracFocus disclosures in 2011 and 2012, 30% or more in 2.2% of counties, and 50% or more in 1.0% of counties (see Table B-2). Consumption estimates follow the same pattern, with higher percentages in each category: hydraulic fracturing water consumption was 10%, 30%, and 50% or more of 2010 total water consumption in 13.5%, 6.2%, and 4.0% of counties with FracFocus disclosures (see Table B-2). Thus, hydraulic fracturing represents a relatively large user and consumer of water in these counties.

Whether water quantity impacts occur from water acquisition for hydraulic fracturing depends on the balance between water withdrawals and availability. From our survey of the literature and our county level assessments, southern and western Texas appear to have the highest potential for impacts, of the areas assessed in this chapter, given the combination of high hydraulic fracturing water use, relatively low water availability, intense periods of drought, and reliance on declining groundwater resources. Importantly, our results do not preclude the possibility of local water quantity impacts in areas with comparatively lower potential, nor do they necessarily mean impacts have occurred in the high potential areas. Our survey provides an indicator of areas with higher *potential* for impacts, and could be used to target resources for future studies.

Local impacts to drinking water resources have occurred in areas with increased hydraulic fracturing activity. In a detailed case study, Scanlon et al. (2014) observed generally adequate water supplies for hydraulic fracturing in the Eagle Ford play in southern Texas, except in specific locations. They found excessive drawdown of groundwater locally, with estimated declines of approximately 100 to 200 ft (30 to 60 m) in a small proportion of the play (~6% of the area) after hydraulic fracturing activity increased in 2009. In 2011, drinking water wells in an area overlapping the Haynesville Shale ran out of water due to higher-than-normal groundwater withdrawals and drought (LA Ground Water Resources Commission, 2012). Hydraulic fracturing water use likely contributed to these conditions, along with other water users and the lack of precipitation. By contrast, two EPA case studies in the Upper Colorado and the Susquehanna River Basins found minimal impacts from current hydraulic fracturing water withdrawals (U.S. EPA, 2015d) (Sections 4.5. and 4.5). These site-specific findings emphasize the need to focus on regional and local dynamics when considering the impacts from hydraulic fracturing water withdrawals.

- *What are the possible impacts of water withdrawals for hydraulic fracturing on water quality?*

Water withdrawals for hydraulic fracturing, similar to all water withdrawals, have the potential to alter the quality of drinking water resources. Groundwater withdrawals exceeding natural recharge rates decrease water storage in aquifers, potentially mobilizing contaminants or allowing the infiltration of lower-quality water from the land surface or adjacent formations. Pumping can also promote changes in reduction-oxidation (redox) conditions and mobilize chemicals from geologic sources (e.g., uranium). Withdrawals can also decrease groundwater discharge to streams, potentially affecting surface water quality. Areas with declining groundwater resources, particularly in drought-prone regions, are most likely to experience water quality impacts from hydraulic fracturing water withdrawals.

Surface water withdrawals also have the potential to affect water quality, particularly in smaller streams. Withdrawals may lower water levels and alter stream flow, decreasing a stream's capacity to dilute contaminants. Studies by the EPA show that streams can be vulnerable to changes in water quality due to water withdrawals, most notably smaller streams or during periods of low flow (U.S. EPA, 2015d). Managing the rate and timing of surface water withdrawals (e.g., passby flows) can help mitigate potential impacts on water quality.

A.2.2. Chemical Mixing

- *What is currently known about the frequency, severity, and causes of spills of hydraulic fracturing fluids and additives?*

There has not been much work on the frequency of spills of hydraulic fracturing fluids and additives. Using spills data from three states (Pennsylvania, Colorado, and North Dakota), there is an estimated median of 2.6 reported spills for every 100 wells, with a range of 0.4 to 12.2. These values are uncertain because these rates used different criteria for including a spill, what the denominator is for the well type (e.g., drilled or finished), and includes more than hydraulic fracturing fluids and additives. Using solely the North Dakota database, we estimate 2.6 reported spills of injected fluid or chemical per 100 wells fractured (Rahm et al., 2015; U.S. EPA, 2015j; Brantley et al., 2014; Gradient, 2013).[1] Estimates of the frequency of on-site spills from hydraulic fracturing operations were unavailable for other areas. It is unknown whether these spill estimates are representative of national occurrences.

The severity of a spill depends on several factors, including: spill amount (mass, volume, concentration), the fate and transport of the spill, if it reaches a water resource, the characteristics of the receiving water resource, and the hazard associated with the chemicals themselves. There is little known on the severity of hydraulic fracturing fluid and additive spills. The reported volume of chemicals or hydraulic fracturing fluid spilled range of 5 to 19,320 gal (19 to 73,130 L), with a median volume of 420 gal (1,600 L) per spill. Spill reports contain little information on chemical-specific spill composition. Spilled fluids were often described by their additive type (e.g., acids, biocides, friction reducers, cross-linkers, gels,) or as a blended hydraulic fracturing fluid. Specific chemicals mentioned in spill reports included hydrochloric acid and potassium chloride.

Spill causes included equipment failure, human error, failure of container integrity, and other (e.g., weather and vandalism). The most common cause was equipment failure. Equipment failure included blowout preventer failure, corrosion, and failed valves. More than 30% of the chemical or hydraulic fracturing fluid spills characterized by the EPA came from fluid storage units (e.g., tanks, totes, and trailers) (U.S. EPA, 2015j).

---

[1] Spill frequency estimates are for a given number of wells over a given period of time. These are not annual estimates nor are they for over a lifetime of the wells.

- *What are the identities and volumes of chemicals used in hydraulic fracturing fluids, and how might this composition vary at a given site and across the country?*

The EPA has identified 1,084 different chemicals used in chemical mixing. A recent study of FracFocus disclosure data reported an additional 263 new CASRNs, increasing the total number of chemicals identified as used by approximately 24% (Konschnik and Dayalu, 2016). Industry use of confidential business information (CBI) is one factor that likely limits the completeness of these chemical lists. The EPA's analysis of disclosures to FracFocus 1.0 found that 11% of ingredients were reported to FracFocus as CBI (U.S. EPA, 2015a), and the more recent analysis by Konschnik and Dayalu (2016) indicated a 5.6% increase in the number of CBI ingredients.

Hydraulic fracturing chemicals cover a wide range of chemical classes and a wide range of physicochemical properties. The chemicals include acids, aromatic hydrocarbons, bases, hydrocarbon mixtures, polymers, and surfactants. Thirty-two chemicals, excluding water, quartz, and sodium chloride, have been reported to be used at 10% or more sites. The ten most common chemicals (excluding quartz) are methanol, hydrotreated light petroleum distillates, hydrochloric acid, isopropanol, ethylene glycol, peroxydisulfuric acid diammonium salt, sodium hydroxide, guar gum, glutaraldehyde, and propargyl alcohol. These chemicals can be present in multiple additives. Methanol, hydrotreated light petroleum distillates, and hydrochloric acid are the three chemicals reported to be used in more than half of all frac jobs, with methanol being used at 72% of all sites. Operators used a median of 14 unique chemicals per well according to the EPA's analysis of disclosures to FracFocus 1.0 (U.S. EPA, 2015a).

The composition of hydraulic fracturing fluids varies by state, by well, and within the same service company and geologic formation. This variability likely results from several factors, including the geology of the formation, production goals, the availability and cost of different chemicals, and operator preference (U.S. EPA, 2015a).

The estimated median volumes of individual chemicals injected per well ranged from a few gallons to thousands of gallons, with a median of 650 gal (2,500 L) per chemical per well (U.S. EPA, 2015c). There is an estimated 9,100 gal (34,000 L) to 30,000 gal (114,000L) of chemicals used per well.

- *What are the chemical, physical, and toxicological properties of hydraulic fracturing chemical additives?*

The EPA identified 1,084 different chemicals reported to be used in hydraulic fracturing fluid from 2005 to 2013. Of these, 455 (more than 40%) were individual organic chemicals with physicochemical properties that vary, from fully miscible to insoluble and from highly hydrophobic to highly hydrophilic. We were able to estimate the physicochemical properties of these 455 chemicals using the EPA's Estimation Program Interface (EPI) Suite™ software. Of the 20 most frequently used chemicals, three have low mobility: (1) distillates, petroleum, hydrotreated light; (2) solvent naphtha, petroleum, heavy aromatic; and, (3) naphthalene. These chemicals have the potential to act as long term sources of contamination if spilled on-site.

The chemicals with determinable physicochemical properties were not necessarily the chemicals most frequently reported as used in hydraulic fracturing fluids or activities. Of the 455 chemicals for which physicochemical properties were available, 18 of the top 20 most mobile chemicals were included in 2% or less of disclosures (U.S. EPA, 2015c). However, two highly mobile chemicals, choline chloride and tetrakis (hydroxymethyl) phosphonium sulfate were reported in 14% and 11% of disclosures, respectively. These two chemicals are relatively more common, and, if spilled, would move quickly through the environment with the flow of water.

Of the 1,084 chemicals identified by the EPA as used in hydraulic fracturing fluids, chronic oral RfVs and/or OSFs from selected federal, state, and international sources were available for 98 (9%) of these chemicals. From the federal sources alone, chronic oral RfVs were available for 81 chemicals (7%), and OSFs were available for 15 chemicals (1%). Chronic oral RfVs and OSFs from these selected sources were not available for the majority of chemicals used in hydraulic fracturing fluid, representing a potential data gap with regard to hazard identification. Of the chemicals that have these selected toxicity values, health effects associated with chronic oral exposure include the potential for carcinogenesis, immune system effects, changes in body weight, changes in blood chemistry, cardiotoxicity, neurotoxicity, liver and kidney toxicity, and reproductive and developmental toxicity.

When considering the hazard evaluation of these chemicals on a nationwide scale, chemicals such as propargyl alcohol stand out for their relatively low RfVs, high frequency of use, and expected transport and mobility in water. However, the EPA's analysis of disclosures to FracFocus 1.0 indicates that most chemicals are used infrequently on a nationwide scale. Potential exposures to the majority of these chemicals are likely to be a local issue, rather than a national one. Accordingly, potential hazard and risk considerations for hydraulic fracturing fluid additives are best made on a site-specific, well-specific basis.

- *If spills occur, how might hydraulic fracturing chemical additives contaminate drinking water resources?*

The potential for spilled fluids to contaminate groundwater or surface water resources depends on the characteristics of the spill, the environmental fate and transport of the spilled fluid, and spill response activities. Spill characteristics (e.g., the volume and chemical composition of the spilled fluid) describe the identity and volume of chemicals that enter the environment due to a spill. The environmental fate and transport of the spilled fluid describes how spilled chemicals move and transform in the environment. Spill response activities include actions designed to remove spilled fluids from the environment. Because all of these factors influence whether spilled fluids reach groundwater and surface water resources, they affect the frequency and severity of potential impacts to drinking water resources from spills during the chemical mixing stage of the hydraulic fracturing water cycle.

The movement of spilled hydraulic fracturing fluids and additives through the environment is difficult to assess, because of the site-specific and chemical-specific nature of spills and because hydraulic fracturing-related spills typically involve complex mixtures of chemicals. In the absence of site-specific studies of actual spills, we relied on fundamental environmental fate and transport principles to describe how hydraulic fracturing fluids and chemicals used in hydraulic fracturing fluids can move through the environment to drinking water resources.

The environmental fate and transport of hydraulic fracturing fluids and chemicals depend on site-specific environmental conditions and the physicochemical properties of the chemicals spilled. Site-specific environmental characteristics can affect how spilled liquids move through soil into the subsurface or over the land surface. Generally, highly permeable soils or preferential flow paths can allow spilled liquids to move quickly into and through the subsurface, limiting the opportunity for spilled liquids to move over land to surface water resources. When spilled liquids move underground, the distance between the land surface and the groundwater resource can affect whether spilled liquids reach groundwater. Large spills volumes are more likely to be able to travel the distance between the land surface and the groundwater resource and impact the latter. In low permeability soils, spilled liquids are less able to move into the subsurface and are more likely to move over the land surface. When spilled liquids move over the land surface, the volume spilled and the distance between the source of the spill and nearby surface water resources can affect whether the spilled liquid reaches surface water.

A.2.3. Well Injection

- *How effective are current well construction practices at containing fluids—both liquids and gases—before, during, and after fracturing?*

A well will be exposed to the highest stress during the relatively brief phase of injection for hydraulic fracturing. If the well cannot withstand these stresses, the casing or cement can fail, resulting in the unintended movement of hydraulic fracturing fluids or naturally-occurring liquids or gases into the surrounding environment and, potentially, an impact on drinking water quality. These failures can be the result of inadequate design and/or construction, or degradation of the casing and/or cement that allows fluid to move laterally from inside the well to the formation or vertically along the wellbore from the production zone to shallower drinking water resources.

The presence of multiple layers of casing strings can isolate and protect geologic zones containing drinking water. Most wells used in hydraulic fracturing operations are designed with one or more of these layers of casing.

Cementing of the surface casing to below the lowest drinking water resource is a key protective measure to prevent hydraulic fracturing fluids, or other fluids, from reaching drinking water resources. Most states require this (GWPC, 2014); however, our data indicate adequate casing and/or cement are not present in all wells. For example, studies in Wyoming and Colorado have documented wells with partially uncemented surface casing (Fleckenstein et al., 2015; WYOGCC, 2014).

The presence of properly placed, adequate cement in those portions of the well that intersect porous or permeable water- and/or hydrocarbon-bearing zones can also prevent fluids from moving into drinking water resources. Wells with cement that does not resist formation or operational stresses have the potential to promote unintended subsurface fluid movement. In Bainbridge Township, Ohio, hydraulic fracturing was performed in a well with improperly emplaced and inadequate cement. This resulted in natural gas movement upward along the wellbore, contamination of the drinking water aquifer, and the loss of 26 private drinking water wells (Bair et al., 2010).

Even in optimally designed and constructed wells, metal casings and cement can degrade over time—either as a result of aging or stresses exerted over years of operations—and affect the integrity of the well. We have limited access to data and information regarding the degree to which the integrity of wells is verified before or after hydraulic fracturing operations.

- *Can subsurface migration of fluids—both liquids and gases—to drinking water resources occur, and what local geologic or artificial features might allow this?*

The presence of artificial penetrations, such as inadequately constructed or degraded offset wells or undetected abandoned wells near the well undergoing hydraulic fracturing, can provide pathways that allow fluid movement to drinking water resources. If the fractures created during hydraulic fracturing intersect a nearby, previously-fractured production well or its fracture network, hydraulic fracturing fluids or other fluids can move to that well in an event known as well communication or a "frac hit" (Jackson et al., 2013a). Instances of well communication have occurred in New Mexico (Vaidyanathan, 2014) and Texas (Craig et al., 2012). Additionally, abandoned wells near a well undergoing hydraulic fracturing can provide a pathway for vertical fluid movement to drinking water resources, if those wells were not properly plugged or the plugs and cement have degraded over time. This can be a significant issue in areas with legacy (i.e., historic) oil and gas exploration and when wells are re-entered and fractured (or re-fractured) to increase production in a reservoir.

Some hydraulic fracturing operations involve the injection of fluids into formations with relatively limited vertical separation from drinking water resources. Where the separation between the production zone and drinking water resource is small, and where natural or induced fractures transecting the layers between these formations are present, there is an increased potential for impacts to drinking water quality.

Hydraulic fracturing is also performed within formations that meet the salinity threshold used in some definitions of a drinking water resource, in addition to the broader definition of a drinking water resource developed for this assessment. By definition, these hydraulic fracturing operations affect the quality of the drinking water resources.

A.2.4. Produced Water Handling

- *What is currently known about the frequency, severity, and causes of spills of flowback and produced water?*

Surface spills of produced water from unconventional oil and gas production have occurred across the country. Some produced water spills have affected drinking water resources, including private drinking water wells. Analysis of data from North Dakota suggests a produced water spill rate of 5 to 7 spills per 100 active production wells. Of these, an estimated 84% are confined to the production or exploration facility and expected to have a lower potential to impact drinking water resources. Half of the spills are estimated to be less than 1,000 gal (3,800 L), but a small number of large spills have occurred. For example, in North Dakota in 2015, there were 12 releases of 21,000 gal (79,000 L) or more out of a total of 609 spills. The largest reported spill was 2.9 million gal (11.0 million L). The causes identified for these spills are container and equipment failures, human error, well communication, blowouts, pipeline leaks, and dumping. Although specific impacts from a few spills have been documented, the severity of most spills is unknown.

- ***What is the composition of hydraulic fracturing flowback and produced water, and what factors might influence this composition?***

The geochemical content of water flowing back initially reflects injected fluids. After initial flowback, returning fluid geochemistry shifts to reflect the geochemistry of formation waters and formation solids. According to the available literature and data, conventional and unconventional produced water content are often similar with respect to the occurrence and concentration of many constituents. Much produced water is generally characterized as saline (with the exception of most coalbed methane produced water) and enriched in major anions, cations, metals, naturally occurring radionuclides, and organics. The composition of produced water must be determined through sampling and analysis, both of which have limitations. Sampling limitations include equipment configurations that make it difficult to access representative fluids. Analytical limitations include identifying target analytes in advance, without sufficient knowledge of the composition of the fluid sampled, as well as the lack of appropriate analytical methods.

Typically, unconventional produced water contains low levels of heavy metals. However, elevated strontium and barium levels are characteristic of Marcellus Shale produced water. Elevated levels of technologically enhanced naturally occurring radioactive materials (TENORM) have also been documented in the Marcellus Shale produced water. Other formations also contain TENORM, but fewer data are available. Composition data were limited, in general. Most of the available data on produced water content were for shale formations and CBM basins, while few data were available for sandstone formations.

Recent published research has identified several hundred organic chemicals in produced water. Many of these are naturally-occurring constituents of petroleum, while fewer are known hydraulic fracturing chemicals. The identification of many organic chemicals in produced water depends on the availability of advanced laboratory analytical methods and equipment. Much less is known about subsurface transformation products and only a few have been identified. Recent research shows that subsurface transformation reactions may reduce concentrations of some hydraulic fracturing additives through oxidation (gelling agents and friction reducers), may create chlorinated and brominated organic compounds, and that surfactants (i.e., glycols) may be resistant to degradation and remain in produced water.

Hydraulic fracturing flowback and produced water composition is influenced by the composition of injected hydraulic fracturing fluids, the targeted geological formation and associated hydrocarbon products, the stratigraphic environment, and subsurface processes and residence time. Spatial variability of produced water content occurs between plays of different rock type (e.g., coal vs. sandstone), between plays of the same rock type (e.g., Barnett Shale vs. Bakken Shale), and within formations of the same source rock (e.g., northeastern vs. southwestern Marcellus Shale).

- *What are the chemical, physical, and toxicological properties of hydraulic fracturing flowback and produced water constituents?*

This assessment identified 599 chemicals that are reported to have been detected in hydraulic fracturing produced water. These include chemicals that are added to hydraulic fracturing fluids during the chemical mixing stage, as well as naturally occurring organic chemicals, metals, naturally occurring radioactive material, and other subterranean chemicals that may be mobilized by the hydraulic fracturing process.

The identified constituents of produced water include inorganic chemicals (cations and anions in the form of metals, metalloids, non-metals, and radioactive materials), organic chemicals and compounds, and unidentified materials measured as TOC (total organic carbon) and DOC (dissolved organic carbon). Some constituents are readily transported with water (i.e., chloride and bromide), while others depend strongly on the geochemical conditions in the receiving water body (i.e., radium and barium), and assessment of their transport is based on site-specific factors. Using the EPA's EPI Suite software, we were able to obtain actual or estimated physicochemical properties for 521 (87%) individual organic chemicals of the 599 chemicals identified in produced water. The EPI Suite™ results are constrained by their applicability to one temperature (25 °C), and salinity (low). Temperature changes impact Henry's law constant, K_{ow}, and solubility, and depend on the characteristics of the chemical and ions present. In some cases, the effect changes exponentially with salinity. Therefore, property values that depart from the EPI Suite™ values are expected for the 599 chemicals identified in produced water at elevated temperature and salinity. Although little is known concerning attenuation of hydraulic fracturing fluid constituents, Kekacs et al. (2015) report that salinity above 40,000 mg/L initially inhibited aerobic degradation of the organic constituents of a synthetic fracturing fluid (for 6.5 days), even though the bacterial communities were pre-acclimated to the salts.

Of the 599 chemicals identified by the EPA as detected in produced water, chronic oral RfVs and/or OSFs from selected federal, state, and international sources were available for 120 (20%) of these chemicals. From the federal sources alone, chronic oral RfVs were available for 97 chemicals (16%), and OSFs were available for 30 chemicals (5%). Of the chemicals that have these selected toxicity values, health effects associated with chronic oral exposure include the potential for carcinogenesis, immune system effects, changes in body weight, changes in blood chemistry, pulmonary toxicity, neurotoxicity, liver and kidney toxicity, and reproductive and developmental toxicity.

In a hazard evaluation of produced water data, chemicals such as benzene, pyridine, and naphthalene stood out for their relatively lower RfVs, high average concentrations, and expected transport and mobility in water. However, the chemicals present in produced water are likely to vary on a regional and well-specific basis as a result of geological differences, as well as differences between hydraulic fracturing fluid formulations. Therefore, potential hazard and risk considerations are best made on a site-specific basis.

- *If spills occur, how might hydraulic fracturing flowback and produced water contaminate drinking water resources?*

Both the scientific literature and published reports have shown that produced water spills have impacted drinking water resources. Spills of produced water may impact drinking water resources if the spill or release is of sufficient volume and duration to reach the resource at a sufficient concentration. During the first few months of production produced water is most likely to contain hydraulic fracturing additives and have low salinity. Later, the composition of shale-gas produced water will be dominated by high salinity. Spilled produced water can flow overland to reach surface water resources. Some of that water might infiltrate to impact soils and groundwater. Which path the spill takes depends on different conditions, such as the distance to a water receptor, spill volume, soil characteristics, and the physicochemical properties of the chemical. Of the produced water spills documents by the EPA, 17 (8%) reached surface water resources and 1 (0.4%) was documented to reach groundwater (U.S. EPA, 2015j), although groundwater impacts from 107 additional spills were unknown. More spills (141 or 63%) impacted soil, and the impacts of 30 spills were unknown.

A.2.5. Wastewater Disposal and Reuse

- *What are the common treatment and disposal methods for hydraulic fracturing wastewater, and where are these methods practiced?*

The majority of hydraulic fracturing wastewater in the United States is disposed of via underground injection wells. As of 2014–2015, most states where hydraulic fracturing occurs have access to an adequate number of Class IID injection wells regulated under the Underground Injection Control (UIC) Program. The Marcellus Shale region, especially the northeastern region, is an exception. Due to the lack of available injection wells, wastewater reuse, with or without treatment beforehand (at centralized waste treatment facilities (CWTs) or mobile facilities), is currently the primary means of wastewater management and may continue to increase in western shale plays as the practice becomes encouraged and economically favorable. Other methods of management used to a lesser degree include evaporation and agricultural use (for low-total dissolved solids (TDS) wastewater), both of which occur in the western United States.

- *How effective are conventional POTWs and commercial treatment systems in removing organic and inorganic contaminants of concern in hydraulic fracturing wastewater?*

Publicly owned treatment works (POTWs) using basic treatment processes cannot effectively reduce TDS concentrations in highly saline hydraulic fracturing wastewater. CWTs that use advanced treatment processes such as mechanical vapor recompression, distillation, and reverse osmosis have been shown to remove TDS constituents with removal efficiencies ranging from 97% to over 99% (Table F-4). These advanced treatment processes can also remove other constituents found in hydraulic fracturing wastewater such as metals, cations, anions, and some organics.

Indirect discharge, where wastewater is pretreated by a CWT and sent to a POTW, may be an effective option for hydraulic fracturing wastewater treatment (with restrictions on contaminant concentrations in the pretreated wastewater). This option would require careful planning to ensure that the pretreated wastewater blended with POTW influent is of appropriate quality to prevent deleterious effects on biological processes in the POTW or the pass-through of contaminants.

Facilities that treat wastewater for reuse and employ only basic treatment are unable to remove all contaminants in hydraulic fracturing wastewater, especially if the CWTs do not include specific processes (e.g., distillation, advanced oxidation, adsorption) that target constituents of concern. Depending on the water quality requirements for a particular site, these lower quality treated waters may be of adequate quality for reuse in subsequent hydraulic fracturing operations (and will be less costly).

- ***What are the potential impacts from surface water disposal of treated hydraulic fracturing wastewater on drinking water treatment facilities?***

Inadequate bromide and iodide removal from treated hydraulic fracturing wastewater has the potential to affect surface water quality and place a burden on downstream drinking water treatment facilities due to the formation of disinfection byproducts (DBPs). This occurs when bromide and iodide react with organic carbon and drinking water disinfectants. Although sampling data are limited both for treated wastewaters and receiving waters, bromide has reached drinking water resources via some discharges. One utility in Pennsylvania found that elevated bromide in their source water led to elevated disinfection byproducts in their treated drinking water.

Ammonium in hydraulic fracturing wastewater could also impact downstream drinking water supplies by altering disinfection chemistry. Other constituents (e.g., including radionuclides, barium, and organic compounds) may impact drinking water resources if they are present in high concentrations in the wastewater and the applied wastewater treatment does not adequately remove them. Constituents such as radium, metals, and organics can also accumulate in sediments downstream of discharge points.

As of 2014–2015, there is a lack of data on the concentrations of most hydraulic fracturing wastewater constituents in the water near drinking water intakes in regions with hydraulic fracturing activity. Therefore, it is not known whether or to what degree these contaminants have affected drinking water systems.

This page is intentionally left blank.

Appendix B. Water Acquisition Supplemental Information

This page is intentionally left blank.

Appendix B. Water Acquisition Supplemental Information

B.1. Supplemental Tables

Table B-1. Average annual hydraulic fracturing water use and consumption in 2011 and 2012 compared to total annual water use and consumption in 2010 by state.
Hydraulic fracturing water use data from the EPA FracFocus 1.0 project database (U.S. EPA, 2015c). Annual total water use data from the U.S. Geological Survey (USGS) Water Census (Maupin et al., 2014). Estimates of consumption were derived from hydraulic fracturing water use and total water use data. States listed in descending order by the volume of hydraulic fracturing water use.

State	Total annual water use in 2010 (millions of gal)[a,b]	Average annual hydraulic fracturing water use in 2011 and 2012 (millions of gal)[c]	Hydraulic fracturing water use compared to total water use (%)[d]	Hydraulic fracturing water consumption compared to total water consumption (%)[d,e]
Texas	9,052,000	19,942	0.2	0.7
Pennsylvania	2,967,450	5,105	0.2	1.4
Arkansas	4,124,500	3,676	0.1	0.1
Colorado	4,015,000	3,277	0.1	0.1
Oklahoma	1,157,050	2,949	0.3	0.8
Louisiana	3,117,100	2,462	0.1	0.4
North Dakota	419,750	2,181	0.5	2.9
West Virginia	1,288,450	657	0.1	0.5
Wyoming	1,715,500	538	<0.1	<0.1
New Mexico	1,153,400	371	<0.1	<0.1
Ohio	3,445,600	273	<0.1	0.1
Utah	1,627,900	251	<0.1	<0.1
Montana	2,792,250	155	<0.1	<0.1
Kansas	1,460,000	66	<0.1	<0.1
California	13,870,000	44	<0.1	<0.1
Michigan	3,942,000	28	<0.1	<0.1
Mississippi	1,434,450	18	<0.1	<0.1

State	Total annual water use in 2010 (millions of gal)[a,b]	Average annual hydraulic fracturing water use in 2011 and 2012 (millions of gal)[c]	Hydraulic fracturing water use compared to total water use (%)[d]	Hydraulic fracturing water consumption compared to total water consumption (%)[d,e]
Alaska[f]	397,850	7	<0.1	<0.1
Virginia	2,792,250	1	<0.1	<0.1
Alabama	3,635,400	1	<0.1	<0.1
Total for all 20 states	64,407,900	42,001	0.1	0.2

[a] Texas, Colorado, Pennsylvania, North Dakota, Oklahoma, and Utah all made some degree of reporting to the FracFocus national registry mandatory rather than voluntary during this time period analyzed, January 1, 2011, to February 28, 2013. Three other states started requiring disclosure to either FracFocus or the state (Louisiana, Montana, and Ohio), and five states required or began requiring disclosure to the state (Arkansas, Michigan, New Mexico, West Virginia, and Wyoming). Alabama, Alaska, California, Kansas, Mississippi, and Virginia did not have reporting requirements during the period of time studied (U.S. EPA, 2015a).

[b] State-level data accessed from the USGS website (http://water.usgs.gov/watuse/data/2010/) on January 27, 2015. Total water withdrawals per day (located in downloaded Table 1) were multiplied by 365 days to estimate total water use for the year (Maupin et al., 2014).

[c] Average of water used for hydraulic fracturing in 2011 and 2012 based on the EPA FracFocus 1,0 project database (U.S. EPA, 2015c).

[d] Percentages were calculated by averaging annual water use for hydraulic fracturing in the EPA FracFocus 1.0 project database in 2011 and 2012 for a given state (U.S. EPA, 2015c), and then dividing by 2010 USGS total water use (Maupin et al., 2014) and multiplying by 100. Note that the annual hydraulic fracturing water use based on the EPA FracFocus 1.0 project database (the numerator) was not added to the 2010 total USGS water use value in the denominator, and the percentage is simply calculated as by dividing annual hydraulic fracturing use by 2010 total water use or consumption. This was done because of the difference in years between the two datasets, and because the USGS 2010 Census (Maupin et al., 2014) already included an estimate of hydraulic fracturing water use in its mining category. This approach is also consistent with that of other literature on this topic; see Nicot and Scanlon (2012).

[e] Consumption values were calculated with use-specific consumption rates predominantly from the USGS, including 19.2% for public supply, 19.2% for domestic use, 60.7% for irrigation, 60.7% for livestock, 14.8% for industrial uses, 14.8% for mining (Solley et al., 1998), and 2.7% for thermoelectric power (Diehl and Harris, 2014). We used a rate of 71.6% for aquaculture (Verdegem and Bosma, 2009) (evaporation per kg fish + infiltration per kg)/(total water use per kg) *100. These rates were multiplied by each USGS water use value (Maupin et al., 2014) to yield a total water consumption estimate. To calculate a consumption amount for hydraulic fracturing, we used a consumption rate of 82.5%. This was calculated by taking the median value for all reported produced water/injected water percentages in Tables 7-1 and 7-2 of this assessment and then subtracting from 100%. If a range of values was given, the midpoint was used. Note that this is likely a low estimate of consumption since much of this return water is not subsequently treated and reused, but rather disposed of in injection wells—see Chapter 8.

[f] All reported hydraulic fracturing disclosures for Alaska passed state locational quality assurance methods, but not county methods (U.S. EPA, 2015c). Thus, only state-level cumulative values were reported here, and no county-level data are provided in subsequent tables.

Table B-2. Average annual hydraulic fracturing water use and consumption in 2011 and 2012 compared to total annual water use and consumption in 2010 by county.

The counties listed contained wells used for hydraulic fracturing based on the EPA FracFocus 1.0 project database (U.S. EPA, 2015c). Annual total water use data from the USGS Water Census (Maupin et al., 2014). Estimates of consumption derived from hydraulic fracturing water use and total water use data.

State	County	Total annual water use in 2010 (millions of gal)[a]	Average annual hydraulic fracturing water use in 2011 and 2012 (millions of gal)[b]	Hydraulic fracturing water use compared to total water use (%)[c]	Hydraulic fracturing water consumption compared to total water consumption (%)[c,d]
Alabama	Jefferson	29,685.5	0.6	<0.1	<0.1
	Tuscaloosa	14,319.0	0.5	<0.1	<0.1
Arkansas	Cleburne	9,471.8	740.9	7.8	32.9
	Conway	10,643.4	798.1	7.5	21.2
	Faulkner	3,204.7	284.0	8.9	13.7
	Independence	57,195.5	80.3	0.1	0.3
	Logan	1,525.7	2.4	0.2	0.3
	Sebastian	1,365.1	0.6	<0.1	<0.1
	Van Buren	1,587.8	899.6	56.7	168.8
	White	32,131.0	869.8	2.7	4.7
California	Yell	1,507.5	<0.1	<0.1	<0.1
	Colusa	304,782.3	<0.1	<0.1	<0.1
	Glenn	221,420.0	<0.1	<0.1	<0.1
	Kern	788,359.9	41.7	<0.1	<0.1
	Los Angeles	1,118,363.7	0.2	<0.1	<0.1
	Sutter	263,511.8	0.2	<0.1	<0.1
	Ventura	262,610.2	1.8	<0.1	<0.1
Colorado	Adams	84,285.8	3.2	<0.1	<0.1
	Arapahoe	68,255.0	4.0	<0.1	<0.1
	Boulder	84,537.7	4.1	<0.1	<0.1
	Broomfield	2,336.0	4.5	0.2	0.4
	Delta	131,221.2	0.5	<0.1	<0.1

State	County	Total annual water use in 2010 (millions of gal)[a]	Average annual hydraulic fracturing water use in 2011 and 2012 (millions of gal)[b]	Hydraulic fracturing water use compared to total water use (%)[c]	Hydraulic fracturing water consumption compared to total water consumption (%)[c,d]
Colorado, cont.	Dolores	2,040.4	0.1	<0.1	<0.1
	El Paso	42,380.2	<0.1	<0.1	<0.1
	Elbert	5,040.7	<0.1	<0.1	<0.1
	Fremont	53,366.7	0.6	<0.1	<0.1
	Garfield	95,436.6	1,804.2	1.9	2.7
	Jackson	126,968.9	1.0	<0.1	<0.1
	La Plata	122,873.6	3.5	<0.1	<0.1
	Larimer	150,690.3	5.4	<0.1	<0.1
	Las Animas	26,911.5	7.9	<0.1	<0.1
	Mesa	275,476.5	122.1	<0.1	0.1
	Moffat	62,093.8	14.5	<0.1	<0.1
	Morgan	67,901.0	3.9	<0.1	<0.1
	Phillips	21,509.5	0.2	<0.1	<0.1
	Rio Blanco	97,513.4	147.3	0.2	0.2
	Routt	74,460.0	0.1	<0.1	<0.1
	San Miguel	13,848.1	0.3	<0.1	<0.1
	Weld	168,677.5	1,149.4	0.7	1.0
	Yuma	80,595.7	0.4	<0.1	<0.1
Kansas	Barber	2,164.5	9.9	0.5	0.7
	Clark	1,898.0	0.8	<0.1	0.1
	Comanche	3,011.3	25.6	0.9	1.2
	Finney	102,685.5	2.4	<0.1	<0.1
	Grant	47,128.8	0.2	<0.1	<0.1
	Gray	69,379.2	3.3	<0.1	<0.1
	Harper	1,357.8	17.3	1.3	2.0
	Haskell	72,496.3	0.1	<0.1	<0.1

State	County	Total annual water use in 2010 (millions of gal)[a]	Average annual hydraulic fracturing water use in 2011 and 2012 (millions of gal)[b]	Hydraulic fracturing water use compared to total water use (%)[c]	Hydraulic fracturing water consumption compared to total water consumption (%)[c,d]
Kansas, cont.	Hodgeman	8,460.7	2.7	<0.1	<0.1
	Kearny	64,134.2	<0.1	<0.1	<0.1
	Lane	5,628.3	0.8	<0.1	<0.1
	Meade	55,958.2	<0.1	<0.1	<0.1
	Morton	17,403.2	<0.1	<0.1	<0.1
	Ness	1,478.3	1.6	0.1	0.2
	Seward	57,443.7	<0.1	<0.1	<0.1
	Sheridan	26,393.2	0.7	<0.1	<0.1
	Stanton	41,420.2	<0.1	<0.1	<0.1
	Stevens	72,124.0	0.1	<0.1	<0.1
	Sumner	3,442.0	0.2	<0.1	<0.1
Louisiana	Allen	8,942.5	0.1	<0.1	<0.1
	Beauregard	10,161.6	2.3	<0.1	0.1
	Bienville	4,810.7	108.9	2.3	10.0
	Bossier	5,599.1	110.1	2.0	4.9
	Caddo	53,644.1	153.6	0.3	1.7
	Calcasieu	81,621.3	0.1	<0.1	<0.1
	Caldwell	1,398.0	<0.1	<0.1	<0.1
	Claiborne	952.7	3.8	0.4	1.1
	DeSoto	13,373.6	1,085.9	8.1	47.4
	East Feliciana	1,350.5	3.7	0.3	0.7
	Jackson	1,456.4	<0.1	<0.1	<0.1
	Lincoln	3,000.3	3.3	0.1	0.3
	Natchitoches	12,530.5	12.7	0.1	0.2
	Rapides	199,976.2	1.7	<0.1	<0.1
	Red River	1,606.0	569.6	35.5	83.2

State	County	Total annual water use in 2010 (millions of gal)[a]	Average annual hydraulic fracturing water use in 2011 and 2012 (millions of gal)[b]	Hydraulic fracturing water use compared to total water use (%)[c]	Hydraulic fracturing water consumption compared to total water consumption (%)[c,d]
Louisiana, cont.	Sabine	1,522.1	395.2	26.0	76.6
	Tangipahoa	7,329.2	1.9	<0.1	0.1
	Union	1,481.9	4.9	0.3	1.0
	Webster	2,664.5	1.2	<0.1	0.1
	West Feliciana	15,191.3	2.3	<0.1	0.1
	Winn	846.8	1.1	0.1	0.4
Michigan	Cheboygan	2,777.7	<0.1	<0.1	<0.1
	Gladwin	850.5	1.1	0.1	0.4
	Kalkaska	1,233.7	24.0	1.9	3.7
	Missaukee	1,423.5	<0.1	<0.1	<0.1
	Ogemaw	1,179.0	<0.1	<0.1	<0.1
	Roscommon	1,000.1	2.4	0.2	0.9
Mississippi	Amite	792.1	14.4	1.8	3.8
	Wilkinson	1,270.2	3.2	0.3	0.4
Montana	Daniels	1,408.9	0.6	<0.1	0.1
	Garfield	1,631.6	0.5	<0.1	<0.1
	Glacier	46,760.2	5.1	<0.1	<0.1
	Musselshell	26,827.5	0.4	<0.1	<0.1
	Richland	94,797.8	83.5	0.1	0.1
	Roosevelt	31,539.7	52.1	0.2	0.2
	Rosebud	71,412.3	3.5	<0.1	<0.1
	Sheridan	7,354.8	9.7	0.1	0.2
New Mexico	Chaves	88,078.2	2.8	<0.1	<0.1
	Colfax	17,450.7	0.7	<0.1	<0.1
	Eddy	70,612.9	225.6	0.3	0.5

State	County	Total annual water use in 2010 (millions of gal)[a]	Average annual hydraulic fracturing water use in 2011 and 2012 (millions of gal)[b]	Hydraulic fracturing water use compared to total water use (%)[c]	Hydraulic fracturing water consumption compared to total water consumption (%)[c,d]
New Mexico, cont.	Harding	1,168.0	0.1	<0.1	<0.1
	Lea	64,057.5	113.7	0.2	0.3
	Rio Arriba	39,080.6	16.5	<0.1	0.1
	Roosevelt	63,367.7	<0.1	<0.1	<0.1
	San Juan	125,432.3	11.6	<0.1	<0.1
	Sandoval	23,922.1	0.4	<0.1	<0.1
North Dakota	Billings	762.9	44.4	5.8	16.2
	Bottineau	1,164.4	0.1	<0.1	<0.1
	Burke	394.2	63.6	16.1	40.8
	Divide	806.7	102.2	12.7	18.6
	Dunn	1,076.8	309.5	28.7	43.1
	Golden Valley	208.1	4.6	2.2	3.8
	Mckenzie	13,753.2	588.4	4.3	6.2
	Mclean	7,873.1	12.2	0.2	0.4
	Mountrail	1,248.3	449.4	36.0	98.3
	Stark	1,168.0	48.0	4.1	8.5
	Williams	7,705.2	558.5	7.2	11.3
Ohio	Ashland	2,033.1	1.5	0.1	0.2
	Belmont	65,528.5	1.9	<0.1	0.1
	Carroll	1,127.9	152.7	13.5	37.3
	Columbiana	3,763.2	30.7	0.8	2.2
	Coshocton	53,775.5	5.4	<0.1	0.1
	Guernsey	2,379.8	8.4	0.4	0.7
	Harrison	481.8	16.5	3.4	7.3
	Jefferson	632,917.3	26.2	<0.1	0.1

State	County	Total annual water use in 2010 (millions of gal)[a]	Average annual hydraulic fracturing water use in 2011 and 2012 (millions of gal)[b]	Hydraulic fracturing water use compared to total water use (%)[c]	Hydraulic fracturing water consumption compared to total water consumption (%)[c,d]
Ohio, cont.	Knox	3,270.4	1.1	<0.1	0.1
	Medina	3,540.5	1.3	<0.1	0.1
	Muskingum	6,018.9	5.1	0.1	0.3
	Noble	478.2	8.3	1.7	3.4
	Portage	18,414.3	3.2	<0.1	0.1
	Stark	16,479.8	2.4	<0.1	<0.1
	Tuscarawas	14,165.7	6.7	<0.1	0.2
	Wayne	6,051.7	1.7	<0.1	0.1
Oklahoma	Alfalfa	2,996.7	182.7	6.1	12.0
	Beaver	15,341.0	23.1	0.2	0.3
	Beckham	4,099.0	108.0	2.6	4.7
	Blaine	3,763.2	203.3	5.4	9.3
	Bryan	5,062.6	10.3	0.2	0.4
	Caddo	24,064.5	25.4	0.1	0.3
	Canadian	5,584.5	441.9	7.9	15.6
	Carter	159,906.5	161.9	0.1	0.5
	Coal	1,193.6	85.9	7.2	21.5
	Custer	3,281.4	19.0	0.6	1.2
	Dewey	10,953.7	162.6	1.5	6.2
	Ellis	8,486.3	184.3	2.2	3.2
	Garvin	16,279.0	15.0	0.1	0.4
	Grady	13,537.9	111.5	0.8	2.3
	Grant	5,569.9	77.8	1.4	5.2
	Harper	3,266.8	8.8	0.3	0.4
	Hughes	3,394.5	30.5	0.9	2.2
	Jefferson	4,496.8	<0.1	<0.1	<0.1

State	County	Total annual water use in 2010 (millions of gal)[a]	Average annual hydraulic fracturing water use in 2011 and 2012 (millions of gal)[b]	Hydraulic fracturing water use compared to total water use (%)[c]	Hydraulic fracturing water consumption compared to total water consumption (%)[c,d]
Oklahoma, cont.	Johnston	1,671.7	32.9	2.0	4.7
	Kay	16,957.9	17.3	0.1	0.4
	Kingfisher	3,744.9	10.2	0.3	0.5
	Kiowa	5,022.4	0.1	<0.1	<0.1
	Latimer	1,062.2	0.6	0.1	0.1
	Le Flore	8,635.9	0.3	<0.1	<0.1
	Logan	4,077.1	4.2	0.1	0.3
	Love	2,011.2	4.4	0.2	0.5
	Major	6,321.8	1.2	<0.1	<0.1
	Marshall	2,613.4	98.4	3.8	7.2
	McClain	2,952.9	2.1	0.1	0.2
	Noble	12,990.4	25.3	0.2	1.8
	Oklahoma	47,836.9	1.2	<0.1	<0.1
	Osage	6,971.5	3.8	0.1	0.2
	Pawnee	4,839.9	15.7	0.3	1.4
	Payne	4,332.6	9.9	0.2	0.6
	Pittsburg	6,314.5	349.0	5.5	16.0
	Roger Mills	2,847.0	235.5	8.3	12.6
	Seminole	124,837.3	0.1	<0.1	<0.1
	Stephens	49,990.4	27.7	0.1	0.3
	Texas	110,208.1	0.1	<0.1	<0.1
	Washita	3,310.6	102.1	3.1	5.4
	Woods	4,139.1	155.1	3.7	10.9

State	County	Total annual water use in 2010 (millions of gal)[a]	Average annual hydraulic fracturing water use in 2011 and 2012 (millions of gal)[b]	Hydraulic fracturing water use compared to total water use (%)[c]	Hydraulic fracturing water consumption compared to total water consumption (%)[c,d]
Pennsylvania	Allegheny	234,140.2	13.6	<0.1	<0.1
	Armstrong	65,853.3	55.7	0.1	1.8
	Beaver	157,793.2	30.5	<0.1	0.2
	Blair	8,303.8	5.9	0.1	0.2
	Bradford	4,354.5	1,059.4	24.3	78.2
	Butler	5,730.5	121.8	2.1	6.0
	Cameron	292.0	6.6	2.3	4.1
	Centre	16,560.1	38.5	0.2	0.5
	Clarion	1,843.3	8.1	0.4	1.4
	Clearfield	111,051.3	111.5	0.1	2.3
	Clinton	6,161.2	94.4	1.5	3.0
	Columbia	3,810.6	5.6	0.1	0.4
	Crawford	5,091.8	2.4	<0.1	0.1
	Elk	7,876.7	37.5	0.5	1.9
	Fayette	16,465.2	120.2	0.7	2.7
	Forest	744.6	7.7	1.0	1.6
	Greene	13,023.2	359.0	2.8	24.7
	Huntingdon	5,121.0	2.7	0.1	0.2
	Indiana	21,819.7	16.2	0.1	0.7
	Jefferson	1,730.1	13.8	0.8	1.7
	Lawrence	36,598.6	27.0	0.1	1.0
	Lycoming	5,854.6	704.6	12.0	33.8
	McKean	4,723.1	60.5	1.3	4.9
	Potter	2,281.3	16.5	0.7	1.0
	Somerset	10,833.2	5.8	0.1	0.2
	Sullivan	222.7	66.5	29.9	79.8

State	County	Total annual water use in 2010 (millions of gal)[a]	Average annual hydraulic fracturing water use in 2011 and 2012 (millions of gal)[b]	Hydraulic fracturing water use compared to total water use (%)[c]	Hydraulic fracturing water consumption compared to total water consumption (%)[c,d]
Pennsylvania, cont.	Susquehanna	1,617.0	751.3	46.5	123.4
	Tioga	2,909.1	566.3	19.5	47.3
	Venango	2,989.4	2.4	0.1	0.3
	Warren	5,099.1	2.3	<0.1	0.2
	Washington	130,535.0	433.7	0.3	4.6
	Westmoreland	14,607.3	207.0	1.4	3.8
	Wyoming	4,788.8	150.0	3.1	15.2
Texas	Andrews	23,363.7	236.2	1.0	2.7
	Angelina	5,540.7	0.8	<0.1	<0.1
	Archer	2,536.8	0.1	<0.1	<0.1
	Atascosa	15,038.0	327.3	2.2	4.0
	Austin	2,555.0	2.1	0.1	0.1
	Bee	3,087.9	20.0	0.6	1.1
	Borden	2,427.3	8.0	0.3	1.0
	Bosque	3,544.2	0.7	<0.1	<0.1
	Brazos	24,790.8	7.7	<0.1	0.1
	Brooks	1,204.5	1.5	0.1	0.3
	Burleson	10,694.5	3.0	<0.1	<0.1
	Cherokee	24,845.6	0.5	<0.1	<0.1
	Clay	1,963.7	<0.1	<0.1	<0.1
	Cochran	24,035.3	3.0	<0.1	<0.1
	Coke	12,713.0	0.3	<0.1	<0.1
	Colorado	52,465.1	0.1	<0.1	<0.1
	Concho	2,832.4	<0.1	<0.1	<0.1
	Cooke	4,533.3	454.3	10.0	29.9

State	County	Total annual water use in 2010 (millions of gal)[a]	Average annual hydraulic fracturing water use in 2011 and 2012 (millions of gal)[b]	Hydraulic fracturing water use compared to total water use (%)[c]	Hydraulic fracturing water consumption compared to total water consumption (%)[c,d]
Texas, cont.	Cottle	733.7	0.3	<0.1	0.1
	Crane	8,566.6	92.3	1.1	5.7
	Crockett	4,281.5	279.0	6.5	29.5
	Crosby	27,261.9	1.3	<0.1	<0.1
	Culberson	14,311.7	37.7	0.3	0.4
	Dallas	112,204.7	5.6	<0.1	<0.1
	Dawson	28,842.3	17.5	0.1	0.1
	DeWitt	2,394.4	546.6	22.8	48.6
	Denton	60,684.9	455.0	0.7	2.3
	Dimmit	4,073.4	1,794.2	44.0	81.3
	Ector	21,958.4	226.5	1.0	4.6
	Edwards	332.2	<0.1	<0.1	<0.1
	Ellis	8,530.1	4.2	<0.1	0.1
	Erath	5,876.5	0.8	<0.1	<0.1
	Fayette	9,008.2	13.7	0.2	1.2
	Fisher	2,854.3	1.8	0.1	0.1
	Franklin	1,956.4	<0.1	<0.1	<0.1
	Freestone	297,861.9	53.9	<0.1	0.5
	Frio	20,589.7	127.5	0.6	0.9
	Gaines	121,778.6	21.6	<0.1	<0.1
	Garza	5,234.1	0.6	<0.1	<0.1
	Glasscock	20,680.9	598.1	2.9	4.2
	Goliad	142,963.2	<0.1	<0.1	<0.1
	Gonzales	7,121.2	577.9	8.1	17.6
	Grayson	8,143.2	9.3	0.1	0.3
	Gregg	33,010.6	9.4	<0.1	0.2

State	County	Total annual water use in 2010 (millions of gal)[a]	Average annual hydraulic fracturing water use in 2011 and 2012 (millions of gal)[b]	Hydraulic fracturing water use compared to total water use (%)[c]	Hydraulic fracturing water consumption compared to total water consumption (%)[c,d]
Texas, cont.	Grimes	112,500.3	15.5	<0.1	0.3
	Hansford	43,643.1	2.9	<0.1	<0.1
	Hardeman	2,230.2	0.4	<0.1	<0.1
	Hardin	2,376.2	0.1	<0.1	<0.1
	Harrison	11,869.8	141.6	1.2	6.0
	Hartley	113,555.2	1.9	<0.1	<0.1
	Haskell	12,143.6	0.1	<0.1	<0.1
	Hemphill	3,150.0	263.9	8.4	16.3
	Hidalgo	171,630.3	8.0	<0.1	<0.1
	Hockley	46,314.9	3.0	<0.1	<0.1
	Hood	9,351.3	76.0	0.8	2.2
	Houston	3,686.5	8.6	0.2	0.6
	Howard	10,811.3	97.6	0.9	2.7
	Hutchinson	34,437.8	0.3	<0.1	<0.1
	Irion	1,335.9	411.4	30.8	74.5
	Jack	2,241.1	14.0	0.6	2.2
	Jefferson	88,585.5	<0.1	<0.1	<0.1
	Jim Hogg	306.6	0.1	<0.1	0.1
	Johnson	9,241.8	582.0	6.3	18.5
	Jones	5,679.4	<0.1	<0.1	<0.1
	Karnes	1,861.5	1,055.2	56.7	120.1
	Kenedy	456.3	0.2	0.1	0.1
	Kent	6,132.0	0.4	<0.1	<0.1
	King	1,485.6	<0.1	<0.1	<0.1
	Kleberg	1,171.7	3.4	0.3	0.5
	Knox	9,800.3	<0.1	<0.1	<0.1

State	County	Total annual water use in 2010 (millions of gal)[a]	Average annual hydraulic fracturing water use in 2011 and 2012 (millions of gal)[b]	Hydraulic fracturing water use compared to total water use (%)[c]	Hydraulic fracturing water consumption compared to total water consumption (%)[c,d]
Texas, cont.	La Salle	2,474.7	1,288.7	52.1	93.7
	Lavaca	3,763.2	45.0	1.2	2.0
	Lee	3,120.8	1.2	<0.1	0.1
	Leon	2,171.8	56.2	2.6	6.6
	Liberty	20,662.7	<0.1	<0.1	<0.1
	Limestone	11,158.1	10.7	0.1	0.9
	Lipscomb	11,015.7	89.0	0.8	1.1
	Live Oak	1,916.3	294.0	15.3	40.1
	Loving	781.1	138.4	17.7	94.1
	Lynn	19,892.5	1.1	<0.1	<0.1
	Madison	1,554.9	45.3	2.9	8.2
	Marion	3,606.2	5.9	0.2	0.9
	Martin	14,063.5	432.0	3.1	4.7
	Maverick	20,498.4	52.4	0.3	0.4
	McMullen	657.0	745.9	113.5	350.4
	Medina	19,228.2	0.2	<0.1	<0.1
	Menard	1,014.7	<0.1	<0.1	<0.1
	Midland	12,891.8	307.4	2.4	3.7
	Milam	16,665.9	4.9	<0.1	0.1
	Mitchell	6,559.1	11.0	0.2	0.3
	Montague	3,989.5	925.3	23.2	77.8
	Montgomery	32,565.3	0.2	<0.1	<0.1
	Moore	57,075.1	<0.1	<0.1	<0.1
	Nacogdoches	5,891.1	271.7	4.6	12.5
	Navarro	18,699.0	4.8	<0.1	0.1
	Newton	2,263.0	0.2	<0.1	<0.1

State	County	Total annual water use in 2010 (millions of gal)[a]	Average annual hydraulic fracturing water use in 2011 and 2012 (millions of gal)[b]	Hydraulic fracturing water use compared to total water use (%)[c]	Hydraulic fracturing water consumption compared to total water consumption (%)[c,d]
Texas, cont.	Nolan	4,124.5	4.5	0.1	0.2
	Nueces	85,767.7	1.0	<0.1	<0.1
	Ochiltree	21,348.9	33.3	0.2	0.2
	Oldham	2,124.3	1.3	0.1	0.1
	Orange	150,128.2	0.3	<0.1	<0.1
	Palo Pinto	18,403.3	9.6	0.1	0.3
	Panola	6,365.6	346.5	5.4	20.7
	Parker	8,241.7	261.7	3.2	9.8
	Pecos	52,954.2	8.2	<0.1	<0.1
	Polk	204,009.5	0.2	<0.1	<0.1
	Potter	2,029.4	0.4	<0.1	<0.1
	Reagan	9,333.1	410.5	4.4	7.8
	Reeves	20,772.2	164.2	0.8	1.1
	Roberts	7,690.6	38.2	0.5	1.2
	Robertson	158,344.3	45.4	<0.1	0.2
	Runnels	2,847.0	<0.1	<0.1	<0.1
	Rusk	582,134.9	65.8	<0.1	0.3
	Sabine	799.4	31.1	3.9	13.9
	San Augustine	1,131.5	182.1	16.1	50.8
	San Patricio	4,172.0	1.1	<0.1	<0.1
	Schleicher	967.3	27.0	2.8	5.0
	Scurry	14,187.6	1.1	<0.1	<0.1
	Shelby	4,920.2	133.6	2.7	8.2
	Sherman	78,073.5	<0.1	<0.1	<0.1
	Smith	11,231.1	0.2	<0.1	<0.1
	Somervell	746,005.3	4.8	<0.1	<0.1

State	County	Total annual water use in 2010 (millions of gal)[a]	Average annual hydraulic fracturing water use in 2011 and 2012 (millions of gal)[b]	Hydraulic fracturing water use compared to total water use (%)[c]	Hydraulic fracturing water consumption compared to total water consumption (%)[c,d]
Texas, cont.	Starr	9,552.1	5.0	0.1	0.1
	Stephens	13,446.6	2.6	<0.1	0.1
	Sterling	719.1	36.6	5.1	11.9
	Stonewall	923.5	0.9	0.1	0.3
	Sutton	1,153.4	1.6	0.1	0.3
	Tarrant	104,430.2	1,443.0	1.4	3.9
	Terrell	543.9	0.1	<0.1	<0.1
	Terry	48,362.5	7.5	<0.1	<0.1
	Tyler	1,872.5	0.1	<0.1	<0.1
	Upshur	8,610.4	0.2	<0.1	<0.1
	Upton	7,975.3	462.6	5.8	14.2
	Van Zandt	4,139.1	0.1	<0.1	<0.1
	Walker	4,478.6	3.4	0.1	0.2
	Waller	9,829.5	0.1	<0.1	<0.1
	Ward	6,909.5	107.3	1.6	4.6
	Washington	2,430.9	2.2	0.1	0.2
	Webb	15,862.9	1,117.8	7.0	18.2
	Wharton	81,606.7	<0.1	<0.1	<0.1
	Wheeler	6,522.6	858.0	13.2	21.5
	Wichita	25,936.9	0.1	<0.1	<0.1
	Wilbarger	12,683.8	0.2	<0.1	<0.1
	Willacy	15,209.6	0.1	<0.1	<0.1
	Wilson	7,843.9	84.5	1.1	1.7
	Winkler	5,274.3	7.7	0.1	0.5
	Wise	24,966.0	529.7	2.1	8.9
	Wood	19,334.1	0.2	<0.1	<0.1

State	County	Total annual water use in 2010 (millions of gal)[a]	Average annual hydraulic fracturing water use in 2011 and 2012 (millions of gal)[b]	Hydraulic fracturing water use compared to total water use (%)[c]	Hydraulic fracturing water consumption compared to total water consumption (%)[c,d]
Texas, cont.	Yoakum	77,325.3	7.5	<0.1	<0.1
	Young	21,162.7	0.1	<0.1	<0.1
	Zapata	2,697.4	1.1	<0.1	0.1
	Zavala	14,410.2	130.0	0.9	1.3
Utah	Carbon	15,067.2	7.3	<0.1	0.1
	Duchesne	119,811.3	85.5	0.1	0.1
	San Juan	10,632.5	0.3	<0.1	<0.1
	Sevier	52,512.6	<0.1	<0.1	<0.1
	Uintah	100,229.0	157.5	0.2	0.2
Virginia	Buchanan	313.9	0.6	0.2	0.3
	Dickenson	1,741.1	0.8	<0.1	0.2
	Wise	1,927.2	0.1	<0.1	<0.1
West Virginia	Barbour	773.8	19.9	2.6	6.9
	Brooke	4,551.6	54.8	1.2	5.1
	Doddridge	405.2	78.5	19.4	69.4
	Hancock	28,718.2	1.2	<0.1	<0.1
	Harrison	20,232.0	40.2	0.2	1.9
	Lewis	901.6	2.4	0.3	0.8
	Marion	5,982.4	70.1	1.2	4.9
	Marshall	158,358.9	84.5	0.1	0.7
	Monongalia	42,102.8	6.8	<0.1	0.1
	Ohio	3,825.2	116.5	3.0	10.4
	Pleasants	24,703.2	<0.1	<0.1	<0.1
	Preston	2,890.8	8.4	0.3	1.4
	Ritchie	587.7	2.8	0.5	1.7
	Taylor	824.9	52.9	6.4	17.6
	Tyler	4,934.8	2.1	<0.1	0.2
	Upshur	1,814.1	34.9	1.9	6.8

State	County	Total annual water use in 2010 (millions of gal)[a]	Average annual hydraulic fracturing water use in 2011 and 2012 (millions of gal)[b]	Hydraulic fracturing water use compared to total water use (%)[c]	Hydraulic fracturing water consumption compared to total water consumption (%)[c,d]
West Virginia, cont.	Webster	1,292.1	2.3	0.2	0.3
	Wetzel	1,467.3	78.2	5.3	11.9
Wyoming	Big Horn	143,368.4	2.9	<0.1	<0.1
	Campbell	44,318.3	11.7	<0.1	0.1
	Carbon	137,130.5	4.5	<0.1	<0.1
	Converse	56,972.9	106.8	0.2	0.3
	Fremont	186,150.0	28.2	<0.1	<0.1
	Goshen	144,248.0	5.8	<0.1	<0.1
	Hot Springs	28,572.2	0.3	<0.1	<0.1
	Johnson	43,205.1	<0.1	<0.1	<0.1
	Laramie	86,297.0	18.3	<0.1	<0.1
	Lincoln	74,562.2	0.8	<0.1	<0.1
	Natrona	62,885.9	1.8	<0.1	<0.1
	Niobrara	25,148.5	0.1	<0.1	<0.1
	Park	111,317.7	0.9	<0.1	<0.1
	Sublette	61,006.1	314.8	0.5	0.7
	Sweetwater	61,699.6	39.4	0.1	0.1
	Uinta	79,518.9	0.6	<0.1	<0.1
	Washakie	60,400.2	1.1	<0.1	<0.1

[a] County-level data accessed from the USGS website (http://water.usgs.gov/watuse/data/2010/) on November 11, 2014. Total daily water withdrawals were multiplied by 365 days to estimate total water use for the year (Maupin et al., 2014).

[b] Average of water used for hydraulic fracturing in 2011 and 2012, based on the EPA FracFocus 1.0 project database (U.S. EPA, 2015c).

[c] Percentages were calculated by averaging annual water use for hydraulic fracturing in the EPA FracFocus 1.0 project database in 2011 and 2012 for a given county (U.S. EPA, 2015c), and then dividing by 2010 USGS total water use for that county (Maupin et al., 2014) and multiplying by 100.

[d] Consumption values were calculated with use-specific consumption rates predominantly from the USGS, including 19.2% for public supply, 19.2% for domestic use, 60.7% for irrigation, 60.7% for livestock, 14.8% for industrial uses, 14.8% for mining (Solley et al., 1998), and 2.7% for thermoelectric power (Diehl and Harris, 2014). We used a rate of 71.6% for aquaculture (Verdegem and Bosma, 2009) (evaporation per kg fish + infiltration per kg)/(total water use per kg)*100. These rates were multiplied by each USGS water use value (Maupin et al., 2014) to yield a total water consumption estimate. To calculate a consumption amount for hydraulic fracturing, we used a consumption rate of 82.5%. This was calculated by taking the median value for all reported produced water/injected water percentages in Tables 7-1 and 7-2 of this assessment and then subtracting from 100%. If a range of values was given, the midpoint was used. Note that this is likely a low estimate of consumption since much of this return water is not subsequently treated and reused, but rather disposed of in injection wells—see Chapter 8.

Table B-3. Comparison of water use per well estimates from the EPA FracFocus 1.0 project database (U.S. EPA, 2015c) and literature sources.

State	Basin[a]	Water use per well (gal) - EPA FracFocus 1.0 project database estimate[b]	Water use per well (gal) - Literature estimate[b,c]	EPA FracFocus 1.0 project database estimate as a percentage of literature estimate (%)
Colorado	Denver	403,686	2,900,000	14
North Dakota	-	2,140,842	2,200,000	97
Oklahoma	-	2,591,778	3,000,000	86
Pennsylvania[d]	-	4,301,701	4,450,000	97
Texas	Fort Worth	3,881,220	4,500,000	86
Texas	Salt	3,139,980	4,000,000	78
Texas	Western Gulf	3,777,648	4,600,000	82
Average[e]	-	-	-	77
Median[e]	-	-	-	86

[a] In cases where a basin is not specified, estimates were for the entire state and not specific to a particular basin. Basin boundaries for the EPA FracFocus 1.0 project database estimates were determined from data from the U.S. EIA (U.S. EPA, 2015b).

[b] The type of literature estimate determined the specific comparison with the EPA FracFocus 1.0 project database. If averages were given in the literature (as for North Dakota and Pennsylvania), those values were compared with EPA FracFocus 1.0 project database averages; where medians were given in the literature (as for Colorado, Oklahoma, and Texas), they were compared with EPA FracFocus 1.0 project database medians.

[c] Literature estimates were from the following sources: Colorado (Goodwin et al., 2014), North Dakota (North Dakota State Water Commission, 2014), Pennsylvania (Mitchell et al., 2013), and Texas (Nicot and Scanlon, 2012)—see far right-column and footnotes in Table B-5 for details on literature estimates. Where the literature provided a range, the mid-point was used. Only literature estimates that were not directly derived from FracFocus were included.

[d] The results from Mitchell et al. (2013) were used for Pennsylvania since they were derived from Pennsylvania Department of Environment Protection (PA DEP) records. Estimates from Hansen et al. (2013) were not included here because they were based on data from the FracFocus national registry.

[e] Average and median percentage calculations were not weighted by the number of wells for a given estimate.

Table B-4. Comparison of well counts from the EPA FracFocus 1.0 project database (U.S. EPA, 2015c) and state databases for North Dakota, Pennsylvania, and West Virginia.

State	EPA FracFocus 1.0 project database well counts[a]			State database well counts			EPA FracFocus 1.0 project database counts as a percentage of state database counts		
	2011	2012	Total	2011	2012	Total	2011	2012	Total
North Dakota[b]	613	1,458	2,071	1,225	1,740	2,965	50%	84%	70%
Pennsylvania[c]	1,137	1,257	2,394	1,963	1,347	3,310	58%	93%	72%
West Virginia[d]	93	176	269	214	251	465	43%	70%	58%
Average	-	-	-	-	-	-	50%	82%	67%

[a] EPA FracFocus 1.0 project database wells counts (U.S. EPA, 2015c).

[b] For North Dakota state well counts, we used a North Dakota Department of Mineral Resources online database containing a list of horizontal wells completed in the Bakken Formation. Data for North Dakota were accessed on July 9, 2014 at https://www.dmr.nd.gov/oilgas/bakkenwells.asp.

[c] For Pennsylvania state well counts, we used completed horizontal wells as a proxy for hydraulically fractured wells in the state. The Pennsylvania Department of Environmental Protection has online databases of permitted and spudded wells, which differentiate between conventional and unconventional wells and can generate summary statistics at both the county and state scale. The number of spudded wells (i.e., wells drilled) provided a better comparison with the number of hydraulically fractured wells in the EPA FracFocus 1.0 project database than that of permitted wells. The number of permitted wells was nearly double that of spudded in 2011 and 2012, indicating that almost half of the wells permitted were not drilled in that same year. Therefore, we used spudded wells here. Data for Pennsylvania were accessed on February 11, 2014 from http://www.depreportingservices.state.pa.us/ReportServer/Pages/ReportViewer.aspx?/Oil_Gas/Spud_External_Data.

[d] For West Virginia state well counts, data on the number of hydraulically fractured wells per year were received from the West Virginia Department of Environmental Protection on February 25, 2014.

Table B-5. Water use per hydraulically fractured well as reported in the EPA FracFocus 1.0 project database (U.S. EPA, 2015c) by state and basin, covering the time period of January 2011 through February 2013.

This table highlights 15 of the 20 states accounting for almost all disclosures reported in the EPA FracFocus 1.0 project database (U.S. EPA, 2015c). All EPA FracFocus 1.0 project database estimates were limited to disclosures with valid state, county, and volume information. Other literature estimates are also included where available. NA indicates other literature estimates were not available.

State	Basin/Total[a]	Number of disclosures	Mean (gal)	Median (gal)	10th percentile (gal)	90th percentile (gal)	Literature estimates
Arkansas	Arkoma	1,423	5,190,254	5,259,965	3,234,963	7,121,249	NA
	Total	1,423	5,190,254	5,259,965	3,234,963	7,121,249	NA
California	San Joaquin	677	131,653	77,238	22,100	285,029	NA
	Other	34	132,391	36,099	13,768	361,192	NA
	Total	711	131,689	76,818	21,462	285,306	NA
Colorado	Denver	3,166	753,887	403,686	143,715	2,588,946	130,000 gal (average)[b]
	Uinta-Piceance	1,520	2,739,523	1,798,414	840,778	5,066,380	2.9 million gal (median, Wattenberg field of Niobrara play)[c]
	Raton	146	108,003	95,974	24,917	211,526	NA
	Other	66	605,740	183,408	34,412	601,816	NA
	Total	4,898	1,348,842	463,462	147,353	3,092,024	NA
Kansas	Total	121	1,135,973	1,453,788	10,836	2,227,926	NA
Louisiana	TX-LA-MS Salt	939	5,289,100	5,116,650	2,851,654	7,984,838	NA
	Other	27	896,899	232,464	87,003	3,562,400	NA
	Total	966	5,166,337	5,077,863	1,812,099	7,945,630	NA
Montana	Williston	187	1,640,085	1,552,596	375,864	3,037,398	NA
	Other	20	945,541	1,017,701	157,639	1,575,197	NA
	Total	207	1,572,979	1,455,757	367,326	2,997,552	NA

Appendix B – Water Acquisition Supplemental Information

State	Basin/Total[a]	Number of disclosures	Mean (gal)	Median (gal)	10th percentile (gal)	90th percentile (gal)	Literature estimates
New Mexico	Permian	732	991,369	426,258	89,895	2,502,923	NA
	San Juan	363	159,680	97,734	27,217	313,919	NA
	Other	50	33,787	8,358	1,100	98,841	NA
	Total	1,145	685,882	175,241	35,638	1,871,666	NA
North Dakota	Williston	2,109	2,140,842	2,022,380	969,380	3,313,482	NA
	Total	2,109	2,140,842	2,022,380	969,380	3,313,482	2.2 million gal (average)[d]
Ohio	Appalachian	146	4,206,955	3,887,499	2,885,568	5,571,027	NA
	Total	146	4,206,955	3,887,499	2,885,568	5,571,027	NA
Oklahoma	Anadarko	935	3,742,703	3,259,774	1,211,700	6,972,652	Many formations reported[e]
	Arkoma	158	6,323,750	6,655,929	172,375	9,589,554	Many formations reported[e]
	Ardmore	98	6,637,332	8,021,559	81,894	8,835,842	Many formations reported[e]
	Other	592	1,963,480	1,866,144	1,319,247	2,785,352	NA
	Total	1,783	3,539,775	2,591,778	1,260,906	7,402,230	3 million gal (median)[e]
Pennsylvania	Appalachian	2,445	4,301,701	4,184,936	2,313,649	6,615,981	4.1-4.6 million gal (average, Marcellus play, Susquehanna River Basin)[f]
	Total	2,445	4,301,701	4,184,936	2,313,649	6,615,981	4.1-4.5[g] and 4.3-4.6[h] million gal (average)
Texas	Permian	8,419	1,068,511	841,134	40,090	1,814,633	Many formations reported[i]
	Western Gulf	4,549	3,915,540	3,777,648	173,832	6,786,052	4.5-4.7 million gal (median, Eagle Ford play)[i]
	Fort Worth	2,564	3,880,724	3,881,220	923,381	6,649,406	4.5 million gal (median, Barnett play)[i]
	TX-LA-MS Salt	626	4,261,363	3,139,980	193,768	10,010,707	6-7.5 million gal (median, Texas-Haynesville play) and 0.5-1 million gal (median, Cotton Valley play)[i]

State	Basin/Total[a]	Number of disclosures	Mean (gal)	Median (gal)	10th percentile (gal)	90th percentile (gal)	Literature estimates
Texas, cont.	Anadarko	604	4,128,702	3,341,310	492,421	8,292,996	Many formations reported[i]
	Other	120	1,601,897	184,239	21,470	5,678,588	NA
	Total	16,882	2,494,452	1,420,613	58,709	6,115,195	Not reported by state[i]
Utah	Uinta-Piceance	1,396	375,852	304,105	77,166	770,699	NA
	Other	10	58,874	56,245	28,745	97,871	NA
	Total	1,406	373,597	302,075	76,286	769,360	NA
West Virginia	Appalachian	273	5,034,217	5,012,238	3,170,210	7,297,080	NA
	Total	273	5,034,217	5,012,238	3,170,210	7,297,080	4.7-6 million gal (average)[g]
Wyoming	Greater Green River	861	841,702	752,979	147,020	1,493,266	NA
	Powder River	351	739,129	5,927	5,353	2,863,182	NA
	Other	193	613,618	41,664	22,105	1,818,606	NA
	Total	1,405	784,746	322,793	5,727	1,837,602	NA

a Basin boundaries for the EPA FracFocus 1.0 project database well locations were determined from data from the U.S. EIA (U.S. EPA, 2015b).

b Literature estimates for California were from a California Council on Science and Technology report using data from FracFocus (CCST, 2014).

c Literature estimates for the Denver Basin were from Goodwin et al. (2014). Goodwin et al. (2014) assessed 200 randomly sampled wells in the Wattenberg Field of the Denver Basin (Niobrara Play), using industry data for wells operated by Noble Energy, drilled between January 1, 2010, and July 1, 2013. Water consumption is reported rather than water use, but Goodwin et al. (2014) assume, based on Noble Energy practices, that water use and water consumption were identical because none of the flowback or produced water is reused for hydraulic fracturing. Goodwin et al. reported drilling water consumed, hydraulic fracturing water consumed, and total water consumed. We present hydraulic fracturing water consumption here (hydraulic fracturing water consumption was approximately 95% of the total).

d Literature estimates for North Dakota were from an informational bulletin from the North Dakota State Water Commission (2014). No further information was available.

e Murray (2013), who assessed water use for oil and gas operations from 2000–2010 for eight formations in Oklahoma using data from the Oklahoma Corporation Commission. It is not possible to extract an estimate corresponding to 2011–2012 from Murray without the raw data, because medians were presented for the 10-year period rather than separated by year.

f The range of average annual water use per hydraulically fractured well in the Susquehanna River Basin for 2011 and 2012, calculated from SRBC (2016).

g Hansen et al. (2013), using data from FracFocus via Skytruth for Pennsylvania as a whole, the range of annual averages is reported for 2011 and 2012. Similarly, for West Virginia, the range of annual averages is reported for 2011 and 2012 (partial year).

[h] Mitchell et al. (2013), using data reported to the Pennsylvania Department of Environmental Protection. Mitchell et al. (2013) reported water use in the Ohio River Basin for 2011 and 2012 (partial year) for horizontal and vertical wells. Here we report results for horizontal wells, which made up the majority of wells over the two-year period (i.e., 93%, 1,191 horizontal wells versus 96 vertical wells). A range is reported as before because the average water use differed between the two years.

[i] Literature estimates for Texas were from Nicot et al. (2012), using proprietary data from IHS. In most cases, Nicot et al. (2012) reported at the play scale or smaller, rather than the EIA basin scale used for the EPA FracFocus 1.0 project database. We reference 2011 and 2012 (partial year) for Nicot et al. (2012) where possible to overlap with the period of study for the EPA FracFocus 1.0 project database, though more years were available for most formations. A range is reported for some medians because median water use was different for the two years. There were five formations reported for the Permian Basin (Wolfberry, Wolfcamp, Canyon, Clearfork, and San Andres-Greyburg). The most active area in the Permian Basin in 2011–2012 was the Wolfberry, which reported a median of 1 to 1.1 million gal (3.8 to 4.2 million L) per well—these were mostly vertical wells. For the TX-LA-MS Salt Basin Nicot et al. (2012) reported two formations (TX-Haynesville and Cotton Valley), with similar levels of activity in 2011-2012. Wells in TX-Haynesville were predominantly horizontal, while those in Cotton Valley were predominantly vertical (though horizontal wells in Cotton Valley were also reported). There were three fields reported in the Anadarko Basin (Granite Wash, Cleveland, and Marmaton). The most active area in the Anadarko Basin in 2011-2012 was the Granite Wash, which reported a median of 3.3 to 5.2 million gal (12 to 20 million L) per well and where wells were mostly horizontal.

Table B-6. Estimated percent domestic use water from groundwater and self-supplied by county in 2010.

Counties listed contained hydraulically fractured wells with valid state, county, and volume information (U.S. EPA, 2015c). Data estimated from the USGS Water Census (Maupin et al., 2014).

State	County	Percent domestic use water from groundwater[a,b]	Percent domestic use water self supplied[a,c]
Alabama	Jefferson	11.9	0.8
	Tuscaloosa	10.7	6.1
Arkansas	Cleburne	0.0	0.0
	Conway	8.6	8.6
	Faulkner	48.0	3.5
	Independence	20.5	9.4
	Logan	0.0	0.0
	Sebastian	0.0	0.0
	Van Buren	6.4	6.4
	White	0.4	0.0
	Yell	1.8	1.8
California	Colusa	97.9	10.3
	Glenn	96.5	21.6
	Kern	74.5	1.7
	Los Angeles	45.0	4.2
	Sutter	19.4	4.6
	Ventura	30.9	3.9
Colorado	Adams	18.1	2.8
	Arapahoe	19.3	1.3
	Boulder	1.7	1.5
	Broomfield	0.0	0.0
	Delta	59.6	28.4
	Dolores	55.2	51.4
	El Paso	19.6	5.1
	Elbert	100.0	75.2
	Fremont	15.6	15.6

State	County	Percent domestic use water from groundwater[a,b]	Percent domestic use water self supplied[a,c]
Colorado, cont.	Garfield	36.7	28.5
	Jackson	84.4	40.7
	La Plata	24.4	11.3
	Larimer	2.3	0.8
	Las Animas	26.3	16.0
	Mesa	7.3	6.2
	Moffat	36.4	25.8
	Morgan	57.9	4.9
	Phillips	100.0	25.3
	Rio Blanco	60.2	32.5
	Routt	22.6	5.9
	San Miguel	71.4	32.5
	Weld	4.7	0.7
	Yuma	100.0	38.1
Kansas	Barber	100.0	19.0
	Clark	100.0	24.2
	Comanche	100.0	19.2
	Finney	100.0	2.1
	Grant	100.0	23.8
	Gray	100.0	36.4
	Harper	100.0	10.3
	Haskell	100.0	35.2
	Hodgeman	100.0	42.3
	Kearny	100.0	14.6
	Lane	100.0	24.1
	Meade	100.0	25.4
	Morton	100.0	21.7
	Ness	100.0	24.2
	Seward	100.0	15.7

State	County	Percent domestic use water from groundwater[a,b]	Percent domestic use water self supplied[a,c]
Kansas, cont.	Sheridan	100.0	44.9
	Stanton	100.0	29.8
	Stevens	100.0	25.9
	Sumner	51.3	0.0
Louisiana	Allen	100.0	7.5
	Beauregard	100.0	20.6
	Bienville	100.0	16.8
	Bossier	29.4	14.6
	Caddo	12.2	8.8
	Calcasieu	98.3	12.7
	Caldwell	100.0	6.5
	Claiborne	100.0	10.4
	DeSoto	55.8	21.8
	East Feliciana	100.0	11.8
	Jackson	100.0	13.8
	Lincoln	100.0	4.2
	Natchitoches	23.2	11.4
	Rapides	100.0	3.3
	Red River	83.2	27.6
	Sabine	67.5	36.2
	Tangipahoa	100.0	26.9
	Union	100.0	11.2
	Webster	100.0	11.3
	West Feliciana	100.0	2.4
	Winn	100.0	16.4
Michigan	Cheboygan	100.0	76.4
	Gladwin	100.0	84.5
	Kalkaska	100.0	89.0
	Missaukee	100.0	90.6

State	County	Percent domestic use water from groundwater[a,b]	Percent domestic use water self supplied[a,c]
Michigan, cont.	Ogemaw	100.0	90.8
	Roscommon	100.0	91.9
Mississippi	Amite	100.0	26.0
	Wilkinson	100.0	11.1
Montana	Daniels	100.0	29.4
	Garfield	100.0	70.0
	Glacier	62.1	17.7
	Musselshell	89.9	54.5
	Richland	100.0	30.8
	Roosevelt	84.2	20.9
	Rosebud	51.3	10.3
	Sheridan	100.0	31.0
New Mexico	Chaves	100.0	11.8
	Colfax	30.7	2.6
	Eddy	100.0	2.2
	Harding	100.0	25.0
	Lea	100.0	17.4
	Rio Arriba	84.0	42.3
	Roosevelt	100.0	8.9
	San Juan	14.6	12.9
	Sandoval	98.9	23.2
North Dakota	Billings	NA	33.3
	Bottineau	100.0	13.7
	Burke	100.0	12.5
	Divide	100.0	12.5
	Dunn	100.0	21.4
	Golden Valley	100.0	7.7
	Mckenzie	75.8	15.7
	Mclean	12.5	9.9

State	County	Percent domestic use water from groundwater[a,b]	Percent domestic use water self supplied[a,c]
North Dakota, cont.	Mountrail	65.7	11.5
	Stark	NA	5.7
	Williams	27.4	7.3
Ohio	Ashland	98.8	57.4
	Belmont	76.4	8.9
	Carroll	96.4	76.4
	Columbiana	63.2	43.2
	Coshocton	99.3	34.9
	Guernsey	37.6	9.5
	Harrison	65.6	45.9
	Jefferson	33.1	10.2
	Knox	99.2	41.1
	Medina	98.4	83.1
	Muskingum	93.4	17.0
	Noble	8.0	8.0
	Portage	32.6	18.3
	Stark	91.2	30.9
	Tuscarawas	94.0	23.5
	Wayne	99.1	49.0
Oklahoma	Alfalfa	100.0	14.6
	Beaver	100.0	47.9
	Beckham	100.0	10.6
	Blaine	100.0	8.8
	Bryan	26.0	7.8
	Caddo	45.4	35.1
	Canadian	100.0	0.0
	Carter	17.5	0.5
	Coal	31.5	27.5
	Custer	70.8	13.2

State	County	Percent domestic use water from groundwater[a,b]	Percent domestic use water self supplied[a,c]
Oklahoma, cont.	Dewey	100.0	22.5
	Ellis	100.0	31.4
	Garvin	41.3	15.8
	Grady	100.0	34.2
	Grant	100.0	13.2
	Harper	100.0	22.6
	Hughes	23.6	6.7
	Jefferson	13.5	1.8
	Johnston	53.4	1.1
	Kay	39.2	4.6
	Kingfisher	100.0	28.3
	Kiowa	10.3	0.0
	Latimer	12.6	12.6
	Le Flore	14.3	13.1
	Logan	61.1	34.6
	Love	100.0	3.8
	Major	100.0	28.1
	Marshall	20.1	4.4
	Mcclain	95.9	23.9
	Noble	23.3	14.3
	Oklahoma	22.0	2.5
	Osage	18.0	14.9
	Pawnee	38.2	27.7
	Payne	47.9	12.6
	Pittsburg	0.6	0.0
	Roger Mills	80.1	19.4
	Seminole	78.8	16.1
	Stephens	99.2	14.9
	Texas	100.0	10.9

State	County	Percent domestic use water from groundwater[a,b]	Percent domestic use water self supplied[a,c]
Oklahoma, cont.	Washita	53.9	18.2
	Woods	100.0	14.7
Pennsylvania	Allegheny	15.7	15.3
	Armstrong	45.3	36.8
	Beaver	54.7	26.8
	Blair	34.9	24.0
	Bradford	100.0	65.2
	Butler	51.8	42.8
	Cameron	29.0	29.0
	Centre	93.1	21.3
	Clarion	61.5	55.8
	Clearfield	38.4	22.7
	Clinton	48.4	38.1
	Columbia	77.5	56.7
	Crawford	97.7	66.0
	Elk	25.3	15.6
	Fayette	19.2	16.1
	Forest	100.0	78.3
	Greene	31.9	31.9
	Huntingdon	73.2	57.8
	Indiana	52.2	49.1
	Jefferson	60.7	46.1
	Lawrence	40.5	38.8
	Lycoming	60.0	29.3
	McKean	56.6	33.3
	Potter	93.7	58.1
	Somerset	42.6	33.5
	Sullivan	100.0	76.9
	Susquehanna	79.9	74.7

State	County	Percent domestic use water from groundwater[a,b]	Percent domestic use water self supplied[a,c]
Pennsylvania, cont.	Tioga	81.3	58.3
	Venango	95.9	32.7
	Warren	96.9	49.4
	Washington	21.6	21.5
	Westmoreland	21.3	19.8
	Wyoming	100.0	70.6
Texas	Andrews	100.0	23.4
	Angelina	100.0	9.8
	Archer	16.9	16.9
	Atascosa	100.0	16.3
	Austin	100.0	55.6
	Bee	100.0	52.5
	Borden	100.0	71.4
	Bosque	88.7	30.3
	Brazos	100.0	2.1
	Brooks	100.0	35.3
	Burleson	100.0	42.9
	Cherokee	87.5	26.1
	Clay	44.6	36.7
	Cochran	100.0	23.3
	Coke	29.0	28.9
	Colorado	100.0	45.4
	Concho	96.8	5.0
	Cooke	75.5	8.9
	Cottle	100.0	21.4
	Crane	100.0	14.3
	Crockett	100.0	42.5
	Crosby	35.6	19.0
	Culberson	100.0	13.8

State	County	Percent domestic use water from groundwater[a,b]	Percent domestic use water self supplied[a,c]
Texas, cont.	Dallas	1.0	0.7
	Dawson	100.0	33.8
	DeWitt	100.0	42.3
	Denton	9.0	3.6
	Dimmit	100.0	30.5
	Ector	100.0	28.3
	Edwards	100.0	42.1
	Ellis	32.2	7.9
	Erath	100.0	43.3
	Fayette	100.0	27.6
	Fisher	NA	36.8
	Franklin	0.9	0.0
	Freestone	100.0	31.2
	Frio	100.0	20.4
	Gaines	100.0	45.5
	Garza	20.1	17.2
	Glasscock	NA	100.0
	Goliad	NA	66.7
	Gonzales	96.8	15.9
	Grayson	56.0	4.2
	Gregg	20.8	14.1
	Grimes	100.0	26.0
	Hansford	100.0	16.4
	Hardeman	87.6	13.3
	Hardin	100.0	29.5
	Harrison	43.8	24.8
	Hartley	100.0	39.7
	Haskell	100.0	15.7
	Hemphill	100.0	27.5

State	County	Percent domestic use water from groundwater[a,b]	Percent domestic use water self supplied[a,c]
Texas, cont.	Hidalgo	9.2	1.6
	Hockley	100.0	27.4
	Hood	70.8	39.8
	Houston	79.7	36.6
	Howard	100.0	19.8
	Hutchinson	27.3	14.9
	Irion	100.0	50.0
	Jack	46.7	43.8
	Jefferson	25.0	5.8
	Jim Hogg	NA	25.0
	Johnson	34.9	6.8
	Jones	60.5	60.5
	Karnes	100.0	17.6
	Kenedy	100.0	25.0
	Kent	100.0	37.5
	King	100.0	33.3
	Kleberg	100.0	1.9
	Knox	86.2	24.2
	La Salle	100.0	43.3
	Lavaca	100.0	56.0
	Lee	100.0	15.9
	Leon	100.0	41.4
	Liberty	98.5	42.5
	Limestone	46.5	32.5
	Lipscomb	100.0	23.5
	Live Oak	32.8	32.1
	Loving	NA	0.0
	Lynn	64.1	32.2
	Madison	100.0	66.9

State	County	Percent domestic use water from groundwater[a,b]	Percent domestic use water self supplied[a,c]
Texas, cont.	Marion	13.7	8.4
	Martin	100.0	48.9
	Maverick	27.6	27.6
	McMullen	100.0	40.0
	Medina	98.0	23.6
	Menard	36.4	36.4
	Midland	100.0	22.1
	Milam	82.5	41.1
	Mitchell	100.0	14.7
	Montague	57.1	49.7
	Montgomery	100.0	26.6
	Moore	100.0	8.1
	Nacogdoches	55.6	21.6
	Navarro	22.0	22.0
	Newton	100.0	63.7
	Nolan	100.0	17.6
	Nueces	5.6	5.6
	Ochiltree	100.0	16.8
	Oldham	100.0	58.8
	Orange	99.1	41.2
	Palo Pinto	11.7	11.7
	Panola	96.6	58.7
	Parker	63.5	41.1
	Pecos	100.0	31.3
	Polk	41.9	41.7
	Potter	100.0	12.6
	Reagan	100.0	16.2
	Reeves	100.0	31.1
	Roberts	100.0	33.3

State	County	Percent domestic use water from groundwater[a,b]	Percent domestic use water self supplied[a,c]
Texas, cont.	Robertson	97.1	22.5
	Runnels	13.5	13.5
	Rusk	90.7	41.8
	Sabine	76.2	69.0
	San Augustine	78.0	74.4
	San Patricio	88.8	21.8
	Schleicher	100.0	40.0
	Scurry	32.5	27.7
	Shelby	66.2	58.2
	Sherman	100.0	33.3
	Smith	48.0	13.7
	Somervell	87.7	69.3
	Starr	23.2	23.2
	Stephens	13.5	13.5
	Sterling	NA	18.8
	Stonewall	NA	40.0
	Sutton	100.0	26.7
	Tarrant	3.7	1.3
	Terrell	100.0	25.0
	Terry	100.0	16.7
	Tyler	100.0	73.6
	Upshur	54.1	23.2
	Upton	100.0	15.2
	Van Zandt	65.7	39.0
	Walker	57.7	30.6
	Waller	100.0	37.2
	Ward	100.0	4.5
	Washington	48.2	36.0
	Webb	99.4	0.5

State	County	Percent domestic use water from groundwater[a,b]	Percent domestic use water self supplied[a,c]
Texas, cont.	Wharton	100.0	45.9
	Wheeler	100.0	31.3
	Wichita	8.8	2.9
	Wilbarger	100.0	11.5
	Willacy	28.4	28.4
	Wilson	100.0	6.9
	Winkler	100.0	3.8
	Wise	51.3	50.4
	Wood	21.3	12.9
	Yoakum	100.0	36.0
	Young	19.3	18.9
	Zapata	13.9	13.9
	Zavala	100.0	15.2
Utah	Carbon	50.0	1.2
	Duchesne	57.1	10.4
	San Juan	68.3	47.5
Utah	Sevier	100.0	10.0
	Uintah	87.7	3.1
Virginia	Buchanan	NA	27.6
	Dickenson	2.5	2.5
	Wise	5.9	2.3
West Virginia	Barbour	24.1	24.8
	Brooke	33.4	6.8
	Doddridge	60.6	62.1
	Hancock	67.7	6.9
	Harrison	8.8	8.9
	Lewis	29.5	30.3
	Marion	5.8	4.9
	Marshall	96.5	12.0

State	County	Percent domestic use water from groundwater[a,b]	Percent domestic use water self supplied[a,c]
West Virginia, cont.	Monongalia	5.3	5.5
	Ohio	5.4	3.4
	Pleasants	100.0	27.9
	Preston	66.1	41.0
	Ritchie	45.2	46.4
	Taylor	14.9	14.9
	Tyler	44.4	39.2
	Upshur	27.3	27.8
	Webster	41.9	43.2
	Wetzel	96.3	28.6
Wyoming	Big Horn	79.4	11.3
	Campbell	100.0	0.6
	Carbon	63.8	6.7
	Converse	96.5	17.0
	Fremont	49.3	23.7
	Goshen	100.0	21.1
	Hot Springs	31.9	8.2
	Johnson	40.8	35.4
	Laramie	38.1	13.0
	Lincoln	82.4	9.0
	Natrona	69.0	6.6
	Niobrara	100.0	16.3
	Park	18.9	13.7
	Sublette	54.6	22.1
	Sweetwater	3.5	0.4
	Uinta	19.5	11.5
	Washakie	100.0	16.0

[a] Data accessed from the USGS website (http://water.usgs.gov/watuse/data/2010/) on November 11, 2014. Domestic water use is water used for indoor household purposes such as drinking, food preparation, bathing, washing clothes and dishes, flushing toilets, and outdoor purposes such as watering lawns and gardens (Maupin et al., 2014).

[b] Percent domestic water use from groundwater estimated with the following equation: (Domestic public supply volume from groundwater + Domestic self-supplied volume from groundwater)/ Domestic total water use volume * 100. Domestic public supply volume from groundwater was estimated by multiplying the volume of domestic water from public supply by the ratio of public supply volume from groundwater to total public supply volume.

[c] Percent domestic water use self-supplied estimated by dividing the volume of domestic water self-supplied by total domestic water use volume.

Table B-7. Projected hydraulic fracturing water use by Texas counties between 2015 and 2060, expressed as a percentage of 2010 total county water use.

Hydraulic fracturing water use data from Nicot et al. (2012). Total water use data from 2010 from the USGS Water Census (Maupin et al., 2014). All 254 Texas counties are listed by descending order of percentages in 2030.

Texas county	Projected hydraulic fracturing water use as a percentage of 2010 total water use[a,b,c]									
	2015	2020	2025	2030	2035	2040	2045	2050	2055	2060
McMullen	126.2	137.0	152.1	165.1	176.7	164.0	145.3	126.6	108.0	89.3
Irion	36.1	59.2	70.5	63.7	53.4	43.1	32.8	22.4	12.1	5.4
La Salle	58.4	58.3	59.7	60.8	61.9	54.6	45.3	36.0	26.7	17.4
San Augustine	60.2	56.2	52.2	48.2	44.2	40.2	36.2	32.1	28.1	24.1
Sterling	12.0	32.0	39.9	40.5	41.0	34.7	28.3	21.9	15.6	10.7
Dimmit	38.2	38.1	38.9	39.0	38.7	33.9	27.9	22.0	16.0	10.1
Sabine	9.6	19.2	28.7	38.3	35.1	31.9	28.7	25.6	22.3	19.2
Leon	9.9	19.3	27.0	34.6	32.9	29.0	25.1	21.2	17.3	13.5
Karnes	48.1	43.0	37.9	32.6	27.2	21.8	16.4	11.0	5.6	0.2
Loving	13.1	17.4	23.4	29.4	28.8	26.2	23.6	20.9	18.3	15.7
Shackelford	0.0	7.9	15.7	23.6	21.2	18.9	16.5	14.1	11.8	9.4
Madison	5.5	11.8	15.7	19.7	17.4	15.2	13.0	10.9	8.7	6.5
Schleicher	10.5	15.8	19.1	19.7	17.1	14.5	11.9	9.3	6.7	4.7
Sutton	0.0	11.0	15.1	19.1	23.2	20.6	18.1	15.5	12.9	10.3
Shelby	11.0	20.4	19.4	18.4	17.4	15.7	14.1	12.5	10.9	9.3
DeWitt	26.9	24.1	21.4	18.4	15.4	12.3	9.3	6.3	3.2	0.2
Hemphill	25.7	23.1	20.5	17.8	15.2	12.6	10.0	7.3	4.7	2.1
Terrell	0.0	9.7	13.2	16.8	20.4	18.2	15.9	13.6	11.3	9.0
Coryell	7.0	24.4	22.8	16.5	10.1	3.8	0.0	0.0	0.0	0.0

Appendix B – Water Acquisition Supplemental Information

Texas county	Projected hydraulic fracturing water use as a percentage of 2010 total water use[a,b,c]									
	2015	2020	2025	2030	2035	2040	2045	2050	2055	2060
Montague	28.6	24.5	20.4	16.3	12.2	8.2	4.1	0.0	0.0	0.0
Crockett	7.6	12.5	14.8	13.4	11.2	9.1	6.9	4.7	2.5	1.1
Upton	12.1	15.2	14.1	12.9	11.7	9.8	7.9	5.9	4.0	2.7
Borden	3.1	8.6	12.0	12.1	12.2	10.3	8.4	6.4	4.5	3.1
Live Oak	13.3	12.4	11.5	11.8	12.2	12.7	13.2	11.7	9.8	7.8
Reagan	11.2	14.0	12.7	11.3	9.9	8.1	6.4	4.6	2.8	1.6
Clay	3.2	5.9	8.6	11.3	10.3	9.4	8.4	7.5	6.6	5.6
Wheeler	17.6	15.3	13.1	10.8	8.6	6.3	4.1	1.8	0.0	0.0
Lavaca	7.9	13.2	12.0	10.7	9.4	8.1	6.7	5.4	4.0	2.7
Washington	0.0	6.7	11.8	10.7	9.6	8.6	7.5	6.4	5.3	4.3
Nacogdoches	7.9	11.4	10.7	10.0	9.2	8.3	7.5	6.6	5.7	4.9
Hill	17.1	14.7	12.2	9.8	7.3	4.9	2.4	0.0	0.0	0.0
Jack	3.5	5.3	7.1	8.8	7.9	7.1	6.2	5.3	4.4	3.5
Panola	7.2	10.2	9.2	8.5	7.7	7.0	6.3	5.5	4.8	4.0
Jim Hogg	4.8	6.4	8.0	8.0	6.9	6.0	4.9	3.9	2.9	1.8
Howard	4.4	7.1	8.5	8.0	6.8	5.6	4.4	3.2	2.1	1.3
Parker	3.7	5.0	6.3	7.6	6.8	6.1	5.3	4.5	3.8	3.0
Hamilton	8.8	10.7	8.9	7.1	5.3	3.5	1.8	0.0	0.0	0.0
Johnson	14.2	11.9	9.5	7.1	4.7	2.4	0.0	0.0	0.0	0.0
Midland	6.7	8.3	7.7	7.1	6.2	5.2	4.1	3.0	2.0	1.2
Kenedy	4.1	5.4	6.8	6.8	5.9	5.1	4.1	3.3	2.4	1.6
Fayette	3.9	8.4	7.6	6.6	5.5	4.4	3.4	2.3	1.2	0.2

B-42

| Texas county | Projected hydraulic fracturing water use as a percentage of 2010 total water use[a,b,c] | | | | | | | | | |
	2015	2020	2025	2030	2035	2040	2045	2050	2055	2060
Lee	2.1	4.1	5.3	6.5	5.8	5.1	4.3	3.6	2.9	2.1
Winkler	2.9	3.8	5.1	6.3	6.0	5.4	4.7	4.1	3.4	2.8
Wilson	6.7	7.7	7.0	6.2	5.4	4.6	3.9	3.1	2.3	1.5
Martin	5.7	7.1	6.5	6.0	5.3	4.4	3.5	2.6	1.8	1.2
Burleson	1.0	2.9	4.3	5.7	5.1	4.5	3.9	3.3	2.6	2.0
Atascosa	6.3	5.7	5.6	5.6	5.6	5.6	5.0	4.2	3.4	2.7
Bosque	1.8	3.0	4.3	5.5	5.1	4.6	4.2	3.7	3.2	2.8
Webb	7.5	7.1	6.3	5.4	4.6	3.8	3.1	2.3	1.4	0.5
Gonzales	8.0	7.1	6.2	5.3	4.4	3.6	2.7	1.8	0.9	0.0
Marion	1.1	2.4	3.8	5.1	5.2	4.7	4.2	3.7	3.2	2.7
Harrison	4.3	6.1	5.5	5.1	4.6	4.2	3.7	3.3	2.9	2.4
Eastland	0.0	3.9	5.9	5.0	4.2	3.3	2.5	1.7	0.8	0.0
Archer	1.0	2.4	3.6	4.9	4.5	4.1	3.7	3.3	2.9	2.5
Zavala	4.7	5.5	5.2	4.9	4.6	4.3	4.0	3.4	2.7	2.0
Roberts	6.9	6.0	5.1	4.2	3.4	2.5	1.6	0.7	0.0	0.0
Maverick	2.5	3.0	3.6	4.2	4.8	4.5	4.0	3.6	3.1	2.6
Cooke	11.9	9.3	6.7	4.1	1.5	0.0	0.0	0.0	0.0	0.0
Ward	2.7	3.2	4.2	4.1	4.0	3.6	3.2	2.7	2.3	1.9
Austin	0.0	1.2	2.5	3.7	3.4	3.0	2.6	2.2	1.9	1.5
Reeves	1.4	1.8	2.7	3.7	3.9	3.6	3.3	3.0	2.6	2.3
Glasscock	3.1	4.1	3.9	3.6	3.1	2.6	2.1	1.5	1.0	0.7
Tyler	1.9	2.6	3.2	3.2	2.8	2.4	2.0	1.6	1.1	0.7

Appendix B – Water Acquisition Supplemental Information

| Texas county | Projected hydraulic fracturing water use as a percentage of 2010 total water use[a,b,c] | | | | | | | | | |
	2015	2020	2025	2030	2035	2040	2045	2050	2055	2060
Hood	1.4	2.0	2.6	3.2	2.9	2.6	2.2	1.9	1.6	1.3
Garza	1.5	2.0	2.5	2.9	2.7	2.4	2.1	1.8	1.5	1.2
Andrews	2.3	3.0	2.9	2.7	2.6	2.3	2.0	1.7	1.4	1.1
Crane	1.3	1.7	2.1	2.6	3.1	2.8	2.5	2.2	1.9	1.7
Erath	0.9	1.4	1.9	2.4	2.2	2.0	1.8	1.6	1.4	1.2
Wise	3.6	3.2	2.8	2.4	2.0	1.6	1.2	0.8	0.4	0.0
Upshur	0.2	0.9	1.7	2.4	2.9	2.6	2.3	2.1	1.8	1.5
Mitchell	1.2	1.6	2.0	2.4	2.1	1.9	1.7	1.4	1.2	0.9
Ector	1.5	2.0	2.1	2.3	2.2	1.9	1.7	1.4	1.2	1.0
Culberson	0.3	0.4	1.3	2.2	2.9	2.6	2.4	2.1	1.9	1.6
Lipscomb	1.7	3.0	2.6	2.1	1.7	1.3	0.8	0.4	0.0	0.0
Angelina	0.4	0.9	1.5	2.1	2.2	2.0	1.8	1.6	1.4	1.2
Houston	2.1	2.7	2.4	2.1	1.8	1.5	1.2	0.9	0.6	0.3
Frio	1.8	1.8	1.9	1.9	1.8	1.8	1.7	1.5	1.2	0.9
Newton	1.8	2.3	2.1	1.8	1.6	1.3	1.0	0.8	0.5	0.3
Kleberg	1.0	1.4	1.7	1.7	1.5	1.3	1.1	0.8	0.6	0.4
Brooks	1.0	1.3	1.7	1.7	1.5	1.2	1.0	0.8	0.6	0.4
Brazos	0.4	0.9	1.2	1.5	1.4	1.2	1.0	0.8	0.7	0.5
Comanche	0.4	0.7	1.0	1.4	1.2	1.1	1.0	0.8	0.7	0.5
Ochiltree	0.6	1.1	1.5	1.2	1.0	0.7	0.5	0.2	0.0	0.0
Palo Pinto	0.3	0.6	0.9	1.2	1.1	1.0	0.8	0.7	0.6	0.5
Limestone	0.9	1.0	1.1	1.2	1.1	1.0	0.8	0.7	0.6	0.4

| Texas county | Projected hydraulic fracturing water use as a percentage of 2010 total water use[a,b,c] | | | | | | | | | |
	2015	2020	2025	2030	2035	2040	2045	2050	2055	2060
Duval	0.7	0.9	1.1	1.1	1.0	0.8	0.7	0.5	0.4	0.3
Stephens	0.1	0.4	0.8	1.1	1.0	0.9	0.8	0.6	0.5	0.4
Dawson	0.5	0.8	1.0	1.1	1.1	1.0	0.8	0.6	0.5	0.3
Scurry	0.0	0.6	0.8	1.0	1.2	1.1	0.9	0.8	0.7	0.5
Bee	0.8	1.1	1.1	1.0	0.9	0.7	0.6	0.4	0.3	0.1
Val Verde	0.0	0.5	0.8	0.9	1.1	1.0	0.9	0.8	0.6	0.5
Colorado	<0.1	0.3	0.6	0.9	0.8	0.7	0.6	0.5	0.4	0.4
Tarrant	2.1	1.7	1.3	0.9	0.4	0.0	0.0	0.0	0.0	0.0
Zapata	0.5	0.7	0.8	0.8	0.7	0.6	0.5	0.4	0.3	0.2
Ellis	0.3	0.5	0.6	0.8	0.7	0.6	0.6	0.5	0.4	0.3
Jim Wells	0.4	0.6	0.7	0.7	0.6	0.5	0.4	0.4	0.3	0.2
Lynn	0.0	0.4	0.6	0.7	0.8	0.8	0.7	0.6	0.5	0.4
Henderson	0.1	0.3	0.5	0.7	0.8	0.7	0.6	0.5	0.4	0.4
Hansford	0.0	0.4	0.8	0.7	0.5	0.4	0.3	0.2	0.1	0
Gaines	0.2	0.3	0.5	0.5	0.5	0.4	0.4	0.3	0.2	0.2
Gregg	0.1	0.2	0.3	0.4	0.4	0.4	0.4	0.3	0.3	0.2
Refugio	0.2	0.3	0.4	0.4	0.3	0.3	0.2	0.2	0.1	0.1
Caldwell	0.4	0.5	0.4	0.4	0.3	0.3	0.2	0.2	0.1	0.1
Pecos	0.1	0.1	0.2	0.4	0.5	0.4	0.4	0.3	0.3	0.2
Anderson	0.1	0.2	0.3	0.4	0.4	0.4	0.4	0.3	0.3	0.2
Young	0.0	0.1	0.2	0.4	0.3	0.3	0.3	0.2	0.2	0.1
San Patricio	0.2	0.3	0.4	0.4	0.3	0.3	0.2	0.2	0.1	0.1

Texas county	Projected hydraulic fracturing water use as a percentage of 2010 total water use[a,b,c]									
	2015	2020	2025	2030	2035	2040	2045	2050	2055	2060
Smith	0.1	0.1	0.2	0.3	0.4	0.3	0.3	0.3	0.2	0.2
Cherokee	0.1	0.2	0.2	0.3	0.4	0.3	0.3	0.2	0.2	0.2
McLennan	0.1	0.1	0.2	0.3	0.3	0.2	0.2	0.2	0.2	0.1
Terry	0.0	0.2	0.2	0.3	0.3	0.3	0.3	0.2	0.2	0.2
Starr	0.2	0.2	0.3	0.3	0.2	0.2	0.2	0.1	0.1	0.1
Cochran	0.1	0.2	0.2	0.2	0.3	0.2	0.2	0.2	0.2	0.1
Jasper	0.2	0.3	0.2	0.2	0.2	0.1	0.1	0.1	0.1	<0.1
Dallas	0.2	0.3	0.2	0.2	0.1	0.1	<0.1	0.0	0.0	0.0
Robertson	0.1	0.2	0.2	0.2	0.2	0.1	0.1	0.1	0.1	0.1
Grimes	<0.1	0.1	0.1	0.2	0.1	0.1	0.1	0.1	0.1	0.1
Yoakum	0.1	0.1	0.2	0.2	0.1	0.1	0.1	0.1	0.1	0.1
Freestone	0.1	0.1	0.1	0.2	0.2	0.1	0.1	0.1	0.1	0.1
Cass	<0.1	0.1	0.1	0.2	0.2	0.2	0.1	0.1	0.1	0.1
Hutchinson	0.0	0.1	0.2	0.1	0.1	0.1	0.1	<0.1	<0.1	0.0
Rusk	<0.1	0.1	0.1	0.1	0.1	0.1	0.1	0.1	0.1	<0.1
Willacy	<0.1	0.1	0.1	0.1	0.1	0.1	0.1	<0.1	0.1	<0.1
Victoria	<0.1	0.1	0.1	0.1	0.1	0.1	<0.1	<0.1	<0.1	<0.1
Sherman	0.0	0.0	<0.1	0.1	0.1	0.1	<0.1	<0.1	<0.1	<0.1
Calhoun	<0.1	0.1	0.1	0.1	0.1	0.1	<0.1	<0.1	<0.1	<0.1
Lubbock	0.0	0.0	<0.1	0.1	0.1	0.1	0.1	0.1	0.1	0.1
Jackson	<0.1	<0.1	0.1	0.1	<0.1	<0.1	<0.1	<0.1	<0.1	<0.1
Matagorda	<0.1	<0.1	<0.1	<0.1	<0.1	<0.1	<0.1	<0.1	<0.1	<0.1

| Texas county | Projected hydraulic fracturing water use as a percentage of 2010 total water use[a,b,c] | | | | | | | | | | |
	2015	2020	2025	2030	2035	2040	2045	2050	2055	2060
Polk	<0.1	<0.1	<0.1	<0.1	<0.1	<0.1	<0.1	<0.1	<0.1	<0.1
Wharton	<0.1	<0.1	<0.1	<0.1	<0.1	<0.1	<0.1	<0.1	<0.1	<0.1
Nueces	<0.1	<0.1	<0.1	<0.1	<0.1	<0.1	<0.1	<0.1	<0.1	<0.1
Hidalgo	<0.1	<0.1	<0.1	<0.1	<0.1	<0.1	<0.1	<0.1	<0.1	<0.1
Cameron	<0.1	<0.1	<0.1	<0.1	<0.1	<0.1	<0.1	<0.1	<0.1	<0.1
Somervell	<0.1	<0.1	<0.1	<0.1	<0.1	<0.1	<0.1	<0.1	<0.1	<0.1
Goliad	<0.1	<0.1	<0.1	<0.1	<0.1	<0.1	<0.1	<0.1	<0.1	<0.1
Brazoria	<0.1	<0.1	<0.1	<0.1	<0.1	<0.1	<0.1	<0.1	<0.1	<0.1
Fort Bend	<0.1	<0.1	<0.1	<0.1	<0.1	<0.1	<0.1	<0.1	<0.1	<0.1
Aransas	0.0	0.0	0.0	0.0	0.0	0.0	0.0	0.0	0.0	0.0
Armstrong	0.0	0.0	0.0	0.0	0.0	0.0	0.0	0.0	0.0	0.0
Bailey	0.0	0.0	0.0	0.0	0.0	0.0	0.0	0.0	0.0	0.0
Bandera	0.0	0.0	0.0	0.0	0.0	0.0	0.0	0.0	0.0	0.0
Bastrop	0.0	0.0	0.0	0.0	0.0	0.0	0.0	0.0	0.0	0.0
Baylor	0.0	0.0	0.0	0.0	0.0	0.0	0.0	0.0	0.0	0.0
Bell	0.0	0.0	0.0	0.0	0.0	0.0	0.0	0.0	0.0	0.0
Bexar	0.0	0.0	0.0	0.0	0.0	0.0	0.0	0.0	0.0	0.0
Blanco	0.0	0.0	0.0	0.0	0.0	0.0	0.0	0.0	0.0	0.0
Bowie	0.0	0.0	0.0	0.0	0.0	0.0	0.0	0.0	0.0	0.0
Brewster	0.0	0.0	0.0	0.0	0.0	0.0	0.0	0.0	0.0	0.0
Briscoe	0.0	0.0	0.0	0.0	0.0	0.0	0.0	0.0	0.0	0.0
Brown	0.0	0.0	0.0	0.0	0.0	0.0	0.0	0.0	0.0	0.0

Appendix B – Water Acquisition Supplemental Information

| Texas county | Projected hydraulic fracturing water use as a percentage of 2010 total water use[a,b,c] | | | | | | | | | |
	2015	2020	2025	2030	2035	2040	2045	2050	2055	2060
Burnet	0.0	0.0	0.0	0.0	0.0	0.0	0.0	0.0	0.0	0.0
Callahan	0.0	0.0	0.0	0.0	0.0	0.0	0.0	0.0	0.0	0.0
Camp	0.0	0.0	0.0	0.0	0.0	0.0	0.0	0.0	0.0	0.0
Carson	0.0	0.0	0.0	0.0	0.0	0.0	0.0	0.0	0.0	0.0
Castro	0.0	0.0	0.0	0.0	0.0	0.0	0.0	0.0	0.0	0.0
Chambers	0.0	0.0	0.0	0.0	0.0	0.0	0.0	0.0	0.0	0.0
Childress	0.0	0.0	0.0	0.0	0.0	0.0	0.0	0.0	0.0	0.0
Coke	0.0	0.0	0.0	0.0	0.0	0.0	0.0	0.0	0.0	0.0
Coleman	0.0	0.0	0.0	0.0	0.0	0.0	0.0	0.0	0.0	0.0
Collin	0.0	0.0	0.0	0.0	0.0	0.0	0.0	0.0	0.0	0.0
Collingsworth	0.0	0.0	0.0	0.0	0.0	0.0	0.0	0.0	0.0	0.0
Comal	0.0	0.0	0.0	0.0	0.0	0.0	0.0	0.0	0.0	0.0
Concho	0.0	0.0	0.0	0.0	0.0	0.0	0.0	0.0	0.0	0.0
Cottle	0.0	0.0	0.0	0.0	0.0	0.0	0.0	0.0	0.0	0.0
Crosby	0.0	0.0	0.0	0.0	0.0	0.0	0.0	0.0	0.0	0.0
Dallam	0.0	0.0	0.0	0.0	0.0	0.0	0.0	0.0	0.0	0.0
Deaf Smith	0.0	0.0	0.0	0.0	0.0	0.0	0.0	0.0	0.0	0.0
Delta	0.0	0.0	0.0	0.0	0.0	0.0	0.0	0.0	0.0	0.0
Denton	1.7	1.1	0.6	0.0	0.0	0.0	0.0	0.0	0.0	0.0
Dickens	0.0	0.0	0.0	0.0	0.0	0.0	0.0	0.0	0.0	0.0
Donley	0.0	0.0	0.0	0.0	0.0	0.0	0.0	0.0	0.0	0.0
Edwards	0.0	0.0	0.0	0.0	0.0	0.0	0.0	0.0	0.0	0.0

| Texas county | Projected hydraulic fracturing water use as a percentage of 2010 total water use[a,b,c] | | | | | | | | | |
	2015	2020	2025	2030	2035	2040	2045	2050	2055	2060
El Paso	0.0	0.0	0.0	0.0	0.0	0.0	0.0	0.0	0.0	0.0
Falls	0.0	0.0	0.0	0.0	0.0	0.0	0.0	0.0	0.0	0.0
Fannin	0.0	0.0	0.0	0.0	0.0	0.0	0.0	0.0	0.0	0.0
Fisher	0.0	0.0	0.0	0.0	0.0	0.0	0.0	0.0	0.0	0.0
Floyd	0.0	0.0	0.0	0.0	0.0	0.0	0.0	0.0	0.0	0.0
Foard	0.0	0.0	0.0	0.0	0.0	0.0	0.0	0.0	0.0	0.0
Franklin	0.0	0.0	0.0	0.0	0.0	0.0	0.0	0.0	0.0	0.0
Galveston	0.0	0.0	0.0	0.0	0.0	0.0	0.0	0.0	0.0	0.0
Gillespie	0.0	0.0	0.0	0.0	0.0	0.0	0.0	0.0	0.0	0.0
Gray	0.0	0.0	0.0	0.0	0.0	0.0	0.0	0.0	0.0	0.0
Grayson	0.0	0.0	0.0	0.0	0.0	0.0	0.0	0.0	0.0	0.0
Guadalupe	0.0	0.0	0.0	0.0	0.0	0.0	0.0	0.0	0.0	0.0
Hale	0.0	0.0	0.0	0.0	0.0	0.0	0.0	0.0	0.0	0.0
Hall	0.0	0.0	0.0	0.0	0.0	0.0	0.0	0.0	0.0	0.0
Hardeman	0.0	0.0	0.0	0.0	0.0	0.0	0.0	0.0	0.0	0.0
Hardin	0.0	0.0	0.0	0.0	0.0	0.0	0.0	0.0	0.0	0.0
Harris	0.0	0.0	0.0	0.0	0.0	0.0	0.0	0.0	0.0	0.0
Hartley	0.0	0.0	0.0	0.0	0.0	0.0	0.0	0.0	0.0	0.0
Haskell	0.0	0.0	0.0	0.0	0.0	0.0	0.0	0.0	0.0	0.0
Hays	0.0	0.0	0.0	0.0	0.0	0.0	0.0	0.0	0.0	0.0
Hockley	0.0	0.0	0.0	0.0	0.0	0.0	0.0	0.0	0.0	0.0
Hopkins	0.0	0.0	0.0	0.0	0.0	0.0	0.0	0.0	0.0	0.0

Appendix B – Water Acquisition Supplemental Information

| Texas county | Projected hydraulic fracturing water use as a percentage of 2010 total water use[a,b,c] | | | | | | | | | |
	2015	2020	2025	2030	2035	2040	2045	2050	2055	2060
Hudspeth	0.0	0.0	0.0	0.0	0.0	0.0	0.0	0.0	0.0	0.0
Hunt	0.0	0.0	0.0	0.0	0.0	0.0	0.0	0.0	0.0	0.0
Jeff Davis	0.0	0.0	0.0	0.0	0.0	0.0	0.0	0.0	0.0	0.0
Jefferson	0.0	0.0	0.0	0.0	0.0	0.0	0.0	0.0	0.0	0.0
Jones	0.0	0.0	0.0	0.0	0.0	0.0	0.0	0.0	0.0	0.0
Kaufman	0.0	0.0	0.0	0.0	0.0	0.0	0.0	0.0	0.0	0.0
Kendall	0.0	0.0	0.0	0.0	0.0	0.0	0.0	0.0	0.0	0.0
Kent	0.0	0.0	0.0	0.0	0.0	0.0	0.0	0.0	0.0	0.0
Kerr	0.0	0.0	0.0	0.0	0.0	0.0	0.0	0.0	0.0	0.0
Kimble	0.0	0.0	0.0	0.0	0.0	0.0	0.0	0.0	0.0	0.0
King	0.0	0.0	0.0	0.0	0.0	0.0	0.0	0.0	0.0	0.0
Kinney	0.0	0.0	0.0	0.0	0.0	0.0	0.0	0.0	0.0	0.0
Knox	0.0	0.0	0.0	0.0	0.0	0.0	0.0	0.0	0.0	0.0
Lamar	0.0	0.0	0.0	0.0	0.0	0.0	0.0	0.0	0.0	0.0
Lamb	0.0	0.0	0.0	0.0	0.0	0.0	0.0	0.0	0.0	0.0
Lampasas	0.0	0.0	0.0	0.0	0.0	0.0	0.0	0.0	0.0	0.0
Liberty	0.0	0.0	0.0	0.0	0.0	0.0	0.0	0.0	0.0	0.0
Llano	0.0	0.0	0.0	0.0	0.0	0.0	0.0	0.0	0.0	0.0
McCulloch	0.0	0.0	0.0	0.0	0.0	0.0	0.0	0.0	0.0	0.0
Mason	0.0	0.0	0.0	0.0	0.0	0.0	0.0	0.0	0.0	0.0
Medina	0.0	0.0	0.0	0.0	0.0	0.0	0.0	0.0	0.0	0.0
Menard	0.0	0.0	0.0	0.0	0.0	0.0	0.0	0.0	0.0	0.0

| Texas county | Projected hydraulic fracturing water use as a percentage of 2010 total water use[a,b,c] | | | | | | | | | |
	2015	2020	2025	2030	2035	2040	2045	2050	2055	2060
Milam	0.0	0.0	0.0	0.0	0.0	0.0	0.0	0.0	0.0	0.0
Mills	0.0	0.0	0.0	0.0	0.0	0.0	0.0	0.0	0.0	0.0
Montgomery	0.0	0.0	0.0	0.0	0.0	0.0	0.0	0.0	0.0	0.0
Moore	0.0	0.0	0.0	0.0	0.0	0.0	0.0	0.0	0.0	0.0
Morris	0.0	0.0	0.0	0.0	0.0	0.0	0.0	0.0	0.0	0.0
Motley	0.0	0.0	0.0	0.0	0.0	0.0	0.0	0.0	0.0	0.0
Navarro	0.0	0.0	0.0	0.0	0.0	0.0	0.0	0.0	0.0	0.0
Nolan	0.0	0.0	0.0	0.0	0.0	0.0	0.0	0.0	0.0	0.0
Oldham	0.0	0.0	0.0	0.0	0.0	0.0	0.0	0.0	0.0	0.0
Orange	0.0	0.0	0.0	0.0	0.0	0.0	0.0	0.0	0.0	0.0
Parmer	0.0	0.0	0.0	0.0	0.0	0.0	0.0	0.0	0.0	0.0
Potter	0.0	0.0	0.0	0.0	0.0	0.0	0.0	0.0	0.0	0.0
Presidio	0.0	0.0	0.0	0.0	0.0	0.0	0.0	0.0	0.0	0.0
Rains	0.0	0.0	0.0	0.0	0.0	0.0	0.0	0.0	0.0	0.0
Randall	0.0	0.0	0.0	0.0	0.0	0.0	0.0	0.0	0.0	0.0
Real	0.0	0.0	0.0	0.0	0.0	0.0	0.0	0.0	0.0	0.0
Red River	0.0	0.0	0.0	0.0	0.0	0.0	0.0	0.0	0.0	0.0
Rockwall	0.0	0.0	0.0	0.0	0.0	0.0	0.0	0.0	0.0	0.0
Runnels	0.0	0.0	0.0	0.0	0.0	0.0	0.0	0.0	0.0	0.0
San Jacinto	0.0	0.0	0.0	0.0	0.0	0.0	0.0	0.0	0.0	0.0
San Saba	0.0	0.0	0.0	0.0	0.0	0.0	0.0	0.0	0.0	0.0
Stonewall	0.0	0.0	0.0	0.0	0.0	0.0	0.0	0.0	0.0	0.0

Appendix B – Water Acquisition Supplemental Information

| Texas county | Projected hydraulic fracturing water use as a percentage of 2010 total water use[a,b,c] | | | | | | | | | |
	2015	2020	2025	2030	2035	2040	2045	2050	2055	2060
Swisher	0.0	0.0	0.0	0.0	0.0	0.0	0.0	0.0	0.0	0.0
Taylor	0.0	0.0	0.0	0.0	0.0	0.0	0.0	0.0	0.0	0.0
Throckmorton	0.0	0.0	0.0	0.0	0.0	0.0	0.0	0.0	0.0	0.0
Titus	0.0	0.0	0.0	0.0	0.0	0.0	0.0	0.0	0.0	0.0
Tom Green	0.0	0.0	0.0	0.0	0.0	0.0	0.0	0.0	0.0	0.0
Travis	0.0	0.0	0.0	0.0	0.0	0.0	0.0	0.0	0.0	0.0
Trinity	0.0	0.0	0.0	0.0	0.0	0.0	0.0	0.0	0.0	0.0
Uvalde	0.0	0.0	0.0	0.0	0.0	0.0	0.0	0.0	0.0	0.0
Van Zandt	0.0	0.0	0.0	0.0	0.0	0.0	0.0	0.0	0.0	0.0
Walker	0.0	0.0	0.0	0.0	0.0	0.0	0.0	0.0	0.0	0.0
Waller	0.0	0.0	0.0	0.0	0.0	0.0	0.0	0.0	0.0	0.0
Wichita	0.0	0.0	0.0	0.0	0.0	0.0	0.0	0.0	0.0	0.0
Wilbarger	0.0	0.0	0.0	0.0	0.0	0.0	0.0	0.0	0.0	0.0
Williamson	0.0	0.0	0.0	0.0	0.0	0.0	0.0	0.0	0.0	0.0
Wood	0.0	0.0	0.0	0.0	0.0	0.0	0.0	0.0	0.0	0.0

[a] Total water use data accessed from the USGS website (http://water.usgs.gov/watuse/data/2010/) on April 21, 2015 (Maupin et al., 2014). Data from Nicot et al. (2012) transcribed.

[b] Percentages calculated by dividing projected hydraulic fracturing water use volumes from Nicot et al. (2012) by 2010 total water use from the USGS (Maupin et al., 2014) and multiplying by 100. Note, the projected hydraulic fracturing water use volume from Nicot et al. (2012) was not added to the 2010 total USGS water use value in the denominator, and is simply expressed as a percentage compared to 2010 total water use. This was done because of the difference in years between the two datasets, and because the USGS 2010 Water Census (Maupin et al., 2014) included hydraulic fracturing water use estimates in their mining category. This approach is consistent with that of other literature on this topic; see Nicot and Scanlon (2012). Estimates of projected hydraulic fracturing water use as a percentage of 2010 total water use exceeded 100% when projected hydraulic fracturing water use exceeded 2010 total water use in that county in 2010.

[c] Percentages less than 0.1 were not rounded and simply noted as "<0.1," but where the percentage was actually zero because there was no projected hydraulic fracturing water use we noted that as "0.0."

B.2. Supplemental Discussion: Potential for Water Acquisition Impacts by Location

This section includes an expanded discussion of the potential for water acquisition impacts by location. This discussion provides further examples of the concepts illustrated in Chapter 4, Section 4.5, and includes a discussion for Oklahoma and Kansas (Section B.2.1) and Utah, New Mexico, and California (Section B.2.2).

B.2.1. Oklahoma and Kansas

Oklahoma had the fifth most disclosures in the EPA FracFocus 1.0 project database (5.0% of disclosures) (Table B-5, Figure 4-4). Three major basins—the Anadarko, which includes the Woodford play; the Arkoma, which includes the Fayetteville play; and the Ardmore, which includes the Woodford play—contain 67% of the disclosures in Oklahoma (Table B-5, Figure B-1). Few wells were reported for Kansas (Kansas disclosures comprise 0.4% of the EPA FracFocus 1.0 project database), but because of the shared geology of the Cherokee Platform across the two states, we group Kansas with Oklahoma. Oklahoma and Kansas were two of the three states where a large fraction of wells were not associated with a basin defined by the U.S. EIA (U.S. EPA, 2015c) (Table B-5).[1]

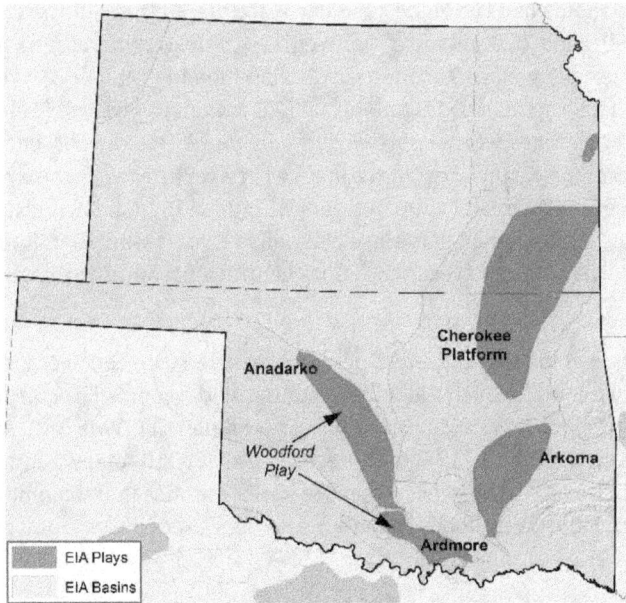

Figure B-1. Major U.S. EIA shale plays and basins for Oklahoma and Kansas.
Source: EIA (2015).

---

[1] Alaska was the other state in the EPA FracFocus 1.0 project database where the U.S. EIA shale basins did not adequately describe well locations, with all 37 wells in Alaska not associated with a U.S. EIA basin. For all other states, U.S. EIA shale basins captured 86%–100% of the wells in the EPA FracFocus 1.0 project database (U.S. EPA, 2015c).

Types of water used: Water for hydraulic fracturing in Oklahoma and Kansas comes from both surface and groundwater (Kansas Water Office, 2014; Taylor, 2012). Data on temporary water use permits in Oklahoma (which make up the majority of water use permits for Oklahoma oil and gas mining) show that, in 2011, approximately 63% and 37% of water for hydraulic fracturing came from surface and groundwater, respectively (Taylor, 2012) (Table 4-1). General water use in Oklahoma follows an east-west divide, with the eastern half dependent on surface sources and the western half relying heavily on groundwater (OWRB, 2014). Water obtained for fracturing is assumed to fit this pattern as well. No data are available on the proportion of hydraulic fracturing water that is sourced from surface versus groundwater resources in Kansas.

For both Oklahoma and Kansas, data are also lacking to describe the extent to which reused wastewater is used as a percentage of total injected volume. However, the quality of Oklahoma's Woodford Shale wastewater has been described as low in TDS, and thus reuse could reduce the demand for fresh water (Kuthnert et al., 2012).

Water use per well: Estimates of median water use per well in Oklahoma include 2.6 million gal (9.8 million L) and 3 million gal (11 million L) (U.S. EPA; Murray, 2013, respectively). Water use for hydraulic fracturing increased from 2000 to 2011, driven by volumes required for fracturing horizontal wells across the state (Murray, 2013). Within the state, there are wide ranges in water use for different formations. According to the EPA FracFocus 1.0 project database, the Ardmore and Arkoma Basins of Oklahoma had the highest median water use in the country, with medians of 8.0 and 6.7 million gal (30.3 and 25.4 million L) per well, respectively; whereas the Anadarko Basin had lower median water use per well (3.3 million gal (12.5 million L) (Table B-5). Wells not associated with a U.S. EIA basin had a median of 1.9 million gal (7.2 million L) per well (Table B-5). It is not clear why lower water volumes were reportedly used in unassociated wells, but Oklahoma has several CBM deposits in the eastern part of the state where very low water use per well has been reported (i.e., less than approximately 300,000 gal (1.1 million L) in the Arbuckle and Hartshorne formations) (Murray, 2013). Median water use per well in Kansas was 1.5 million gal (5.7 million L), focused mostly in a five-county area in the south-central and southwest portions of the state (Table B-5).

Water use/consumption at the county scale: Operators reported using an average of 71.9 million gal (272.2 million L) of water annually in Oklahoma counties with reported fracturing activity in 2011 and 2012; in Kansas, this value was 3.5 million gal (13.2 million L) (Table B-2). Average hydraulic fracturing water use in 2011 and 2012 did not exceed 10% of 2010 total water use in any county in Oklahoma or Kansas (Table B-2). However, there were six counties in Oklahoma (Alfalfa, Canadian, Coal, Pittsburg, Rogers Mills, and Woods) where fracturing water consumption exceeded 10% of 2010 total county water consumption.

Potential for impacts: The potential for impacts on drinking water resources appears to be low in Oklahoma and Kansas at the county scale, since hydraulic fracturing water use and consumption are generally low as a percentage of total water use, consumption, and availability at this scale (Text Box 4-2, Figure 4-6a,b). If local impacts to water quantity or quality do occur, they are more likely to happen in western Oklahoma than in the eastern half of the state or Kansas. Of the six Oklahoma counties where fracturing consumption exceeded 10% of 2010 water consumption, three (Alfalfa, Canadian, and Roger Mills) are in the western half of the state where surface water availability is lowest (Figure 4-7a). Surface water is fully allocated in the Panhandle and West Central regions, encompassing much of the state's northwestern quadrant (OWRB, 2014). As a result, residents generally rely on groundwater in western Oklahoma (Table B-6), and it is likely that fracturing does as well.

Projecting out to 2060, Oklahoma's Water Plan concludes that aquifer storage depletions are likely in the Panhandle and West Central regions due to over-pumping, particularly for irrigation (OWRB, 2014). Groundwater depletions are anticipated to be small relative to storage, but will be the largest in summer months and may lead to higher pumping costs, the need for deeper water wells, lower water yields, and detrimental effects on water quality (OWRB, 2014). Drought conditions are likely to exacerbate this problem, and Oklahoma's Water Plan raises the potential for climate change to affect future water supplies in the state (OWRB, 2014). In the adjacent Texas Panhandle, future irrigation needs may go unmet (TWDB, 2012), and this may be the case in western Oklahoma as well.

Aquifer depletions in western Oklahoma may be associated with groundwater quality degradation, particularly under drought conditions. The central portion of the Ogallala aquifer underlying the Oklahoma Panhandle and western Oklahoma contains elevated levels of some constituents (e.g., nitrate) due to over-pumping, although generally it is of better quality than the southern portion of the aquifer (Gurdak et al., 2009). Additional groundwater withdrawals for hydraulic fracturing in western Oklahoma may add to these water quality issues, particularly in combination with other substantial water uses (e.g., irrigation) (Gurdak et al., 2009).

B.2.2. Utah, New Mexico, and California

Together, Utah, New Mexico, and California accounted for approximately 9% of disclosures in the EPA FracFocus 1.0 project database (3.8%, 3.1% and 1.9% of disclosures, respectively) (Table B-5, Figure 4-4). Almost all reported hydraulic fracturing in Utah and California was in the Uinta-Piceance Basin (99%) and San Joaquin Basin (95%), respectively. Activity in New Mexico mostly occurs in the Permian and San Juan Basins, which together comprised 96% of reported disclosures in that state (Figure B-2).

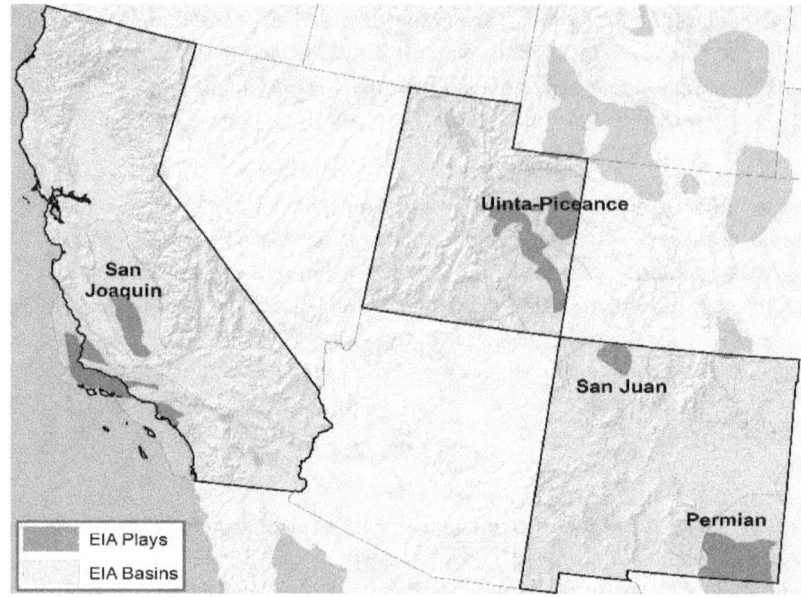

Figure B-2. Major U.S. EIA shale plays and basins for Utah, New Mexico, and California.
Source: EIA (2015).

Types of water used: Of these three states, California has the most information available on the sources of water used for hydraulic fracturing. Most current and proposed fracturing activity occurs in Kern County in the San Joaquin Basin, where operators depend mainly on surface water purchased from nearby irrigation districts (CCST, 2014). California irrigation districts receive water allocated by the State Water Project, and deliveries may be restricted or eliminated during drought years (CCST, 2014).[1] In addition to publicly-supplied surface water, operators may also self-supply a smaller proportion of water from on-site groundwater wells (CCST, 2014). Most water used for hydraulic fracturing in California is fresh (91% of annual water used in well stimulation) (CCST, 2015a). Approximately 13% of water demand for hydraulic fracturing is offset by the reuse of wastewater, according to well stimulation records (CCST, 2015a) (Table 4-2).

The source, quality, and provisioning of water used for hydraulic fracturing in Utah and New Mexico are not as well characterized. A 2010 New Mexico water use report summarizes withdrawals for a variety of water use categories, and 26% and 74% of mining water use (which includes water used for oil and gas production) came from surface and groundwater withdrawals, respectively (NM OSE, 2013). If hydraulic fracturing water use in New Mexico follows the same pattern as other mining uses (e.g., for metals, coal, geothermal), then it is likely that groundwater is the primary source. To our knowledge, no data are available to characterize the source of water for

---

[1] The California State Water Project is a water storage and distribution system maintained by the California Department of Water Resources, which provides water for urban and agricultural water suppliers in Northern California, the San Francisco Bay Area, the San Joaquin Valley, the Central Coast, and Southern California (California Department of Water Resources, 2015).

hydraulic fracturing operations in Utah. In addition, data are lacking on the reuse of wastewater as a proportion of total water injected for both Utah and New Mexico.

Water use per well: Median water use per well in Utah, New Mexico, and California is lower than in other states in the EPA FracFocus 1.0 project database: Utah ranks 13th (approximately 302,000 gal or 1.14 million L), New Mexico ranks 14th (approximately 175,000 gal or 662,000 L), and California ranks 15th (approximately 77,000 gal or 291,000 L) out of the 15 states (Table B-5). A possible explanation for the low water use per well in Utah and New Mexico is the presence of CBM in the Uinta (Utah) and San Juan (New Mexico) Basins. Low water use per well in California is attributed to the prevalence of vertical wells and the use of crosslinked gels. Vertical wells dominate because the complex geology precludes long horizontal drilling and fracturing (CCST, 2014).

For California, the California Council on Science and Technology (CCST) reports average water use per well of 130,000 gal (490,000 L), which agrees with the state average of approximately 131,700 gal (498,500 L) according to the EPA FracFocus 1.0 project database (CCST, 2014) (Table B-5); this is to be expected, because estimates from CCST are also based on data submitted to FracFocus.

Water use/consumption at the county scale: Hydraulic fracturing in Utah, New Mexico, and California uses relatively small amounts of water at the county scale compared to most other states (Table B-1). Only in four counties (Duchesne and Uintah Counties in Utah, and Eddy and Lea Counties in New Mexico) did hydraulic fracturing operators use more than 50 million gal (189 million L) annually in 2011 and 2012 (Table B-2). Fracturing water use and consumption did not exceed 1% of 2010 total water use and consumption in any county.

Potential for impacts: At present, hydraulic fracturing does not use or consume much water compared to other users or consumers in Utah, New Mexico, and California at the county scale (Figure 4-2a,b). Likewise, it also does not use much water compared to county level water availability estimates (Text Box 4-2, Figure 4-6a,b). In general, however, Utah, New Mexico, and California have low surface water availability (Figure 4-7a), high groundwater dependence (Figure 4-7b), and have experienced frequent periods of drought over the last decade (National Drought Mitigation Center, 2015). All of these factors increase the potential for localized impacts. In California, two recent studies conclude changes in water quantity or quality are possible in the San Joaquin Basin due to hydraulic fracturing withdrawals, especially within Kern County where oil and gas activities are concentrated and fresh water is in limited supply (Tiedeman et al., 2016; CCST, 2015a). The combination of factors also suggest future problems could arise if hydraulic fracturing water withdrawals increase substantially in these states beyond present levels, without commensurate steps to reduce fresh water demand.

This page is intentionally left blank.

Appendix C. Chemical Mixing Supplemental Information

This page is intentionally left blank.

Appendix C. Chemical Mixing Supplemental Information

C.1. Most Frequently Reported Chemicals in Gas- and Oil-Producing Wells

Table C-1. Chemicals reported in 10% or more of disclosures in the EPA FracFocus 1.0 project database for gas-producing wells, with the number of disclosures (for reported chemicals), percentage of disclosures, and the median maximum concentration (% by mass) of that chemical in hydraulic fracturing fluid.
Chemicals ranked by frequency of occurrence (U.S. EPA, 2015c). See Text Box 5-2 for more information.

Chemical name	CASRN	Number of disclosures	Percentage of disclosures	Median maximum concentration in hydraulic fracturing fluid (% by mass)
Hydrochloric acid	7647-01-0	12,351	72.8%	15%
Methanol	67-56-1	12,269	72.3%	30%
Distillates, petroleum, hydrotreated light	64742-47-8	11,897	70.1%	30%
Isopropanol	67-63-0	8,008	47.2%	30%
Water	7732-18-5	7,998	47.1%	63%
Ethanol	64-17-5	6,325	37.3%	5%
Propargyl alcohol	107-19-7	5,811	34.2%	10%
Glutaraldehyde	111-30-8	5,635	33.2%	30%
Ethylene glycol	107-21-1	5,493	32.4%	35%
Citric acid	77-92-9	4,832	28.5%	60%
Sodium hydroxide	1310-73-2	4,656	27.4%	5%
Peroxydisulfuric acid, diammonium salt	7727-54-0	4,618	27.2%	100%
Quartz	14808-60-7	3,758	22.1%	10%
2,2-Dibromo-3-nitrilopropionamide	10222-01-2	3,668	21.6%	100%
Sodium chloride	7647-14-5	3,608	21.3%	30%
Guar gum	9000-30-0	3,586	21.1%	60%
Acetic acid	64-19-7	3,563	21.0%	50%
2-Butoxyethanol	111-76-2	3,325	19.6%	10%
Naphthalene	91-20-3	3,294	19.4%	5%
Solvent naphtha, petroleum, heavy arom.	64742-94-5	3,287	19.4%	30%

Chemical name	CASRN	Number of disclosures	Percentage of disclosures	Median maximum concentration in hydraulic fracturing fluid (% by mass)
Quaternary ammonium compounds, benzyl-C12-16-alkyldimethyl, chlorides	68424-85-1	3,259	19.2%	7%
Potassium hydroxide	1310-58-3	2,843	16.8%	15%
Ammonium chloride	12125-02-9	2,483	14.6%	10%
Choline chloride	67-48-1	2,477	14.6%	75%
Poly(oxy-1,2-ethanediyl)-nonylphenyl-hydroxy (mixture)	127087-87-0	2,455	14.5%	5%
Sodium chlorite	7758-19-2	2,372	14.0%	10%
1,2,4-Trimethylbenzene	95-63-6	2,229	13.1%	1%
Carbonic acid, dipotassium salt	584-08-7	2,154	12.7%	60%
Methenamine	100-97-0	2,134	12.6%	1%
Formic acid	64-18-6	2,118	12.5%	60%
Didecyl dimethyl ammonium chloride	7173-51-5	2,063	12.2%	10%
N,N-Dimethylformamide	68-12-2	1,892	11.2%	13%
Phenolic resin	9003-35-4	1,852	10.9%	5%
Thiourea polymer	68527-49-1	1,702	10.0%	30%
Polyethylene glycol	25322-68-3	1,696	10.0%	60%

Table C-2. Chemicals reported in 10% or more of disclosures in the EPA FracFocus 1.0 project database for oil-producing wells, with the number of disclosures (for reported chemicals), percentage of disclosures, and the median maximum concentration (% by mass) of that chemical in hydraulic fracturing fluid.

Chemicals ranked by frequency of occurrence (U.S. EPA, 2015c).

Chemical name	CASRN	Number of disclosures	Percentage of disclosures	Median maximum concentration in hydraulic fracturing fluid (% by mass)
Methanol	67-56-1	12,484	71.8%	30%
Distillates, petroleum, hydrotreated light	64742-47-8	10,566	60.8%	40%
Peroxydisulfuric acid, diammonium salt	7727-54-0	10,350	59.6%	100%
Ethylene glycol	107-21-1	10,307	59.3%	30%
Hydrochloric acid	7647-01-0	10,029	57.7%	15%
Guar gum	9000-30-0	9,110	52.4%	50%
Sodium hydroxide	1310-73-2	8,609	49.5%	10%
Quartz	14808-60-7	8,577	49.4%	2%
Water	7732-18-5	8,538	49.1%	67%
Isopropanol	67-63-0	8,031	46.2%	15%
Potassium hydroxide	1310-58-3	7,206	41.5%	15%
Glutaraldehyde	111-30-8	5,927	34.1%	15%
Propargyl alcohol	107-19-7	5,599	32.2%	5%
Acetic acid	64-19-7	4,623	26.6%	30%
2-Butoxyethanol	111-76-2	4,022	23.1%	10%
Solvent naphtha, petroleum, heavy arom.	64742-94-5	3,821	22.0%	5%
Sodium chloride	7647-14-5	3,692	21.2%	25%
Ethanol	64-17-5	3,536	20.3%	45%
Citric acid	77-92-9	3,310	19.0%	60%
Phenolic resin	9003-35-4	3,109	17.9%	5%
Naphthalene	91-20-3	3,060	17.6%	5%
Nonyl phenol ethoxylate	9016-45-9	2,829	16.3%	20%
Diatomaceous earth, calcined	91053-39-3	2,655	15.3%	100%

Chemical name	CASRN	Number of disclosures	Percentage of disclosures	Median maximum concentration in hydraulic fracturing fluid (% by mass)
Methenamine	100-97-0	2,559	14.7%	1%
Tetramethylammonium chloride	75-57-0	2,428	14.0%	1%
Carbonic acid, dipotassium salt	584-08-7	2,402	13.8%	60%
Ethoxylated propoxylated C12-14 alcohols	68439-51-0	2,342	13.5%	2%
Choline chloride	67-48-1	2,264	13.0%	75%
Boron sodium oxide	1330-43-4	2,228	12.8%	30%
Tetrakis(hydroxymethyl)phosphonium sulfate	55566-30-8	2,130	12.3%	50%
1,2,4-Trimethylbenzene	95-63-6	2,118	12.2%	1%
Boric acid	10043-35-3	2,070	11.9%	25%
Polyethylene glycol	25322-68-3	2,025	11.7%	5%
2-Mercaptoethanol	60-24-2	2,012	11.6%	100%
2,2-Dibromo-3-nitrilopropionamide	10222-01-2	1,988	11.4%	98%
Formic acid	64-18-6	1,948	11.2%	60%
Sodium persulfate	7775-27-1	1,914	11.0%	100%
Phosphonic acid	13598-36-2	1,865	10.7%	1%
Sodium tetraborate decahydrate	1303-96-4	1,862	10.7%	30%
Potassium metaborate	13709-94-9	1,682	9.7%	60%
Ethylenediaminetetraacetic acid tetrasodium salt hydrate	64-02-8	1,676	9.6%	0%
Poly(oxy-1,2-ethanediyl)-nonylphenyl-hydroxy (mixture)	127087-87-0	1,668	9.6%	5%

C.2. Most Frequently Reported Chemicals for Each State

Table C-3a. Chemicals most frequently reported in disclosures in the EPA FracFocus 1.0 project database for each state and number (and percentage) of disclosures where a chemical is reported for that state, Alabama to Montana.

The 20 most frequently reported hydraulic fracturing fluid chemicals were identified for the 20 states that reported in disclosures in the EPA FracFocus 1.0 project database, resulting in a total of 93 chemicals. The chemicals were ranked by counting the number of states where that chemical was in the top 20; chemicals used most widely among the most states come first. For example, methanol is reported in 19 of 20 states, so methanol is ranked first (U.S. EPA, 2015c).

Chemical name	CASRN	Alabama	Alaska	Arkansas	California	Colorado	Kansas	Louisiana	Michigan	Mississippi	Montana
Methanol	67-56-1	55 (100%)		1333 (99.7%)	228 (39.0%)	2883 (63.3%)	77 (79.4%)	596 (59.2%)	13 (92.9%)	3 (75%)	121 (62.7%)
Distillates, petroleum, hydrotreated light	64742-47-8		9 (45%)	743 (55.6%)	322 (55.0%)	3358 (73.7%)	87 (89.7%)	844 (83.9%)	14 (100%)	4 (100%)	115 (59.6%)
Ethylene glycol	107-21-1	55 (100%)	20 (100%)	291 (21.8%)	350 (59.8%)		61 (62.9%)	341 (33.9%)	10 (71.4%)	3 (75%)	95 (49.2%)
Isopropanol	67-63-0	55 (100%)	13 (65%)	586 (43.9%)		2586 (56.8%)	24 (24.7%)	515 (51.2%)	11 (78.6%)		123 (63.7%)
Quartz	14808-60-7		20 (100%)		519 (88.7%)	1048 (23.0%)	22 (22.7%)	377 (37.5%)		2 (50%)	124 (64.2%)
Sodium hydroxide	1310-73-2		20 (100%)	285 (21.3%)	403 (68.9%)	996 (21.9%)	27 (27.8%)	535 (53.2%)		2 (50%)	105 (54.4%)
Ethanol	64-17-5			603 (45.1%)		2258 (49.6%)	78 (80.4%)	420 (41.7%)		4 (100%)	
Guar gum	9000-30-0		10 (50%)		545 (93.2%)			494 (49.1%)		2 (50%)	83 (43.0%)
Hydrochloric acid	7647-01-0	55 (100%)		1330 (99.5%)		2408 (52.9%)	82 (84.5%)	569 (56.6%)			45 (23.3%)
Peroxydisulfuric acid, diammonium salt	7727-54-0		10 (50%)		484 (82.7%)		21 (21.6%)	273 (27.2%)	8 (57.1%)		119 (61.7%)
Propargyl alcohol	107-19-7			813 (60.8%)			69 (71.1%)	299 (29.7%)	5 (35.7%)		

Chemical name	CASRN	Alabama	Alaska	Arkansas	California	Colorado	Kansas	Louisiana	Michigan	Mississippi	Montana
Glutaraldehyde	111-30-8			737 (55.1%)			73 (75.3%)	364 (36.3%)		2 (50%)	
Naphthalene	91-20-3	55 (100%)				1363 (29.9%)	41 (42.3%)	293 (29.2%)	12 (85.7%)		95 (49.2%)
2-Butoxyethanol	111-76-2	55 (100%)	20 (100%)						11 (78.6%)		
Citric acid	77-92-9						45 (46.4%)				
Saline	7647-14-5					1574 (34.5%)		408 (40.6%)		2 (50%)	
Solvent naphtha, petroleum, heavy arom.	64742-94-5			375 (28.0%)		1507 (33.1%)	42 (43.3%)				135 (70.0%)
Quaternary ammonium compounds, benzyl-C12-16-alkyldimethyl, chlorides	68424-85-1						52 (53.6%)			2 (50%)	
2,2-Dibromo-3-nitrilopropionamide	10222-01-2	55 (100%)				2215 (48.6%)			10 (71.4%)		70 (36.3%)
Potassium hydroxide	1310-58-3							340 (33.8%)		4 (100%)	115 (59.6%)
Choline chloride	67-48-1					1235 (27.1%)					
Polyethylene glycol	25322-68-3	55 (100%)							7 (50%)		69 (35.8%)
1,2,4-Trimethylbenzene	95-63-6					1211 (26.63%)	39 (40.2%)				
Ammonium chloride	12125-02-9			277 (20.7%)		1280 (28.0%)					

Chemical name	CASRN	Alabama	Alaska	Arkansas	California	Colorado	Kansas	Louisiana	Michigan	Mississippi	Montana
Diatomaceous earth, calcined	91053-39-3		20 (100%)		417 (71.3%)						
Didecyl dimethyl ammonium chloride	7173-51-5			317 (23.7%)						2 (50%)	
Sodium chlorite	7758-19-2							352 (35.0%)		4 (100%)	
Sodium erythorbate	6381-77-7			435 (32.5%)			29 (29.9%)				
N,N-Dimethylformamide	68-12-2										
Nonyl phenol ethoxylate	9016-45-9										
Poly(oxy-1,2-ethanediyl)-nonylphenyl-hydroxy (mixture)	127087-87-0				1150 (25.2%)	39 (40.2%)					
Sodium persulfate	7775-27-1									4 (100%)	
Tetramethylammonium chloride	75-57-0										85 (44.0%)
1,2-Propylene glycol	57-55-6								10 (71.4%)		
5-Chloro-2-methyl-3(2H)-isothiazolone	26172-55-4		20 (100%)		389 (66.5%)						
Acetic acid	64-19-7					959 (21.0%)		284 (28.2%)			
Ammonium acetate	631-61-8									2 (50%)	
Boric acid	10043-35-3		3 (15%)								

Appendix C – Chemical Mixing Supplemental Information

Chemical name	CASRN	Alabama	Alaska	Arkansas	California	Colorado	Kansas	Louisiana	Michigan	Mississippi	Montana
Carbonic acid, dipotassium salt	584-08-7					1159 (25.4%)					
Cristobalite	14464-46-1		20 (100%)		389 (66.5%)						
Formic acid	64-18-6	55 (100%)						293 (29.1%)			
Hemicellulase enzyme	9012-54-8										
Hemicellulase enzyme concentrate	9025-56-3				395 (67.5%)						
Iron(II) sulfate heptahydrate	7782-63-0								7 (50%)		
Magnesium chloride	7786-30-3		20 (100%)		389 (66.5%)						
Magnesium nitrate	10377-60-3		20 (100%)		389 (66.5%)						
Phenolic resin	9003-35-4										
Sodium hypochlorite	7681-52-9					1046 (23.0%)					
Sodium tetraborate decahydrate	1303-96-4		14 (70%)								
Solvent naphtha, petroleum, heavy aliph.	64742-96-7								7 (50%)	2 (50%)	
1-Butoxy-2-propanol	5131-66-8				315 (53.8%)						
1-Propanol	71-23-8					1232 (27.0%)					

Chemical name	CASRN	Alabama	Alaska	Arkansas	California	Colorado	Kansas	Louisiana	Michigan	Mississippi	Montana
1,2-Ethanediaminium, N,N'-bis[2-[bis(2-hydroxyethyl)methyl ammonio]ethyl]-N,N'bis(2-hydroxyethyl)-N,N'-dimethyl-,tetrachloride	138879-94-4			343 (58.6%)							
2-bromo-3-nitrilopropionamide	1113-55-9										
2-Ethylhexanol	104-76-7										83 (43.0052%)
2-Methyl-3(2H)-isothiazolone	2682-20-4		20 (100%)		389 (66.5%)						
2-Propenoic acid, polymer with 2-propenamide	9003-06-9										
Alkenes, C>10 .alpha.-	64743-02-8			241 (18.0%)							
Benzene, 1,1'-oxybis-, tetrapropylene derivs., sulfonated	119345-03-8										50 (25.9%)
Benzenesulfonic acid, dodecyl-, compd. with N1-(2-aminoethyl)-1,2-ethanediamine (1:?)	40139-72-8										48 (24.9%)
Benzyldimethyldodec ylammonium chloride	139-07-1			268 (20.0%)							
Benzylhexadecyldime thylammonium chloride	122-18-9			268 (20.0%)							

Appendix C – Chemical Mixing Supplemental Information

Chemical name	CASRN	Alabama	Alaska	Arkansas	California	Colorado	Kansas	Louisiana	Michigan	Mississippi	Montana
Boron sodium oxide	1330-43-4				361 (61.7%)						
C10-C16 ethoxylated alcohol	68002-97-1		3 (15%)								
Calcium chloride	10043-52-4		20 (100%)								
Carbon dioxide	124-38-9								7 (50%)		
Cinnamaldehyde (3-phenyl-2-propenal)	104-55-2	55 (100%)									
Diethylene glycol	111-46-6										
Diethylene glycol monobutyl ether	112-34-5								7 (50%)		
Diethylenetriamine	111-40-0										55 (28.5%)
Distillates, petroleum, hydrotreated light paraffinic	64742-55-8				314 (53.7%)						
Distillates, petroleum, hydrotreated middle	64742-46-7		3 (15%)								
Ethoxylated C12-16 alcohols	68551-12-2										
Ethoxylated C14-15 alcohols	68951-67-7			241 (18.0%)							
Formic acid, potassium salt	590-29-4										
Glycerin, natural	56-81-5								7 (50%)		

Chemical name	CASRN	Alabama	Alaska	Arkansas	California	Colorado	Kansas	Louisiana	Michigan	Mississippi	Montana
Isotridecanol, ethoxylated	9043-30-5				312 (53.3%)						
Methenamine	100-97-0							298 (29.6%)			
Naphtha, petroleum, hydrotreated heavy	64742-48-9										
Poly(oxy-1,2-ethanediyl), .alpha.,.alpha.'-[[(9Z)-9-octadecenylimino]di-2,1-ethanediyl]bis[,omega.-hydroxy-	26635-93-8								9 (64.3%)		
Potassium chloride	7447-40-7								7 (50%)		
Sodium bromate	7789-38-0								7 (50%)		
Sodium perborate tetrahydrate	10486-00-7										
Sulfamic acid	5329-14-6									2 (50%)	
Terpenes and Terpenoids, sweet orange-oil	68647-72-3									2 (50%)	
Tetradecyl dimethyl benzyl ammonium chloride	139-08-2			268 (20.0%)							
Tetrakis(hydroxymethyl)phosphonium sulfate	55566-30-8										
Thiourea polymer	68527-49-1			384 (28.7%)							

Appendix C – Chemical Mixing Supplemental Information

Chemical name	CASRN	Alabama	Alaska	Arkansas	California	Colorado	Kansas	Louisiana	Michigan	Mississippi	Montana
Tri-n-butyl tetradecyl phosphonium chloride	81741-28-8										
Trisodium phosphate	7601-54-9						19 (19.6%)				

Table C-3b. Chemicals most frequently reported in disclosures in the EPA FracFocus 1.0 project database for each state and number (and percentage) of disclosures where a chemical is reported for that state, New Mexico to Wyoming.

The 20 most frequently reported hydraulic fracturing fluid chemicals were identified for the 20 states that reported in disclosures in the EPA FracFocus 1.0 project database, resulting in a total of 93 chemicals. The chemicals were ranked by counting the number of states where that chemical was in the top 20; chemicals used most widely among the most states come first. For example, methanol is reported in 19 of 20 states, so methanol is ranked first (U.S. EPA, 2015c).

Chemical name	CASRN	New Mexico	North Dakota	Ohio	Oklahoma	Pennsylvania	Texas	Utah	Virginia	West Virginia	Wyoming
Methanol	67-56-1	1012 (90.8%)	1059 (53.3%)	76 (52.1%)	1270 (70.3%)	1633 (68.6%)	12664 (78.5%)	984 (78.5%)	48 (60.8%)	153 (64.0%)	460 (38.4%)
Distillates, petroleum, hydrotreated light	64742-47-8	699 (62.7%)	943 (47.5%)	122 (83.6%)	1270 (70.3%)	1434 (60.2%)	10677 (66.1%)	934 (74.5%)		196 (82.0%)	612 (51.1%)
Ethylene glycol	107-21-1	503 (45.1%)	724 (36.4%)	83 (56.8%)	843 (46.7%)	807 (33.9%)	9591 (59.4%)	1065 (85.0%)	22 (27.8%)	141 (59.0%)	
Isopropanol	67-63-0	695 (62.3%)	739 (37.2%)	71 (48.6%)	764 (42.28%)	735 (30.9%)	7731 (47.9%)	661 (52.8%)	43 (54.4%)	74 (31.0%)	516 (43.1%)
Quartz	14808-60-7	762 (68.3%)	920 (46.3%)	66 (45.2%)	491 (27.2%)		6869 (42.6%)	503 (40.1%)		53 (22.2%)	356 (29.7%)
Sodium hydroxide	1310-73-2	329 (29.5%)	1028 (51.7%)		490 (27.1%)	406 (17.0%)	7371 (45.7%)	466 (37.2%)			688 (57.4%)
Ethanol	64-17-5	529 (47.4%)	545 (27.4%)	87 (59.6%)	838 (46.4%)	388 (16.3%)	3439 (21.3%)		50 (63.3%)	130 (54.3%)	298 (24.9%)
Guar gum	9000-30-0	702 (63.0%)	1094 (55.1%)	74 (50.7%)	457 (25.3%)	538 (22.6%)	6863 (42.5%)	538 (42.9%)		55 (23.0%)	823 (68.7%)
Hydrochloric acid	7647-01-0	880 (78.9%)		145 (99.3%)	1372 (75.9%)	2279 (95.7%)	11424 (70.8%)	1064 (84.9%)	68 (86.1%)	229 (95.8%)	
Peroxydisulfuric acid, diammonium salt	7727-54-0	836 (75.0%)	1089 (54.8%)	93 (63.7%)	713 (39.5%)		8666 (53.7%)	483 (38.5%)		128 (53.6%)	771 (64.4%)
Propargyl alcohol	107-19-7	760 (68.2%)		72 (49.3%)	732 (40.5%)	1371 (57.6%)	6269 (38.8%)	456 (36.4%)	22 (27.8%)	138 (57.7%)	

Chemical name	CASRN	New Mexico	North Dakota	Ohio	Oklahoma	Pennsyl-vania	Texas	Utah	Virginia	West Virginia	Wyoming
Glutaraldehyde	111-30-8	632 (56.7%)		105 (71.9%)	989 (54.7%)	819 (34.4%)	6470 (40.1%)			169 (70.7%)	260 (21.7%)
Naphthalene	91-20-3		864 (43.5%)		448 (24.8%)			478 (38.1%)	7 (8.9%)		
2-Butoxyethanol	111-76-2	412 (37.0%)				498 (20.9%)	3898 (24.1%)	663 (52.9%)	70 (88.6%)	62 (25.9%)	
Citric acid	77-92-9	447 (40.1%)		96 (65.8%)	644 (35.6%)	701 (29.4%)	3820 (23.7%)	992 (79.2%)	63 (79.8%)	98 (41.0%)	
Saline	7647-14-5		491 (24.7%)				3462 (21.4%)		7 (8.9%)	53 (22.2%)	274 (22.9%)
Solvent naphtha, petroleum, heavy arom.	64742-94-5		981 (49.4%)		557 (30.8%)		2751 (17.0%)		7 (8.9%)		415 (34.6%)
Quaternary ammonium compounds, benzyl-C12-16-alkyldimethyl, chlorides	68424-85-1			54 (37.0%)	597 (33.0%)	373 (15.7%)				53 (22.2%)	
2,2-Dibromo-3-nitrilopropionamide	10222-01-2					804 (33.8%)			22 (27.8%)		
Potassium hydroxide	1310-58-3		1176 (59.2%)	106 (72.6%)			6369 (39.5%)				
Choline chloride	67-48-1	384 (34.4%)		55 (37.7%)				649 (51.8%)	45 (57.0%)		
Polyethylene glycol	25322-68-3		567 (28.5%)			688 (28.9%)					
1,2,4-Trimethylbenzene	95-63-6		496 (25.0%)						7 (8.9%)		
Ammonium chloride	12125-02-9					732 (30.7%)				50 (20.9%)	

Chemical name	CASRN	New Mexico	North Dakota	Ohio	Oklahoma	Pennsylvania	Texas	Utah	Virginia	West Virginia	Wyoming
Diatomaceous earth, calcined	91053-39-3	419 (37.6%)						435 (34.7%)			
Didecyl dimethyl ammonium chloride	7173-51-5			46 (31.6%)						49 (20.5%)	
Sodium chlorite	7758-19-2		482 (24.3%)								271 (22.6%)
Sodium erythorbate	6381-77-7								10 (12.7%)	76 (31.8%)	
N,N-Dimethylformamide	68-12-2			68 (46.6%)	355 (19.6%)			410 (32.7%)			
Nonyl phenol ethoxylate	9016-45-9	333 (29.9%)						447 (35.7%)	25 (31.6%)		
Poly(oxy-1,2-ethanediyl)-nonylphenyl-hydroxy (mixture)	127087-87-0								7 (8.9%)		
Sodium persulfate	7775-27-1					373 (15.7%)					308 (25.7%)
Tetramethylammonium chloride	75-57-0		579 (29.1%)								315 (26.3%)
1,2-Propylene glycol	57-55-6								22 (27.8%)		
5-Chloro-2-methyl-3(2H)-isothiazolone	26172-55-4										
Acetic acid	64-19-7										
Ammonium acetate	631-61-8										323 (27.0%)
Boric acid	10043-35-3			82 (56.2%)							

Appendix C – Chemical Mixing Supplemental Information

Chemical name	CASRN	New Mexico	North Dakota	Ohio	Oklahoma	Pennsyl-vania	Texas	Utah	Virginia	West Virginia	Wyoming
Carbonic acid, dipotassium salt	584-08-7		482 (24.2%)								
Cristobalite	14464-46-1										
Formic acid	64-18-6										
Hemicellulase enzyme	9012-54-8					367 (15.4%)			11 (13.9%)		
Hemicellulase enzyme concentrate	9025-56-3	331 (29.7%)									
Iron(II) sulfate heptahydrate	7782-63-0								22 (27.8%)		
Magnesium chloride	7786-30-3										
Magnesium nitrate	10377-60-3										
Phenolic resin	9003-35-4	419 (37.6%)					2903 (18.0%)				
Sodium hypochlorite	7681-52-9										282 (23.5%)
Sodium tetraborate decahydrate	1303-96-4										265 (22.1%)
Solvent naphtha, petroleum, heavy aliph.	64742-96-7										
1-Butoxy-2-propanol	5131-66-8										
1-Propanol	71-23-8										

Chemical name	CASRN	New Mexico	North Dakota	Ohio	Oklahoma	Pennsyl-vania	Texas	Utah	Virginia	West Virginia	Wyoming
1,2-Ethanediaminium, N,N'-bis[2-[bis(2-hydroxyethyl) methylammonio]ethyl]-N,N'bis(2-hydroxyethyl)-N,N'-dimethyl-, tetrachloride	138879-94-4										
2-Bromo-3-nitrilopropionamide	1113-55-9								11 (13.9%)		
2-Ethylhexanol	104-76-7										
2-Methyl-3(2H)-isothiazolone	2682-20-4										
2-Propenoic acid, polymer with 2-propenamide	9003-06-9							486 (38.8%)			
Alkenes, C>10 .alpha.-	64743-02-8										
Benzene, 1,1'-oxybis-, tetrapropylene derivs., sulfonated	119345-03-8										
Benzenesulfonic acid, dodecyl-, compd. with N1-(2-aminoethyl)-1,2-ethanediamine (1:?)	40139-72-8										
Benzyldimethyldodecylam monium chloride	139-07-1										
Benzylhexadecyldimethyla mmonium chloride	122-18-9										
Boron sodium oxide	1330-43-4										
C10-C16 ethoxylated alcohol	68002-97-1										

Appendix C – Chemical Mixing Supplemental Information

Chemical name	CASRN	New Mexico	North Dakota	Ohio	Oklahoma	Pennsyl-vania	Texas	Utah	Virginia	West Virginia	Wyoming
Calcium chloride	10043-52-4										
Carbon dioxide	124-38-9										
Cinnamaldehyde (3-phenyl-2-propenal)	104-55-2										
Diethylene glycol	111-46-6			45 (30.8%)							
Diethylene glycol monobutyl ether	112-34-5										
Diethylenetriamine	111-40-0										
Distillates, petroleum, hydrotreated light paraffinic	64742-55-8										
Distillates, petroleum, hydrotreated middle	64742-46-7										
Ethoxylated C12-16 alcohols	68551-12-2									57 (23.8%)	
Ethoxylated C14-15 alcohols	68951-67-7										
Formic acid, potassium salt	590-29-4										361 (30.1%)
Glycerin, natural	56-81-5										
Isotridecanol, ethoxylated	9043-30-5										
Methenamine	100-97-0										
Naphtha, petroleum, hydrotreated heavy	64742-48-9										384 (32.1%)

Chemical name	CASRN	New Mexico	North Dakota	Ohio	Oklahoma	Pennsylvania	Texas	Utah	Virginia	West Virginia	Wyoming
Poly(oxy-1,2-ethanediyl), .alpha.,.alpha.'-[[(9Z)-9-octadecenylimino]di-2,1-ethanediyl]bis[.omega.-hydroxy-	26635-93-8										
Potassium chloride	7447-40-7										
Sodium bromate	7789-38-0										
Sodium perborate tetrahydrate	10486-00-7				351 (19.4%)						
Sulfamic acid	5329-14-6										
Terpenes and terpenoids, sweet orange-oil	68647-72-3										
Tetradecyl dimethyl benzyl ammonium chloride	139-08-2										
Tetrakis(hydroxymethyl)phosphonium sulfate	55566-30-8							945 (75.4%)			
Thiourea polymer	68527-49-1										
Tri-n-butyl tetradecyl phosphonium chloride	81741-28-8					350 (14.7%)					
Trisodium phosphate	7601-54-9										

C.3. Estimating Volume and Mass for 74 Chemicals Reported in Disclosures in the EPA FracFocus 1.0 Project Database

Volume and mass were estimated using the chemical data reported in the disclosures in the EPA FracFocus 1.0 project database. The total hydraulic fracturing fluid volume reported was used to calculate the total fluid mass by assuming the fluid has a density of 1 g/mL. This is a simplifying assumption based on the fact that more than 93% of disclosures are inferred to use water as a base fluid. Water had a median concentration of 88% (by mass) in the fracturing fluid, with 10th and 90th percentiles of 77% and 95%. Roughly 2% of disclosures reported the use of non-aqueous base fluids, which contained roughly 60% (median) water (U.S. EPA, 2015c). The use of non-aqueous base fluids would introduce additional error in our calculations. We made the simplifying assumption that this error is negligible. Some disclosures reported using brine, which has a density between 1.0 and 1.1 g/mL. This would introduce at most an error of 10% for the fluid calculation (the difference of a chemical being present at 10 versus 9 gal, 1,000 versus 900 gal). We also assume that the mass of chemicals present in calculating the total fluid mass is negligible. Given that ≤2% of the fluid volume are chemicals, and assuming the density of which is 3 mg/L, the error introduced is approximately 6%. For reference, for the chemicals we are calculating volumes, chlorine dioxide is the densest at 2.757 mg/L. Chemical with densities less than 1 mg/L introduce approximately <1% error.

Next, the mass of each chemical per disclosure was calculated. Each chemical is reported in disclosures to FracFocus 1.0 as a maximum concentration by mass in the hydraulic fracturing fluid. This introduces error, as we only know that it is equal to or less than this mass fraction. In EPA's analysis of the EPA FracFocus 1.0 project database (U.S. EPA, 2015a), an example additive is comprised of three chemicals with maximum ingredient concentration of 60% in the additive and a maximum concentration of 0.22% in the hydraulic fracturing fluid. Each of the three chemicals cannot be present at 60%. Because of how chemical information was reported to FracFocus 1.0, we have no way to know the actual proportions of each chemical in the additive and thus must calculate chemical mass based on the given information. Therefore, our calculations likely overestimate actual volumes. However, in some cases, the concentration in the additive that is given is less than 100% and only one chemical is listed in the additive. In these cases, it appears that the disclosure is reporting the concentration of that chemical in water. Hydrogen chloride (HCl) is listed as the sole ingredient in the acid additive, and the maximum concentration is 40% by mass. In this case, the HCl is diluted down to 40%, so the total volume would be underestimated.

After all the chemical masses are calculated, the volume is calculated by dividing chemical mass by density.

Given the limited information available, due to the limits of the FracFocus 1.0 chemical reporting and general lack of publicly available data, and despite the errors associated with these calculations, these calculations provide context for the general magnitude of volumes for each of the chemicals used on-site. These calculations are used to calculate median volumes for each chemical on a per well basis. These volume calculations are for the chemicals themselves, not the additives.

The analysis considered 34,495 disclosures and 672,358 ingredient records that met selected quality assurance criteria, including: completely parsed; unique combination of fracture date and API well number; fracture date between January 1, 2011, and February 28, 2013; criteria for water volumes; valid CASRN; and valid concentrations. Disclosures that did not meet quality assurance criteria (4,035) or other, query-specific criteria were excluded from our analysis.

Density data were gathered from Reaxys® and other sources as noted. Reaxys® (http://www.elsevier.com/online-tools/reaxys) is an online database of chemistry literature and data. Direct density source, as provided by Reaxys®, is provided in Table C-7.

Reporting hydraulic fracturing well records to FracFocus 1.0 was required in six of the 20 states with data in FracFocus between January 1, 2011, and February 28, 2013. An additional three states required disclosure to either FracFocus or the state, and five states required reporting to the state. Reporting to FracFocus 1.0 was optional in other states. Some states changed their reporting requirements during the course of the study. The EPA FracFocus 1.0 project database, developed using data directly from FracFocus 1.0, therefore does not encompass all data on chemicals used in hydraulic fracturing. As stated in Text Box 4-2, this mix of voluntary and mandatory disclosure requirements limits the completeness of data included in the EPA FracFocus 1.0 project database for estimating hydraulic fracturing fluid compositions and volumes. According to a comparison between the EPA FracFocus 1.0 project database reported fluid volumes and literature values, water use per well was reported to be about 86% of the literature values (median of estimated values; see Chapter 4, Text Box 4-1). If the fluid volume is underreported, then estimated chemical volumes based on fluid volume would be similarly underestimated. Using the underreporting of 86%, then the estimated median chemical volume would be 760 gal (2,900 L).

Table C-4. Estimated mean, median, 5th percentile, and 95th percentile volumes in gallons for chemicals reported in 100 or more disclosures in the EPA FracFocus 1.0 project database, where density information was available.

Chemicals are listed in alphabetical order. Density information came from Reaxys® and other sources. All density sources are referenced in Table C-7.

Name	CASRN	Volume (gal)			
		Mean	Median	5th Percentile	95th Percentile
(4R)-1-methyl-4-(prop-1-en-2-yl)cyclohexene	5989-27-5	2,702	406	0	19,741
1-Butoxy-2-propanol	5131-66-8	167	21	5	654
1-Decanol	112-30-1	28	4	0	33
1-Octanol	111-87-5	5	4	0	10
1-Propanol	71-23-8	128	55	6	367
1,2-Propylene glycol	57-55-6	13,105	72	4	61,071
1,2,4-Trimethylbenzene	95-63-6	38	6	0	43
2-Butoxyethanol	111-76-2	385	26	0	1,811
2-Ethylhexanol	104-76-7	100	11	0	292
2-Mercaptoethanol	60-24-2	1,175	445	0	4,194
2,2-Dibromo-3-nitrilopropionamide	10222-01-2	183	5	0	341
Acetic acid	64-19-7	646	47	0	1,042
Acetic anhydride	108-24-7	239	50	3	722
Acrylamide	79-06-1	95	3	0	57
Adipic acid	124-04-9	153	0	0	109
Aluminum chloride	7446-70-0	2	0	0	0
Ammonia	7664-41-7	44	35	2	138
Ammonium acetate	631-61-8	839	117	0	1,384
Ammonium chloride	12125-02-9	526	58	3	548
Ammonium hydroxide	1336-21-6	7	2	0	14
Benzyl chloride	100-44-7	52	0	0	40
Carbonic acid, dipotassium salt	584-08-7	467	113	0	1,729
Chlorine dioxide	10049-04-4	31	11	0	28
Choline chloride	67-48-1	2,131	290	28	4,364

Name	CASRN	Volume (gal)			
		Mean	Median	5th Percentile	95th Percentile
Cinnamaldehyde (3-phenyl-2-propenal)	104-55-2	68	3	0	697
Citric acid	77-92-9	163	20	1	269
Dibromoacetonitrile	3252-43-5	22	13	1	45
Diethylene glycol	111-46-6	168	16	0	102
Diethylenetriamine	111-40-0	92	21	0	207
Dodecane	112-40-3	190	31	0	151
Ethanol	64-17-5	831	121	1	2,645
Ethanolamine	141-43-5	70	30	0	283
Ethyl acetate	141-78-6	0	0	0	0
Ethylene glycol	107-21-1	614	184	4	2,470
Ferric chloride	7705-08-0	0	0	0	0
Formalin	50-00-0	200	0	0	8
Formic acid	64-18-6	501	38	1	1,229
Fumaric acid	110-17-8	2	0	0	12
Glutaraldehyde	111-30-8	1,313	122	2	1,165
Glycerin, natural	56-81-5	413	109	10	911
Glycolic acid	79-14-1	38	10	4	94
Hydrochloric acid	7647-01-0	28,320	3,110	96	26,877
Isopropanol	67-63-0	2,095	55	0	1,264
Isopropylamine	75-31-0	83	121	0	172
Magnesium chloride	7786-30-3	14	0	0	2
Methanol	67-56-1	1,218	110	2	3,731
Methenamine	100-97-0	3,386	100	0	3,648
Methoxyacetic acid	625-45-6	36	4	2	115
N,N-Dimethylformamide	68-12-2	119	10	0	216
Naphthalene	91-20-3	72	12	0	204
Nitrogen, liquid	7727-37-9	41,841	26,610	3,091	108,200
Ozone	10028-15-6	15,844	15,473	8,785	26,063

Name	CASRN	Volume (gal)			
		Mean	Median	5th Percentile	95th Percentile
Peracetic acid	79-21-0	300	268	50	663
Phosphonic acid	13598-36-2	1,201	0	0	3
Phosphoric acid Divosan X-Tend formulation	7664-38-2	13	4	0	15
Potassium acetate	127-08-2	209	1	0	994
Propargyl alcohol	107-19-7	183	2	0	51
Saline	7647-14-5	876	85	0	1,544
Saturated sucrose	57-50-1	1	1	0	2
Silica, amorphous	7631-86-9	6,877	8	0	38,371
Sodium carbonate	497-19-8	228	16	0	1,319
Sodium formate	141-53-7	0	0	0	0
Sodium hydroxide	1310-73-2	551	38	0	1,327
Sulfur dioxide	7446-09-5	0	0	0	0
Sulfuric acid	7664-93-9	3	0	0	3
tert-Butyl hydroperoxide (70% solution in Water)	75-91-2	156	64	0	557
Tetramethylammonium chloride	75-57-0	970	483	2	3,508
Thioglycolic acid	68-11-1	55	7	2	229
Toluene	108-88-3	18	0	0	11
Tridecane	629-50-5	190	31	0	190
Triethanolamine	102-71-6	846	60	0	2,264
Triethyl phosphate	78-40-0	55	1	0	533
Triethylene glycol	112-27-6	5,198	116	28	945
Triisopropanolamine	122-20-3	46	4	1	330
Trimethyl borate	121-43-7	83	40	4	283
Undecane	1120-21-4	273	29	0	1,641

Table C-5. Estimated mean, median, 5th percentile, and 95th percentile volumes in liters for chemicals reported in 100 or more disclosures in the EPA FracFocus 1.0 project database, where density information was available.

Chemicals are listed in alphabetical order. Density information came from Reaxys® and other sources. All density sources are referenced in Table C-7.

		Volume (L)			
Name	CASRN	Mean	Median	5th Percentile	95th Percentile
(4R)-1-methyl-4-(prop-1-en-2-yl)cyclohexene	5989-27-5	10,229	1,536	0	74,729
1-Butoxy-2-propanol	5131-66-8	631	80	18	2,475
1-Decanol	112-30-1	107	14	1	123
1-Octanol	111-87-5	21	14	1	39
1-Propanol	71-23-8	483	208	22	1,391
1,2-Propylene glycol	57-55-6	49,607	274	15	231,179
1,2,4-Trimethylbenzene	95-63-6	145	24	0	165
2-Butoxyethanol	111-76-2	1,459	98	0	6,856
2-Ethylhexanol	104-76-7	377	40	1	1,106
2-Mercaptoethanol	60-24-2	4,449	1,685	0	15,878
2,2-Dibromo-3-nitrilopropionamide	10222-01-2	692	18	0	1,292
Acetic acid	64-19-7	2,446	176	0	3,945
Acetic anhydride	108-24-7	906	189	12	2,734
Acrylamide	79-06-1	361	10	0	216
Adipic acid	124-04-9	578	0	0	414
Aluminum chloride	7446-70-0	6	0	0	0
Ammonia	7664-41-7	166	134	7	523
Ammonium acetate	631-61-8	3,177	444	0	5,238
Ammonium chloride	12125-02-9	1,992	218	12	2,074
Ammonium hydroxide	1336-21-6	27	6	1	52
Benzyl chloride	100-44-7	196	1	0	151
Carbonic acid, dipotassium salt	584-08-7	1,769	429	0	6,544
Chlorine dioxide	10049-04-4	117	43	1	106
Choline chloride	67-48-1	8,068	1,096	107	16,521

Name	CASRN	Volume (L)			
		Mean	Median	5th Percentile	95th Percentile
Cinnamaldehyde (3-phenyl-2-propenal)	104-55-2	258	12	0	2,638
Citric acid	77-92-9	618	77	5	1,019
Dibromoacetonitrile	3252-43-5	82	50	4	170
Diethylene glycol	111-46-6	636	61	1	384
Diethylenetriamine	111-40-0	347	80	0	785
Dodecane	112-40-3	719	117	0	572
Ethanol	64-17-5	3,144	458	6	10,011
Ethanolamine	141-43-5	264	112	0	1,070
Ethyl acetate	141-78-6	0	0	0	0
Ethylene glycol	107-21-1	2,324	697	14	9,349
Ferric chloride	7705-08-0	0	0	0	0
Formalin	50-00-0	756	2	0	31
Formic acid	64-18-6	1,896	144	2	4,653
Fumaric acid	110-17-8	9	0	0	46
Glutaraldehyde	111-30-8	4,972	462	6	4,409
Glycerin, natural	56-81-5	1,565	412	38	3,447
Glycolic acid	79-14-1	146	39	14	356
Hydrochloric acid	7647-01-0	107,204	11,772	362	101,741
Isopropanol	67-63-0	7,932	210	1	4,786
Isopropylamine	75-31-0	314	458	0	652
Magnesium chloride	7786-30-3	52	0	0	8
Methanol	67-56-1	4,609	416	6	14,125
Methenamine	100-97-0	12,817	378	0	13,810
Methoxyacetic acid	625-45-6	136	17	8	436
N,N-Dimethylformamide	68-12-2	449	38	2	819
Naphthalene	91-20-3	271	44	0	774
Nitrogen, liquid	7727-37-9	158,384	100,731	11,700	409,583
Ozone	10028-15-6	59,976	58,570	33,254	98,658

Name	CASRN	Volume (L)			
		Mean	Median	5th Percentile	95th Percentile
Peracetic acid	79-21-0	1,137	1,016	190	2,511
Phosphonic acid	13598-36-2	4,547	2	0	11
Phosphoric acid Divosan X-Tend formulation	7664-38-2	51	15	0	57
Potassium acetate	127-08-2	790	3	0	3,762
Propargyl alcohol	107-19-7	693	9	0	193
Saline	7647-14-5	3,317	321	0	5,844
Saturated sucrose	57-50-1	5	2	0	6
Silica, amorphous	7631-86-9	26,031	32	0	145,251
Sodium carbonate	497-19-8	862	62	0	4,991
Sodium formate	141-53-7	1	1	0	1
Sodium hydroxide	1310-73-2	2,087	144	1	5,024
Sulfur dioxide	7446-09-5	2	0	0	0
Sulfuric acid	7664-93-9	10	0	0	12
tert-Butyl hydroperoxide (70% solution in Water)	75-91-2	591	242	0	2,109
Tetramethylammonium chloride	75-57-0	3,672	1,830	8	13,279
Thioglycolic acid	68-11-1	208	28	6	868
Toluene	108-88-3	69	0	0	41
Tridecane	629-50-5	721	118	0	721
Triethanolamine	102-71-6	3,203	228	0	8,570
Triethyl phosphate	78-40-0	209	6	0	2,019
Triethylene glycol	112-27-6	19,676	439	106	3,579
Triisopropanolamine	122-20-3	174	16	4	1,249
Trimethyl borate	121-43-7	314	152	16	1,072
Undecane	1120-21-4	1,035	111	0	6,212

Table C-6. Calculated mean, median, 5th percentile, and 95th percentile chemical masses reported in 100 or more disclosures in the EPA FracFocus 1.0 project database, where density information was available.

Density information came from Reaxys® and other sources. All density sources are referenced in Table C-7. Number of disclosures reported for each chemical is also included.

Name	CASRN	Mass (kg)				Disclosures
		Mean	Median	5th Percentile	95th Percentile	
(4R)-1-methyl-4-(prop-1-en-2-yl)cyclohexene	5989-27-5	8,593	1,290	0	62,772	578
1-Butoxy-2-propanol	5131-66-8	555	71	16	2,178	773
1-Decanol	112-30-1	89	12	1	102	434
1-Octanol	111-87-5	17	12	1	32	434
1-Propanol	71-23-8	386	167	18	1,113	1,481
1,2-Propylene glycol	57-55-6	51,095	282	15	238,114	1,023
1,2,4-Trimethylbenzene	95-63-6	126	21	0	143	3,976
2-Butoxyethanol	111-76-2	1,313	88	0	6,170	6,778
2-Ethylhexanol	104-76-7	313	34	0	918	1,291
2-Mercaptoethanol	60-24-2	489	185	0	1,747	2,051
2,2-Dibromo-3-nitrilopropionamide	10222-01-2	1,660	44	0	3,102	4,927
Acetic acid	64-19-7	2,544	183	0	4,103	7,643
Acetic anhydride	108-24-7	969	203	12	2,925	1,377
Acrylamide	79-06-1	408	11	0	244	251
Adipic acid	124-04-9	785	0	0	564	233
Aluminum chloride	7446-70-0	15	0	0	0	122
Ammonia	7664-41-7	111	90	4	351	398
Ammonium acetate	631-61-8	3,718	520	0	6,129	1,504
Ammonium chloride	12125-02-9	2,530	277	16	2,633	3,288
Ammonium hydroxide	1336-21-6	48	11	2	94	1,173
Benzyl chloride	100-44-7	214	1	0	165	1,833
Carbonic acid, dipotassium salt	584-08-7	4,298	1,042	0	15,902	4,093
Chlorine dioxide	10049-04-4	321	117	3	291	331
Choline chloride	67-48-1	9,440	1,282	125	19,329	4,241

Name	CASRN	Mass (kg)				Disclosures
		Mean	Median	5th Percentile	95th Percentile	
Cinnamaldehyde (3-phenyl-2-propenal)	104-55-2	284	13	0	2,902	1,377
Citric acid	77-92-9	989	123	8	1,630	7,503
Dibromoacetonitrile	3252-43-5	193	118	11	403	272
Diethylene glycol	111-46-6	712	68	1	430	1,732
Diethylenetriamine	111-40-0	330	76	0	746	784
Dodecane	112-40-3	539	88	0	429	131
Ethanol	64-17-5	2,484	361	4	7,908	9,233
Ethanolamine	141-43-5	267	113	0	1,081	585
Ethyl acetate	141-78-6	0	0	0	0	110
Ethylene glycol	107-21-1	2,557	767	15	10,283	14,767
Ferric chloride	7705-08-0	0	0	0	0	118
Formalin	50-00-0	816	2	0	34	456
Formic acid	64-18-6	2,313	176	2	5,677	3,781
Fumaric acid	110-17-8	15	0	0	75	224
Glutaraldehyde	111-30-8	4,972	462	6	4,409	10,963
Glycerin, natural	56-81-5	1,972	519	47	4,343	1,829
Glycolic acid	79-14-1	217	58	21	530	595
Hydrochloric acid	7647-01-0	107,204	11,772	362	101,741	20,996
Isopropanol	67-63-0	6,187	163	1	3,733	15,058
Isopropylamine	75-31-0	213	311	0	444	255
Magnesium chloride	7786-30-3	120	1	0	18	1,113
Methanol	67-56-1	3,641	329	5	11,159	23,225
Methenamine	100-97-0	15,380	454	0	16,572	4,412
Methoxyacetic acid	625-45-6	161	20	9	514	584
N,N-Dimethylformamide	68-12-2	422	36	2	770	2,972
Naphthalene	91-20-3	220	35	0	627	5,945
Nitrogen, liquid	7727-37-9	129,875	82,599	9,594	335,858	713
Ozone	10028-15-6	129	126	71	212	209

Name	CASRN	Mass (kg)				Disclosures
		Mean	Median	5th Percentile	95th Percentile	
Peracetic acid	79-21-0	1,251	1,117	209	2,762	221
Phosphonic acid	13598-36-2	7,730	3	0	18	2,216
Phosphoric acid Divosan X-Tend formulation	7664-38-2	48	14	0	54	315
Potassium acetate	127-08-2	1,216	5	0	5,793	325
Propargyl alcohol	107-19-7	658	9	0	183	10,771
Saline	7647-14-5	7,197	696	0	12,682	6,673
Saturated sucrose	57-50-1	6	2	0	7	125
Silica, amorphous	7631-86-9	57,267	71	0	319,553	2,423
Sodium carbonate	497-19-8	2,191	158	0	12,678	396
Sodium formate	141-53-7	2	1	1	2	204
Sodium hydroxide	1310-73-2	4,445	306	2	10,701	12,585
Sulfur dioxide	7446-09-5	2	0	0	0	224
Sulfuric acid	7664-93-9	18	0	0	22	402
tert-Butyl hydroperoxide (70% solution in water)	75-91-2	532	218	0	1,898	814
Tetramethylammonium chloride	75-57-0	4,296	2,141	10	15,537	3,162
Thioglycolic acid	68-11-1	277	37	8	1,155	156
Toluene	108-88-3	59	0	0	35	214
Tridecane	629-50-5	541	88	0	541	132
Triethanolamine	102-71-6	3,588	255	0	9,599	1,498
Triethyl phosphate	78-40-0	222	6	0	2,140	991
Triethylene glycol	112-27-6	22,038	491	119	4,008	528
Triisopropanolamine	122-20-3	177	17	4	1,274	251
Trimethyl borate	121-43-7	292	141	14	997	294
Undecane	1120-21-4	766	82	0	4,597	241

Table C-7. Associated chemical densities and references used to calculate chemical mass and estimate chemical volume.

Name	CASRN	Density (g/mL)	Reference
(4R)-1-methyl-4-(prop-1-en-2-yl)cyclohexene	5989-27-5	0.84	Dejoye Tanzi et al. (2012)
1-Butoxy-2-propanol	5131-66-8	0.88	Pal et al. (2013)
1-Decanol	112-30-1	0.83	Faria et al. (2013)
1-Octanol	111-87-5	0.82	Dubey and Kumar (2013)
1-Propanol	71-23-8	0.8	Rani and Maken (2013)
1,2-Propylene glycol	57-55-6	1.03	Moosavi et al. (2013)
1,2,4-Trimethylbenzene	95-63-6	0.87	He et al. (2008)
2-Butoxyethanol	111-76-2	0.9	Dhondge et al. (2010)
2-Ethylhexanol	104-76-7	0.83	Laavi et al. (2012)
2-Mercaptoethanol	60-24-2	0.11	Rawat et al. (1976)
2,2-Dibromo-3-nitrilopropionamide	10222-01-2	2.4	Fels (1900)
Acetic acid	64-19-7	1.04	Thalladi et al. (2000)
Acetic anhydride	108-24-7	1.07	Radwan and Hanna (1976)
Acrylamide	79-06-1	1.13	Carpenter and Davis (1957)
Adipic acid	124-04-9	1.36	Thalladi et al. (2000)
Aluminum chloride	7446-70-0	2.48	Sigma-Aldrich (2015a)
Ammonia	7664-41-7	0.67	Harlow et al. (1997)
Ammonium acetate	631-61-8	1.17	Biltz and Balz (1928)
Ammonium chloride	12125-02-9	1.27	Haynes (2014)
Ammonium hydroxide	1336-21-6	1.8	Xiao et al. (2013)
Benzyl chloride	100-44-7	1.09	Sarkar et al. (2012)
Carbonic acid, dipotassium salt	584-08-7	2.43	Sigma-Aldrich (2014b)
Chlorine dioxide	10049-04-4	2.757	Haynes (2014)
Choline chloride	67-48-1	1.17	Shanley and Collin (1961)
Cinnamaldehyde (3-phenyl-2-propenal)	104-55-2	1.1	Masood et al. (1976)
Citric acid	77-92-9	1.6	Bennett and Yuill (1935)
Dibromoacetonitrile	3252-43-5	2.37	Wilt (1956)
Diethylene glycol	111-46-6	1.12	Chasib (2013)
Diethylenetriamine	111-40-0	0.95	Dubey and Kumar (2011)
Dodecane	112-40-3	0.75	Baragi et al. (2013)
Ethanol	64-17-5	0.79	Kiselev et al. (2012)
Ethanolamine	141-43-5	1.01	Blanco et al. (2013)

Name	CASRN	Density (g/mL)	Reference
Ethyl acetate	141-78-6	0.89	Laavi et al. (2013)
Ethylene glycol	107-21-1	1.1	Rodnikova et al. (2012)
Ferric chloride	7705-08-0	2.9	Haynes (2014)
Formalin	50-00-0	1.08	Alfa Aesar (2015)
Formic acid	64-18-6	1.22	Casanova et al. (1981)
Fumaric acid	110-17-8	1.64	Huffman and Fox (1938)
Glutaraldehyde	111-30-8	1	Oka (1962)
Glycerin, natural	56-81-5	1.26	Egorov et al. (2013)
Glycolic acid	79-14-1	1.49	Pijper (1971)
Hydrochloric acid	7647-01-0	1	Steinhauser et al. (1990)
Isopropanol	67-63-0	0.78	Zhang et al. (2013)
Isopropylamine	75-31-0	0.68	Sarkar and Roy (2009)
Magnesium chloride	7786-30-3	2.32	Haynes (2014)
Methanol	67-56-1	0.79	Kiselev et al. (2012)
Methenamine	100-97-0	1.2	Mak (1965)
Methoxyacetic acid	625-45-6	1.18	Haynes (2014)
N,N-Dimethylformamide	68-12-2	0.94	Smirnov and Badelin (2013)
Naphthalene	91-20-3	0.81	Dyshin et al. (2008)
Nitrogen, liquid	7727-37-9	0.82	finemech (2012)
Ozone	10028-15-6	0.002144	Haynes (2014)
Peracetic acid	79-21-0	1.1	Sigma-Aldrich (2015b)
Phosphonic acid	13598-36-2	1.7	Sigma-Aldrich (2014a)
Phosphoric acid Divosan X-Tend formulation	7664-38-2	0.94	Fadeeva et al. (2004)
Potassium acetate	127-08-2	1.54	Haynes (2014)
Propargyl alcohol	107-19-7	0.95	Vijaya Kumar et al. (1996)
Saline	7647-14-5	2.17	Sigma-Aldrich (2010)
Saturated sucrose	57-50-1	1.13	Hagen and Kaatze (2004)
Silica, amorphous	7631-86-9	2.2	Fujino et al. (2004)
Sodium carbonate	497-19-8	2.54	Haynes (2014)
Sodium formate	141-53-7	1.97	Fuess et al. (1982)
Sodium hydroxide	1310-73-2	2.13	Haynes (2014)
Sulfur dioxide	7446-09-5	1.3	Sigma-Aldrich (2015c)
Sulfuric acid	7664-93-9	1.83	Sigma-Aldrich (2015d)

Name	CASRN	Density (g/mL)	Reference
tert-Butyl hydroperoxide (70% solution in water)	75-91-2	0.9	Sigma-Aldrich (2007)
Tetramethylammonium chloride	75-57-0	1.17	Haynes (2014)
Thioglycolic acid	68-11-1	1.33	Biilmann (1906)
Toluene	108-88-3	0.86	Martinez-Reina et al. (2012)
Tridecane	629-50-5	0.75	Zhang et al. (2011)
Triethanolamine	102-71-6	1.12	Blanco et al. (2013)
Triethyl phosphate	78-40-0	1.06	Krakowiak et al. (2001)
Triethylene glycol	112-27-6	1.12	Afzal et al. (2009)
Triisopropanolamine	122-20-3	1.02	IUPAC (2014)
Trimethyl borate	121-43-7	0.93	Sigma-Aldrich (2015e)
Undecane	1120-21-4	0.74	de Oliveira et al. (2011)

C.4.   Estimating Spill Rates Based on State Spill Report Data

Several studies have provided estimates for the frequency of hydraulic fracturing-related spills. This section compiles analyses for three states: Pennsylvania, Colorado, and North Dakota (Table C-8).

In Pennsylvania, spills related to hydraulic fracturing activity are estimated to occur at a rate between 0.4 to 12.2 reported spills per 100 wells installed in the Marcellus Shale. Three studies (Rahm et al., 2015; Brantley et al., 2014; Gradient, 2013) calculated a spill rate for the Marcellus Shale in Pennsylvania using reports from the Pennsylvania Department of Environmental Protection (PA DEP) Oil and Gas Compliance Report Database. The PA DEP database provides a searchable format based on Notices of Violations from routine inspections or investigations of spill reports or complaints. Each study had different criteria for inclusion, presented in Table C-8, resulting in a range of rates even when using the same data source. Spill estimates include different criteria for how the rates were calculated. All three of these sources consider spills that occur during hydraulic fracturing activity. These include produced water, hydraulic fracturing chemicals, and diesel. Brantley et al. (2014) present data for major spills (> 400 gal or 1,514 L) that reached a water body, which would be a low-end estimate of the total number of spills occuring on site.

In Colorado, there is an estimated average of 1.3 reported spills on or near the well pad for every 100 hydraulically fractured wells, based on spill reports from the Colorado Oil and Gas Conservation Commission (COGCC) Information System. In its study of spills related to hydraulic fracturing, the EPA determined that Colorado spill reports were the most detailed spill reports from among the nine state data sources investigated and generally provided more of the information needed to determine whether a spill was related to hydraulic fracturing (U.S. EPA, 2015j). Here, we estimate the spill rate in Colorado by dividing the number of hydraulic fracturing-related spills identified by the EPA (U.S. EPA, 2015j, Appendix B) by the number of wells hydraulically fractured in Colorado for specific time periods between January 2006 and April 2012. We used three data sources to estimate the number of wells: (1) there were 172 reported spills in Colorado for the 15,000 wells fractured from January 2006 to April 2012 (DrillingInfo, 2012), (2) there were 50 reported spills in Colorado for the 3,559 wells fractured from January 2011 to April 2012 (U.S. EPA, 2015c), and (3) there were 41 reported spills in Colorado for the 3,000 wells fractured from September 2009 to October 2010 (U.S. EPA, 2013a). These data give an estimated average of 1.3 reported spills on or near the well pad for every 100 hydraulically fractured wells (Table C-8).

In North Dakota, using the North Dakota spills database, there were an estimated 2.6 reported spills of hydraulic fracturing fluids and chemicals per 100 wells fractured in 2015 (North Dakota Department of Health, 2015; see Appendix E). There were 22 reported spills of injection fluid and 17 spills of injection chemical. In 2015, there were 1490 wells fractured (North Dakota Department of Mineral Resources, 2016). Due to including only spills of fluids and chemicals, this estimate may fall on the low side.

The spill rates presented in Table C-8 are based on spill reports found in three state data sources and are limited by both the spills reported in the state data sources and the inclusion criteria defined by each of the studies. Spills identified from state data sources are likely a subset of the total number of spills that occurred within a state for a specified time period. Some spills may not be recorded in state data sources, because they do not meet the spill reporting requirements in place at the time of the spill. Additionally, the PA DEP Notices of Violation may include spills not specifically related to hydraulic fracturing, such as spills of drilling fluids.

The inclusion criteria used by each of the studies affects which spills are used to calculate a spill rate. More restrictive criteria, such as only counting spills that were greater than 400 gal (1,514 L), results in a lower number of spills being used for estimating spill rates, while less restrictive criteria, such as all spills from wells marked unconventional in the PA DEP database, results in a greater number of spills being used for estimating spill rates. Rahm et al. applied the least restrictive criteria of the four studies (i.e., spills from unconventional wells) when identifying spills, while Brantley et al. applied more restrictive criteria (i.e., spills of > 400 gal or 1,514 L in which spilled fluids reached a surface water body). This would contribute to the different spill rates calculated by these two studies.

Based on previous studies and the analysis here, hydraulic fracturing-related spills rates in Pennsylvania, Colorado, and North Dakota range from 0.4 to 12.2 reported spills per 100 wells, with a median rate of 2.6 reported spills for every 100 wells. These numbers may not be representative of national spill rates or rates in other regions.

Table C-8. Estimations of spill rates.

Spill rates from four different sources. Each source used different criteria to identify and include spills in their analysis.

Spill rate[a]	Data source	Time period	Inclusion criteria	Information source
0.4[b], 0.8[c]	PA DEP[d]	2008 - 2013	Volume spilled > 400 gal; all spills reported to reach water body.[e]	Brantley et al. (2014)
3.3[f]	PA DEP[d]	2009 - 2012	"Unconventional" well; spills with unknown volumes not included. Includes any spill during HF activities	Gradient (2013)
12.2[g], 11.6[h]	PA DEP[d]	2007 - July 2013	"Unconventional" well based on environmental violation rates.	Rahm et al. (2015)
1.3[i]	COGCC[j]	Jan 2006 - May 2012	Specifically related to hydraulic fracturing on or near well pad	U.S. EPA (2013a)
2.6[i]	ND	2015	Spills reported as injection fluid (22) or injection chemical (17)	North Dakota Department of Health (2015)
Median Spill Rate: 2.6 reported spills per 100 wells				

[a] Spill rate is the number of reported spills per 100 wells.

[b] Spill rate is calculated as the number of spills per 100 wells spudded.

[c] Spill rate is calculated as number of spills per 100 wells completed.

[d] PA DEP (2016).

[e] 32 spills >400 gal: 9 were brine (e.g., produced water), 7 were gel or hydraulic fracturing fluids, 5 were hydrostatic test waters or sediments, 2 were unknown, 1 was diesel.

[f] Spill rate is calculated as the number of spills per 100 wells installed.

[g] Mean spill rate is calculated as the number of spills per 100 wells drilled.

[h] Median spill rate is calculated as the number of spills per 100 drilled.

[i] Spill rate is calculated as the number of spills per 100 wells fractured.

[j] COGCC (2016).

C.5. Selected Physicochemical Properties of Organic Chemicals Used in Hydraulic Fracturing Fluids

Table C-9. Selected physicochemical properties of organic chemicals reported as used in hydraulic fracturing fluids.
Properties are provided for chemicals, where available from EPI Suite™ version 4.1 (U.S. EPA, 2012b). Selected physicochemical properties of organic chemicals reported as used in hydraulic fracturing fluids. In the table, "--" indicates no information is available.

Chemical name	CASRN	Log K_{ow}		Water solubility estimate from log K_{ow} (mg/L at 25°C)	Henry's law constant (atm-m³/mol at 25°C)		
		Estimated	Measured		Bond method	Group method 25	Measured
(13Z)-N,N-bis(2-hydroxyethyl)-N-methyldocos-13-en-1-aminium chloride	120086-58-0	4.38	--	0.3827	3.32×10^{-15}	--	--
(2,3-Dihydroxypropyl)trimethyl ammonium chloride	34004-36-9	-5.8	--	1.00×10^{6}	9.84×10^{-18}	--	--
(E)-Crotonaldehyde	123-73-9	0.6	--	4.15×10^{4}	5.61×10^{-5}	1.90×10^{-5}	1.94×10^{-5}
[Nitrilotris(methylene)]tris-phosphonic acid pentasodium salt	2235-43-0	-5.45	-3.53	1.00×10^{6}	1.65×10^{-34}	--	--
1-(1-Naphthylmethyl)quinolinium chloride	65322-65-8	5.57	--	0.02454	1.16×10^{-7}	--	--
1-(Alkyl* amino)-3-aminopropane *(42%C12, 26%C18, 15%C14, 8%C16, 5%C10, 4%C8)	68155-37-3	4.74	--	23.71	6.81×10^{-8}	2.39×10^{-8}	--
1-(Phenylmethyl)pyridinium Et Me derivatives, chlorides	68909-18-2	4.1	--	14.13	1.78×10^{-5}	--	--
1,2,3-Trimethylbenzene	526-73-8	3.63	3.66	75.03	7.24×10^{-3}	6.58×10^{-3}	4.36×10^{-3}
1,2,4-Trimethylbenzene	95-63-6	3.63	3.63	79.59	7.24×10^{-3}	6.58×10^{-3}	6.16×10^{-3}
1,2-Benzisothiazolin-3-one	2634-33-5	0.64	--	2.14×10^{4}	6.92×10^{-9}	--	--
1,2-Dibromo-2,4-dicyanobutane	35691-65-7	1.63	--	424	3.94×10^{-10}	--	--

Chemical name	CASRN	Log K_{ow}		Water solubility estimate from log K_{ow} (mg/L at 25°C)	Henry's law constant (atm-m^3/mol at 25°C)		
		Estimated	Measured		Bond method	Group method 25	Measured
1,2-Dimethylbenzene	95-47-6	3.09	3.12	224.1	6.56×10^{-3}	6.14×10^{-3}	5.18×10^{-3}
1,2-Ethanediaminium, N,N′-bis[2-[bis(2-hydroxyethyl)methylammonio]ethyl]-N,N′-bis(2-hydroxyethyl)-N,N′-dimethyl-, tetrachloride	138879-94-4	-23.19	--	1.00×10^6	2.33×10^{-35}	--	--
1,2-Propylene glycol	57-55-6	-0.78	-0.92	8.11×10^5	1.74×10^{-7}	1.31×10^{-10}	1.29×10^{-8}
1,2-Propylene oxide	75-56-9	0.37	0.03	1.29×10^5	1.60×10^{-4}	1.23×10^{-4}	6.96×10^{-5}
1,3,5-Triazine	290-87-9	-0.2	0.12	1.03×10^5	1.21×10^{-6}	--	--
1,3,5-Triazine-1,3,5(2H,4H,6H)-triethanol	4719-04-4	-4.67	--	1.00×10^6	1.08×10^{-11}	--	--
1,3,5-Trimethylbenzene	108-67-8	3.63	3.42	120.3	7.24×10^{-3}	6.58×10^{-3}	8.77×10^{-3}
1,3-Butadiene	106-99-0	2.03	1.99	792.3	7.79×10^{-2}	7.05×10^{-2}	7.36×10^{-2}
1,3-Dichloropropene	542-75-6	2.29	2.04	1,994	2.45×10^{-2}	3.22×10^{-3}	3.55×10^{-3}
1,4-Dioxane	123-91-1	-0.32	-0.27	2.14×10^5	5.91×10^{-6}	1.12×10^{-7}	4.80×10^{-6}
1,6-Hexanediamine	124-09-4	0.35	--	5.34×10^5	3.21×10^{-9}	7.05×10^{-10}	--
1,6-Hexanediamine dihydrochloride	6055-52-3	0.35	--	5.34×10^5	3.21×10^{-9}	7.05×10^{-10}	--
1-[2-(2-Methoxy-1-methylethoxy)-1-methylethoxy]-2-propanol	20324-33-8	-0.2	--	1.96×10^5	2.36×10^{-11}	4.55×10^{-13}	--
1-Amino-2-propanol	78-96-6	-1.19	-0.96	1.00×10^6	4.88×10^{-10}	2.34×10^{-10}	--
1-Benzylquinolinium chloride	15619-48-4	4.4	--	6.02	1.19×10^{-6}	--	--
1-Butanol	71-36-3	0.84	0.88	7.67×10^4	9.99×10^{-6}	9.74×10^{-6}	8.81×10^{-6}
1-Butoxy-2-propanol	5131-66-8	0.98	--	4.21×10^4	1.30×10^{-7}	4.88×10^{-8}	--

Chemical name	CASRN	Log K_{ow}		Water solubility estimate from log K_{ow} (mg/L at 25°C)	Henry's law constant (atm-m³/mol at 25°C)		
		Estimated	Measured		Bond method	Group method 25	Measured
1-Decanol	112-30-1	3.79	4.57	28.21	5.47×10^{-5}	7.73×10^{-5}	3.20×10^{-5}
1-Dodecyl-2-pyrrolidinone	2687-96-9	5.3	4.2	5.862	7.12×10^{-7}	--	--
1-Eicosene	3452-07-1	10.03	--	1.26×10^{-5}	1.89×10^{1}	6.74×10^{1}	--
1-Ethyl-2-methylbenzene	611-14-3	3.58	3.53	96.88	8.71×10^{-3}	9.52×10^{-3}	5.53×10^{-3}
1-Hexadecene	629-73-2	8.06	--	0.001232	6.10	1.69×10^{1}	--
1-Hexanol	111-27-3	1.82	2.03	6,885	1.76×10^{-5}	1.94×10^{-5}	1.71×10^{-5}
1-Methoxy-2-propanol	107-98-2	-0.49	--	1.00×10^{6}	5.56×10^{-8}	1.81×10^{-8}	9.20×10^{-7}
1-Octadecanamine, acetate (1:1)	2190-04-7	7.71	--	0.04875	9.36×10^{-4}	2.18×10^{-3}	--
1-Octadecanamine, N,N-dimethyl-	124-28-7	8.39	--	0.008882	4.51×10^{-3}	3.88×10^{-2}	--
1-Octadecene	112-88-9	9.04	--	1.256×10^{-4}	10.7	3.38×10^{1}	--
1-Octanol	111-87-5	2.81	3	814	3.10×10^{-5}	3.88×10^{-5}	2.45×10^{-5}
1-Pentanol	71-41-0	1.33	1.51	2.09×10^{4}	1.33×10^{-5}	1.38×10^{-5}	1.30×10^{-5}
1-Propanaminium, 3-chloro-2-hydroxy-N,N,N-trimethyl-, chloride	3327-22-8	-4.48	--	1.00×10^{6}	9.48×10^{-17}		--
1-Propanesulfonic acid	5284-66-2	-1.4	--	1.00×10^{6}	2.22×10^{-8}		--
1-Propanol	71-23-8	0.35	0.25	2.72×10^{5}	7.52×10^{-6}	6.89×10^{-6}	7.41×10^{-6}
1-Propene	115-07-1	1.68	1.77	1,162	1.53×10^{-1}	1.58×10^{-1}	1.96×10^{-1}
1-tert-Butoxy-2-propanol	57018-52-7	0.87	--	5.24×10^{4}	1.30×10^{-7}	5.23×10^{-8}	--
1-Tetradecene	1120-36-1	7.08	--	0.01191	3.46	8.48	--

Chemical name	CASRN	Log K_{ow}		Water solubility estimate from log K_{ow} (mg/L at 25°C)	Henry's law constant (atm-m³/mol at 25°C)		
		Estimated	Measured		Bond method	Group method 25	Measured
1-Tridecanol	112-70-9	5.26	--	4.533	1.28×10^{-4}	2.18×10^{-4}	--
1-Undecanol	112-42-5	4.28	--	43.04	7.26×10^{-5}	1.09×10^{-4}	--
2-(2-Butoxyethoxy)ethanol	112-34-5	0.29	0.56	7.19×10^{4}	1.52×10^{-9}	4.45×10^{-11}	7.20×10^{-9}
2-(2-Ethoxyethoxy)ethanol	111-90-0	-0.69	-0.54	8.28×10^{5}	8.63×10^{-10}	2.23×10^{-11}	2.23×10^{-8}
2-(2-Ethoxyethoxy)ethyl acetate	112-15-2	0.32	--	3.09×10^{4}	5.62×10^{-8}	7.22×10^{-10}	2.29×10^{-8}
2-(Dibutylamino)ethanol	102-81-8	2.01	2.65	3,297	9.70×10^{-9}	1.02×10^{-8}	--
2-(Hydroxymethylamino)ethanol	34375-28-5	-1.53	--	1.00×10^{6}	1.62×10^{-12}	--	--
2-(Thiocyanomethylthio)benzothiazole	21564-17-0	3.12	3.3	41.67	6.49×10^{-12}	--	--
2,2'-(Diazene-1,2-diyldiethane-1,1-diyl)bis-4,5-dihydro-1H-imidazole dihydrochloride	27776-21-2	2.12	--	193.3	3.11×10^{-14}	--	--
2,2'-(Octadecylimino)diethanol	10213-78-2	6.85	--	0.08076	1.06×10^{-8}	7.39×10^{-12}	--
2,2'-[Ethane-1,2-diylbis(oxy)]diethanamine	929-59-9	-2.17	--	1.00×10^{6}	2.50×10^{-13}	8.10×10^{-16}	--
2,2'-Azobis(2-amidinopropane) dihydrochloride	2997-92-4	-3.28	--	1.00×10^{6}	1.21×10^{-14}	--	--
2,2-Dibromo-3-nitrilopropionamide	10222-01-2	1.01	0.82	2,841	6.16×10^{-14}	--	1.91×10^{-8}
2,2-Dibromopropanediamide	73003-80-2	0.37	--	1.00×10^{4}	3.58×10^{-14}	--	--
2,4-Hexadienoic acid, potassium salt, (2E,4E)-	24634-61-5	1.62	1.33	1.94×10^{4}	5.72×10^{-7}	4.99×10^{-8}	--
2,6,8-Trimethyl-4-nonanol	123-17-1	4.48	--	24.97	9.63×10^{-5}	4.45×10^{-4}	--

Chemical name	CASRN	Log K_{ow}		Water solubility estimate from log K_{ow} (mg/L at 25°C)	Henry's law constant (atm-m³/mol at 25°C)		
		Estimated	Measured		Bond method	Group method 25	Measured
2-Acrylamido-2-methyl-1-propanesulfonic acid	15214-89-8	-2.19	--	1.00×10^6	5.18×10^{-15}	--	--
2-Amino-2-methylpropan-1-ol	124-68-5	-0.74	--	1.00×10^6	6.48×10^{-10}	--	--
2-Aminoethanol hydrochloride	2002-24-6	-1.61	-1.31	1.00×10^6	3.68×10^{-10}	9.96×10^{-11}	--
2-Bromo-3-nitrilopropionamide	1113-55-9	-0.31	--	3,274	5.35×10^{-13}	--	--
2-Butanone oxime	96-29-7	1.69	0.63	3.66×10^4	1.04×10^{-5}	--	--
2-Butoxy-1-propanol	15821-83-7	0.98	--	4.21×10^4	1.30×10^{-7}	4.88×10^{-8}	--
2-Butoxyethanol	111-76-2	0.57	0.83	6.45×10^4	9.79×10^{-8}	2.08×10^{-8}	1.60×10^{-6}
2-Dodecylbenzenesulfonic acid- N-(2-aminoethyl)ethane-1,2-diamine(1:1)	40139-72-8	4.78	--	0.7032	6.27×10^{-8}	--	--
2-Ethoxyethanol	110-80-5	-0.42	-0.32	7.55×10^5	5.56×10^{-8}	1.04×10^{-8}	4.70×10^{-7}
2-Ethoxynaphthalene	93-18-5	3.74	--	38.32	4.13×10^{-5}	4.06×10^{-4}	--
2-Ethyl-1-hexanol	104-76-7	2.73	--	1,379	3.10×10^{-5}	4.66×10^{-5}	2.65×10^{-5}
2-Ethyl-2-hexenal	645-62-5	2.62	--	548.6	2.06×10^{-4}	4.88×10^{-4}	--
2-Ethylhexyl benzoate	5444-75-7	5.19	--	1.061	2.52×10^{-4}	2.34×10^{-4}	--
2-Hydroxyethyl acrylate	818-61-1	-0.25	-0.21	5.07×10^5	4.49×10^{-9}	7.22×10^{-10}	--
2-Hydroxyethylammonium hydrogen sulphite	13427-63-9	-1.61	-1.31	1.00×10^6	3.68×10^{-10}	9.96×10^{-11}	--
2-Hydroxy-N,N-bis(2-hydroxyethyl)-N-methylethanaminium chloride	7006-59-9	-6.7	--	1.00×10^6	4.78×10^{-19}	--	--
2-Mercaptoethanol	60-24-2	-0.2	--	1.94×10^5	1.27×10^{-7}	3.38×10^{-8}	1.80×10^{-7}

Chemical name	CASRN	Log K_{ow}		Water solubility estimate from log K_{ow} (mg/L at 25°C)	Henry's law constant (atm-m³/mol at 25°C)		
		Estimated	Measured		Bond method	Group method 25	Measured
2-Methoxyethanol	109-86-4	-0.91	-0.77	1.00×10^{6}	4.19×10^{-8}	7.73×10^{-9}	3.30×10^{-7}
2-Methyl-1-propanol	78-83-1	0.77	0.76	9.71×10^{4}	9.99×10^{-6}	1.17×10^{-5}	9.78×10^{-6}
2-Methyl-2,4-pentanediol	107-41-5	0.58	--	3.26×10^{4}	4.06×10^{-7}	3.97×10^{-10}	--
2-Methyl-3(2H)-isothiazolone	2682-20-4	-0.83	--	5.37×10^{5}	4.96×10^{-8}	--	--
2-Methyl-3-butyn-2-ol	115-19-5	0.45	0.28	2.40×10^{5}	1.04×10^{-6}	--	3.91×10^{-6}
2-Methylbutane	78-78-4	2.72	--	184.6	1.29	1.44	1.40
2-Methylquinoline hydrochloride	62763-89-7	2.69	2.59	498.5	7.60×10^{-7}	2.13×10^{-6}	--
2-Phosphono-1,2,4-butanetricarboxylic acid	37971-36-1	-1.66	--	1.00×10^{6}	1.17×10^{-26}	--	--
2-Phosphonobutane-1,2,4-tricarboxylic acid, potassium salt (1:x)	93858-78-7	-1.66	--	1.00×10^{6}	1.17×10^{-26}	--	--
2-Propenoic acid, 2-(2-hydroxyethoxy)ethyl ester	13533-05-6	-0.52	-0.3	3.99×10^{5}	6.98×10^{-11}	1.54×10^{-12}	--
3-(Dimethylamino)propylamine	109-55-7	-0.45	--	1.00×10^{6}	6.62×10^{-9}	4.45×10^{-9}	--
3,4,4-Trimethyloxazolidine	75673-43-7	0.13	--	8.22×10^{5}	6.63×10^{-6}	--	--
3,5,7-Triazatricyclo(3.3.1.13,7)decane, 1-(3-chloro-2-propenyl)-, chloride, (Z)-	51229-78-8	-5.92	--	1.00×10^{6}	1.76×10^{-8}	--	--
3,7-Dimethyl-2,6-octadienal	5392-40-5	3.45	--	84.71	3.76×10^{-4}	4.35×10^{-5}	--
3-Hydroxybutanal	107-89-1	-0.72	--	1.00×10^{6}	4.37×10^{-9}	2.28×10^{-9}	--
3-Methoxypropylamine	5332-73-0	-0.42	--	1.00×10^{6}	1.56×10^{-7}	1.94×10^{-8}	--
3-Phenylprop-2-enal	104-55-2	1.82	1.9	2,150	1.60×10^{-6}	3.38×10^{-7}	--

Chemical name	CASRN	Log K_{ow} Estimated	Log K_{ow} Measured	Water solubility estimate from log K_{ow} (mg/L at 25°C)	Henry's law constant (atm-m³/mol at 25°C) Bond method	Group method 25	Measured
4,4-Dimethyloxazolidine	51200-87-4	-0.08	--	1.00×10^6	3.02×10^{-6}	--	--
4,6-Dimethyl-2-heptanone	19549-80-5	2.56	--	528.8	2.71×10^{-4}	4.55×10^{-4}	--
4-[Abieta-8,11,13-trien-18-yl(3-oxo-3-phenylpropyl)amino]butan-2-one hydrochloride	143106-84-7	7.72	--	0.002229	2.49×10^{-12}	1.20×10^{-14}	--
4-Ethyloct-1-yn-3-ol	5877-42-9	2.87	--	833.9	4.27×10^{-6}	--	--
4-Hydroxy-3-methoxybenzaldehyde	121-33-5	1.05	1.21	6,875	8.27×10^{-11}	2.81×10^{-9}	2.15×10^{-9}
4-Methoxybenzyl formate	122-91-8	1.61	--	2,679	1.15×10^{-6}	2.13×10^{-6}	--
4-Methoxyphenol	150-76-5	1.59	1.58	1.65×10^4	3.32×10^{-8}	5.35×10^{-7}	--
4-Methyl-2-pentanol	108-11-2	1.68	--	1.38×10^4	1.76×10^{-5}	3.88×10^{-5}	4.45×10^{-5}
4-Methyl-2-pentanone	108-10-1	1.16	1.31	8,888	1.16×10^{-4}	1.34×10^{-4}	1.38×10^{-4}
4-Nonylphenol	104-40-5	5.99	5.76	1.57	5.97×10^{-6}	1.23×10^{-5}	3.40×10^{-5}
5-Chloro-2-methyl-3(2H)-isothiazolone	26172-55-4	-0.34	--	1.49×10^5	3.57×10^{-8}		--
Acetaldehyde	75-07-0	-0.17	-0.34	2.57×10^5	6.78×10^{-5}	6.00×10^{-5}	6.67×10^{-5}
Acetic acid	64-19-7	0.09	-0.17	4.76×10^5	5.48×10^{-7}	2.94×10^{-7}	1.00×10^{-7}
Acetic acid, C6-8-branched alkyl esters	90438-79-2	3.25	--	117.8	9.60×10^{-4}	1.07×10^{-3}	--
Acetic acid, hydroxy-, reaction products with triethanolamine	68442-62-6	-2.48	-1	1.00×10^6	4.18×10^{-12}	3.38×10^{-19}	7.05×10^{-13}
Acetic acid, mercapto-, monoammonium salt	5421-46-5	0.03	0.09	2.56×10^5	1.94×10^{-8}	--	--
Acetic anhydride	108-24-7	-0.58	--	3.59×10^5	3.57×10^{-5}	--	5.71×10^{-6}

Chemical name	CASRN	Log K_{ow}		Water solubility estimate from log K_{ow} (mg/L at 25°C)	Henry's law constant (atm-m³/mol at 25°C)		
		Estimated	Measured		Bond method	Group method 25	Measured
Acetone	67-64-1	-0.24	-0.24	2.20×10^5	4.96×10^{-5}	3.97×10^{-5}	3.50×10^{-5}
Acetonitrile, 2,2',2''-nitrilotris-	7327-60-8	-1.39	--	1.00×10^6	2.61×10^{-15}	--	--
Acetophenone	98-86-2	1.67	1.58	4,484	9.81×10^{-6}	1.09×10^{-5}	1.04×10^{-5}
Acetyltriethyl citrate	77-89-4	1.34	--	688.2	6.91×10^{-11}	--	--
Acrolein	107-02-8	0.19	-0.01	1.40×10^5	3.58×10^{-5}	1.94×10^{-5}	1.22×10^{-4}
Acrylamide	79-06-1	-0.81	-0.67	5.04×10^5	5.90×10^{-9}	--	1.70×10^{-9}
Acrylic acid	79-10-7	0.44	0.35	1.68×10^5	2.89×10^{-7}	1.17×10^{-7}	3.70×10^{-7}
Acrylic acid, with sodium-2-acrylamido-2-methyl-1-propanesulfonate and sodium phosphinate	110224-99-2	-2.19	--	1.00×10^6	5.18×10^{-15}	--	--
Alcohols, C10-12, ethoxylated	67254-71-1	5.47	--	0.9301	1.95×10^{-2}	2.03×10^{-2}	--
Alcohols, C11-14-iso-, C13-rich	68526-86-3	5.19	--	5.237	1.28×10^{-4}	2.62×10^{-4}	--
Alcohols, C11-14-iso-, C13-rich, ethoxylated	78330-21-9	4.91	--	5.237	1.25×10^{-6}	7.73×10^{-7}	--
Alcohols, C12-13, ethoxylated	66455-14-9	5.96	--	0.2995	2.58×10^{-2}	2.87×10^{-2}	--
Alcohols, C12-14, ethoxylated propoxylated	68439-51-0	6.67	--	0.02971	7.08×10^{-4}	1.23×10^{-4}	--
Alcohols, C12-14-secondary	126950-60-5	5.19	--	5.237	1.28×10^{-4}	3.62×10^{-4}	--
Alcohols, C12-16, ethoxylated	68551-12-2	6.45	--	0.09603	3.43×10^{-2}	4.06×10^{-2}	--
Alcohols, C14-15, ethoxylated	68951-67-7	7.43	--	0.009765	6.04×10^{-2}	8.10×10^{-2}	--
Alcohols, C6-12, ethoxylated	68439-45-2	4.49	--	8.832	1.10×10^{-2}	1.02×10^{-2}	--

Chemical name	CASRN	Log K_{ow}		Water solubility estimate from log K_{ow} (mg/L at 25°C)	Henry's law constant (atm-m³/mol at 25°C)		
		Estimated	Measured		Bond method	Group method 25	Measured
Alcohols, C7-9-iso-, C8-rich, ethoxylated	78330-19-5	2.46	--	1,513	3.04×10^{-7}	1.38×10^{-7}	--
Alcohols, C9-11, ethoxylated	68439-46-3	4.98	--	2.874	1.47×10^{-2}	1.44×10^{-2}	--
Alcohols, C9-11-iso-, C10-rich, ethoxylated	78330-20-8	4.9	--	3.321	1.47×10^{-2}	2.39×10^{-2}	--
Alkanes, C12-14-iso-	68551-19-9	6.65	--	0.03173	1.24×10^{1}	2.28×10^{1}	--
Alkanes, C13-16-iso-	68551-20-2	7.63	--	0.003311	2.19×10^{1}	4.55×10^{1}	--
Alkenes, C>10 alpha-	64743-02-8	8.55	--	0.0003941	8.09	2.39×10^{1}	--
Alkyl* dimethyl ethylbenzyl ammonium chloride *(50%C12, 30%C14, 17%C16, 3%C18)	85409-23-0_1	3.97	--	3.23	1.11×10^{-11}	--	--
Alkyl* dimethyl ethylbenzyl ammonium chloride *(60%C14, 30%C16, 5%C12, 5%C18)	68956-79-6	4.95	--	0.3172	1.96×10^{-11}	--	--
Alkylbenzenesulfonate, linear	42615-29-2	4.71	--	0.8126	6.27×10^{-8}	--	--
alpha-Lactose monohydrate	5989-81-1	-5.12	--	1.00×10^{6}	4.47×10^{-22}	9.81×10^{-45}	--
alpha-Terpineol	98-55-5	3.33	2.98	371.7	1.58×10^{-5}	3.15×10^{-6}	1.22×10^{-5}
Amaranth	915-67-3	1.63	--	1.789	1.49×10^{-30}	--	--
Aminotrimethylene phosphonic acid	6419-19-8	-5.45	-3.53	1.00×10^{6}	1.65×10^{-34}	--	--
Ammonium acetate	631-61-8	0.09	-0.17	4.76×10^{5}	5.48×10^{-7}	2.94×10^{-7}	1.00×10^{-7}
Ammonium acrylate	10604-69-0	0.44	0.35	1.68×10^{5}	2.89×10^{-7}	1.17×10^{-7}	3.70×10^{-7}
Ammonium citrate (1:1)	7632-50-0	-1.67	-1.64	1.00×10^{6}	8.33×10^{-18}	--	4.33×10^{-14}

Chemical name	CASRN	Log K_{ow}		Water solubility estimate from log K_{ow} (mg/L at 25°C)	Henry's law constant (atm-m³/mol at 25°C)		
		Estimated	Measured		Bond method	Group method 25	Measured
Ammonium citrate (2:1)	3012-65-5	-1.67	-1.64	1.00×10^6	8.33×10^{-18}	--	4.33×10^{-14}
Ammonium dodecyl sulfate	2235-54-3	2.42	--	163.7	1.84×10^{-7}	--	--
Ammonium hydrogen carbonate	1066-33-7	-0.46	--	8.42×10^5	6.05×10^{-9}	--	--
Ammonium lactate	515-98-0	-0.65	-0.72	1.00×10^6	1.13×10^{-7}	--	8.13×10^{-8}
Anethole	104-46-1	3.39	--	98.68	2.56×10^{-4}	2.23×10^{-3}	--
Aniline	62-53-3	1.08	0.9	2.08×10^4	1.90×10^{-6}	2.18×10^{-6}	2.02×10^{-6}
Benactyzine hydrochloride	57-37-4	2.89	--	292.1	2.07×10^{-10}	--	--
Benzamorf	12068-08-5	4.71	--	0.8126	6.27×10^{-8}	--	--
Benzene	71-43-2	1.99	2.13	2,000	5.39×10^{-3}	5.35×10^{-3}	5.55×10^{-3}
Benzene, C10-16-alkyl derivatives	68648-87-3	8.43	9.36	0.0002099	1.78×10^{-1}	3.97×10^{-1}	--
Benzenesulfonic acid	98-11-3	-1.17	--	6.90×10^5	2.52×10^{-9}	--	--
Benzenesulfonic acid, (1-methylethyl)-,	37953-05-2	0.29	--	2.46×10^4	4.89×10^{-9}	--	--
Benzenesulfonic acid, (1-methylethyl)-, ammonium salt	37475-88-0	0.29	--	2.46×10^4	4.89×10^{-9}	--	--
Benzenesulfonic acid, (1-methylethyl)-, sodium salt	28348-53-0	0.29	--	2.46×10^4	4.89×10^{-9}	--	--
Benzenesulfonic acid, C10-16-alkyl derivatives, compounds with cyclohexylamine	255043-08-4	4.71	--	0.8126	6.27×10^{-8}	--	--

Chemical name	CASRN	Log K_{ow}		Water solubility estimate from log K_{ow} (mg/L at 25°C)	Henry's law constant (atm-m³/mol at 25°C)		
		Estimated	Measured		Bond method	Group method 25	Measured
Benzenesulfonic acid, C10-16-alkyl derivatives, compounds with triethanolamine	68584-25-8	5.2	--	0.255	8.32×10^{-8}	--	--
Benzenesulfonic acid, C10-16-alkyl derivatives, potassium salts	68584-27-0	5.2	--	0.255	8.32×10^{-8}	--	--
Benzenesulfonic acid, dodecyl-, branched, compounds with 2-propanamine	90218-35-2	4.49	--	1.254	6.27×10^{-8}	--	--
Benzenesulfonic acid, mono-C10-16-alkyl derivatives, sodium salts	68081-81-2	4.22	--	2.584	4.72×10^{-8}	--	--
Benzoic acid	65-85-0	1.87	1.87	2,493	1.08×10^{-7}	4.55×10^{-8}	3.81×10^{-8}
Benzyl chloride	100-44-7	2.79	2.3	1,030	2.09×10^{-3}	3.97×10^{-4}	4.12×10^{-4}
Benzyldimethyldodecylammonium chloride	139-07-1	2.93	--	36.47	7.61×10^{-12}	--	--
Benzylhexadecyldimethylammonium chloride	122-18-9	4.89	--	0.3543	2.36×10^{-11}	--	--
Benzyltrimethylammonium chloride	56-93-9	-2.47	--	1.00×10^{6}	3.37×10^{-13}	--	--
Bicine	150-25-4	-3.27	--	3.52×10^{5}	1.28×10^{-14}	--	--
Bis(1-methylethyl)naphthalenesulfonic acid, cyclohexylamine salt	68425-61-6	2.92	--	43.36	9.29×10^{-10}	--	--
Bis(2-chloroethyl) ether	111-44-4	1.56	1.29	6,435	1.89×10^{-4}	4.15×10^{-7}	1.70×10^{-5}
Bisphenol A	80-05-7	3.64	3.32	172.7	9.16×10^{-12}	--	--
Bronopol	52-51-7	-1.51	--	8.37×10^{5}	6.35×10^{-21}	--	--

Chemical name	CASRN	Log K_{ow}		Water solubility estimate from log K_{ow} (mg/L at 25°C)	Henry's law constant (atm-m³/mol at 25°C)		
		Estimated	Measured		Bond method	Group method 25	Measured
Butane	106-97-8	2.31	2.89	135.6	9.69×10^{-1}	8.48×10^{-1}	9.50×10^{-1}
Butanedioic acid, sulfo-, 1,4-bis(1,3-dimethylbutyl) ester, sodium salt	2373-38-8	3.98	--	0.1733	1.61×10^{-12}	--	--
Butene	25167-67-3	2.17	2.4	354.8	2.03×10^{-1}	2.68×10^{-1}	2.33×10^{-1}
Butyl glycidyl ether	2426-08-6	1.08	0.63	2.66×10^{4}	4.37×10^{-6}	5.23×10^{-7}	2.47×10^{-5}
Butyl lactate	138-22-7	0.8	--	5.30×10^{4}	8.49×10^{-5}	--	1.92×10^{-6}
Butyryl trihexyl citrate	82469-79-2	8.21	--	5.56×10^{-5}	3.65×10^{-9}	--	--
C.I. Acid Red 1	3734-67-6	0.51	--	6.157	3.73×10^{-29}	--	--
C.I. Acid Violet 12, disodium salt	6625-46-3	0.59	--	3.379	2.21×10^{-30}	--	--
C.I. Pigment Red 5	6410-41-9	7.65	--	4.38×10^{-5}	4.36×10^{-21}	--	--
C.I. Solvent Red 26	4477-79-6	9.27	--	5.68×10^{-5}	5.48×10^{-13}	4.66×10^{-13}	--
C10-16-Alkyldimethylamines oxides	70592-80-2	2.87	--	89.63	1.14×10^{-13}	--	--
C10-C16 Ethoxylated alcohol	68002-97-1	4.99	--	4.532	1.25×10^{-6}	4.66×10^{-7}	--
C12-14 tert-Alkyl ethoxylated amines	73138-27-9	3.4	--	264.2	1.29×10^{-10}	--	--
Calcium dodecylbenzene sulfonate	26264-06-2	4.71	--	0.8126	6.27×10^{-8}	--	--
Camphor	76-22-2	3.04	2.38	339.1	7.00×10^{-5}	--	8.10×10^{-5}
Carbon dioxide	124-38-9	0.83	0.83	2.57×10^{4}	1.52×10^{-2}	--	1.52×10^{-2}
Carbonic acid, dipotassium salt	584-08-7	-0.46	--	8.42×10^{5}	6.05×10^{-9}	--	--
Chloromethane	74-87-3	1.09	0.91	2.26×10^{4}	8.20×10^{-3}	8.88×10^{-3}	8.82×10^{-3}

Chemical name	CASRN	Log K_{ow}		Water solubility estimate from log K_{ow} (mg/L at 25°C)	Henry's law constant (atm-m³/mol at 25°C)		
		Estimated	Measured		Bond method	Group method 25	Measured
Chlorobenzene	108-90-7	2.64	2.84	400.5	3.99×10^{-3}	4.55×10^{-3}	3.11×10^{-3}
Choline bicarbonate	78-73-9	-5.16	--	1.00×10^{6}	2.03×10^{-16}	--	--
Choline chloride	67-48-1	-5.16	--	1.00×10^{6}	2.03×10^{-16}	--	--
Citric acid	77-92-9	-1.67	-1.64	1.00×10^{6}	8.33×10^{-18}	--	4.33×10^{-14}
Citronellol	106-22-9	3.56	3.91	105.5	5.68×10^{-5}	2.13×10^{-5}	--
Coconut trimethylammonium chloride	61789-18-2	1.22	--	2,816	9.42×10^{-11}	--	--
Coumarin	91-64-5	1.51	1.39	5,126	6.95×10^{-6}	--	9.92×10^{-8}
Cumene	98-82-8	3.45	3.66	75.03	1.05×10^{-2}	1.23×10^{-2}	1.15×10^{-2}
Cyclohexane	110-82-7	3.18	3.44	43.02	2.55×10^{-1}	1.94×10^{-1}	1.50×10^{-1}
Cyclohexanol	108-93-0	1.64	1.23	3.37×10^{4}	4.90×10^{-6}	3.70×10^{-6}	4.40×10^{-6}
Cyclohexanone	108-94-1	1.13	0.81	2.41×10^{4}	5.11×10^{-5}	1.28×10^{-5}	9.00×10^{-6}
Cyclohexylamine sulfate	19834-02-7	1.63	1.49	6.40×10^{4}	1.38×10^{-5}	--	4.16×10^{-6}
D&C Red no. 28	18472-87-2	9.62	--	1.64×10^{-8}	6.37×10^{-21}	--	--
D&C Red no. 33	3567-66-6	0.48	--	11.87	1.15×10^{-26}	--	--
Daidzein	486-66-8	2.55	--	568.4	3.91×10^{-16}	--	--
Dapsone	80-08-0	0.77	0.97	3,589	3.11×10^{-14}	--	--
Dazomet	533-74-4	0.94	0.63	1.94×10^{4}	2.84×10^{-3}	--	4.98×10^{-10}
Decyldimethylamine	1120-24-7	4.46	--	82.23	4.68×10^{-4}	2.45×10^{-3}	--
D-Glucitol	50-70-4	-3.01	-2.2	1.00×10^{6}	7.26×10^{-13}	2.94×10^{-29}	--

Chemical name	CASRN	Log K_{ow}		Water solubility estimate from log K_{ow} (mg/L at 25°C)	Henry's law constant (atm-m³/mol at 25°C)		
		Estimated	Measured		Bond method	Group method 25	Measured
D-Gluconic acid	526-95-4	-1.87	--	1.00×10^6	4.74×10^{-13}	--	--
D-Glucopyranoside, methyl	3149-68-6	-2.5	--	1.00×10^6	1.56×10^{-14}	2.23×10^{-24}	--
D-Glucose	50-99-7	-2.89	-3.24	1.00×10^6	9.72×10^{-15}	1.62×10^{-26}	--
Di(2-ethylhexyl) phthalate	117-81-7	8.39	7.6	0.001132	1.18×10^{-5}	1.02×10^{-5}	2.70×10^{-7}
Dibromoacetonitrile	3252-43-5	0.47	--	9,600	4.06×10^{-7}	--	--
Dichloromethane	75-09-2	1.34	1.25	1.10×10^4	9.14×10^{-3}	3.01×10^{-3}	3.25×10^{-3}
Didecyldimethylammonium chloride	7173-51-5	4.66	--	0.9	6.85×10^{-10}	--	--
Diethanolamine	111-42-2	-1.71	-1.43	1.00×10^6	3.92×10^{-11}	3.46×10^{-15}	3.87×10^{-11}
Diethylbenzene	25340-17-4	4.07	3.72	58.86	1.16×10^{-2}	1.47×10^{-2}	2.61×10^{-3}
Diethylene glycol	111-46-6	-1.47	--	1.00×10^6	2.03×10^{-9}	1.20×10^{-13}	--
Diethylene glycol monomethyl ether	111-77-3	-1.18	--	1.00×10^6	6.50×10^{-10}	1.65×10^{-11}	--
Diethylenetriamine	111-40-0	-2.13	--	1.00×10^6	3.10×10^{-13}	1.09×10^{-14}	--
Diisobutyl ketone	108-83-8	2.56	--	528.8	2.71×10^{-4}	4.55×10^{-4}	1.17×10^{-4}
Diisopropanolamine	110-97-4	-0.88	-0.82	1.00×10^6	6.91×10^{-11}	1.90×10^{-14}	--
Diisopropylnaphthalene	38640-62-9	6.08	--	0.2421	1.99×10^{-3}	1.94×10^{-3}	--
Dimethyl adipate	627-93-0	1.39	1.03	7,749	9.77×10^{-7}	1.28×10^{-7}	2.31×10^{-6}
Dimethyl glutarate	1119-40-0	0.9	0.62	2.02×10^4	7.36×10^{-7}	9.09×10^{-8}	6.43×10^{-7}
Dimethyl succinate	106-65-0	0.4	0.35	3.96×10^4	5.54×10^{-7}	6.43×10^{-8}	--
Dimethylaminoethanol	108-01-0	-0.94	--	1.00×10^6	1.77×10^{-9}	1.77×10^{-9}	3.73×10^{-7}

Chemical name	CASRN	Log K_{ow}		Water solubility estimate from log K_{ow} (mg/L at 25°C)	Henry's law constant (atm-m^3/mol at 25°C)		
		Estimated	Measured		Bond method	Group method 25	Measured
Dimethyldiallylammonium chloride	7398-69-8	-2.49	--	1.00×10^6	7.20×10^{-12}	--	--
Diphenyl oxide	101-84-8	4.05	4.21	15.58	1.18×10^{-4}	2.81×10^{-4}	2.79×10^{-4}
Dipropylene glycol	25265-71-8	-0.64	--	3.11×10^5	3.58×10^{-9}	6.29×10^{-10}	--
Di-sec-butylphenol	31291-60-8	5.41	--	3.723	3.74×10^{-6}	6.89×10^{-6}	--
Disodium dodecyl(sulphonatophenoxy)benzenesulphonate	28519-02-0	5.05	--	0.0353	6.40×10^{-16}	--	--
Disodium ethylenediaminediacetate	38011-25-5	-4.79	--	1.00×10^6	1.10×10^{-16}	--	--
Disodium ethylenediaminetetraacetate dihydrate	6381-92-6	-3.86	--	2.28×10^5	1.17×10^{-23}	--	5.77×10^{-16}
D-Lactic acid	10326-41-7	-0.65	-0.72	1.00×10^6	1.13×10^{-7}	--	8.13×10^{-8}
D-Limonene	5989-27-5	4.83	4.57	4.581	3.80×10^{-1}	--	3.19×10^{-2}
Docusate sodium	577-11-7	6.1	--	0.001227	5.00×10^{-12}	--	--
Dodecane	112-40-3	6.23	6.1	0.1099	9.35	1.34×10^1	8.18
Dodecylbenzene	123-01-3	7.94	8.65	0.001015	1.34×10^{-1}	2.81×10^{-1}	--
Dodecylbenzenesulfonic acid	27176-87-0	4.71	--	0.8126	6.27×10^{-8}	--	--
Dodecylbenzenesulfonic acid, monoethanolamine salt	26836-07-7	4.71	--	0.8126	6.27×10^{-8}	--	--
Epichlorohydrin	106-89-8	0.63	0.45	5.06×10^4	5.62×10^{-5}	2.62×10^{-6}	3.04×10^{-5}
Ethanaminium, N,N,N-trimethyl-2-[(1-oxo-2-propenyl)oxy]-, chloride	44992-01-0	-3.1	--	1.00×10^6	6.96×10^{-15}	--	--

Chemical name	CASRN	Log K_{ow}		Water solubility estimate from log K_{ow} (mg/L at 25°C)	Henry's law constant (atm-m³/mol at 25°C)		
		Estimated	Measured		Bond method	Group method 25	Measured
Ethane	74-84-0	1.32	1.81	938.6	5.50×10^{-1}	4.25×10^{-1}	5.00×10^{-1}
Ethanol	64-17-5	-0.14	-0.31	7.92×10^{5}	5.67×10^{-6}	4.88×10^{-6}	5.00×10^{-6}
Ethanol, 2,2',2''-nitrilotris-, tris(dihydrogen phosphate) (ester), sodium salt	68171-29-9	-3.13	--	1.00×10^{6}	3.08×10^{-36}	--	--
Ethanol, 2-[2-[2-(tridecyloxy)ethoxy]ethoxy]-, hydrogen sulfate, sodium salt	25446-78-0	2.09	--	42	9.15×10^{-13}	--	--
Ethanolamine	141-43-5	-1.61	-1.31	1.00×10^{6}	3.68×10^{-10}	9.96×10^{-11}	--
Ethoxylated dodecyl alcohol	9002-92-0	4.5	--	14.19	9.45×10^{-7}	3.30×10^{-7}	--
Ethyl acetate	141-78-6	0.86	0.73	2.99×10^{4}	2.33×10^{-4}	1.58×10^{-4}	1.34×10^{-4}
Ethyl acetoacetate	141-97-9	-0.2	0.25	5.62×10^{4}	1.57×10^{-7}	--	1.20×10^{-6}
Ethyl benzoate	93-89-0	2.32	2.64	421.5	4.61×10^{-5}	2.45×10^{-5}	7.33×10^{-5}
Ethyl lactate	97-64-3	-0.18	--	4.73×10^{5}	4.82×10^{-5}	--	5.83×10^{-7}
Ethyl salicylate	118-61-6	3.09	2.95	737.1	6.04×10^{-6}	3.01×10^{-9}	--
Ethylbenzene	100-41-4	3.03	3.15	228.6	7.89×10^{-3}	8.88×10^{-3}	7.88×10^{-3}
Ethylene	74-85-1	1.27	1.13	3,449	9.78×10^{-2}	1.62×10^{-1}	2.28×10^{-1}
Ethylene glycol	107-21-1	-1.2	-1.36	1.00×10^{6}	1.31×10^{-7}	5.60×10^{-11}	6.00×10^{-8}
Ethylene oxide	75-21-8	-0.05	-0.3	2.37×10^{5}	1.20×10^{-4}	5.23×10^{-5}	1.48×10^{-4}
Ethylenediamine	107-15-3	-1.62	-2.04	1.00×10^{6}	1.03×10^{-9}	1.77×10^{-10}	1.73×10^{-9}
Ethylenediaminetetraacetic acid	60-00-4	-3.86	--	2.28×10^{5}	1.17×10^{-23}	--	5.77×10^{-16}

Chemical name	CASRN	Log K_{ow}		Water solubility estimate from log K_{ow} (mg/L at 25°C)	Henry's law constant (atm-m³/mol at 25°C)		
		Estimated	Measured		Bond method	Group method 25	Measured
Ethylenediaminetetraacetic acid tetrasodium salt	64-02-8	-3.86	--	2.28×10^5	1.17×10^{-23}	--	5.77×10^{-16}
Ethylenediaminetetraacetic acid, disodium salt	139-33-3	-3.86	--	2.28×10^5	1.17×10^{-23}	--	5.77×10^{-16}
Ethyne	74-86-2	0.5	0.37	1.48×10^4	2.40×10^{-2}	2.45×10^{-2}	2.17×10^{-2}
Fatty acids, C18-unsaturated, dimers	61788-89-4	14.6	--	2.31×10^{-10}	4.12×10^{-8}	9.74×10^{-9}	--
FD&C Blue no. 1	3844-45-9	-0.15	--	0.2205	2.25×10^{-35}	--	--
FD&C Yellow no. 5	1934-21-0	-1.82	--	7.388	1.31×10^{-28}	--	--
FD&C Yellow no. 6	2783-94-0	1.4	--	242.7	3.26×10^{-23}	--	--
Formaldehyde	50-00-0	0.35	0.35	5.70×10^4	9.29×10^{-5}	6.14×10^{-5}	3.37×10^{-7}
Formamide	75-12-7	-1.61	-1.51	1.00×10^6	1.53×10^{-8}	--	1.39×10^{-9}
Formic acid	64-18-6	-0.46	-0.54	9.55×10^5	7.50×10^{-7}	5.11×10^{-7}	1.67×10^{-7}
Formic acid, potassium salt	590-29-4	-0.46	-0.54	9.55×10^5	7.50×10^{-7}	5.11×10^{-7}	1.67×10^{-7}
Fumaric acid	110-17-8	0.05	-0.48	1.04×10^5	1.35×10^{-12}	8.48×10^{-14}	--
Furfural	98-01-1	0.83	0.41	5.36×10^4	1.34×10^{-5}	--	3.77×10^{-6}
Furfuryl alcohol	98-00-0	0.45	0.28	2.21×10^5	2.17×10^{-7}	--	7.86×10^{-8}
Galantamine hydrobromide	69353-21-5	2.29	--	1,606	1.70×10^{-13}	--	--
Gluconic acid	133-42-6	-1.87	--	1.00×10^6	4.74×10^{-13}	--	--
Glutaraldehyde	111-30-8	-0.18	--	1.67×10^5	1.10×10^{-7}	2.39×10^{-8}	--
Glycerol	56-81-5	-1.65	-1.76	1.00×10^6	6.35×10^{-9}	1.51×10^{-15}	1.73×10^{-8}

Chemical name	CASRN	Log K_{ow}		Water solubility estimate from log K_{ow} (mg/L at 25°C)	Henry's law constant (atm-m³/mol at 25°C)		
		Estimated	Measured		Bond method	Group method 25	Measured
Glycine, N-(carboxymethyl)-N-(2-hydroxyethyl)-, disodium salt	135-37-5	-3.04	--	1.90×10^5	3.90×10^{-17}	--	--
Glycine, N-(hydroxymethyl)-, monosodium salt	70161-44-3	-3.41	--	7.82×10^5	1.80×10^{-12}	--	--
Glycine, N,N-bis(carboxymethyl)-, trisodium salt	5064-31-3	-3.81	--	7.39×10^5	1.19×10^{-16}	--	--
Glycine, N-[2-[bis(carboxymethyl)amino]ethyl]-N-(2-hydroxyethyl)-, trisodium salt	139-89-9	-4.09	--	4.31×10^5	3.81×10^{-24}	--	--
Glycolic acid	79-14-1	-1.07	-1.11	1.00×10^6	8.54×10^{-8}	6.29×10^{-11}	--
Glycolic acid sodium salt	2836-32-0	-1.07	-1.11	1.00×10^6	8.54×10^{-8}	6.29×10^{-11}	--
Glyoxal	107-22-2	-1.66	--	1.00×10^6	3.70×10^{-7}	--	3.33×10^{-9}
Glyoxylic acid	298-12-4	-1.4	--	1.00×10^6	2.98×10^{-9}	--	--
Heptane	142-82-5	3.78	4.66	3.554	2.27	2.39	2.00
Hexadecyltrimethylammonium bromide	57-09-0	3.18	--	28.77	2.93×10^{-10}	--	--
Hexane	110-54-3	3.29	3.9	17.24	1.71	1.69	1.80
Hexanedioic acid	124-04-9	0.23	0.08	1.67×10^5	9.53×10^{-12}	8.10×10^{-13}	4.71×10^{-12}
Hydroxyvalerenic acid	1619-16-5	3.31	--	282.1	--	--	--
Indole	120-72-9	2.05	2.14	1,529	8.86×10^{-7}	1.99×10^{-6}	5.28×10^{-7}
Isoascorbic acid	89-65-6	-1.88	-1.85	1.00×10^6	4.07×10^{-8}	--	--
Isobutane	75-28-5	2.23	2.76	175.1	9.69×10^{-1}	1.02	1.19

Chemical name	CASRN	Log K_{ow}		Water solubility estimate from log K_{ow} (mg/L at 25°C)	Henry's law constant (atm-m^3/mol at 25°C)		
		Estimated	Measured		Bond method	Group method 25	Measured
Isobutene	115-11-7	2.23	2.34	399.2	2.40×10^{-1}	2.34×10^{-1}	2.18×10^{-1}
Isooctanol	26952-21-6	2.73	--	1,379	3.10×10^{-5}	4.66×10^{-5}	9.21×10^{-5}
Isopentyl alcohol	123-51-3	1.26	1.16	4.16×10^4	1.33×10^{-5}	1.65×10^{-5}	1.41×10^{-5}
Isopropanol	67-63-0	0.28	0.05	4.02×10^5	7.52×10^{-6}	1.14×10^{-5}	8.10×10^{-6}
Isopropanolamine dodecylbenzene	42504-46-1	7.94	8.65	0.001015	1.34×10^{-1}	2.81×10^{-1}	--
Isopropylamine	75-31-0	0.27	0.26	8.38×10^5	1.34×10^{-5}	--	4.51×10^{-5}
Isoquinoline	119-65-3	2.14	2.08	1,551	6.88×10^{-7}	4.15×10^{-7}	--
Isoquinoline, reaction products with benzyl chloride and quinoline	68909-80-8	2.14	2.08	1,551	6.88×10^{-7}	4.15×10^{-7}	--
Isoquinolinium, 2-(phenylmethyl)-, chloride	35674-56-7	4.4	--	6.02	1.19×10^{-6}	--	--
Lactic acid	50-21-5	-0.65	-0.72	1.00×10^6	1.13×10^{-7}	--	8.13×10^{-8}
Lactose	63-42-3	-5.12	--	1.00×10^6	4.47×10^{-22}	9.81×10^{-45}	--
Lauryl hydroxysultaine	13197-76-7	-1.3	--	7.71×10^4	1.04×10^{-21}	--	--
L-Dilactide	4511-42-6	1.65	--	3,165	1.22×10^{-5}	--	--
L-Glutamic acid	56-86-0	-3.83	-3.69	9.42×10^5	1.47×10^{-14}	--	--
L-Lactic acid	79-33-4	-0.65	-0.72	1.00×10^6	1.13×10^{-7}	--	8.13×10^{-8}
Methane	74-82-8	0.78	1.09	2,610	4.14×10^{-1}	6.58×10^{-1}	6.58×10^{-1}
Methanol	67-56-1	-0.63	-0.77	1.00×10^6	4.27×10^{-6}	3.62×10^{-6}	4.55×10^{-6}
Methenamine	100-97-0	-4.15	--	1.00×10^6	1.63×10^{-1}	--	1.64×10^{-9}

Chemical name	CASRN	Log K_{ow}		Water solubility estimate from log K_{ow} (mg/L at 25°C)	Henry's law constant (atm-m³/mol at 25°C)		
		Estimated	Measured		Bond method	Group method 25	Measured
Methoxyacetic acid	625-45-6	-0.68	--	1.00×10^6	4.54×10^{-8}	8.68×10^{-9}	6.42×10^{-9}
Methyl salicylate	119-36-8	2.6	2.55	1,875	4.55×10^{-6}	2.23×10^{-9}	9.81×10^{-5}
Methyl vinyl ketone	78-94-4	0.41	--	6.06×10^4	2.61×10^{-5}	1.38×10^{-5}	4.65×10^{-5}
Methylcyclohexane	108-87-2	3.59	3.61	28.4	3.39×10^{-1}	3.30×10^{-1}	4.30×10^{-1}
Methylene bis(thiocyanate)	6317-18-6	0.62	--	2.72×10^4	2.61×10^{-8}	--	--
Methylenebis(5-methyloxazolidine)	66204-44-2	-0.58	--	1.00×10^6	1.07×10^{-7}	--	--
Morpholine	110-91-8	-0.56	-0.86	1.00×10^6	1.14×10^{-7}	3.22×10^{-9}	1.16×10^{-6}
Morpholinium, 4-ethyl-4-hexadecyl-, ethyl sulfate	78-21-7	4.54	--	0.9381	2.66×10^{-12}	--	--
N-(2-Acryloyloxyethyl)-N-benzyl-N,N-dimethylammonium chloride	46830-22-2	-1.39	--	4.42×10^5	5.62×10^{-16}	--	--
N-(3-Chloroallyl)hexaminium chloride	4080-31-3	-5.92	--	1.00×10^6	1.76×10^{-8}	--	--
N,N,N-Trimethyl-3-((1-oxooctadecyl)amino)-1-propanaminium methyl sulfate	19277-88-4	4.38	--	0.7028	2.28×10^{-16}	--	--
N,N,N-Trimethyloctadecan-1-aminium chloride	112-03-8	4.17	--	2.862	5.16×10^{-10}	--	--
N,N'-Dibutylthiourea	109-46-6	2.57	2.75	2,287	4.17×10^{-6}	--	--
N,N-Dimethyldecylamine oxide	2605-79-0	1.4	--	2,722	4.88×10^{-14}	--	--
N,N-Dimethylformamide	68-12-2	-0.93	-1.01	9.78×10^5	7.38×10^{-8}	--	7.39×10^{-8}

Appendix C – Chemical Mixing Supplemental Information

Chemical name	CASRN	Log K_{ow}		Water solubility estimate from log K_{ow} (mg/L at 25°C)	Henry's law constant (atm-m³/mol at 25°C)		
		Estimated	Measured		Bond method	Group method 25	Measured
N,N-Dimethylmethanamine hydrochloride	593-81-7	0.04	0.16	1.00×10^6	3.65×10^{-5}	1.28×10^{-4}	1.04×10^{-4}
N,N-Dimethyl-methanamine-N-oxide	1184-78-7	-3.02	--	1.00×10^6	3.81×10^{-15}	--	--
N,N-dimethyloctadecylamine hydrochloride	1613-17-8	8.39	--	0.008882	4.51×10^{-3}	3.88×10^{-2}	--
N,N'-Methylenebisacrylamide	110-26-9	-1.52	--	7.01×10^4	1.14×10^{-9}	--	--
Naphthalene	91-20-3	3.17	3.3	142.1	5.26×10^{-4}	3.70×10^{-4}	4.40×10^{-4}
Naphthalenesulfonic acid, bis(1-methylethyl)-	28757-00-8	2.92	--	43.36	9.29×10^{-10}	--	--
Naphthalenesulphonic acid, bis (1-methylethyl)-methyl derivatives	99811-86-6	4.02	--	3.45	1.13×10^{-9}	--	--
Naphthenic acid ethoxylate	68410-62-8	3.41	--	112.5	3.62×10^{-8}	2.74×10^{-9}	--
Nitrilotriacetamide	4862-18-4	-4.75	--	1.00×10^6	1.61×10^{-18}	--	--
Nitrilotriacetic acid	139-13-9	-3.81	--	7.39×10^5	1.19×10^{-16}	--	--
Nitrilotriacetic acid trisodium monohydrate	18662-53-8	-3.81	--	7.39×10^5	1.19×10^{-16}	--	--
N-Methyl-2-pyrrolidone	872-50-4	-0.11	-0.38	2.48×10^5	3.16×10^{-8}	--	3.20×10^{-9}
N-Methyldiethanolamine	105-59-9	-1.5	--	1.00×10^6	8.61×10^{-11}	2.45×10^{-14}	3.14×10^{-11}
N-Methylethanolamine	109-83-1	-1.15	-0.94	1.00×10^6	8.07×10^{-10}	2.50×10^{-10}	--
N-Methyl-N-hydroxyethyl-N-hydroxyethoxyethylamine	68213-98-9	-1.78	--	1.00×10^6	1.34×10^{-12}	5.23×10^{-17}	--

Chemical name	CASRN	Log K_{ow}		Water solubility estimate from log K_{ow} (mg/L at 25°C)	Henry's law constant (atm-m³/mol at 25°C)		
		Estimated	Measured		Bond method	Group method 25	Measured
N-Oleyl diethanolamide	13127-82-7	6.63	--	0.1268	9.35×10^{-9}	1.94×10^{-12}	--
Oleic acid	112-80-1	7.73	7.64	0.01151	4.48×10^{-5}	1.94×10^{-5}	--
Pentaethylenehexamine	4067-16-7	-3.67	--	1.00×10^{6}	8.36×10^{-24}	2.56×10^{-27}	--
Pentane	109-66-0	2.8	3.39	49.76	1.29	1.20	1.25
Pentyl acetate	628-63-7	2.34	2.3	996.8	5.45×10^{-4}	4.45×10^{-4}	3.88×10^{-4}
Pentyl butyrate	540-18-1	3.32	--	101.9	9.60×10^{-4}	8.88×10^{-4}	--
Peracetic acid	79-21-0	-1.07	--	1.00×10^{6}	1.39×10^{-6}	--	2.14×10^{-6}
Phenanthrene	85-01-8	4.35	4.46	0.677	5.13×10^{-5}	2.56×10^{-5}	4.23×10^{-5}
Phenol	108-95-2	1.51	1.46	2.62×10^{4}	5.61×10^{-7}	6.58×10^{-7}	3.33×10^{-7}
Phosphonic acid (dimethylamino(methylene))	29712-30-9	-1.9	--	1.00×10^{6}	1.00×10^{-24}	--	--
Phosphonic acid, (((2-hydroxyethyl)(phosphonomethyl)amino)ethyl)imino]bis(methylene)]bis-, compd. with 2-aminoethanol	129828-36-0	-6.73	--	1.00×10^{6}	5.29×10^{-42}	--	--
Phosphonic acid, (1-hydroxyethylidene)bis-, potassium salt	67953-76-8	-0.01	--	1.34×10^{5}	9.79×10^{-26}	--	--
Phosphonic acid, (1-hydroxyethylidene)bis-, tetrasodium salt	3794-83-0	-0.01	--	1.34×10^{5}	9.79×10^{-26}	--	--
Phosphonic acid, [[(phosphonomethyl)imino]bis[2,1-ethanediylnitrilobis(methylene)]]tetrakis-	15827-60-8	-9.72	--	1.00×10^{6}	--	--	--

Chemical name	CASRN	Log K_{ow}		Water solubility estimate from log K_{ow} (mg/L at 25°C)	Henry's law constant (atm-m³/mol at 25°C)		
		Estimated	Measured		Bond method	Group method 25	Measured
Phosphonic acid, [[(phosphonomethyl)imino]bis[2,1-ethanediylnitrilobis(methylene)]]tetrakis-, ammonium salt (1:x)	70714-66-8	-9.72	--	1.00×10^6	--	--	--
Phosphonic acid, [[(phosphonomethyl)imino]bis[2,1-ethanediylnitrilobis(methylene)]]tetrakis-, sodium salt	22042-96-2	-9.72	--	1.00×10^6	--	--	--
Phosphonic acid, [[(phosphonomethyl)imino]bis[6,1-hexanediylnitrilobis(methylene)]]tetrakis-	34690-00-1	-5.79	--	1.00×10^6	--	--	--
Phthalic anhydride	85-44-9	2.07	1.6	3,326	6.35×10^{-6}	--	1.63×10^{-8}
Poly(oxy-1,2-ethanediyl), .alpha.-(octylphenyl)-.omega.-hydroxy-, branched	68987-90-6	5.01	--	3.998	1.24×10^{-7}	1.07×10^{-6}	--
Potassium acetate	127-08-2	0.09	-0.17	4.76×10^5	5.48×10^{-7}	2.94×10^{-7}	1.00×10^{-7}
Potassium oleate	143-18-0	7.73	7.64	0.01151	4.48×10^{-5}	1.94×10^{-5}	--
Propane	74-98-6	1.81	2.36	368.9	7.30×10^{-1}	6.00×10^{-1}	7.07×10^{-1}
Propanol, 1(or 2)-(2-methoxymethylethoxy)-	34590-94-8	-0.27	--	4.27×10^5	1.15×10^{-9}	1.69×10^{-9}	--
Propargyl alcohol	107-19-7	-0.42	-0.38	9.36×10^5	5.88×10^{-7}		1.15×10^{-6}
Propylene carbonate	108-32-7	0.08	-0.41	2.58×10^5	3.63×10^{-4}	--	3.45×10^{-8}
Propylene pentamer	15220-87-8	6.28	--	0.05601	3.92×10^{-1}	1.09×10^{-3}	--
p-Xylene	106-42-3	3.09	3.15	228.6	6.56×10^{-3}	6.14×10^{-3}	6.90×10^{-3}

Chemical name	CASRN	Log K_{ow}		Water solubility estimate from log K_{ow} (mg/L at 25°C)	Henry's law constant (atm-m³/mol at 25°C)		
		Estimated	Measured		Bond method	Group method 25	Measured
Pyrimidine	289-95-2	-0.06	-0.4	2.87×10^5	2.92×10^{-6}	--	--
Pyrrole	109-97-7	0.88	0.75	3.12×10^4	9.07×10^{-6}	7.73×10^{-6}	1.80×10^{-5}
Quaternary ammonium compounds, di-C8-10-alkyldimethyl, chlorides	68424-95-3	2.69	--	90.87	2.20×10^{-10}	--	--
Quinaldine	91-63-4	2.69	2.59	498.5	7.60×10^{-7}	2.13×10^{-6}	--
Quinoline	91-22-5	2.14	2.03	1,711	6.88×10^{-7}	1.54×10^{-6}	1.67×10^{-6}
Rhodamine B	81-88-9	6.03	--	0.0116	--	--	--
Sodium 1-octanesulfonate	5324-84-5	1.06	--	5,864	9.15×10^{-8}	--	--
Sodium 2-mercaptobenzothiolate	2492-26-4	2.86	2.42	543.4	3.63×10^{-8}	--	--
Sodium acetate	127-09-3	0.09	-0.17	4.76×10^5	5.48×10^{-7}	2.94×10^{-7}	1.00×10^{-7}
Sodium benzoate	532-32-1	1.87	1.87	2,493	1.08×10^{-7}	4.55×10^{-8}	3.81×10^{-8}
Sodium bicarbonate	144-55-8	-0.46	--	8.42×10^5	6.05×10^{-9}	--	--
Sodium bis(tridecyl) sulfobutanedioate	2673-22-5	11.15	--	7.46×10^{-9}	8.51×10^{-11}	--	--
Sodium C14-16 alpha-olefin sulfonate	68439-57-6	4.36	--	2.651	4.95×10^{-7}	--	--
Sodium caprylamphopropionate	68610-44-6	-0.26	--	615.1	1.19×10^{-9}	2.45×10^{-10}	--
Sodium carbonate	497-19-8	-0.46	--	8.42×10^5	6.05×10^{-9}	--	--
Sodium chloroacetate	3926-62-3	0.34	0.22	1.95×10^5	1.93×10^{-7}	8.88×10^{-8}	9.26×10^{-9}
Sodium decyl sulfate	142-87-0	1.44	--	1,617	1.04×10^{-7}	--	--
Sodium D-gluconate	527-07-1	-1.87	--	1.00×10^6	4.74×10^{-13}	--	--

Appendix C – Chemical Mixing Supplemental Information

Chemical name	CASRN	Log K_{ow} Estimated	Log K_{ow} Measured	Water solubility estimate from log K_{ow} (mg/L at 25°C)	Henry's law constant (atm-m³/mol at 25°C) Bond method	Henry's law constant (atm-m³/mol at 25°C) Group method 25	Henry's law constant (atm-m³/mol at 25°C) Measured
Sodium diacetate	126-96-5	0.09	-0.17	4.76×10^5	5.48×10^{-7}	2.94×10^{-7}	1.00×10^{-7}
Sodium dichloroisocyanurate	2893-78-9	1.28	--	3,613	3.22×10^{-12}	--	--
Sodium dl-lactate	72-17-3	-0.65	-0.72	1.00×10^6	1.13×10^{-7}	--	8.13×10^{-8}
Sodium dodecyl sulfate	151-21-3	2.42	--	163.7	1.84×10^{-7}	--	--
Sodium erythorbate (1:1)	6381-77-7	-1.88	-1.85	1.00×10^6	4.07×10^{-8}	--	--
Sodium ethasulfate	126-92-1	0.38	--	1.82×10^4	5.91×10^{-8}	--	--
Sodium formate	141-53-7	-0.46	-0.54	9.55×10^5	7.50×10^{-7}	5.11×10^{-7}	1.67×10^{-7}
Sodium hydroxymethanesulfonate	870-72-4	-3.85	--	1.00×10^6	4.60×10^{-13}	--	--
Sodium l-lactate	867-56-1	-0.65	-0.72	1.00×10^6	1.13×10^{-7}	--	8.13×10^{-8}
Sodium maleate (1:x)	18016-19-8	0.05	-0.48	1.04×10^5	1.35×10^{-12}	8.48×10^{-14}	--
Sodium N-methyl-N-oleoyltaurate	137-20-2	4.43	--	0.4748	1.00×10^{-12}	--	--
Sodium octyl sulfate	142-31-4	0.46	--	1.58×10^4	5.91×10^{-8}	--	--
Sodium salicylate	54-21-7	2.24	2.26	3,808	1.42×10^{-8}	5.60×10^{-12}	7.34×10^{-9}
Sodium sesquicarbonate	533-96-0	-0.46	--	8.42×10^5	6.05×10^{-9}	--	--
Sodium thiocyanate	540-72-7	0.58	--	4.36×10^4	1.46×10^{-4}	--	--
Sodium trichloroacetate	650-51-1	1.44	1.33	1.20×10^4	2.39×10^{-8}	--	1.35×10^{-8}
Sodium xylenesulfonate	1300-72-7	-0.07	--	5.89×10^4	3.06×10^{-9}	--	--
Sorbic acid	110-44-1	1.62	1.33	1.94×10^4	5.72×10^{-7}	4.99×10^{-8}	--
Sorbitan sesquioleate	8007-43-0	14.32	--	2.31×10^{-11}	7.55×10^{-12}	1.25×10^{-16}	--

Chemical name	CASRN	Log K_{ow}		Water solubility estimate from log K_{ow} (mg/L at 25°C)	Henry's law constant (atm-m³/mol at 25°C)		
		Estimated	Measured		Bond method	Group method 25	Measured
Sorbitan, mono-(9Z)-9-octadecenoate	1338-43-8	5.89	--	0.01914	1.42×10^{-12}	5.87×10^{-20}	--
Sorbitan, monooctadecanoate	1338-41-6	6.1	--	0.01218	1.61×10^{-12}	2.23×10^{-19}	--
Sorbitan, tri-(9Z)-9-octadecenoate	26266-58-0	22.56	--	1.12×10^{-19}	4.02×10^{-11}	2.68×10^{-13}	--
Styrene	100-42-5	2.89	2.95	343.7	2.76×10^{-3}	2.81×10^{-3}	2.75×10^{-3}
Sucrose	57-50-1	-4.27	-3.7	1.00×10^{6}	4.47×10^{-22}	--	--
Sulfan blue	129-17-9	-1.34	--	50.67	1.31×10^{-26}	--	--
Sulfuric acid, mono-C12-18-alkyl esters, sodium salts	68955-19-1	3.9	--	5.165	4.29×10^{-7}	--	--
Sulfuric acid, mono-C6-10-alkyl esters, ammonium salts	68187-17-7	0.46	--	1.58×10^{4}	5.91×10^{-8}	--	--
Symclosene	87-90-1	0.94	--	4,610	6.19×10^{-11}	--	--
tert-Butyl hydroperoxide	75-91-2	0.94	--	1.97×10^{4}	1.60×10^{-5}	--	--
tert-Butyl perbenzoate	614-45-9	2.89	--	159.2	2.06×10^{-4}	--	--
Tetradecane	629-59-4	7.22	7.2	0.009192	1.65×10^{1}	2.68×10^{1}	9.20
Tetradecyldimethylbenzylammonium chloride	139-08-2	3.91	--	3.608	1.34×10^{-11}	--	--
Tetraethylene glycol	112-60-7	-2.02	--	1.00×10^{6}	4.91×10^{-13}	5.48×10^{-19}	--
Tetraethylenepentamine	112-57-2	-3.16	--	1.00×10^{6}	2.79×10^{-20}	4.15×10^{-23}	--
Tetrakis(hydroxymethyl)phosphonium sulfate	55566-30-8	-5.03	--	1.00×10^{6}	9.17×10^{-13}	--	--
Tetramethylammonium chloride	75-57-0	-4.18	--	1.00×10^{6}	4.17×10^{-12}	--	--

Chemical name	CASRN	Log K_{ow}		Water solubility estimate from log K_{ow} (mg/L at 25°C)	Henry's law constant (atm-m³/mol at 25°C)		
		Estimated	Measured		Bond method	Group method 25	Measured
Thiamine hydrochloride	67-03-8	0.95	--	3,018	8.24×10^{-17}	--	--
Thiocyanic acid, ammonium salt	1762-95-4	0.58	--	4.36×10^4	1.46×10^{-4}	--	--
Thioglycolic acid	68-11-1	0.03	0.09	2.56×10^5	1.94×10^{-8}	--	--
Thiourea	62-56-6	-1.31	-1.08	5.54×10^5	1.58×10^{-7}	--	1.98×10^{-9}
Toluene	108-88-3	2.54	2.73	573.1	5.95×10^{-3}	5.73×10^{-3}	6.64×10^{-3}
Tributyl phosphate	126-73-8	3.82	4	7.355	3.19×10^{-6}	--	1.41×10^{-6}
Tributyltetradecylphosphonium chloride	81741-28-8	11.22	--	7.90×10^{-7}	2.61×10^{-1}	--	--
Tridecane	629-50-5	6.73	--	0.02746	1.24×10^1	1.90×10^1	2.88
Triethanolamine	102-71-6	-2.48	-1	1.00×10^6	4.18×10^{-12}	3.38×10^{-19}	7.05×10^{-13}
Triethanolamine hydrochloride	637-39-8	-2.48	-1	1.00×10^6	4.18×10^{-12}	3.38×10^{-19}	7.05×10^{-13}
Triethanolamine hydroxyacetate	68299-02-5	-2.97	--	1.00×10^6	6.28×10^{-11}		--
Triethyl citrate	77-93-0	0.33	--	2.82×10^4	6.39×10^{-10}	--	3.84×10^{-9}
Triethyl phosphate	78-40-0	0.87	0.8	1.12×10^4	5.83×10^{-7}	--	3.60×10^{-8}
Triethylene glycol	112-27-6	-1.75	-1.75	1.00×10^6	3.16×10^{-11}	2.56×10^{-16}	--
Triethylenetetramine	112-24-3	-2.65	--	1.00×10^6	9.30×10^{-17}	6.74×10^{-19}	--
Triisopropanolamine	122-20-3	-1.22	--	1.00×10^6	9.77×10^{-12}	4.35×10^{-18}	--
Trimethanolamine	14002-32-5	-3.95	--	1.00×10^6	1.42×10^{-8}	--	--
Trimethylamine	75-50-3	0.04	0.16	1.00×10^6	3.65×10^{-5}	1.28×10^{-4}	1.04×10^{-4}
Tripotassium citrate monohydrate	6100-05-6	-1.67	-1.64	1.00×10^6	8.33×10^{-18}	--	4.33×10^{-14}

Chemical name	CASRN	Log K_{ow}		Water solubility estimate from log K_{ow} (mg/L at 25°C)	Henry's law constant (atm-m³/mol at 25°C)		
		Estimated	Measured		Bond method	Group method 25	Measured
Tripropylene glycol monomethyl ether	25498-49-1	–0.2	--	1.96×10^5	2.36×10^{-11}	4.55×10^{-13}	--
Trisodium citrate	68-04-2	–1.67	–1.64	1.00×10^6	8.33×10^{-18}	--	4.33×10^{-14}
Trisodium citrate dihydrate	6132-04-3	–1.67	–1.64	1.00×10^6	8.33×10^{-18}	--	4.33×10^{-14}
Trisodium ethylenediaminetetraacetate	150-38-9	–3.86	--	2.28×10^5	1.17×10^{-23}	--	5.77×10^{-16}
Trisodium ethylenediaminetriacetate	19019-43-3	–4.32	--	1.00×10^6	3.58×10^{-20}	--	--
Tromethamine	77-86-1	–1.56	--	1.00×10^6	8.67×10^{-13}	--	--
Undecane	1120-21-4	5.74	--	0.2571	7.04	9.52	1.93
Urea	57-13-6	–1.56	–2.11	4.26×10^5	3.65×10^{-10}	--	1.74×10^{-12}
Xylenes	1330-20-7	3.09	3.2	207.2	6.56×10^{-3}	6.14×10^{-3}	7.18×10^{-3}

C.6. Details on the EPI (Estimation Programs Interface) Suite™

The EPI (Estimation Programs Interface) Suite™ (U.S. EPA, 2012b) is an open-source, Windows®-based suite of physicochemical property and environmental fate estimation programs developed by the EPA's Office of Pollution Prevention and Toxics and Syracuse Research Corporation. More information on EPI Suite™ is available at http://www.epa.gov/oppt/exposure/pubs/episuite.htm.

Although only physicochemical properties from EPI Suite™ are provided here, other sources of information were also consulted. QikProp (Schrodinger, 2012) and LeadScope (Leadscope Inc., 2012) are commercial products designed primarily as drug development and screening tools. Properties generated by QikProp and LeadScope are generally more relevant to drug development than to environmental assessment.

QikProp is specifically focused on drug discovery and provides predictions for physically significant descriptors and pharmaceutically (and toxicologically) relevant properties useful in predicting ADME (adsorption, distribution, metabolism, and excretion) characteristics of drug candidates. QikProp's use of whole-molecule descriptors that have a straightforward physical interpretation (as opposed to fragment-based descriptors).

LeadScope is a program designed for interpreting chemical and biological screening data that can assist pharmaceutical scientists in finding promising drug candidates. The software organizes the chemical data by structural features familiar to medicinal chemists. Graphs are used to summarize the data, and structural classes are highlighted that are statistically correlated with biological activity. It incorporates chemically-based data mining, visualization, and advanced informatics techniques (e.g., prediction tools, scaffold generators).

Physicochemical properties of chemicals were generated from the two-dimensional (2-D) chemical structures from the EPA National Center for Computational Toxicology's Distributed Structure-Searchable Toxicity (NCCT DSSTox) Database Network in structure-data file (SDF) format. For EPI Suite™ properties, both the desalted and non-desalted 2-D files were run using the program's batch mode (i.e., processing many molecules at once) to calculate environmentally-relevant, chemical property descriptors. The chemical descriptors in QikProp require 3-D chemical structures. For these calculations, the 2-D desalted chemical structures were converted to 3-D using the Rebuild3D function in the Molecular Operating Environment software (CCG, 2011). All computed physicochemical properties are added into the structure-data file prior to assigning toxicological properties.

Both LeadScope and Qikprop software require input of desalted structures. Therefore, the structures were desalted, a process where salts and complexes are simplified to the neutral, uncomplexed form of the chemical, using the "Desalt Batch" option in the ACD Labs ChemFolder. All LeadScope general chemical descriptors (Parent Molecular Weight, AlogP, Hydrogen Bond Acceptors, Hydrogen Bond Donors, Lipinski Score, Molecular Weight, Parent Atom Count, Polar Surface Area, and Rotatable Bonds) were calculated by default.

C.7.   Top 20 lists for most mobile and least mobile chemicals

Table C-10 and Table C-11 present the 20 highest and lowest log K_{ow} (approximate surrogate for most mobile and least mobile) chemicals, known to be used in hydraulic fracturing fluids, respectively, as ranked by log K_{ow}. These were taken from the list of 917 chemicals with estimated values for physicochemical properties. These tables also include values for aqueous solubility and Henry's law constant, as well as frequency of use, based on chemical information reported in disclosures in the EPA FracFocus 1.0 project database (U.S. EPA, 2015a, c).

Table C-10 shows the chemicals that have the *lowest* log K_{ow} and are, thus, the *most* mobile. These chemicals are fully miscible (i.e., they will mix completely with water), which means they may move through the environment at high concentrations, leading to greater severity of impact. These chemicals generally have low volatility, based on their negative log Henry's law constants (i.e., will remain in water and will not be lost to the air). These chemicals will dissolve in water and move rapidly through the environment (e.g., via infiltration into the subsurface or via overland flow to surface waters). Chemicals exhibiting this combination of properties have greater potential to cause immediate impacts to drinking water resources. Most of the chemicals in the table were infrequently reported (≤2% of wells) in the EPA FracFocus 1.0 project database (U.S. EPA, 2015a). However, choline chloride (14% of wells), used for clay control, and tetrakis(hydroxymethyl)phosphonium sulfate (11% of wells), a biocide, were more commonly reported.

Table C-11 shows the chemicals that have the *highest* log K_{ow} and are, thus, the least mobile. The estimated aqueous solubilities for some of these chemicals are extremely low, with highest solubilities of <10 µg/L. Therefore, the concentration of these chemicals dissolved in water will be low. The estimated Henry's law constants are more variable for these low-mobility chemicals. Chemicals with high log K_{ow} values (>0) and high Henry's law constants will sorb strongly to organic phases and solids and may volatilize. However, their strong preference for the organic or solid phase may slow or reduce volatilization. The chemicals with low Henry's law constants will readily sorb to organic phases and solids. Less mobile chemicals will move slowly through the soil and have potentially delayed and longer-term impacts to drinking water resources. Seven of the chemicals in were reported in disclosures in the EPA FracFocus 1.0 project database (U.S. EPA, 2015c). Five were reported infrequently (<1% of wells). Tri-n-butyltetradecylphosphonium chloride (6% of wells), used as a biocide, and C>10-alpha-alkenes (8% of wells), a mixture of alpha-olefins with carbon numbers greater than 10 used as a corrosion inhibitor, were more commonly reported. The least mobile organic chemical is sorbitan, tri-(9Z)-9-octadecenoate, a mineral oil co-emulsifier (0.05% of wells), with an estimated log K_{ow} of 22.56.[1]

---

[1] Sorbitan, tri-(9Z)-9-octadecenoate, CASRN 26266-58-0, is soluble in hydrocarbons and insoluble in water, listed as an effective coupling agent and co-emulsifier for mineral oil (Santa Cruz Biotechnology, 2015; ChemicalBook, 2010).

Table C-10. Ranking of the 20 most mobile organic chemicals, as determined by the largest log K_{ow}, with CASRN, percent of wells where the chemical is reported from January 1, 2011 to February 28, 2013 (U.S. EPA, 2015c), and physicochemical properties (log K_{ow}, solubility, and Henry's law constant) as estimated by EPI Suite™.

For organic salts, parameters are estimated using the desalted form.

Rank	Chemical name	CASRN	Percent of wells (U.S. EPA, 2015c)[a]	Estimated log K_{ow} (unitless)[b]	Estimated water solubility (mg/L @ 25°C)[c]	Estimated Henry's law constant (atm m³/mole @ 25°C)[d]
1	1,2-Ethanediaminium, N,N'-bis[2-[bis(2-hydroxyethyl)methylammonio]ethyl]-N,N'-bis(2-hydroxyethyl)-N,N'-dimethyl-, tetrachloride	138879-94-4	2%	-23.19	1.00×10^6	2.33×10^{-35}
2	Phosphonic acid, [[[(phosphonomethyl)imino]bis [2,1-ethanediylnitrilobis(methylene)]]tetrakis-	15827-60-8	0.2%	-9.72	1.00×10^6	NA
3	Phosphonic acid, [[[(phosphonomethyl)imino]bis [2,1-ethanediylnitrilobis(methylene)]]tetrakis-, sodium salt	22042-96-2	0.07%	-9.72	1.00×10^6	NA
4	Phosphonic acid, [[[(phosphonomethyl)imino]bis [2,1-ethanediylnitrilobis(methylene)]]tetrakis-, ammonium salt (1:x)	70714-66-8	NA	-9.72	1.00×10^6	NA
5	Phosphonic acid, ((((2-[(2-hydroxyethyl) (phosphonomethyl) amino)ethyl)imino]bis(methylene))bis-, compd. with 2-aminoethanol	129828-36-0	NA	-6.73	1.00×10^6	5.29×10^{-42}
6	2-Hydroxy-N,N-bis(2-hydroxyethyl)-N-methylethanaminium chloride	7006-59-9	NA	-6.7	1.00×10^6	4.78×10^{-19}
7	N-(3-Chloroallyl)hexaminium chloride	4080-31-3	0.02%	-5.92	1.00×10^6	1.76×10^{-8}
8	3,5,7-Triazatricyclo(3.3.1.1 (superscript 3,7))decane, 1-(3-chloro-2-propenyl)-, chloride, (Z)-	51229-78-8	NA	-5.92	1.00×10^6	1.76×10^{-8}
9	(2,3-dihydroxypropyl)trimethylammonium chloride	34004-36-9	NA	-5.8	1.00×10^6	9.84×10^{-18}

Rank	Chemical name	CASRN	Percent of wells (U.S. EPA, 2015c)[a]	Estimated log K_{ow} (unitless)[b]	Estimated water solubility (mg/L @ 25°C)[c]	Estimated Henry's law constant (atm m³/mole @ 25°C)[d]
10	Phosphonic acid, [[(phosphonomethyl)imino]bis [6,1-hexanediylnitrilobis(methylene)]]tetrakis-	34690-00-1	0.006%	-5.79	1.00×10^6	NA
11	[Nitrilotris(methylene)]tris-phosphonic acid pentasodium salt	2235-43-0	0.5%	-5.45	1.00×10^6	1.65×10^{-34}
12	Aminotrimethylene phosphonic acid	6419-19-8	2%	-5.45	1.00×10^6	1.65×10^{-34}
13	Choline chloride	67-48-1	14%	-5.16	1.00×10^6	2.03×10^{-16}
14	Choline bicarbonate	78-73-9	0.2%	-5.16	1.00×10^6	2.03×10^{-16}
15	alpha-Lactose monohydrate	5989-81-1	NA	-5.12	1.00×10^6	4.47×10^{-22}
16	Lactose	63-42-3	NA	-5.12	1.00×10^6	4.47×10^{-22}
17	Tetrakis(hydroxymethyl)phosphonium sulfate	55566-30-8	11%	-5.03	1.00×10^6	9.17×10^{-13}
18	Disodium ethylenediaminediacetate	38011-25-5	0.6%	-4.79	1.00×10^6	1.10×10^{-16}
19	Nitrilotriacetamide	4862-18-4	NA	-4.75	1.00×10^6	1.61×10^{-18}
20	1,3,5-Triazine-1,3,5(2H,4H,6H)-triethanol	4719-04-4	0.2%	-4.67	1.00×10^6	1.08×10^{-11}

[a] Some of the chemicals in these tables have NA (not available) listed as the number of wells, which means that these chemicals have been used in hydraulic fracturing, but they were not reported to disclosures in the EPA FracFocus 1.0 project database for the time period of the study (January 1, 2011, to February 28, 2013) (U.S. EPA, 2015c). Analysis considered 34,675 disclosures and 676,376 ingredient records that met selected quality assurance criteria, including: completely parsed; unique combination of fracture date and API well number; fracture date between January 1, 2011, and February 28, 2013; valid CASRN; and valid concentrations. Disclosures that did not meet our quality assurance criteria (3,855) or other, query-specific criteria were excluded from our analysis.

[b] Log K_{ow} is estimated using the KOWWIN™ model, which uses an atom/fragment contribution method.

[c] Water solubility is estimated using the WSKOWWIN™ model, which estimates a chemical's solubility from K_{ow} and any applicable correction factors.

[d] Henry's law constant is estimated using the HENRYWIN™ model using the bond contribution method.

Table C-11. Ranking of the 20 least mobile organic chemicals, as determined by the largest log Kow, with CASRN, percent of wells where the chemical is reported from January 1, 2011 to February 28, 2013 (U.S. EPA, 2015c), and physicochemical properties (log Kow, solubility, and Henry's law constant) as estimated by EPI Suite™.

For organic salts, parameters are estimated using the desalted form.

Rank	Chemical name	CASRN	Percent of wells (U.S. EPA, 2015c)[a]	Estimated log K_{ow} (unitless)[b]	Estimated water solubility (mg/L @ 25°C)[c]	Estimated Henry's law constant (atm m³/mole @ 25°C)[d]
1	Sorbitan, tri-(9Z)-9 octadecenoate	26266-58-0	0.05%	22.56	1.12×10^{-19}	4.02×10^{-11}
2	Fatty acids, C18-unsatd., dimers	61788-89-4	NA	14.6	2.31×10^{-10}	4.12×10^{-08}
3	Sorbitan sesquioleate	8007-43-0	0.02%	14.32	2.31×10^{-11}	7.55×10^{-12}
4	Tri-n-butyltetradecyl-phosphonium chloride	81741-28-8	6%	11.22	7.90×10^{-7}	2.61×10^{-1}
5	Sodium bis(tridecyl) sulfobutanedioate	2673-22-5	NA	11.15	7.46×10^{-9}	8.51×10^{-11}
6	1-Eicosene	3452-07-1	NA	10.03	1.26×10^{-5}	1.89×10^{1}
7	D&C Red 28	18472-87-2	NA	9.62	1.64×10^{-8}	6.37×10^{-21}
8	C.I. Solvent Red 26	4477-79-6	NA	9.27	5.68×10^{-5}	5.48×10^{-13}
9	1-Octadecene	112-88-9	NA	9.04	1.256×10^{-4}	1.07×10^{1}
10	Alkenes, C>10 alpha-	64743-02-8	8%	8.55	3.941×10^{-4}	8.09×10^{0}
11	Dioctyl phthalate	117-84-0	NA	8.54	4.236×10^{-4}	1.18×10^{-5}
12	Benzene, C10-16-alkyl derivs.	68648-87-3	0.5%	8.43	2.099×10^{-4}	1.78×10^{-1}
13	Di(2-ethylhexyl) phthalate	117-81-7	NA	8.39	1.132×10^{-3}	1.18×10^{-5}
14	1-Octadecanamine, N,N-dimethyl-	124-28-7	NA	8.39	8.882×10^{-3}	4.51×10^{-3}
15	N,N-dimethyloctadecylamine hydrochloride	1613-17-8	NA	8.39	8.882×10^{-3}	4.51×10^{-3}

Rank	Chemical name	CASRN	Percent of wells (U.S. EPA, 2015c)[a]	Estimated log K_{ow} (unitless)[b]	Estimated water solubility (mg/L @ 25°C)[c]	Estimated Henry's law constant (atm m³/mole @ 25°C)[d]
16	Butyryl trihexyl citrate	82469-79-2	0.03%	8.21	5.56×10^{-5}	3.65×10^{-9}
17	1-Hexadecene	629-73-2	NA	8.06	1.232×10^{-3}	6.10×10^{0}
18	Benzo(g,h,i)perylene	191-24-2	NA	7.98	7.321×10^{-4}	1.26×10^{-2}
19	Dodecylbenzene	123-01-3	NA	7.94	1.015×10^{-3}	1.34×10^{-1}
20	Isopropanolamine dodecylbenzene	42504-46-1	0.02%	7.94	1.015×10^{-3}	1.34×10^{-1}

[a] Some of the chemicals in these tables have NA (not available) listed as the number of wells, which means that these chemicals have been used in hydraulic fracturing, but they were not reported in disclosures in the EPA FracFocus 1.0 project databases for the time period of the study (January 1, 2011, to February 28, 2013) (U.S. EPA, 2015c). Analysis considered 34,675 disclosures and 676,376 ingredient records that met selected quality assurance criteria, including: completely parsed; unique combination of fracture date and API well number; fracture date between January 1, 2011, and February 28, 2013; valid CASRN; and valid concentrations. Disclosures that did not meet these quality assurance criteria (3,855) or other, query-specific criteria were excluded from our analysis.

[b] Log K_{ow} is estimated using the KOWWIN™ model, which uses an atom/fragment contribution method.

[c] Water solubility is estimated using the WSKOWWIN™ model, which estimates a chemical's solubility from K_{ow} and any applicable correction factors.

[d] Henry's law constant is estimated using the HENRYWIN™ model using the bond contribution method.

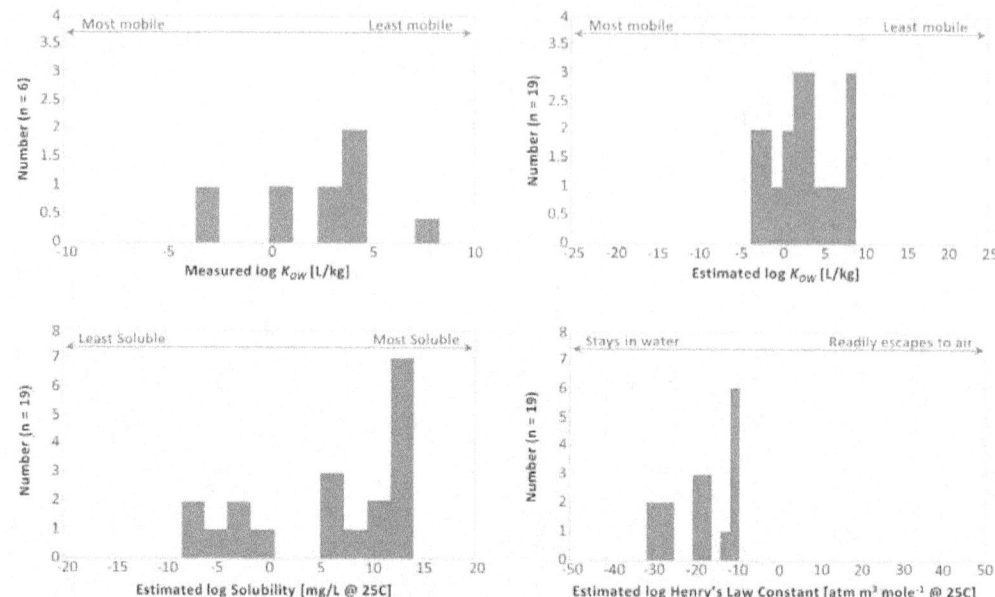

Figure C-1. Histograms of physicochemical properties organic chemicals claimed as confidential by industry that were used in the hydraulic fracturing process.
Measured values of log K_{ow} (upper left). Estimated physicochemical properties for log K_{ow} (upper right), log solubility (lower left), and log Henry's law constant (lower right) for all chemicals. Physicochemical properties (log K_{ow}, solubility, and Henry's law constant) estimated by EPI Suite™. Source: U.S. EPA (2013a).

Appendix D. Well Injection Supplemental Information

This page is intentionally left blank.

Appendix D. Well Injection Supplemental Information

This appendix presents the goals for the design and construction of oil and gas production wells, the well components used to achieve those goals, and methods for testing well integrity to help verify that the goals for well performance are achieved. This information provides additional background for the well component discussions presented in Chapter 6. Information on the pathways associated with the well that can cause fluid movement into drinking water resources is presented in Chapter 6.

D.1. Design Goals for Well Construction

Simply stated, production wells are designed to move oil and gas from the production zone (within the oil and gas reservoir) into the well and then through the well to the surface. There are typically a variety of goals for well design (Renpu, 2011), but the main purposes are facilitating the flow of oil and gas from the hydrocarbon reservoirs to the well (production management) while isolating that oil and gas and the hydrocarbon reservoirs from nearby groundwater resources (zonal isolation).

To achieve these goals, operators design and construct wells to have and maintain mechanical integrity throughout the life of the well. A properly designed and constructed well has two types of mechanical integrity: internal and external. Internal mechanical integrity refers to the absence of significant leakage within the production tubing, casing, or packer. External mechanical integrity refers to the absence of significant leakage along the well outside of the casing.

Achieving mechanical integrity involves designing the well components to resist the stresses they will encounter. Each well component must be designed to withstand all of the stresses to which the well will be subjected, including burst pressure, collapse, tensile, compression (or bending), and cyclic stresses (see Section 6.2.1 for additional information on these stresses). Well materials should also be compatible with the fluids (including liquids or gases) with which they come into contact to prevent leaks caused by corrosion.

These goals are accomplished by the use of one or more layers of casing, cement, and mechanical devices (such as packers), which provide the main barrier preventing migration of fluids from the well into drinking water resources. It should be noted that design conditions will change depending on the specific geology of the site. Technology in the field of hydraulic fracturing is also rapidly evolving with new technologies and techniques being continually developed. Therefore while the following sections outline basic design goals and concepts, they cannot anticipate all possible design conditions.

D.2. Well Components

Casing and cement are used in the design and construction of wells to achieve the goals of mechanical integrity and zonal isolation. Several industry-developed specifications and best practices for well construction have been established to guide well operators in the construction process; see Text Box D-1.[1] The sections below describe options available for casing, cement, and other well components.

Text Box D-1. Selected Industry-Developed Specifications and Recommended Practices for Well Construction in North America.

American Petroleum Institute (API)

- API Guidance Document HF1—Hydraulic Fracturing Operations—Well Construction and Integrity Guidelines (API, 2009a)
- API RP 10B-2—Recommended Practice for Testing Well Cements (API, 2013)
- API RP 10D-2—Recommended Practice for Centralizer Placement and Stop Collar Testing (API, 2004)
- API RP 5C1—Recommended Practices for Care and Use of Casing and Tubing (API, 1999)
- API RP 65-2—Isolating Potential Flow Zones during Well Construction (API, 2010a)
- API Specification 10A—Specification on Cements and Materials for Well Cementing (API, 2010b)
- API Specification 11D1—Packers and Bridge Plugs (API, 2009b)
- API Specification 5CT—Specification for Casing and Tubing (API, 2011)
- API RP 100-1 – Hydraulic Fracturing Well Integrity and Fracture Containment 1st Edition (API, 2015)

Canadian Association of Petroleum Producers (CAPP) and Enform

- Hydraulic Fracturing Operating Practices: Wellbore Construction and Quality Assurance (CAPP, 2013)
- Interim Industry Recommended Practice Volume #24—Fracture Stimulation: Inter-wellbore Communication (Enform, 2013)

Marcellus Shale Coalition (MSC)

- Recommended Practices—Drilling and Completions (MSC, 2013)

D.2.1. Casing

Casing is steel pipe that is placed into the wellbore (the cylindrical hole drilled through the subsurface rock formation) to maintain the stability of the wellbore, to transport the hydrocarbons from the subsurface to the surface, and to prevent intrusion of other fluids into the well and wellbore. Up to four types of casing may be present in a well, including (from largest to smallest-diameter): conductor casing, surface casing, intermediate casing, and production casing. Each is described below.

D.2.1.1. Types of Casing

The **conductor casing** is the largest diameter string of casing. It is typically in the range of 30 in. (76 cm) to 42 in. (107 cm) in diameter (Hyne, 2012). Its main purpose is to prevent unconsolidated material, such as sand, gravel, and soil, from collapsing into the wellbore. Therefore, the casing is

---

[1] Information is not available to determine how often these practices are used or how well they prevent the development of pathways for fluid movement to drinking water resources.

typically installed from the surface to the top of the bedrock or other consolidated formations. The conductor casing may or may not be cemented in place.

The next string of casing is the **surface casing**. A typical surface casing diameter is 13.75 in. (34.93 cm), but diameter can vary (Hyne, 2012). The surface casing's main purposes are to isolate any groundwater resources that are to be protected by preventing fluid migration along the wellbore once the casing is cemented and to provide a sturdy structure to which blow-out prevention equipment can be attached. For these reasons, the surface casing most commonly extends from the surface to some distance beneath the lowermost geologic formation containing groundwater resources to be protected. The specific depth to which the surface casing is set is often governed by the depth of the groundwater resource as defined and identified for protection in state regulations.

Intermediate casing is typically used in wells to control pressure in an intermediate-depth formation. It may be used to reduce or prevent exposure of weak formations to pressure from the weight of the drilling fluid or cement or to allow better control of over-pressured formations. The intermediate casing extends from the surface through the formation of concern. There may be more than one string of concentric intermediate casing present or none at all, depending on the subsurface geology. Intermediate casing may be cemented, especially through over-pressured zones; however, it is not always cemented to the surface. Intermediate casing, when present, is often 8.625 in. (21.908 cm) in diameter but can vary (Hyne, 2012).

Production casing extends from the surface into the production zone. The main purposes of the production casing are to isolate the hydrocarbon product from fluids in surrounding formations and to transport the product to the surface. It can also be used to inject hydraulic fracturing fluids, receive produced water during hydraulic fracturing operations (e.g., if tubing or a temporary fracturing string is not present), and prevent other fluids from mixing with and diluting the produced hydrocarbons. The production casing is generally cemented to some point above the production zone. Production casing is often 5.5 in. (14.0 cm) in diameter but can vary (Hyne, 2012).

Liners are another type of metal tubular (casing-like) well component that can be used to fulfill the same purposes as intermediate and production casing in the production zone. Like casing, they are steel pipe, but differ in that they do not extend from the production zone to the surface. Rather, they are connected to the next largest string of casing by a hanger that is attached to the casing. A frac sleeve is a specialized type of liner that is used during fracturing. It has plugs that can be opened and closed by dropping balls from the surface (see the discussion of well completions below for additional information on the use of frac sleeves).

Production tubing is the smallest, innermost steel pipe in the well and is distinguished from casing by not being cemented in place. It is used to transport the hydrocarbons to the surface. Fracturing may be done through the tubing if present, or through the production casing. Because casing cannot be replaced, tubing is often used, especially if the hydrocarbons contain corrosive substances such as hydrogen sulfide or carbon dioxide. Tubing may not be used in high-volume production wells. Typical tubing diameter is between 1.25 in. (3.18 cm) and 4.5 in. (11.4 cm) (Hyne, 2012).

D.2.1.2. Casing Design Considerations

The stresses that the casing will experience are key factors to consider in designing the casing. If the casing is not designed with sufficient resistance to the stresses it will face, it can fail. Stresses that may cause failure of casing include: pressure exerted during hydraulic fracturing operations, cyclic pressure from multi-stage fracturing, pressure from the formation, and stresses encountered during installation of the casing especially around bends (King and Valencia, 2016; Cheremisinoff and Davletshin, 2015). Maximum values for each of these stresses can be calculated, and the casing can be designed to resist them. Generally, the inner layers such as tubing are designed to collapse before the outer casing will burst (King and Valencia, 2016). Casing strength can be improved by choosing stronger materials or by increasing casing thickness.

Another factor to consider in casing design is corrosion. The casing may be exposed to corrosive substances such as carbon dioxide, hydrogen sulfide, natural brines, and acids used during fracturing. Corrosion resistance may be achieved by using corrosion resistant alloys or by lining the casing (King and Valencia, 2016; Syed and Cutler, 2010). Abrasion from proppant during fracturing can also lead to casing erosion problems (King and Valencia, 2016).

Joint design and installation are equally important in casing design as they are a frequent location of casing leaks (King and Valencia, 2016). Joint failure can occur due to poor design, installation errors, and stress corrosion.

D.2.2. Cement

Cement is the main barrier preventing fluid movement along the wellbore outside the casing. It also lends mechanical strength to the well and protects the casing from corrosion by naturally occurring formation fluids. Cement is placed in the annulus, which is the space between two adjacent casings or the space between the outermost casing and the rock formation through which the wellbore was drilled. The sections below describe considerations for selecting cement and additives, as well as cementing procedures and techniques.

D.2.2.1. Considerations for Cementing

The length and location of the casing section to be cemented and the composition of the cement can vary based on numerous factors, including the presence and locations of weak formations, over- or under-pressured formations, or formations containing fluids; formation permeability; and temperature. State requirements for oil and gas production well construction and the relative costs of well construction options are also factors.

Improper cementing can lead to the formation of channels (small connected voids) in the cement, which can—if they extend across multiple formations or connect to other existing channels or fractures—present pathways for fluid migration. This section describes some of the considerations and concerns for proper cement placement and techniques and materials that are available to address these concerns. Careful selection of cements (and additives) and design of the cementing job can avoid integrity problems related to cement.

To select the appropriate cement type, properties, and additives, operators consider the required strength needed to withstand downhole conditions and compatibility with subsurface chemistry, as described below:

- The cement design needs to **achieve the strength** required under the measured or anticipated downhole conditions. Factors that are taken into account to achieve proper strength can include density, thickening time, the presence of free water, compressive strength, and formation permeability (Renpu, 2011). Commonly, cement properties are varied during the process, with a "weaker" (i.e., less dense) lead cement, followed by a "stronger" (denser) tail cement. The lead cement is designed with a lower density to reduce pressure on the formation and better displace drilling fluid without a large concern for strength. The stronger tail cement provides greater strength for the deeper portions of the well the operator considers as requiring greater strength.

- The **compatibility of the cement** with the chemistry of formation fluids, hydrocarbons, and hydraulic fracturing fluids is important for maintaining well integrity through the life of the well. Most oil and gas wells are constructed using some form of Portland cement. Portland cement is a specific type of cement consisting primarily of calcium silicates with additional iron and aluminum. Industry specifications for recommended cements are determined by the downhole pressure, temperature, and chemical compatibility required.

There are a number of considerations in the design and execution of a cement job. Proper centralization of the casing within the wellbore is one of the more important considerations. Others include the potential for lost cement, gas invasion, cement shrinkage, incomplete removal of drilling mud, settling of solids in the wellbore, and water loss into the formation while curing. These concerns, and techniques available to address them, include the following:

- **Improper centralization of the casing within the wellbore** can lead to preferential flow of cement on the side of the casing with the larger space and little to no cement on the side closer to the formation. If the casing is not centered in the wellbore, cement will flow unevenly during the cement job, leading to the formation of cement channels. Kirksey (2013) notes that, if the casing is off-center by just 25%, the cement job is almost always inadequate. Centralizers are used to keep the casing in the center of the hole and allow an even cement job. To ensure proper centralization, centralizers are placed at regular intervals along the casing (API, 2010a). Centralizer use is especially key in horizontal wells, as the casing will tend to settle (due to gravity) to the bottom of the wellbore if the casing is not centered (Sabins, 1990), leading to inadequate cement on the lower side. Although some operators have avoided using centralizers on horizontal wells because of problems with stuck pipe, improved centralizer designs have allowed increased use of centralizers in horizontal wells (Landry et al., 2015).

- **Lost cement** (sometimes referred to as lost returns) refers to cement that moves out of the wellbore and into the formation instead of filling up the annulus between the casing and the formation. Lost cement can occur in weak formations that fail (fracture) under pressure of the cement or in particularly porous, permeable, or naturally fractured formations. Lost cement can result in lack of adequate cement across a water- or brine-bearing zone. To avoid inadequate placement of cement due to lost cement, records of nearby wells can be examined to determine zones where lost cement returns occur (API, 2009a). If records from nearby wells are not available, cores and logs may be used to identify any high-permeability or mechanically weak formations that might lead to lost cement. Steps can then be taken to eliminate or reduce loss of cement to the formation. Staged cementing (see below) can reduce the hydrostatic pressure on the formation and may avoid fracturing weak formations (Lyons and Pligsa, 2004). Additives such as cellulose or polymers are also available that will lessen the flow of cement into highly porous formations (API, 2010a; Ali et al., 2009).

- **Gas invasion and cement shrinkage** during cement setting can also cause channels and poor bonding. As cement sets, it begins to lose the ability to transmit pressure to the surrounding formation. During the cementing process, the hydrostatic pressure from the cement column keeps formation gas from entering the cement. As the cement sets (hardens), the hydrostatic pressure decreases; if it becomes less than the formation pressure, gas can enter the cement, leading to channels. Cement shrinkage occurs as the cement sets under a high pressure; shrinking can be made worse by left over drilling mud or too large of a space between the casing and formation (Oyarhossein and Dusseault, 2015). Such shrinkage can lead to channels or microannuli along the cement column. These problems can be avoided by using cement additives that increase setting time or expand to offset shrinkage (McDaniel et al., 2014; Wojtanowicz, 2008; Dusseault et al., 2000). Foamed cement can help alleviate problems with shrinkage, although care needs to be taken in cement design to ensure the proper balance of pressure between the cement column and formation (API, 2010a). Cement additives such as latex are also available that will expand upon contact with certain fluids such as hydrocarbons. These cements, termed self-healing cements, are relatively new but have shown early promise in some fields (Ali et al., 2009). Self-healing cements have been found to increase the compressive strength of the cement by 10%, tensile strength by 48%, Poisson's ratio by 66%, and Young's modulus by 56% (Shadravan and Amani, 2015). Rotating the casing during cementing will also delay cement setting by agitating the cement. Another technique called pulsation, where pressure pulses are applied to the cement while it is setting, also can delay cement setting and loss of hydrostatic pressure until the cement is strong enough to resist gas penetration (Stein et al., 2003).

- Another important issue is **removal of drilling mud**. Inadequate removal of drilling mud can prevent cement from filling the entire space between the casing and the formation, resulting in channels in the cement after the mud is eroded away by formation fluids (Jackson and Dussealt, 2014). If drilling mud is not completely removed, it can gather on one side of the wellbore and prevent that portion of the wellbore from being adequately cemented. The drilling mud can then be eroded away after the cement sets, leaving a channel. Drilling mud can be removed by circulating a denser fluid (spacer fluid) to flush the drilling mud out (Kirksey, 2013; Brufatto et al., 2003). Mechanical devices called scratchers can also be attached to the casing, and the casing rotated or reciprocated to scrape drilling mud from the wellbore (Hyne, 2012; Crook, 2008). The spacer fluid, which is circulated prior to the cement to wash the drilling fluid out of the wellbore, must be designed with the appropriate properties and pumped in such a way that it displaces the drilling fluid without mixing with the cement (Kirksey, 2013; API, 2010a; Brufatto et al., 2003).

- Also of concern in horizontal wells is the possibility of **solids settling** at the bottom of the wellbore and free water collecting at the top of the wellbore. This can lead to channels and poor cement bonding. The cement slurry must be properly designed for horizontal wells to minimize free water and solids settling.

- If there is free water in the cement, pressure can cause **water loss into the formation,** leaving behind poor cement or channels (Jiang et al., 2012). In horizontal wells, free water can also accumulate at the top of the wellbore, forming a channel (Sabins, 1990). Minimizing free water in the cement design and using fluid loss control additives can help control the loss of water (Ross and King, 2007).

- Fracturing in stages can lead to **cyclic stresses** being exerted on the cement (King and Valencia, 2016). During fracturing, the cooler temperature fluids are injected into the well at high pressure, resulting in temperature and pressure changes downhole. When injection stops, the temperature returns to the higher reservoir temperatures and pressure returns to normal. One study has found such cycling can lead to temperature changes of as much as 176°F (80°C) (Tian et al., 2015). Exposing cement to several cycles of temperature and pressure variation can lead to a number of problems. Stress may cause cracks in the cement, especially at locations of existing defects in the cement (De Andrade et al., 2015; Syed and Cutler, 2010). Differences between the rates at which steel and cement expand can lead to debonding between the cement and casing. Contraction of fluids at lower temperatures can also create vacuums in some situations, which can stress the casing and cement (Tian et al., 2015). Using cement with lower anelastic strength and higher tensile and impact strength may help alleviate problems caused by cyclic stresses (McDaniel et al., 2014). Self-sealing cements, as described above, may also seal cracks that are initiated during cycling. Some studies have found the ability of such cements to seal flow through cracks in as little as 30 minutes (Cavanagh et al., 2007). Foamed cements have also been found to hold up better to pressure cycles than standard cement slurries (Spaulding, 2015).

D.2.2.2. Cement Placement Techniques

The primary cement job is most commonly conducted by pumping the cement down the inside of the casing, then out the bottom of the casing where it is then forced up the space between the outside of the casing and the formation. (The cement can also be placed in the space between two casings.) If **continuous cement** (i.e., a sheath of cement placed along the entire wellbore) is desired, cement is circulated through the annulus until cement that is pumped down the central casing flows out of the annulus at the surface. A spacer fluid is often pumped ahead of cement to remove any excess drilling fluid left in the wellbore; even if the operator does not plan to circulate cement to the surface, the spacer fluid will still return to the surface, as this is necessary to remove the drilling mud from the annulus. If neither the spacer fluid nor the cement returns to the surface, this indicates that fluids are being lost into the formation.

Staged cementing is a technique that reduces pressure on the formation by decreasing the height (and therefore the weight) of the cement column. This may be necessary if the estimated weight and pressure associated with standard cement emplacement could damage zones where the formation intersected is weak. The reduced hydrostatic pressure at the bottom of the cement column can also reduce the loss of water to permeable formations, improving the quality of the cement job. In multiple-stage cementing, cement is circulated to just below a cement collar placed between two sections of casing. A cement collar will have been placed between two sections of casing, just above, with ports that can be opened by dropping a weighted tool. Two plugs—which are often referred to as bombs or darts because of their shape—are then dropped. The first plug is dropped once the desired cement for the first stage has been pushed out of the casing by a spacer fluid. It closes the section of the well below the cement collar and stops cement from flowing into the lower portion of the well. The second plug (or opening bomb) opens the cement ports in the collar, allowing cement to flow into the annulus between the casing and formation. Cement is then circulated down the wellbore, out the cement ports, into the annulus, and up to the surface. Once cementing is complete, a third plug is dropped to close the cement ports, preventing the newly pumped cement from flowing back into the well (Lyons and Pligsa, 2004); see Figure D-1.

Another less commonly used primary cementing technique is **reverse circulation cementing.** This technique has been developed to decrease the force exerted on weak formations. In reverse circulation cementing, the cement is pumped down the annulus directly between the outside of the outermost casing and the formation. This essentially allows use of lower density cement and lower pumping pressures. With reverse circulation cementing, greater care must be taken in calculating the required cement, ensuring proper cement circulation, and locating the beginning and end of the cemented portion.

Another method used to cement specific portions of the well without circulating cement along the entire wellbore length is to use a **cement basket**. A cement basket is a device that attaches to the well casing. It is made of flexible material such as canvas or rubber that can conform to the shape of the wellbore. The cement basket acts as a one-way barrier to cement flow. Cement can be circulated up the wellbore past the cement basket, but when circulation stops the basket prevents the cement from falling back down the wellbore. Cement baskets can be used to isolate weak formations or formations with voids. They can also be placed above large voids such as mines or caverns with staged cementing used to cement the casing above the void.

If any deficiencies are identified, **remedial cementing** may be performed. The techniques available to address deficiencies in the primary cement job including cement squeezes or top-job cementing. A cement squeeze injects cement under high pressure to fill in voids or spaces in the primary cement job caused by high pressure, failed formations, or improper removal of drilling mud. Although cement squeezes can be used to fix deficiencies in the primary cement job, they require the well to be perforated, which can weaken the well and make it susceptible to degradation by pressure and temperature cycling as would occur during fracturing (Crescent, 2011). Another method of secondary cementing is the top job. In a top job, cement is pumped down the annulus directly to fill the remaining uncemented space when cement fails to circulate to the surface.

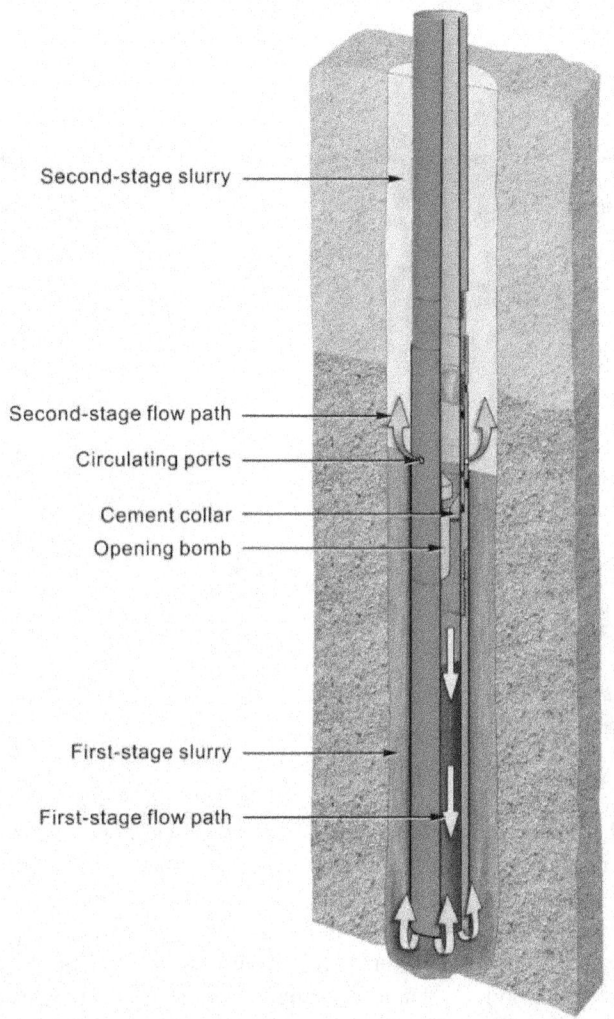

Second-stage slurry

Second-stage flow path

Circulating ports

Cement collar

Opening bomb

First-stage slurry

First-stage flow path

Note: *Figure not to scale*

Two-Stage Cementing Process

Figure D-1. A typical staged cementing process.

D.3. Well Completions

Completion refers to how the well is prepared for production and how flow is established between the formation and the surface. Figure D-2 presents examples of well completion types, including cased, formation packer, and open hole completion.

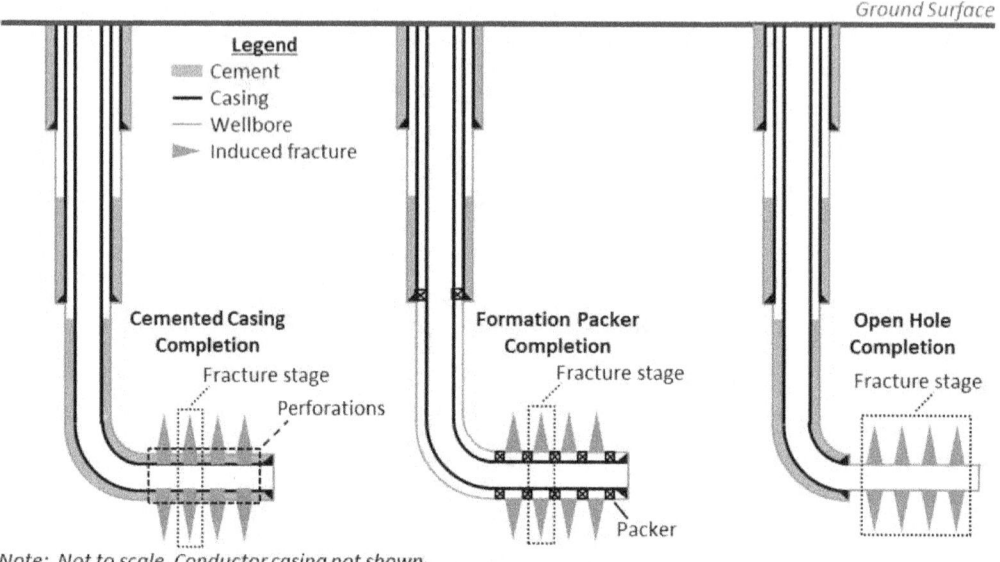

Note: Not to scale. Conductor casing not shown.

Figure D-2. Examples of well completion types.
Configurations shown include cased, formation packer, and open hole completion. From <u>U.S. EPA (2015k)</u>.

A **cased completion**, where the casing extends to the end of the wellbore and is cemented in place, is the most common configuration of the well in the production zone (<u>U.S. EPA, 2015k</u>). Perforations are made through the casing and cement and into the formation using tools called "perf guns" that deliver small explosive charges or other devices, such as sand jets. Hydraulic fracturing then is conducted through the perforations. This is a common technique in wells that produce from several different depths and in low-permeability formations that are fractured (<u>Renpu, 2011</u>). While perforations do control the initiation point of the fracture, this can be a disadvantage if the perforations are not properly aligned with the local stress field. If the perforations are not aligned, the fractures will twist to align with the stress field, leading to tortuosity in the fractures and making fluid movement through them more difficult (<u>Cramer, 2008</u>). Fracturing stages can be isolated from each other using various mechanisms such as plugs or baffle rings, which close off a section of the well when a ball of the correct size is dropped down the well.

A **packer** is a mechanical device used to selectively seal off certain sections of the wellbore. Packers can be used to seal the space between the tubing and casing, between two casings, or between the production casing and formation. The packer has one or more rubber elements that can be manipulated downhole to increase in diameter and make contact with the inner wall of the next-largest casing or the formation, effectively sealing the annulus created between the outside of the tubing and the inside of the casing. Packers vary in how they are constructed and how they are set, based on the downhole conditions in which they are used. There are two types of packers: internal packers and formation packers. Internal packers are used to seal the space between the casing and tubing or between two different casings. They prevent fluid movement into the annulus by isolating the outer casing layers from produced or formation fluids. Formation packers seal the space between the casing and the formation and are often used to isolate fracture stages; they can be used to separate an open hole completion into separate fracture stages. Packers can seal an annulus by several different mechanisms. Mechanical packers expand mechanically against the formation and can exert a significant force on the formation (McDaniel and Rispler, 2009). They are typically less than 5 ft (1.5 m) long and can be used in wells with tighter doglegs (Senters et al., 2016). Swellable packers have elastomer sealing elements that swell when they come into contact with a triggering fluid such as water or hydrocarbons. They exert less force on the formation and can seal larger spaces but take some time to fully swell (McDaniel and Rispler, 2009). Swellable packers are longer and can be affected by thermal changes during fracturing. Cyclic stresses during fracturing can also cause packer failure (Senters et al., 2016). Internal mechanical integrity tests such as pressure tests can verify that the packer is functioning as designed and has not corroded or deteriorated.

In an **open hole completion**, the production casing extends just into the production zone and the entire length of the wellbore through the production zone is left uncased. This is only an option in formations where the wellbore is stable enough to not collapse into the wellbore. In formations that are unstable, a slotted liner may be used in open hole completions to control sand production (Renpu, 2011). Perforations are not needed in an open hole completion, since the production zone is not cased. An open hole completion can be fractured in a single stage or in multiple stages.

If formations are to be fractured in stages, additional completion methods are needed to separate the stages from each other and control the location of the fractures. One possibility is use of a liner with formation packers to isolate each stage. The liner is equipped with sliding sleeves that can be opened by dropping balls down the casing to open each stage. Fracturing typically occurs from the end of the well and continues toward the beginning of the production zone.

D.4. Mechanical Integrity Testing

While proper design and construction of the well's casing and cement are important, it is also important to verify the well was constructed and is performing as designed. Mechanical integrity tests (MITs) can verify that the well was constructed as planned and can detect damage to the production well that occurs during operations, including hydraulic fracturing activities. Verifying that a well has mechanical integrity can prevent potential impacts to drinking water resources or loss of hydrocarbon products by providing early warning of a problem with the well or cement and allowing repairs.

It is important to note that if a well fails an MIT, this does not mean the well has failed or that an impact on drinking water resources has occurred. An MIT failure is a warning that one or more components of the well are not performing as designed and is an indication that corrective actions are necessary. If well remediation is not performed, a loss of well integrity could occur, which could result in fluid movement from the well.

D.4.1. Internal Mechanical Integrity

Internal mechanical integrity is an absence of significant leakage in the tubing, casing, or packers within the well system. Loss of internal mechanical integrity is usually due to corrosion or mechanical failure of the well's tubular and mechanical components.

Internal mechanical integrity can be tested by the use of pressure testing, annulus pressure monitoring, ultrasonic monitoring, and casing inspection logs or caliper logs:

- **Pressure testing** involves raising the pressure in the wellbore to a set level and shutting in the well. If the well has internal mechanical integrity, the pressure should remain constant with only small changes due to temperature fluctuation. Typically, the well is shut in (i.e., production is stopped and the wellhead valves closed) for a time prescribed by regulation, and if the pressure remains within a given percent of the original reading, the well is considered to have passed the test. Usually, the well is pressure tested to the maximum expected pressure; for a well to be used for hydraulic fracturing, this would be the pressure applied during hydraulic fracturing. Performing a pressure test on each casing before the next casing is drilled ensures the casing can withstand subsequent stresses and allows repairs if necessary before problems can develop (Cheremisinoff and Davletshin, 2015). Pressure tests, however, can cause debonding of the cement from the casing, so test length is often limited to reduce this effect (API, 2010a).

- If the annulus between the tubing and casing is sealed by a packer, **annulus pressure monitoring** can give an indication of the integrity of the tubing and casing. If the tubing, casing, and packer all have mechanical integrity, the pressure in the annulus should not change except for small changes in response to temperature fluctuations. The annulus can be filled with a non-corrosive liquid and the level of the liquid can be used as another indication of the integrity of the casing, tubing, and packer. The advantage of monitoring the tubing/production casing annulus is that it can give a continuous, real-time indication of the internal integrity of the well. This is the only MIT test likely to detect problems during normal well operations. Even if the annulus is not filled with a fluid, monitoring its pressure can indicate leaks. If pressure builds up in the annulus and then recovers quickly after having bled off, that condition is referred to as sustained casing pressure or surface casing vent flow and is a sign of a leak in the tubing or casing (Watson and Bachu, 2009). Monitoring of annuli between other sets of casings can also provide information on the integrity of those casings. It can also provide information on external mechanical integrity for annuli open to the formation (see Section D.4.2 for additional information on external MITs). Jackson et al. (2013b) also note that monitoring annular pressure allows the operator to vent gas before it accumulates enough pressure to cause migration into drinking water resources. Measuring annulus flow rate also allows detection of gas flowing into the annulus (Arthur, 2012).

- A newer tool uses **ultrasonic monitors** to detect leaks in casing and other equipment. It measures the attenuation of an ultrasonic signal as it is transmitted through the wellbore. The tool measures transmitted ultrasonic signals as it is lowered down the wellbore. The tool can pick up ultrasonic signals created by a leak, similar to noise logs. The tool only has a range of a few feet but is claimed to detect leaks as small as half a cup per minute (Julian et al., 2007).

- **Caliper logs** have mechanical fingers that extend from a central tool and measure the distance from the center of the wellbore to the side of the casing. Running a caliper log can identify areas where corrosion has altered the diameter of the casing or where holes have formed in the casing. Caliper logs may also detect debris or obstructions in the well. Casing inspection and caliper logs are primarily used to determine the condition of the casing. Regular use of them may identify problems such as corrosion and allow mitigation before they cause a loss of integrity to the casing. To run these logs in a producing well, the tubing must first be pulled.

- **Casing inspection logs** are instruments lowered into the casing to inspect the casing for signs of wear or corrosion. One type of casing log uses video equipment to detect corrosion or holes. Another type uses electromagnetic pulses to detect variations in metal thickness. Running these logs in a producing well requires the tubing to be pulled.

If an internal mechanical integrity problem is detected, the location of the problem must be found. Caliper or casing inspection logs can detect locations of holes in casing. Locations of leaks can also be detected by sealing off different sections of the well using packers and performing pressure tests on each section until the faulty section is located. If the leaks are in the tubing or a packer, the problem may be remedied by replacing the well component. Casing leaks may be remedied by performing a cement squeeze (Section D.2.2).

D.4.2. External Mechanical Integrity

External well mechanical integrity is demonstrated by establishing the absence of significant fluid movement along the outside of the casing, either between the outer casing and cement or between the cement and the wellbore. Failure of an external MIT can indicate improper cementing or degradation of the cement emplaced in the annular space between the outside of the casing and the wellbore. This type of failure can lead to movement of fluids out of intended production zones and toward drinking water resources.

Several types of logs are available to evaluate external mechanical integrity, including temperature logs, noise logs, oxygen activation logs, radioactive tracer logs, and cement evaluation logs.

- **Temperature logs** measure the temperature in the wellbore, and are capable of measuring small changes in temperature. They can be performed using instruments that are lowered down the well on a wireline, or they can be done using fiber optic sensors permanently installed in the well. When performed immediately after cementing, they can detect the heat from the cement setting and determine the location of the top of cement. After the cement has set, temperature logs can sense the difference in temperatures between formation fluids and injected or produced fluids. They may also detect temperature changes due to cooling or warming caused by flow. In this way, temperature logs may detect movement of fluid outside the casing in the wellbore (Arthur, 2012). Temperature logs require interpretation of the causes of temperature changes and are therefore subject to varying results among different users.

- **Noise logs** are sensitive microphones that are lowered down the well on a wireline. They are capable of detecting small noises caused by flowing fluids, such as fluids flowing through channels in the cement (Arthur, 2012). They are most effective at detecting fast-moving gas leaks and less successful with more slowly moving liquid migration.

- **Oxygen activation logs** consist of a neutron source and one or more detectors that are lowered on a wireline. The neutron source bombards oxygen molecules surrounding the wellbore and converts them into unstable nitrogen molecules that rapidly decay back to oxygen, emitting gamma radiation in the process. Gamma radiation detectors above or below the neutron source measure how quickly the oxygen molecules are moving away from the source, thereby determining flow associated with water.

- **Radioactive tracer logs** involve release of a radioactive tracer and then passing a detector up or down the wellbore to measure the path the tracers have taken. They can be used to determine if fluid is flowing up the wellbore. Tracer logs can be very sensitive but may be limited in the range over which leaks can be detected.

- **Cement evaluation logs** (also known as cement bond logs) are acoustic logs consisting of an instrument that sends out acoustic signals along with receivers, separated by some distance, that record the acoustic signals. As the acoustic signals pass through the casing, they will be attenuated to an extent, depending on whether the pipe is free or is bonded to cement. By analyzing the return acoustic signal, the degree of cement bonding with the casing can be determined. The cement evaluation log measures the sound attenuation as sound waves passing through the cement and casing. There are different types of cement evaluation logs available. Some instruments can only return an average value over the entire wellbore. Other instruments are capable of measuring the cement bond radially. Newer acoustic logging techniques with features such as flexural attenuation and acoustic impedance maps can identify channels as small as an inch (2.5 cm) in diameter (Landry et al., 2015). Cement logs do not actually determine whether fluid movement through the annulus is occurring. They only can determine whether cement is present in the annulus and in some cases can give a qualitative assessment of the quality of the cement in the annulus. Cement evaluation logs are used to calculate a bond index which varies between 0 and 1, with 1 representing the strongest bond and 0 representing the weakest bond. It should be noted that these type of tests cannot detect whether or not fluid migration is occurring. They only indicate whether cement is present and give a qualitative indication of the degree of bonding of any cement present. Because interpretation of these logs are qualitative, there is also a great deal of subjectivity in their results.

If the well fails an external MIT, damaged or missing cement may be repaired using a cement squeeze (Wojtanowicz, 2008). A cement squeeze involves injection of cement slurry into voids behind the casing or into permeable formations. Different types of cement squeezes are available depending on the location of the void needing to be filled and well conditions (Kirksey, 2013). Cement squeezes are not always successful, however, and may need to be repeated to successfully seal off flow (Wojtanowicz, 2008).

This page is intentionally left blank.

Appendix E. Produced Water Handling Supplemental Information

This page is intentionally left blank.

Appendix E. Produced Water Handling Supplemental Information

E.1. Specific Definitions of the Terms "Produced Water" and "Flowback"

Various organizations have used different definitions of the terms "produced water" and "flowback." Several examples follow:

E.1.1. Produced Water

The American Petroleum Institute (API): "Produced water is any of the many types of water produced from oil and gas wells" (API, 2010c).

The U.S. Department of Energy (DOE): "Produced water is water trapped in underground formations that is brought to the surface along with oil or gas" (Veil et al., 2004).

The American Water Works Association (AWWA): "Produced water is the combination of flowback and formation water that returns to the surface along with the oil and natural gas" (AWWA, 2013).

E.1.2. Flowback

API: "The fracture fluids that return to the surface after a hydraulic fracture is completed" (API, 2010c).

AWWA: "Fracturing fluids that return to the surface through the wellbore after hydraulic fracturing is complete" (AWWA, 2013).

Other definitions include production of hydrocarbons from the well (Barbot et al., 2013; U.S. EPA, 2012e), or specify a time period (USGS, 2014; Haluszczak et al., 2013; Warner et al., 2013b; Hayes and Severin, 2012a; Hayes, 2009).

E.2. Produced Water Volumes

The EPA (U.S. EPA, 2015m) estimates of flowback volumes and long-term produced water volumes used to generate the summaries appearing in Table 7-3 of Chapter 7 appear below in Table E-1.

Table E-1. Produced water characteristics for wells by basin, formation, and resource type.
Source: U.S. EPA (2016b).

Basin	Formation	Resource type	Well type	Fracturing fluid (Mgal)			Flowback (% of fracturing fluid returned)			Long-term produced water rates (gpd)		
				Weighted average[a]	Range[b]	Data points[c]	Weighted average[a]	Range[b]	Data points[c]	Weighted average[a]	Range[a]	Data points[c]
Anadarko	Caney	Shale	H	8.1	4.4 – 12	11	-	-	0	-	-	0
	Cleveland	Tight	H	1.7	0.2 – 4	928	-	12 – 40	2	410	59 – 2,000	1,160
			V	0.18	0.033 – 3	15	50	50 – 50	1	66	56 – 400	130
	Granite Wash	Tight	H	4.9	0.2 – 8.3	924	-	6.5 – 22	2	980	10 – 2,400	762
			V	0.53	0.085 – 3	72	50	50 – 50	1	520	330 – 790	1,397
			D	-	-	0	-	-	0	480	160 – 940	83
	Mississippi Lime	Tight	H	2	1.3 – 5	3,301	50	50 – 50	1	-	37,000 – 120,000	4
			V	0.34	0.016 – 0.71	59	-	-	0	10	0.71 – 38	16
	Woodford	Shale	H	5.2	1 – 12	3,243	34	20 – 50	3	5,500	3,200 – 6,400	198
			V	0.36	0.015 – 1.6	11	-	-	0	-	-	0
			D	1.6	0.21 – 1.9	10	-	-	0	-	-	0
	Clinton-Medina	Tight	V	-	-	0	-	-	0	7.9	7.3 – 11	551

Basin	Formation	Resource type	Well type	Fracturing fluid (Mgal)			Flowback (% of fracturing fluid returned)			Long-term produced water rates (gpd)		
				Weighted average[a]	Range[b]	Data points[c]	Weighted average[a]	Range[b]	Data points[c]	Weighted average[a]	Range[a]	Data points[c]
Appalachian	Devonian	Shale	V	-	-	0	-	-	0	13	4.8 – 19	197
	Marcellus	Shale	H	4.6	0.9 – 11	17,316	7.1	4 – 47	4,374	820	54 – 13,000	6,494
		Shale	V	0.25	0.11 – 5.4	116	40	21 – 60	7	200	94 – 1,000	741
			D	0.16	0.092 – 0.17	6	-	-	0	-	-	0
	Utica	Shale	H	6.8	1 – 13	1,108	2.5	0.66 – 27	684	800	420 – 1,700	764
Arkoma	Fayetteville	Shale	H	5	1.7 – 11	3,014	-	10 – 20	2	430	150 – 2,300	2,305
Denver-Julesburg	Codell	Tight	H	3.5	2.4 – 7.1	234	16	-	36	400	110 – 1,100	179
			V	0.23	0.11 – 0.46	97	0	0 – 4	13	59	47 – 120	158
			D	0.26	0.14 – 0.5	362	0	0 – 3	8	46	18 – 71	667
	Codell-Niobrara	Tight	H	2.8	2.7 – 5.4	65	7.2	7.2 – 7.2	32	75	19 – 560	38
			V	0.3	0.15 – 0.4	490	2.8	-	21	33	13 – 65	2,113
			D	0.4	0.2 – 0.46	806	0	0 – 5	11	45	28 – 70	1,853

Basin	Formation	Resource type	Well type	Fracturing fluid (Mgal)			Flowback (% of fracturing fluid returned)			Long-term produced water rates (gpd)		
				Weighted average[a]	Range[b]	Data points[c]	Weighted average[a]	Range[b]	Data points[c]	Weighted average[a]	Range[a]	Data points[c]
Denver Julesburg, cont.	Muddy J.	Tight	H	1.4	0.44 – 2.6	6	-	-	0	860	220 – 1,100	6
			V	0.27	0.12 – 0.45	139	0.09	-	15	120	52 – 550	340
			D	0.42	0.17 – 0.62	758	0	0 – 0	11	63	39 – 110	1,106
	Niobrara	Shale	H	2.9	1.9 – 5.1	1,435	16	1.8 – 100	173	760	120 – 1,300	1,213
			V	0.24	0.015 – 0.31	455	33	1.6 – 90	29	330	15 – 600	5,808
			D	0.36	0.13 – 2.9	25	-	-	0	41	8.1 – 590	38
Fort Worth	Barnett	Shale	H	3.7	1 – 7.3	26,495	30	21 – 40	11	530	240 – 4,200	11,957
			V	1.3	0.38 – 1.9	3,773	-	-	0	230	140 – 390	2,416
			D	1.2	0.48 – 1.6	96	-	-	0	210	79 – 410	481
Green River	Hilliard-Baxter-Mancos	Shale	H	1.7	1 – 5.6	2	-	-	0	-	-	0
	Lance	Tight	H	-	-	0	-	-	0	730	350 – 1,100	6

Basin	Formation	Resource type	Well type	Fracturing fluid (Mgal)			Flowback (% of fracturing fluid returned)			Long-term produced water rates (gpd)		
				Weighted average[a]	Range[b]	Data points[c]	Weighted average[a]	Range[b]	Data points[c]	Weighted average[a]	Range[a]	Data points[c]
Green River, cont.	Lance	Tight	V	1.5	0.82 – 3.9	37	3.3	0.88 – 50	38	610	410 – 840	61
			D	.97	0.65 – 2.1	881	12	1.8 – 40	187	650	420 – 1,100	2,787
	Mancos	Shale	H	15	1.8 – 24	24	3.1	0.063 – 17	8	770	-	26
			D	5.4	0.12 – 20	10	-	-	0	140	0.83 – 1,400	36
	Mesaverde	Tight	H	-	-	0	-	-	0	220	130 – 480	5
			V	0.16	0.13 – 0.22	21	18	6.3 – 43	15	440	120 – 780	33
			D	0.19	0.11 – 0.3	448	9.3	0.7 – 36	94	380	150 – 610	856
Illinois	New Albany	Shale	H	-	-	0	-	-	0	2,940	2,940 – 2,940	1
Michigan	Antrim	Shale	V	0.05	0.05 – 0.05	1	-	25 – 75	2	1,300	530 – 4,600	7
Permian	Avalon & Bone Spring	Shale	H	2.3	1.2 – 5.7	965	19	4.9 – 40	48	2,700	2,100 – 5,700	1,171
			V	0.4	0.07 – 1.3	21	-	-	0	2,000	1,000 – 4,800	68
			D	1.8	1.2 – 3.4	40	33	12 – 57	36	1,300	800 – 3,300	94

Appendix E – Produced Water Handling Supplemental Information

Appendix E – Produced Water Handling Supplemental Information

Basin	Formation	Resource type	Well type	Fracturing fluid (Mgal)			Flowback (% of fracturing fluid returned)			Long-term produced water rates (gpd)		
				Weighted average[a]	Range[b]	Data points[c]	Weighted average[a]	Range[b]	Data points[c]	Weighted average[a]	Range[a]	Data points[c]
Permian, cont.	Barnett-Woodford	Shale	H	2.1	0.5 – 4.5	2	-	-	0	-	-	0
	Delaware	Shale	H	1.3	0.42 – 3	85	79	9.7 – 230	20	9,400	5,000 – 29,000	232
			V	0.19	0.044 – 0.38	141	210	84 – 580	19	1,600	1,100 – 3,800	412
			D	0.26	0.15 – 0.4	47	-	-	0	4,500	2,400 – 5,700	90
	Devonian (TX)	Shale	H	0.47	0.091 – 5.5	43	-	-	0	1,700	630 – 2,700	325
			V	0.14	0.075 – 1	187	-	-	0	3,700	1,400 – 5,400	306
			D	0.11	0.037 – 0.13	11	-	-	0	2,400	250 – 12,000	40
	Morrow	Tight	V	-	-	0	-	-	0	130	41 – 290	7
		Tight	D	-	-	0	-	-	0	140	34 – 2,200	66
	Spraberry	Tight	H	1.3	0.069 – 6.5	29	-	-	0	1,000	420 – 3,800	41
			V	0.91	0.071 – 1.6	449	-	-	0	1,000	670 – 1,500	936
			D	1	0.06 – 1.5	16	-	-	0	1,200	660 – 2,500	42

Basin	Formation	Resource type	Well type	Fracturing fluid (Mgal)			Flowback (% of fracturing fluid returned)			Long-term produced water rates (gpd)		
				Weighted average[a]	Range[b]	Data points[c]	Weighted average[a]	Range[b]	Data points[c]	Weighted average[a]	Range[a]	Data points[c]
Permian, cont.	Trend Area	Tight	H	8.3	2.4 – 12	991	-	-	0	890	530 – 3,900	457
			V	1.1	0.58 – 1.9	8,733	-	-	0	780	690 – 920	15,494
			D	1	0.4 – 1.7	41	-	-	0	620	370 – 1,500	50
	Wolfcamp	Shale	H	6.7	1.4 – 12	1,775	16	12 – 23	12	3,500	450 – 15,000	1,237
			V	1.6	0.18 – 2.3	383	-	-	0	780	460 – 1,400	1,142
			D	1.8	0.17 – 3	12	-	-	0	1,700	750 – 3,600	170
Piceance & Uinta	Mesaverde	Tight	D	--	--	0	--	--	0	510	130 – 700	52
	Hermosa	Shale	D	--	--	0	--	--	0	47	27 – 260	21
Powder River	Mowry	Shale	H	2.5	0.76 – 7.4	15	15	4.3 – 580	14	450	61 – 2,100	16
San Juan	Dakota	Tight	V	0.16	0.061 – 0.34	85	1.6	-	22	75	35 – 490	81
			D	0.12	0.063 – 0.32	136	4.1	1.1 – 60	29	230	53 – 950	511
	Mesaverde	Tight	V	--	--	0	--	--	0	43	14 – 560	5
			D	--	--	0	--	--	0	21	15 – 180	49

Basin	Formation	Resource type	Well type	Fracturing fluid (Mgal)			Flowback (% of fracturing fluid returned)			Long-term produced water rates (gpd)		
				Weighted average[a]	Range[b]	Data points[c]	Weighted average[a]	Range[b]	Data points[c]	Weighted average[a]	Range[a]	Data points[c]
San Juan, cont.	Pictured Cliffs	Tight	H	--	--	0	--	--	0	370	190 – 720	7
			D	--	--	0	--	--	0	4,700	1,200 – 8,200	6
TX-LA-MS	Bossier	Shale	H	3.8	2.6 – 5.4	12	--	--	0	37	5.6 – 370	47
			V	0.61	0.22 – 1.7	82	--	--	0	230	4.8 – 480	1,143
			D	0.55	0.18 – 1.1	48	--	--	0	150	1.2 – 300	304
	Cotton Valley	Tight	H	4.4	0.25 – 8.5	433	60	60 – 60	1	710	410 – 2,600	689
			V	0.27	0.018 – 1.4	355	60	60 – 60	1	700	490 – 890	9,267
			D	0.45	0.046 – 4	79	60	60 – 60	1	620	240 – 980	1,912
	Haynesville	Shale	H	5.7	0.95 – 15	3,855	5.2	5.2 – 30	3	910	84 – 1,200	2,575
			V	0.9	0.2 – 2.5	2	--	--	0	330	210 – 560	230
			D	3.9	1.9 – 7.3	35	--	--	0	660	130 – 1,200	204
	Travis Peak	Tight	H	3	0.25 – 6	2	--	--	0	710	110 – 4,200	7
			V	0.17	0.032 – 4	36	--	--	0	630	270 – 930	1,046
			D	--	--	0	--	--	0	520	140 – 800	134

Basin	Formation	Resource type	Well type	Fracturing fluid (Mgal)			Flowback (% of fracturing fluid returned)			Long-term produced water rates (gpd)		
				Weighted average[a]	Range[b]	Data points[c]	Weighted average[a]	Range[b]	Data points[c]	Weighted average[a]	Range[a]	Data points[c]
TX-LA-MS, cont.	Tuscaloosa	Shale	H	11	6.1 – 14	28	--	--	0	--	--	0
Western Gulf	Austin Chalk	Tight	V	13	4.7 – 19	11	--	--	0	7,400	220 – 51,000	64
			H	1.7	0.83 – 5.4	134	--	--	0	2,200	980 – 5,100	752
			V	--	--	0	--	--	0	97	21 – 1,500	51
	Eagle Ford	Shale	H	4.8	1 – 14	12,810	4.2	2.1 – 8.4	1,800	1,900	88 – 6,200	7,971
			V	0.94	0.23 – 2	8	--	--	0	1,200	510 – 2,300	12
			D	--	--	0	--	--	0	4,300	3,000 – 5,600	5
	Edwards	Tight	H	--	--	0	--	--	0	2,300	1,000 – 24,000	266
			V	--	--	0	--	--	0	560	150 – 2,100	32
			D	--	--	0	--	--	0	160	69 – 290	6
	Olmos	Tight	H	1.9	0.37 – 6	246	--	--	0	180	13 – 700	229
			V	0.11	0.078 – 0.21	50	--	--	0	78	52 – 370	1,120
			D	--	--	--	--	--	0	51	15 – 470	16
	Pearsall	Shale	H	3.5	1.6 – 5.6	47	--	--	0	160	53 – 1,500	51

Basin	Formation	Resource type	Well type	Fracturing fluid (Mgal)			Flowback (% of fracturing fluid returned)			Long-term produced water rates (gpd)		
				Weighted average[a]	Range[b]	Data points[c]	Weighted average[a]	Range[b]	Data points[c]	Weighted average[a]	Range[a]	Data points[c]
Western Gulf, cont.	Vicksburg	Tight	V	0.21	0.072 – 0.61	158	--	--	0	700	330 – 990	702
			D	0.23	0.11 – 0.63	40	--	--	0	830	390 – 1,400	193
	Wilcox Lobo	Tight	H	0.33	0.082 – 2.4	8	--	--	0	370	250 – 610	84
			V	0.1	0.042 – 0.6	56	--	--	0	650	400 – 940	1,084
			D	0.094	0.058 – 0.16	14	--	--	0	500	300 – 4,200	395
Williston	Bakken	Shale	H	2.4	0.35 – 10	8,103	19	5 – 47	225	910	500 – 3,800	7,309
			V	0.16	0.04 – 2.7	6	--	--	0	2,400	150 – 5,100	5

"–" indicates no data; H, horizontal well; D, directional well; V, vertical well.

[a] For some formations, if only one data point was reported, the EPA reported it in the range column and did not report a median value.

[b] For some formations, the number of data points was not reported in the data source. In these instances, the EPA reported the number of data points as equal to one, even if the source reported a range and median value.

[c] For some formations, the number of data points was not reported in the data source. In these instances, this table reports that number as 1, except if the source reported a range in which case this table reports the number of data points as 2.

E.2.1.   Summary of Results from Produced Water Studies

Data were collected from six vertical and eight horizontal wells in the Marcellus Shale of Pennsylvania and West Virginia (Hayes, 2009). The author collected samples of flowback after one, five, and 14 days after hydraulic fracturing was completed, as well as a produced water sample 90 days after completion of the wells. Both the vertical and horizontal wells showed their largest volume of flowback between one and five days after fracturing, as shown in Figure E-1.

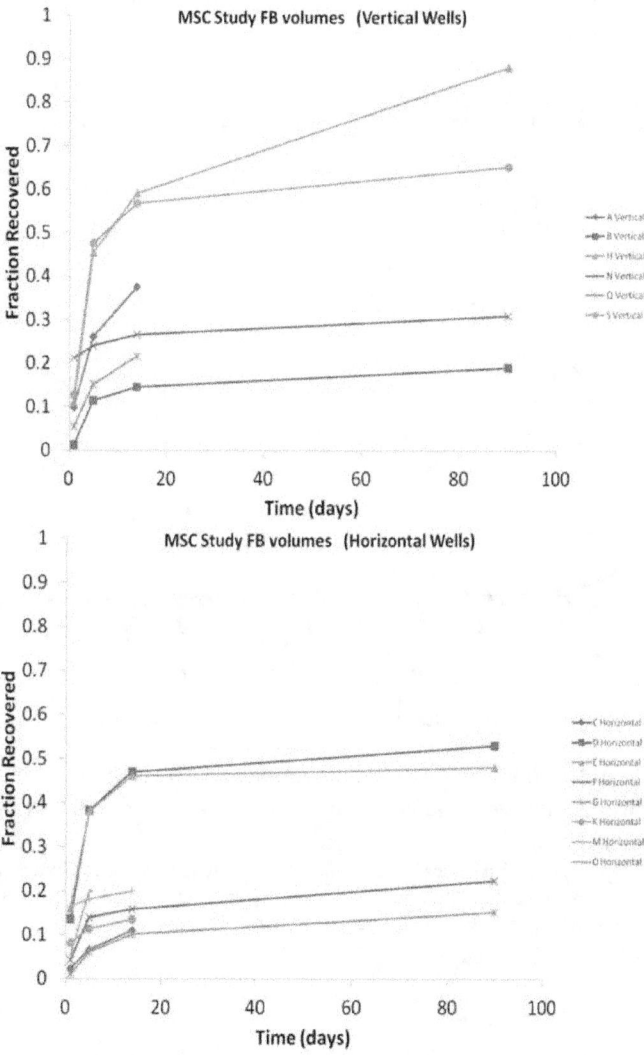

Figure E-1. Fraction of injected hydraulic fracturing fluid recovered from six vertical (top) and eight horizontal (bottom) wells completed in the Marcellus Shale.
Data used with permission from Hayes (2009).

The wells continued to produce water, and at 90 days, samples were available from four each of the horizontal and vertical wells. The vertical wells produced on average 7,600 gal/day (29,000 L/day) and the horizontal wells a similar 8,400 gal/day (32,000 L/day). Results from one Marcellus Shale study were fitted to a power curve (Ziemkiewicz et al., 2014) (Figure E-2). These and the Hayes (2009) data show decreasing rates of flowback with time. In West Virginia, water recovered at the surface within 30 days following injection or before 50% of the hydraulic fracturing fluid volume is returned to the surface is reported as flowback. Data from wells in the Marcellus Shale in West Virginia (Hansen et al., 2013) reveal the variability of recovery from wells in the same formation and that the amount of hydraulic fracturing fluid recovered was estimated to be less than 15% from over 80% of the wells (Figure E-3).

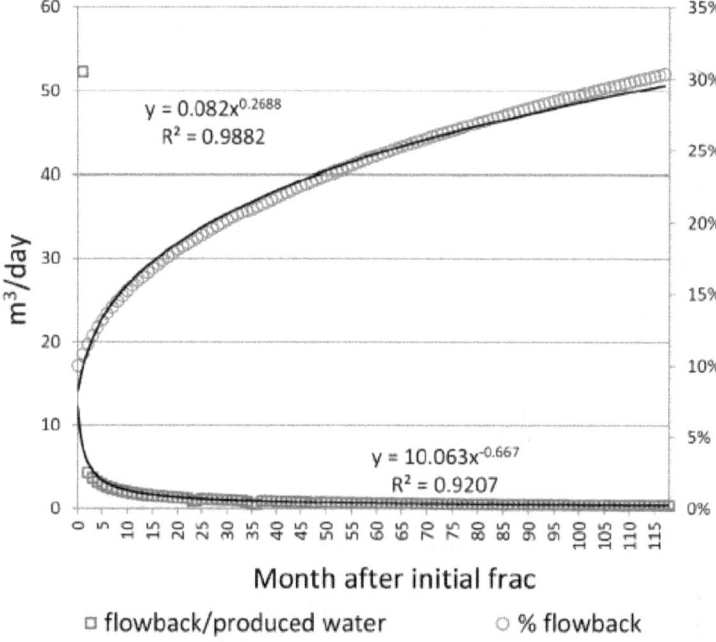

Figure E-2. Example of flowback and produced water from the Marcellus Shale, illustrating rapid decline in water production and cumulative return of approximately 30% of the volume of hydraulic fracturing fluid.

Source: Ziemkiewicz et al. (2014). Ziemkiewicz, P; Quaranta, JD; McCawley, M. (2014). Practical measures for reducing the risk of environmental contamination in shale energy production. Environ. Sci.: Processes & Impacts 16: 1692-1699. Reproduced with permission from The Royal Society of Chemistry. http:// dx.doi.org/10.1039/C3EM00510K.

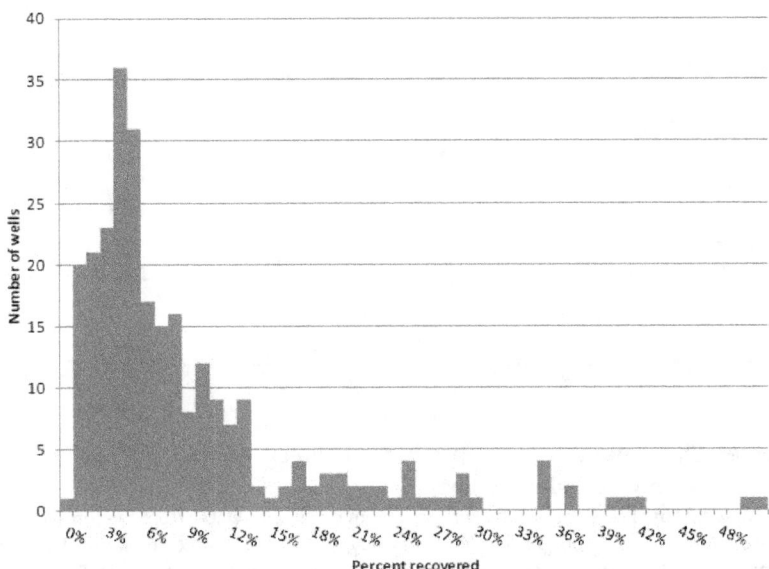

Figure E-3. Percent of hydraulic fracturing fluid recovered for Marcellus Shale wells in West Virginia (2010 – 2012).
One data point showing 98% recovery omitted. Source: Hansen et al. (2013). Reprinted with permission from Downstream Strategies, San Jose State University, and Earthworks Oil & Gas Accountability Project.

Nicot et al. (2014) show a counter-example where the produced water exceeded the amount of hydraulic fracturing fluid injected. When the produced water data were presented as the percentage of hydraulic fracturing fluid, the median exceeded 100% at around 36 months (Figure E-4 and Figure E-5). This means that roughly 50% of the wells were producing more water than was used in stimulating production. Nicot et al. (2014) did not identify the source or mechanism for the excess water. Systematic breaching of the underlying karstic Ellenburger Formation was not believed likely; nor was operator efficiency or skill. A number of geologic factors that could impact water migration were identified by DOE (2011) in the Barnett Shale, including fracture height, aperture size, and density, fracture mineralization, the presence of karst chimneys underlying parts of the Barnett Shale, and others, but the impact of these on water migration was not determined.

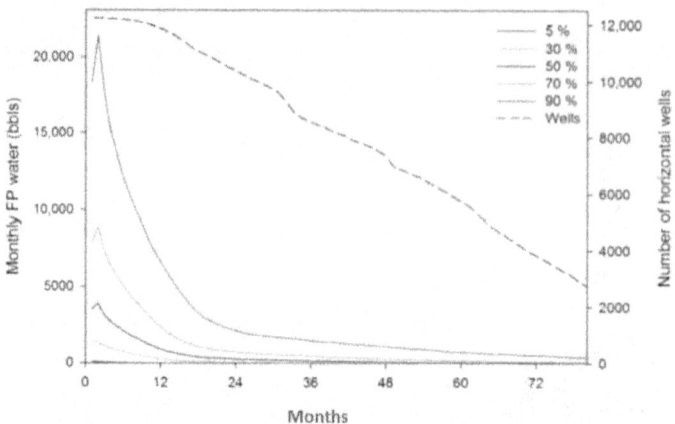

Figure E-4. Barnett Shale monthly water-production percentiles (5th, 30th, 50th, 70th, and 90th) and number of wells with data (dashed line).

FP is the amount of water that flows back to the surface, commingled with water from the formation. Reprinted with permission from Nicot, JP; Scanlon, BR; Reedy, RC; Costley, RA. (2014). Source and fate of hydraulic fracturing water in the Barnett Shale: A historical perspective [Supplemental Information]. Environ Sci Technol 48: 2464-2471. Copyright 2014 American Chemical Society.

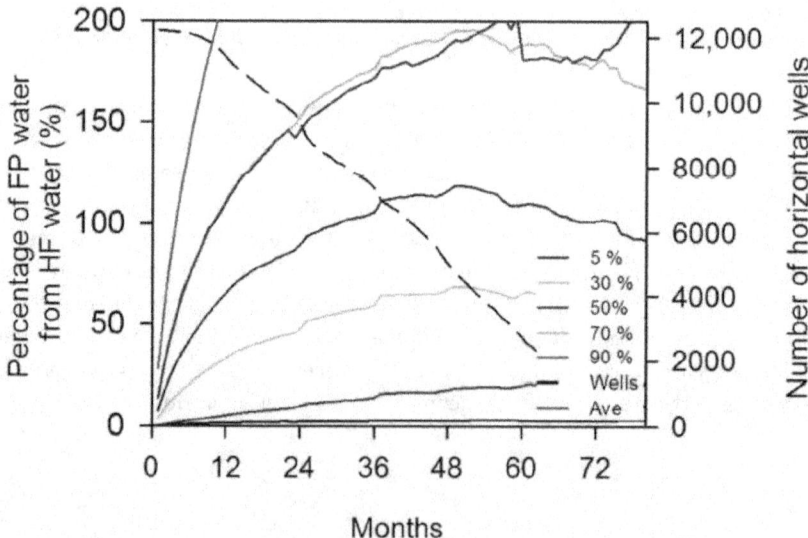

Figure E-5. Barnett Shale production data for approximately 72 months.

Flowback and produced water are reported as the percentage of hydraulic fracturing fluid. The dashed line shows the number of horizontal wells included. Data for each percentile show declining production with time, but the median production exceeds 100% of the hydraulic fracturing fluid. FP is the amount of water that flows back to the surface, commingled with water from the formation. Reprinted with permission from Nicot, JP; Scanlon, BR; Reedy, RC; Costley, RA. (2014). Source and fate of hydraulic fracturing water in the Barnett Shale: A historical perspective. Environ Sci Technol 48: 2464-2471. Copyright 2014 American Chemical Society.

E.3.   Chemical Content of Produced Water

In the main text of Chapter 7, we describe aspects of flowback and produced water composition, including temporal changes in water quality parameters of flowback (Section 7.3.3) and major classes of compounds in produced water (Section 7.3.4). In Section 7.3.4.2, we describe variability as occurring on three levels: between different rock types (e.g., coal vs. sandstone), between formations composed of the same rock types (e.g., Barnett Shale vs. Bakken Shale), and within formations of the same rock type (e.g., northeastern vs. southwestern Marcellus Shale). In this appendix, we present data from the literature that illustrate the differences among these three.

E.3.1.   General Water Quality Parameters

As noted in Section 7.3.4.3, the EPA identified data characterizing the content of flowback and produced water from unconventional reservoirs including 12 shale and tight formations and coalbed methane (CBM) basins. These formations and basins span 18 states. Note that in this subsection we treat all fluids as produced water. As a consequence, the variability of reported concentrations is likely higher than if the data could be standardized to a specific point on the flowback-to-produced water continuum. Table E-2 and Table E-3 provide supporting data on general water quality parameters of produced water in shale, tight formations, and coal seams for 12 formations.

Table E-2. Reported concentrations of general water quality parameters in produced water for unconventional shale and tight formations, presented as: average (minimum-maximum) or *median* (minimum-maximum).
Both averages and medians are reported because this table summarizes published information and authors differed in their use of averages or medians.

Parameter	Units	Shales					Tight formations			
		Bakken[a]	Barnett[b]	Fayetteville[c]	Marcellus		Cotton Valley Group[f]	Devonian Sandstone[g]	Mesaverde[f]	Oswego[f]
States	n/a	MT, ND	TX	AR	PA[d]	PA, WV[e]	LA, TX	PA	CO, NM, UT, WY	OK
Acidity	mg/L	–	NC (ND – ND)	–	NC (<5 – 473)	**162** (5 – 925)		–	–	–
Alkalinity	mg/L	–	**725** (215 – 1,240)	1,347 (5,507 – 1,896)	165 (8 – 577)	**99.8** (7.5 – 577)	–	99 (43 – 194)	–	582 (207 – 1,220)
Ammonium	mg/L	–	–	–	–	–	89 (40 – 131)	–	–	–
Bicarbonate	mg/L	291 (122 – 610)	–	–	–	–	–	524 (ND – 8,440)	2,230 (1,281 – 13,650)	–
Biochemical oxygen demand (BOD)	mg/L	–	**582** (101 – 2,120)	–	–	**141** (2.8 – 12,400)	–	–	–	–
Carbonate	mg/L	–	–	–	–	–	–	–	227 (ND–1,680)	–
Chloride	mg/L	119,000 (90,000 – 133,000)	**34,700** (9,600 – 60,800)	9,156 (5,507 – 12,287)	57,447 (64 – 196,000)	**49,000** (64.2 – 196,000)	101,332 (3,167 – 221,498.7)	132,567 (58,900 – 207,000)	4,260 (8 – 75,000)	44,567 (23,000 – 75,000)
Chemical oxygen demand	mg/L	–	**2,945** (927 – 3,150)	–	15,358 (195 – 36,600)	**4,670** (195 – 36,600)	–	–	–	–

Parameter	Units	Shales			Marcellus		Tight formations			
		Bakken[a]	Barnett[b]	Fayetteville[c]	PA[d]	PA, WV[e]	Cotton Valley Group[f]	Devonian Sandstone[g]	Mesaverde[f]	Oswego[f]
States	n/a	MT, ND	TX	AR	PA[d]	PA, WV[e]	LA, TX	PA	CO, NM, UT, WY	OK
DO	mg/L	-	-	-	-	-	-	0.8 (0.2 – 2.5)	-	-
DOC	mg/L	-	11.2 (5.5 – 65.3)	-	-	117 (3.3 – 5,960)	-	-	-	-
Hardness as CaCO₃	mg/L	-	**5,800** (3,500 – 21,000)	-	34,000 (630 – 95,000)	**25,000** (156 – 106,000)	-	-	-	-
Oil and grease	mg/L	-	**163.5** (88.2 – 1,430)	-	74 (5 – 802)	**16.85** (4.7 – 802)	-	-	-	-
pH	U	5.87 (5.47 – 6.53)	**7.05** (6.5 – 7.2)	-	6.6 (5.1 – 8.4)	**6.5** (4.9 – 7.9)	-	6.3 (5.5 – 6.8)	8 (5.8 – 11.62)	6.3 (6.1 – 6.4)
Specific conductivity	µS/cm	213,000 (205,000 – 220,800)	**111,500** (34,800 – 179,000)	-	-	**183,000** (479 – 763,000)	-	184,800 (118,000 – 211,000)	-	-
Specific gravity	--	1.13 (1.0961 – 1.155)	-	-	-	-	-	-	-	-
TDS	mg/L	196,000 (150,000 – 219,000)	**50,550** (16,400 – 97,800)	13,290 (9,972 – 15,721)	106,390 (680 – 345,000)	**87,800** (680 – 345,000)	164,683 (5,241 – 356,666)	235,125 (106,000 – 354,000)	15,802 (1,032 – 125,304)	73,082 (56,541 – 108,813)

Appendix E – Produced Water Handling Supplemental Information

			Shales					Tight formations			
		Bakken[a]	Barnett[b]	Fayetteville[c]	Marcellus		Cotton Valley Group[f]	Devonian Sandstone[g]	Mesaverde[f]	Oswego[f]
Parameter	Units									
States	n/a	MT, ND	TX	AR	PA[d]	PA, WV[e]	LA, TX	PA	CO, NM, UT, WY	OK
Total Kjeldahl nitrogen	mg/L	-	*171* (26 – 298)	-	-	*94.9* (5.6 – 312)	-	-	-	-
TOC	mg/L	-	*9.75* (6.2 – 36.2)	-	160 (1.2 – 1,530)	*89.2* (1.2 – 5,680)	198 (184 – 212)	-	-	-
Total suspended solids	mg/L	-	*242* (120 – 535)	-	352 (4 – 7,600)	*127* (6.8 – 3,220)	-	-	-	-
Turbidity	NTU	-	*239* (144 – 314)	-	-	126 (2.3 – 1,540)	-	-	-	-

n/a, not applicable; -, no value available; NC, not calculated; ND, not detected; SU= standard units; ***bolded italic*** numbers are medians

[a] Stepan et al. (2010). *n* = 3. Concentrations were calculated based on Stepan et al.'s raw data. Samples had charge balance errors of 1.74, -0.752, and -0.220%

[b] Hayes and Severin (2012a). *n* = 16. This data source reported concentrations without direct presentation of raw data.

[c] Warner et al. (2013a). *n* = 6. Concentrations were calculated based on Warner et al.'s raw data. Both flowback and produced water included.

[d] Barbot et al. (2013). n = 134 – 159. This data source reported concentrations without direct presentation of raw data.

[e] Hayes (2009). *n* = 31 –67. Concentrations were calculated based on Hayes's raw data. Both flowback produced water included. Non-detects and contaminated blanks omitted.

[f] Blondes et al. (2014). Cotton Valley Group, *n*=2; Mesa Verde, *n* = 1 – 407; Oswego, *n* = 4 – 30. Concentrations were calculated based on raw data presented in the U.S. Geological Survey (USGS) National Produced Water Database v2.0.

[g] Dresel and Rose (2010). *n* = 3 – 15. Concentrations were calculated based on Dresel and Rose's raw data.

Table E-3. Reported concentrations of general water quality parameters in produced water for coalbed basins, presented as: average (minimum-maximum).

Parameter	Units	Black Warrior[a]	Powder River[b]	Raton[b]	San Juan[b]
States	n/a	AL, MS	MT, WY	CO, NM	AZ, CO, NM, UT
Alkalinity	mg/L	355 (3 – 1,600)	1,384 (653 – 2,672)	1,107 (130 – 2,160)	3,181 (51 – 11,400)
Ammonium	mg/L	3.60 (0.16 – 8.91)	-	-	-
Bicarbonate	mg/L	427 (2 – 1,922)	1,080 (236 – 3,080)	1,124 (127 – 2,640)	3,380 (117 – 13,900)
Carbonate	mg/L	3 (0 – 64)	2.17 (0.00 – 139.0)	51.30 (1.30 – 316.33)	40.17 (0.00 – 1,178)
Chloride	mg/L	9,078 (11 – 42,800)	21 (BDL – 282)	787 (4.8 – 8,310)	624 (BDL – 20,100)
Chemical oxygen demand	mg/L	830 (0 – 10,500)	-	-	-
Dissolved oxygen	mg/L	-	1.07 (0.11 – 3.48)	0.39 (0.01 – 3.52)	0.51 (0.04 – 1.69)
DOC	mg/L	3.37 (0.53 – 61.41)	3.18 (1.09 – 8.04)	1.26 (0.30 – 8.54)	3.21 (0.89 – 11.41)
Hardness as CaCO$_3$	mg/L	871 (3 – 6,150)	-	-	-
Hydrogen sulfide	mg/L	-	-	4.41 (BDL – 190.0)	23.00 (23.00 – 23.00)
Oil and grease	mg/L	-	-	9.10 (0.60 – 17.6)	-
pH	SU	7.5 (5.3 – 9.0)	7.71 (6.86 – 9.16)	8.19 (6.90 – 9.31)	7.82 (5.40 – 9.26)
Phosphate	mg/L	0.435 (0.026 – 3.570)	BDL (BDL – BDL)	0.04 (BDL – 1.00)	1.89 (BDL – 9.42)
Specific conductivity	μS/cm	20,631 (718 – 97,700)	1,598 (413 – 4,420)	3,199 (742 – 11,550)	5,308 (232 – 18,066)
TDS	mg/L	14,319 (589 – 61,733)	997 (252 – 2,768)	2,512 (244 – 14,800)	4,693 (150 – 39,260)
Total Kjeldahl nitrogen	mg/L	6.08 (0.15 – 38.40)	0.48 (BDL – 4.70)	2.61 (BDL – 26.10)	0.46 (BDL – 3.76)
TOC	mg/L	6.03 (0.00 – 103.00)	3.52 (2.07 – 6.57)	1.74 (0.25 – 13.00)	2.91 (0.95 – 9.36)
Total suspended solids	mg/L	78 (0 – 2,290)	11.0 (1.4 – 72.7)	32.3 (1.0 – 580.0)	47.2 (1.4 – 236.0)
Turbidity	NTU	74 (0 – 539)	8.2 (0.7 – 57.0)	4.5 (0.3 – 25.0)	61.6 (0.8 – 810.0)

n/a, not applicable; -, no value available; BDL, below detection limit.

[a] DOE (2014). *n* = 206. Concentrations were calculated based on raw data presented in the reference.

[b] Dahm et al. (2011). Powder River, *n* = 31; Raton, *n* = 40; San Juan, *n* = 20. This data source reported concentrations without presentation of raw data.

E.3.2. Salinity and Inorganics

Table E-4 and Table E-5 provide supporting data on salinity and inorganic constituents of produced water for 12 formations.

E.3.2.1. Processes Controlling Salinity and Inorganics Concentrations

Multiple mechanisms likely control elevated salt concentrations in flowback and produced water and are largely dependent upon post-injection fluid interactions and the formation's stratigraphic and hydrogeologic environment (Barbot et al., 2013). High inorganic ionic loads observed in flowback and produced water are expressed as TDS.

Subsurface brines or formation waters are saline fluids associated with the targeted formation. Shale and sandstone brines are typically much more saline than coalbed waters. After hydraulic fracturing fluids are injected into the subsurface, the hydraulic fracturing fluids (which are typically not sources of high TDS) contact in-situ brines, which typically contain high ionic loads (Haluszczak et al., 2013).

Deep brines, present in over- or underlying strata, may naturally migrate into targeted formations over geologic time or artificially intrude if a saline aquifer is breached during hydraulic fracturing (Chapman et al., 2012; Maxwell, 2011; Blauch et al., 2009). Whether it is through natural or induced intrusion, saline fluids may contact the producing formation and introduce novel salinity sources to the produced water (Chapman et al., 2012). Despite the general use of fresh water for hydraulic fracturing fluid, some elevated salts in produced water may result from the use of reused saline flowback or produced water as a hydraulic fracturing base fluid (Hayes, 2009).

Table E-4. Reported concentrations (mg/L) of inorganic constituents contributing to salinity in produced water from unconventional reservoirs (including shale and tight formations), presented as: average (minimum-maximum) or *median* (minimum-maximum).

Both averages and medians are reported because this table summarizes published information and authors differed in their use of averages or medians.

Parameter	Shale					Tight Formations			
	Bakken[a]	Barnett[b]	Fayetteville[c]	Marcellus		Cotton Valley Group[f]	Devonian Sandstone[g]	Mesaverde[f]	Oswego[f]
				PA[d]	PA, WV[e]				
States	MT, ND	TX	AR			LA, TX	PA	CO, NM, UT, WY	OK
Bromide	-	589 (117 – 798)	111 (96 – 144)	511 (0.2 – 1,990)	*512 (15.8 – 1,990)*	498 (32 – 1,338)	1,048 (349 – 1,350)	-	-
Calcium	9,680 (7,540 – 13,500)	*1,600 (1,110 – 6,730)*	317 (221 – 386)	7,220 (38 – 41,000)	7,465 (173 – 33,000)	19,998 (181 – 51,400)	20,262 (8,930 – 34,400)	212 (1.01 – 4,580)	5,903 (3,609 – 8,662)
Chloride	119,000 (90,000 – 133,000)	*34,700 (9,600 – 60,800)*	9,156 (5,507 – 12,287)	57,447 (64 – 196,000)	*49,000 (64.2 – 196,000)*	101,332 (3,167 – 221,498.7)	132,567 (58,900 – 207,000)	4,260 (8 – 75,000)	44,567 (23,000 – 75,000)
Fluoride	-	*3.8 (3.5 – 12.8)*	-	-	*0.975 (0.077 – 32.9)*	-	-		-
Iodine	-	-	-	-	-	20 (1 – 36)	39 (11 – 56)	1.01 (1.01 – 1.01)	-
Nitrate as N	-	-	NC (ND – ND)	-	1.7 (0.65 – 15.9)	-	-	0.6 (0.6 – 0.6)	-
Nitrite as N	-	4.7 (3.5 – 38.1)	-	-	11.8 (1.1 – 146)	-	-	-	-
Phosphorus	NC (ND – 0.03)	*0.395 (0.19 – 0.7)*	-	-	*0.3 (0.08 – 21.8)*	-	-	-	-
Potassium	2,970 (0 – 5,770)	316 (80 – 750)	-	-	337 (38 – 3,950)	1,975 (8 – 7,099)	858 (126 – 3,890)	160 (4 – 2,621)	-

Appendix E – Produced Water Handling Supplemental Information

Parameter		Shale					Tight Formations			
		Bakken[a]	Barnett[b]	Fayetteville[c]	Marcellus		Cotton Valley Group[f]	Devonian Sandstone[g]	Mesaverde[f]	Oswego[f]
States		MT, ND	TX	AR	PA[d]	PA, WV[e]	LA, TX	PA	CO, NM, UT, WY	OK
Silica		7 (6.41 – 7)	-	52 (13 – 160)	-	-	4 (4 – 4)	-	-	-
Sodium		61,500 (47,100 – 74,600)	18,850 (4,370 – 28,200)	3,758 (3,152 – 4,607)	21,123 (69 – 117,000)	21,650 (63.8 – 95,500)	39,836 (1,320 – 85,623.24)	58,160 (24,400 – 83,300)	5,828 (132 – 48,817)	19,460 (13,484 – 31,328)
Sulfate		660 (300 – 1,000)	709 (120 – 1,260)	NC (ND – 3)	71 (0 – 763)	58.9 (2.4 – 348)	407 (ND – 2,200.46)	20 (1 – 140)	837 (ND – 14,612)	183 (120 – 271)
Sulfide		-	NC (ND – ND)	-	-	3.2 (1.6 – 5.6)	-	0.7 (0.1 – 2.5)	-	-
Sulfite		-	-	-	-	12.4 (5.2 – 73.6)	-	-	-	-
TDS		196,000 (150,000 – 219,000)	50,550 (16,400 – 97,800)	13,290 (9,972 – 15,721)	106,390 (680 – 345,000)	87,800 (680 – 345,000)	164,683 (5,241 – 356,666)	235,125 (106,000 – 354,000)	15,802 (1,032 – 125,304)	73,082 (56,541 – 108,813)

-, no value available; NC, not calculated; ND, not detected. Bolded italic numbers are medians.

[a] Stepan et al. (2010). n = 3. Concentrations were calculated based on Stepan et al.'s raw data. Samples had charge balance errors of 1.74, -0.752, and -0.220% without presentation of raw data.

[b] Hayes and Severin (2012a). n = 16. This data source reported concentrations without presentation of raw data.

[c] Warner et al. (2013b). n = 6. Concentrations were calculated based on Warner et al.'s raw data. Both flowback and produced water included.

[d] Barbot et al. (2013). n = 95 – 159. This data source reported concentrations without presentation of raw data.

[e] Hayes (2009). n = 8-65. Concentrations were calculated based on Hayes's raw data. Both flowback and produced water included. Non-detects and contaminated blanks omitted.

[f] Blondes et al. (2014) Cotton Valley Group, n = 2; Mesa Verde, n = 1 – 407; Oswego, n = 4 – 30. Concentrations were calculated based on raw data presented in the USGS National Produced Water Database v2.0.

[g] Dresel and Rose (2010). n = 3 – 15. Concentrations were calculated based on Dresel and Rose's raw data.

Table E-5. Reported concentrations (mg/L) of inorganic constituents contributing to salinity in produced water for coalbed methane basins, presented as: average (minimum-maximum).

Parameter	Black Warrior[a]	Powder River[b]	Raton[b]	San Juan[b]
State	AL, MS	MT, WY	CO, NM	AZ, CO, NM, UT
Barium	45.540 (0.136 – 352)	0.61 (0.14 – 2.47)	1.67 (BDL – 27.40)	10.80 (BDL – 74.0)
Boron	0.185 (0 – 0.541)	0.17 (BDL – 0.39)	0.36 (BDL – 4.70)	1.30 (0.21 – 3.45)
Bromide	-	0.09 (BDL – 0.26)	4.86 (0.04 – 69.60)	9.77 (BDL – 43.48)
Calcium	218 (0 – 1,640)	32.09 (2.00 – 154.0)	14.47 (0.81 – 269.0)	53.29 (1.00 – 5,530)
Chloride	9,078 (11 – 42,800)	21 (BDL – 282)	787 (4.8 – 8,310)	624 (BDL – 20,100)
Fluoride	6.13 (0.00 – 22.60)	1.57 (0.40 – 4.00)	4.27 (0.59 – 20.00)	1.76 (0.58 – 10.00)
Magnesium	68.12 (0.18 – 414.00)	14.66 (BDL – 95.00)	3.31 (0.10 – 56.10)	15.45 (BDL – 511.0)
Nitrate	8.70 (0.00 – 127.50)	-	-	-
Nitrite	0.03 (0.00 – 2.08)	-	-	-
Phosphorus	0.32 (0.00 – 5.76)	-	-	-
Potassium	12.02 (0.46 – 74.00)	11.95 (BDL – 44.00)	6.37 (BDL – 29.40)	26.99 (BDL – 970.0)
Silica	8.66 (1.04 – 18.10)	6.46 (4.40 – 12.79)	7.05 (4.86 – 10.56)	12.37 (3.62 – 37.75)
Sodium	4,353 (126 – 16,700)	356 (12 – 1,170)	989 (95 – 5,260)	1,610 (36 – 7,834)
Strontium	11.354 (0.015 – 142.000)	0.60 (0.10 – 1.83)	5.87 (BDL – 47.90)	5.36 (BDL – 27.00)
Sulfate	5.83 (0.00 – 302.00)	5.64 (BDL – 300.0)	14.75 (BDL – 253.00)	25.73 (BDL – 1,800)
TDS	14,319 (589 – 61,733)	997 (252 – 2,768)	2,512 (244 – 14,800)	4,693 (150 – 39,260)

-, no value available; BDL, below detection limit.

[a] DOE (2014). n = 206. Concentrations were calculated based on the authors' raw data.

[b] Dahm et al. (2011). Powder River, n = 31; Raton, n = 40; San Juan, n = 20. This data source reported concentrations without presentation of raw data.

E.3.3. Metals and Metalloids

Table E-6 and Table E-7 provide supporting data on metal constituents of produced water for 12 formations.

E.3.3.1. Processes Controlling Mineral Precipitation and Dissolution

Hydraulic fracturing treatments introduce fluids into the subsurface that are not in equilibrium with respect to formation mineralogy. Subsurface geochemical equilibrium modeling and saturation indices are therefore used to assess the solution chemistry of produced water from unconventional reservoirs and the subsequent likelihood of precipitation and dissolution reactions (Engle and Rowan, 2014; Barbot et al., 2013). Dissolution and precipitation reactions between fracturing fluids, formation solids, and formation water contribute to the chemistry of flowback and produced water.

Depending upon the formation chemistry and composition of the hydraulic fracturing fluid, the hydraulic fracturing fluid may initially have a lower ionic strength than existing formation fluids. Consequently, salts, carbonate, sulfate, and silicate minerals may undergo dissolution or precipitation. Proppants may also undergo dissolution or serve as nucleation sites for precipitation (McLin et al., 2011).

Currently, relatively little literature quantitatively explores subsurface dissolution and precipitation reactions between hydraulic fracturing fluids and formation solids and water. However, the processes that take place will likely be a function of the solubilities of the minerals, the chemistry of the fluid, pH, redox conditions, and temperature.

Documented dissolution processes in unconventional reservoirs include the dissolution of feldspar followed by sodium enrichment in coalbed produced water (Rice et al., 2008). Dissolution of barium-rich minerals (barite ($BaSO_4$) and witherite ($BaCO_3$)), and strontium-rich minerals (celestite ($SrSO_4$) and strontianite ($SrCO_3$)) are known to enrich shale produced waters in barium and strontium (Chapman et al., 2012).

Known precipitation processes in unconventional reservoirs include the precipitation of carbonate and subsequent reduction of calcium and magnesium concentrations in coalbed produced water (Rice et al., 2008). Additionally, calcium carbonate precipitation is suspected to cause declines in pH and alkalinity levels in shale produced water (Barbot et al., 2013).

The subsurface processes associated with fluid-rock interactions take place over a scale of weeks to months through the generation of flowback and produced water. Note that the types and extent of subsurface dissolution and precipitation reactions change with time, from injection through flowback and production. For instance, Engle and Rowan (2014) found that early Marcellus Shale flowback was under-saturated with respect to gypsum ($CaSO_4 \cdot 2H_2O$), halite (NaCl), celestite, strontianite, and witherite, indicating that these minerals would dissolve in the subsurface. Fluids were oversaturated with respect to barite. Saturation indices for gypsum, halite, celestite, and barite all increased during production. Knowing when dissolution and precipitation will likely occur is important, because dissolution and precipitation of minerals change formation permeability and porosity, which can affect production (André et al., 2006).

Table E-6. Reported concentrations (mg/L) of metals and metalloids from produced water from unconventional reservoirs (including shale and tight formations), presented as: average (minimum-maximum) or *median* (minimum-maximum).

Both averages and medians are reported because this table summarizes published information and authors differed in their use of averages or medians. Note that calcium, potassium, and sodium appear in Table E-4.

Parameter	Shale					Tight Formation			
	Bakken[a]	Barnett[b]	Fayetteville[c]	Marcellus		Cotton Valley Group[f]	Devonian Sandstone[g]	Mesaverde[f]	Oswego[f]
				PA[d]	PA, WV[e]				
States	MT, ND	TX	AR	-	PA, WV	LA, TX	PA	CO, NM, UT, WY	OK
Aluminum	-	*0.43* (0.37 – 2.21)	-	-	*2.57* (0.22 – 47.2)	-	-	-	-
Antimony	-	NC (ND – ND)	-	-	*0.028* (0.018 – 0.038)	-	-	-	-
Arsenic	-	NC (ND – ND)	-	-	*0.101* (0.013 – 0.124)	-	-	-	-
Barium	10 (0 – 24.6)	*3.6* (0.93 – 17.9)	4 (3 – 5)	2,224 (0.24 – 13,800)	*542.5* (2,590 – 13,900)	160 (ND – 400.52)	1,488 (7 – 4,370)	139 (4 – 257)	-
Beryllium	-	NC (ND – ND)	-	-	-	-	-	-	-
Boron	116 (39.9 – 192)	*30.3* (7.0 – 31.9)	4.800 (2.395 – 21.102)	-	*12.2* (0.808 – 145)	37 (2 – 100)	-	10 (1 – 14.2)	-
Cadmium	-	NC (ND – ND)	-	-	-	-	-	-	-
Chromium	-	*0.03* (0.01 – 0.12)	-	-	*0.079* (0.011 – 0.567)	-	-	-	-
Cobalt	-	*0.01* (0.01 – 0.01)	-	-	-	-	-	-	-

Appendix E – Produced Water Handling Supplemental Information

Parameter		Shale					Tight Formation			
		Bakken[a]	Barnett[b]	Fayetteville[c]	Marcellus		Cotton Valley Group[f]	Devonian Sandstone[g]	Mesaverde[f]	Oswego[f]
					PA[d]	PA, WV[e]				
States		MT, ND	TX	AR			LA, TX	PA	CO, NM, UT, WY	OK
Copper		NC (ND – 0.21)	0.29 (0.06 – 0.52)	-	-	0.506 (0.253 – 4.150)	0.7 (0.48 – 1)	0.04 (0.01 – 0.13)	-	-
Iron		96 (ND – 120)	24.9 (12.1 – 93.8)	7 (1 – 13)	-	53.65 (2.68 – 574)	-	188 (90 – 458)	9 (1 – 29)	61 (41 – 78)
Lead		-	0.02 (0.01 – 0.02)	-	-	0.066 (0.003 – 0.970)	-	0.02 (0.01 – 0.04)	-	-
Lithium		-	19.0 (2.56 – 37.4)	9.825 (2.777 – 28.145)	-	53.85 (3.410 – 323)	23 (1 – 53)	97.8 (20.2 – 315)	3 (1 – 33)	-
Magnesium		1,270 (630 – 1,750)	255 (149 – 755)	61 (47 – 75)	632 (17 – 2,550)	678 (40.8 – 2,020)	1,363 (27 – 3,712.98)	2,334 (797 – 3,140)	74 (1 – 2,394)	753 (486 – 1,264)
Manganese		7 (4 – 10.2)	0.86 (0.25 – 2.20)	2 (2 – 3)	-	2.825 (0.369 – 18.600)	30.33 (30.33 – 30.33)	19 (5.6 – 68)	-	-
Mercury		-	NC (ND – ND)	-	-	0.00024	-	-	-	-
Molybdenum		NC (ND – <0.2)	0.02 (0.02 – 0.03)	-	-	-	-	-	-	-
Nickel		-	0.04 (0.03 – 0.05)	-	-	0.419 (0.068 – 0.769)	-	-	-	-
Selenium		-	0.03 (0.03 – 0.04)	-	-	0.004	-	-	-	-

Parameter	Shale					Tight Formation			
	Bakken[a]	Barnett[b]	Fayetteville[c]	Marcellus		Cotton Valley Group[f]	Devonian Sandstone[g]	Mesaverde[f]	Oswego[f]
				PA[d]	PA, WV[e]				
States	MT, ND	TX	AR	PA[d]	PA, WV[e]	LA, TX	PA	CO, NM, UT, WY	OK
Silver	-	-	-	-	**4** (3 – 6)	-	-	-	-
Strontium	764 (518 – 1,010)	**529** (48 – 1,550)	27 (14 – 49)	1,695 (0.6 – 8,460)	**1,240** (0.580 – 8,020)	2,312 (39 – 9,770)	3,890 (404 – 13,100)	-	-
Thallium	-	NC (ND – 0.14)	-	-	**0.168**	-	-	-	-
Tin	-	NC (ND – ND)	-	-	-	-	-	-	-
Titanium	-	**0.02** (0.02 – 0.03)	-	-	-	-	-	-	-
Zinc	7 (2 – 11.3)	**0.15** (0.10 – 0.36)	-	-	**0.391** (0.087 – 247)	-	0.20 (0.03 – 1.26)	-	-

-, no value available; NC, not calculated; ND, not detected; BDL, below detection limit. ***Bolded Italic*** numbers are medians.

[a] Stepan et al. (2010). n = 3. Concentrations were calculated based on Stepan et al.'s raw data.

[b] Hayes and Severin (2012a). n = 16. This data source reported concentrations without presentation of raw data.

[c] Warner et al. (2013a). n = 6. Concentrations were calculated based on Warner et al.'s raw data. Both flowback and produced water included.

[d] Barbot et al. (2013). n = 151 – 159. This data source reported concentrations without presentation of data.

[e] Hayes (2009). n = 48. Concentrations were calculated based on Hayes's raw data. Both flowback and produced water included. Non-detects and contaminated blanks omitted.

[f] Blondes et al. (2014). Cotton Valley Group, n = 2; Mesa Verde, n = 1 – 407; Oswego, n = 2; Mesa Verde, n = 2; Oswego, n = 4 – 30. Concentrations were calculated based on raw data presented in the USGS National Produced Water Database v2.0.

[g] Dresel and Rose (2010). n = 3 – 15. Concentrations were calculated based on Dresel and Rose's raw data.

Table E-7. Reported concentrations (mg/L) of metals and metalloids from produced water from coalbed methane, presented as: average (minimum-maximum).

Parameter	Black Warrior[a]	Powder River[b]	Raton[b]	San Juan[b]
States	AL, MS	MT, WY	CO, NM	AZ, CO, NM, UT
Aluminum	0.037 (0 – 0.099)	0.018 (BDL – 0.124)	0.193 (BDL – 2,900)	0.069 (BDL – 0.546)
Antimony	0.006 (0.00 – 0.022)	BDL (BDL – BDL)	BDL (BDL – BDL)	BDL (BDL – BDL)
Arsenic	0.002 (0.0 – 0.085)	0.001 (BDL – 0.004)	0.010 (BD – 0.060)	0.001 (BDL – 0.020)
Barium	45.540 (0.136 – 352)	0.61 (0.14 – 2.47)	1.67 (BDL – 27.40)	10.80 (BDL – 74.0)
Beryllium	0.0 (0.0 – 0.008)	BDL (BDL – BDL)	BDL (BDL – BDL)	BDL (BDL – BDL)
Boron	0.185 (0 – 0.541)	0.17 (BDL – 0.39)	0.36 (BDL – 4.70)	1.30 (0.21 – 3.45)
Cadmium	0.001 (0.00 – 0.015)	BDL (BDL – 0.002)	0.002 (BDL – 0.003)	0.002 (BDL – 0.006)
Calcium	218 (0 – 1,640)	32.09 (2.00 – 154.0)	14.47 (0.81 – 269.0)	53.29 (1.00 – 5,530)
Cesium	0.011 (0.0 – 0.072)	-	-	-
Chromium	0.002 (0.0 – 0.351)	0.012 (BDL – 0.250)	0.105 (BDL – 3.710)	0.002 (BDL – 0.023)
Cobalt	0.023 (0.00 – 0.162)	BDL (BDL – BDL)	0.001 (BDL – 0.018)	0.001 (BDL – 0.017)
Copper	0.001 (0.0 – 0.098)	0.078 (BDL – 1.505)	0.091 (BDL – 4.600)	0.058 (BDL – 0.706)
Iron	8.956 (0.045 – 93.100)	1.55 (BDL – 190.0)	7.18 (0.09 – 95.90)	6.20 (BDL – 258.0)
Lead	0.008 (0.00 – 0.250)	BDL (BDL – BDL)	0.023 (BDL – 0.233)	0.023 (BDL – 0.390)
Lithium	1.157 (0 – 8.940)	0.13 (BDL – 0.34)	0.32 (0.01 – 1.00)	1.61 (0.21 – 4.73)
Magnesium	68.12 (0.18 – 414.00)	14.66 (BDL – 95.00)	3.31 (0.10 – 56.10)	15.45 (BDL – 511.0)
Manganese	0.245 (0.006 – 4.840)	0.02 (BDL – 0.16)	0.11 (0.01 – 2.00)	0.19 (BDL – 1.34)
Mercury	0.000 (0.000 – 0.000)	-	-	-
Molybdenum	0.002 (0 – 0.083)	0.005 (BDL – 0.029)	0.002 (BDL – 0.035)	0.020 (BDL – 0.040)
Nickel	0.015 (0.0 – 0.358)	0.141 (BDL – 2.61)	0.015 (0.004 – 0.11)	0.020 (BDL – 0.13)
Potassium	12.02 (0.46 – 74.00)	11.95 (BDL – 44.00)	6.37 (BDL – 29.40)	26.99 (BDL – 970.0)
Rubidium	0.013 (0.0 – 0.114)	-	-	-
Selenium	0.002 (0.00 – 0.063)	0.006 (BDL – 0.046)	0.017 (BDL – 0.100)	0.018 (BDL – 0.067)
Silver	0.015 (0.0 – 0.565)	0.003 (0.003 – 0.003)	0.015 (BDL – 0.140)	BDL (BDL – BDL)
Sodium	4,353 (126 – 16,700)	356 (12 – 1,170)	989 (95 – 5,260)	1,610 (36 – 7,834)
Strontium	11.354 (0.015 – 142.000)	0.60 (0.10 – 1.83)	5.87 (BDL – 47.90)	5.36 (BDL – 27.00)
Thallium	-	-	-	-

Parameter	Black Warrior[a]	Powder River[b]	Raton[b]	San Juan[b]
States	AL, MS	MT, WY	CO, NM	AZ, CO, NM, UT
Tin	0.00 (0.00 – 0.009)	0.006 (BDL – 0.028)	0.008 (BDL – 0.021)	0.017 (BDL – 0.039)
Titanium	0.003 (0.0 – 0.045)	BDL (BDL – 0.002)	BDL (BDL – 0.002)	0.004 (BDL – 0.020)
Vanadium	0.001 (0.0 – 0.039)	BDL (BDL – BDL)	0.001 (BDL – 0.013)	BDL (BDL – BDL)
Zinc	0.024 (0.0 – 0.278)	0.063 (BDL – 0.390)	0.083 (0.010 – 3.900)	0.047 (0.005 – 0.263)

-, no value available; BDL, below detection limit.

[a] DOE (2014). n = 206. Concentrations were calculated based on the authors' raw data.

[b] Dahm et al. (2011). Powder River, n = 31; Raton, n = 40; San Juan, n = 20. This data source reported concentrations without presentation of raw data.

E.3.4. Naturally Occurring Radioactive Material (NORM) and Technically Enhanced Naturally Occurring Radioactive Material (TENORM)

E.3.4.1. Produced Water Levels of TENORM

Background data on TENORM in the Marcellus Shale and Devonian sandstones are given in Table E-8.

E.3.4.2. Mobilization of Naturally Occurring Radioactive Material

In oil and gas production in both conventional and unconventional reservoirs, radionuclides native to the targeted formation return to the surface with produced water. The principal radionuclides found in oil and gas produced waters include radium-226 of the uranium-238 decay series and radium-228 of the thorium-232 decay series (White, 1992). Levels of TENORM in produced water are controlled by geologic and geochemical interactions between injected and formation fluids, and the targeted formation (Bank, 2011). Mechanisms controlling NORM mobilization into produced water include (1) the TENORM content of the targeted formation; (2) factors governing the release of radionuclides, particularly radium, from the reservoir matrix; and (3) the geochemistry of the produced water (Choppin, 2007, 2006; Fisher, 1998).

Elevated uranium levels in formation solids have been used to identify potential areas of natural gas production for decades (Fertl and Chilingar, 1988). Marine black shales are estimated to contain 3 – 250 ppm uranium depending on depositional conditions (USGS, 1961). Shales that bear significant levels of uranium include the Barnett in Texas, the Woodford in Oklahoma, the New Albany in the Illinois Basin, the Chattanooga Shale in the southeastern United States, and a group of black shales in Kansas and Oklahoma (Swanson, 1955).

Bank et al. (2012) identified Marcellus samples with uranium ranging from 4 – 72 ppm, with an average of 30 ppm. Chermak and Schreiber (2014) compiled mineralogy and trace element data available in the literature for nine U.S. hydrocarbon-producing shales. In this combined data set, uranium levels among different shale plays were found to vary over three orders of magnitude, with samples of the Utica Shale containing approximately 0 – 5 ppm uranium and samples of the Woodford Shale containing uranium in the several-hundred-ppm range.

Table E-8. Reported concentrations (in pCi/L) of radioactive constituents in produced water in unconventional reservoirs (including shale and tight sandstones), presented as: average (minimum-maximum) or *median* (minimum-maximum).
Both averages and medians are reported because this table summarizes published information and authors differed in their use of averages or medians.

Parameter		Marcellus						Devonian Sandstone[a]
			PA NORM Study (PA DEP, 2015)					
States	NY, PA[b]	Fracturing Fluid[c]	Flowback[d]	Produced Water, Conventional Reservoirs[e]	Produced Water, Unconventional Reservoirs[f]	WV[g]		PA
Gross alpha	*6,845* (ND – 123,000)	5,020 (0.695 – 54,100)	*10,700* (288 – 71,000)	*1,835* (465 – 2,570)	*11,300* (2,400 – 41,700)	5,866 (1.84 – 20,920)		-
Gross beta	*1,170* (ND – 12,000)	*1,010* (0.815 – 14,900)	*2,400* (742 – 21,300)	*909* (402 – 1,140)	*3,445* (1,500 – 7,600)	1,172 (9.6 – 4,664)		-
Radium-226	*1,869* (ND – 16,920)	2,160 (64.0 – 21,000)	*4,500* (551 – 25,500)	*243* (81 – 819)	*6,300* (1,700 – 26,600)	358 (15.4 – 1,194)		*2,367* (200 – 5,000)
Radium-228	*557* (ND – 2,589)	218 (4.5 – 1,640)	*633* (248 – 1,740)	*128* (26 – 896)	*941* (366 – 1,900)	94.6 (4.99 – 216)		-
Total Radium	*2,530* (0.192 – 18,045)	283 (10.5 – 456)	-	*371* (107 – 1,715)	*7,180* (2,336 – 28,500)			-
Potassium^{40}			*461* (88.5 – 2,630)			62.44 (nd – 221)		
Thorium^{230}						2.13 (0 – 9.37)		
Thorium^{232}						0.07 (0 – 0.38)		
Uranium^{235}	1 (ND – 20)			-	-			-
Uranium^{238}	42 (ND – 497)			-	-	0.34		-

n/a, not applicable; -, no value available; BDL, below detection limit. ***Bolded italic*** numbers are medians.
[a] Dresel and Rose (2010). n = 3. Concentrations presented were calculated based on Dresel and Rose's raw data.
[b] Rowan et al. (2011). n = 51 total radium; n = 30 gross beta. Concentrations presented were calculated based on Rowan et al.'s raw data for Marcellus samples. Uranium data from Barbot et al. (2013) n = 14.
[c] PA DEP (2015). n = 11. Data reported in Table 3-13 of the referenced paper.
[d] PA DEP (2015). n = 9. Data reported in Table 3-14 of the referenced paper.
[e] PA DEP (2015). n = 9. Values calculated from Table 3-15 for unfiltered samples of the referenced paper.
[f] PA DEP (2015). n = 4. Values calculated from Table 3-15 for unfiltered samples of the referenced paper.
[g] Ziemkiewicz and He (2015). n = 5. Data reported in Table 1 of the referenced paper.

Vine (1956) reported that the principal uranium-bearing coal deposits of the United States are found in Cretaceous and Tertiary formations in the northern Great Plains and Rocky Mountains; in some areas of the West, coal deposits have been found with uranium concentrations in the range of thousands of ppm or greater. In contrast, most Mississippian, Pennsylvanian, and Permian coals in the north-central and eastern United States contain less than 10 ppm uranium, rarely containing 50 ppm or more.

Organic-rich shales and coals are enriched in uranium, thorium, and other trace metals in concentrations above those seen in typical shales or sedimentary rocks (Diehl et al., 2004; USGS, 1997; Wignall and Myers, 1988; Tourtelot, 1979; Vine and Tourtelot, 1970). Unlike shales and coals, sandstones are generally not organic-rich source rocks themselves. Instead, hydrocarbons migrate into these formations over long periods of time (Clark and Veil, 2009). Since TENORM and organic contents are typically positively correlated due to the original, reduced depositional environment (Fertl and Chilingar, 1988), it is unlikely that sandstones would be enriched in TENORM to the same extent as oil- and gas-bearing shales and coals. Therefore, concern related to TENORM within produced water is focused on operations targeting shales and coalbeds.

Radium is most soluble and mobile in chloride-rich, high-TDS, reducing environments (Sturchio et al., 2001; Zapecza and Szabo, 1988; Langmuir and Riese, 1985). In formation fluids with high TDS, calcium, potassium, magnesium, and sodium compete with dissolved radium for sorption sites, limiting radium sorption onto solids and allowing it to accumulate in solution at higher concentrations (Fisher, 1998; Webster et al., 1995). The positive correlation between TDS and radium is well established and TDS is a useful indicator of radium and TENORM activity within produced water, especially in lithologically homogenous reservoirs (Rowan et al., 2011; Sturchio et al., 2001; Fisher, 1998; Kraemer and Reid, 1984).

Uranium and thorium are poorly soluble under reducing conditions and are therefore more concentrated in formation solids than in solution (Fisher, 1998; Kraemer and Reid, 1984; Langmuir and Herman, 1980). However, because uranium becomes more soluble in oxidizing environments, the introduction of relatively oxygen-rich fracturing fluids may promote the temporary mobilization of uranium during hydraulic fracturing and early flowback. In addition, the physical act of hydraulic fracturing creates fresh fractures and exposes organic-rich and highly reduced surfaces from which radionuclides could be released from the rock into formation fluids.

Produced water geochemistry determines, in part, the fate of subsurface radionuclides, particularly radium. Radium may remain in the host mineral or it may be released into formation fluids, where it can remain in solution as the dissolved Ra^{2+} ion, be adsorbed onto oxide grain coatings or clay particles by ion exchange, substitute for other cations during the precipitation of minerals, or form complexes with chloride, sulfate, and carbonate ions (Rowan et al., 2011; Sturchio et al., 2001; Langmuir and Riese, 1985). Uranium- and thorium-containing materials with a small grain size, a large surface-to-volume ratio, and the presence of uranium and thorium near grain surfaces promote the escape of radium into formation fluids. Vinson et al. (2009) point to alpha decay along fracture surfaces as a primary control on radium mobilization in crystalline bedrock aquifers.

Radium may also occur in formation fluids due to other processes, such as the decay of dissolved parent isotopes and adsorption-desorption reactions on formation surfaces (Sturchio et al., 2001).

Preliminary results from fluid-rock interaction studies (Bank, 2011) indicate that a significant percentage of uranium in the Marcellus Shale may be subject to mobilization by hydrochloric acid, which is used as a fracturing fluid additive. More complete understanding these processes will determine the extent to which such processes might influence the TENORM content of flowback and produced water.

E.3.5. Organics

Background data on organics in seven formations is given in Table E-9. Classes of organic compounds identified in produced water are given in Table E-10. Tables H-4 and H-5 give the entire list of chemicals identified as components of produced water. Along with the organic chemicals appearing in Table E-9, Table E-10a presents additional organic chemicals with measured concentrations in produced water. Table E-11 presents data from two studies of the Marcellus Shale. Table E-12 presents data from CBM produced water, while Table E-13 presents data on organics identified in shale and CBM water.

Several classes of naturally occurring organic chemicals are present in produced waters in conventional and unconventional reservoirs, with large concentration ranges (Lee and Neff, 2011). These organic classes include total organic carbon (TOC); saturated hydrocarbons; BTEX (benzene, toluene, ethylbenzene, and xylenes); and polyaromatic hydrocarbons (PAHs) (Table E-10). While TOC concentrations in produced water are detected at the milligrams to grams per liter level, concentrations of individual organic compounds are typically detected at the micrograms to milligrams per liter level.

TOC indicates the level of dissolved and undissolved organics in produced water, including non-volatile and volatile organics (Acharya et al., 2011). TOC concentrations in conventional produced water vary widely from less than 0.1 mg/L to more than 11,000 mg/L. Average TOC concentrations in produced water in unconventional reservoirs range from less than 2.00 mg/L in the Raton CBM basin to approximately 200 mg/L in the Cotton Valley Group sandstones, although individual measurements have exceeded 5,000 mg/L in the Marcellus Shale (Table E-9).

Dissolved organic carbon (DOC) is a general indicator of organic loading and is the fraction of organic carbon available for complexing with metals and supporting microbial growth. DOC values in produced water in unconventional reservoirs range from less than 1.50 mg/L (average) in the Raton Basin to more than 115 mg/L (median) in the Marcellus Shale (Table E-9). Individual DOC concentrations in the Marcellus Shale produced water approach 6,000 mg/L. For comparison, DOC levels in fresh water systems are typically below 5 mg/L.

Biochemical oxygen demand (BOD) is a conventional pollutant under the U.S. Clean Water Act. It is an indirect measure of biodegradable organics in produced water and an estimate of the oxygen demand on a receiving water. Median BOD levels for Barnett and Marcellus Shales produced water exceed 30 mg/L, and both reported maximum concentrations exceeding 12,000 mg/L (Table E-9).

In some circumstances wide variation in produced water median BOD levels may be reflective of flowback reuse in fracturing fluids (Hayes, 2009).

Lastly, BTEX is associated with petroleum. Benzene was found in produced water from several basins: average produced water benzene concentration from the Barnett Shale was 680 µg/L, from the Marcellus Shale was 220 µg/L (median), and from the San Juan Basin was 150 µg/L (Table E-9). Total BTEX concentrations for conventional produced water vary widely from less than 100 µg/L to nearly 580,000 µg/L. For comparison, average total BTEX concentrations in produced water in unconventional reservoirs range from 20 µg/L in the Raton Basin to nearly 3,000 µg/L in the Marcellus (Table E-9). From these data, average total BTEX levels in shale produced water are one to two orders of magnitude higher than those in CBM produced water.

In addition to BTEX, a variety of volatile and semi-volatile organic compounds have been detected in shale and coalbed produced water. Shale produced water contains naphthalene, alkylated toluenes, and methylated aromatics in the form of several benzene and phenol compounds, as shown in Table E-11. Like BTEX, naphthalene, methylated phenols, and acetophenone are associated with petroleum. Detected shale produced water organics such as acetone, 2-butanone, carbon disulfide, and pyridine are potential remnants of additives used as friction reducers or industrial solvents (Hayes, 2009).

Hayes (2009) characterized the content of Marcellus Shale produced water including organics (Table E-11). The author tested for the majority of VOCs and SVOCs, pesticides and PCBs, based on the recommendation of the Pennsylvania and West Virginia Departments of Environmental Protection. Less than 0.5% of VOCs and 0.03% of SVOCs in the produced water were detected above 1 mg/L. More than 96% of VOCs, 98% of SVOCs, and virtually all pesticides and PCBs were at nondetectable levels.

Orem et al. (2014) provided a list of classes of organic compounds in coalbed methane and gas shale produced and formation water (Table E-10). As described in the main text of Chapter 7, these included aromatics, polyaromatic hydrocarbons, heterocyclic compounds, aromatic amines, phenols, phthalates, aliphatic alcohols, fatty acids and nonaromatic compounds. Many of these are naturally occurring components of petroleum hydrocarbons, but the list also contains chemicals that have been used as hydraulic fracturing fluid additives, namely, hexahydro-1,3,5-trimethyl-1,3,5-triazine-2-thione (a biocide), ethylene glycol, dibutyl phthalate, quinoline, and naphthalene, to list a few. See Table H-2.

The organic profile of CBM produced water is characterized by high levels of aromatic and halogenated compounds compared to other produced water in unconventional reservoirs (Sirivedhin and Dallbauman, 2004). PAHs and phenols are the most common organic compounds found in coalbed produced water. Produced water from coalbeds in the Black Warrior Basin mainly contains phenols, multiple naphthalic PAHs, and various decanoic and decenoic fatty acids (Table E-12). CBM-associated organics are also known to include biphenyls, alkyl aromatics, hydroxypyridines, aromatic amines, and nitrogen-, oxygen-, and sulfur-bearing heterocyclics (Orem et al., 2014; Pashin et al., 2014; Benko and Drewes, 2008; Orem et al., 2007; Fisher and Santamaria, 2002).

Table E-9. Concentrations of select organic parameters in produced water from unconventional reservoirs (including shale, a tight formation, and coalbed methane), presented as: average (minimum-maximum) or *median* (minimum-maximum).
Both averages and medians are reported because this table summarizes published information and authors differed in their use of averages or medians.

Parameter	Unit	Shale			Tight Formation	Coal			
		Barnett[a]	Marcellus		Cotton Valley Group[d]	Powder River[e]	Raton[e]	San Juan[e]	Black Warrior[f]
States	n/a	TX	PA[b]	PA, WV[c]	LA, TX	MT, WY	CO, NM	AZ, CO, NM, UT	AL, MS
TOC	mg/L	*9.75* (6.2 – 36.2)	160 (1.2 – 1,530)	*89.2* (1.2 – 5680)	198 (184 – 212)	3.52 (2.07 – 6.57)	1.74 (0.25 – 13.00)	2.91 (0.95 – 9.36)	6.03 (0.00 – 103.00)
DOC	mg/L	*11.2* (5.5 – 65.3)		*117* (3.3 – 5,960)	-	3.18 (1.09 – 8.04)	1.26 (0.30 – 8.54)	3.21 (0.89 – 11.41)	3.37 (0.53 – 61.41)
BOD	mg/L	*582* (101 – 2,120)	-	*141* (2.8 – 12,400)	-	-	-	-	-
Oil and grease	mg/L	*163.5* (88.2 – 1,430)	74 (5 – 802)	*16.9* (4.7 – 802)	-	-	9.10 (0.60 – 17.6)	-	-
Benzene	µg/L	680 (49 – 5,300)	-	*220* (5.8 – 2,000)	-	-	4.7 (BDL – 220.0)	149.7 (BDL – 500.0)	-
Toluene	µg/L	760 (79 – 8,100)	-	*540* (5.1 – 6,200)	-	-	4.7 (BDL – 78.0)	1.7 (BDL – 6.2)	-
Ethylbenzene	µg/L	29 (2.2 – 670)	-	*42* (7.6 – 650)	-	-	0.8 (BDL – 18.0)	10.5 (BDL – 24.0)	-
Xylenes	µg/L	360 (43 – 1,400)	-	*300* (15 – 6,500)	-	-	9.9 (BDL – 190.0)	121.2 (BDL – 327.0)	-

Parameter	Unit	Shale			Tight Formation	Coal			
		Barnett[a]	Marcellus		Cotton Valley Group[d]	Powder River[e]	Raton[e]	San Juan[e]	Black Warrior[f]
			PA[b]	PA, WV[c]					
States	n/a	TX	PA	PA, WV	LA, TX	MT, WY	CO, NM	AZ, CO, NM, UT	AL, MS
Average total BTEX[g]	µg/L	1,829	2,910	1,102	-	-	20.1	283.1	-

n/a, not applicable; -, no value available; BDL, below detection limit. **Bolded italic** numbers are medians.

[a] Hayes and Severin (2012a). n = 16. This data source reported concentrations without presentation of raw data.

[b] Barbot et al. (2013). n = 55 for TOC; n = 62 for oil and grease; no presentation of raw data.

[c] Hayes (2009). n = 13-67. Concentrations were calculated based on Hayes' raw data. Both flowback and produced water included. Non-detects and contaminated blanks omitted.

[d] Blondes et al. (2014). n = 2. Concentrations were calculated based on raw data presented in the USGS National Produced Water Database v2.0.

[e] Dahm et al. (2011). Powder River, n = 31; Raton, n = 40; San Juan, n = 20. This data source reported concentrations without presentation of raw data.

[f] DOE (2014). n = 206. Concentrations were calculated based on the authors' raw data.

[g] Average total BTEX was calculated by summing the average/median concentrations of benzene, toluene, ethylbenzene, and xylenes for a unique formation or basin. Minimum to maximum ranges were not calculated due to inaccessible raw data.

Table E-10. Classes of organic compounds and representative example compounds found in coal bed methane and gas shale formations (Orem et al., 2014).

Compounds also identified as having been used in hydraulic fracturing fluids (Table H-2) are given in bold and italic type.

Extractable hydrocarbons identified in CBM and shale produced and formation water			
Type	**Location**	**Compound classes**	**Representative example compounds**
CBM	Powder River Basin Wyoming	PAHs	Dimethylnaphthalene tetramethylphenanthrene phenanthrenone pyrene
		Heterocyclic compounds	Benzisothiazole 3,4-dihydro1,9(2H,10H)Acridinedione 2(3H)-Benzothiazolone
		Aromatic amines	Dioctyldiphenylamine diphenylamine 2-methyl-N-phenyl Benzenamine
		Phenols	Nonylphenols 4,40-(1-methylethylidene)bis-phenol methoxy-methylphenol
		Other aromatics	Trimethyl benzene 2,4-dimethyl-1-(1-methylpropyl)-benzene
		Phthalates	Diethylphthalate dibutyl phthalate benzyl butyl phthalate didecyl phthalate
		Fatty acids	Dodecanoic acid n-hexadecanoic acid tetradecanoic acid
		Nonaromatic compounds	Kaur-16-ene (a diterpene) 2-[2-[4-(1,1,3,3-tetramethylbutyl)phenoxy]ethoxy]-ethanol
	Tongue River Basin Montana	PAHs	1-Methyl-7-(1-methylethyl)phenanthrene 1-methylnaphthalene 2-methylnaphthalene
		Heterocyclic compounds	Benzothiazole
		Aromatic amines	Diethyltoluamide
		Phenols	2,4-Bis(1,1-dimethylethyl)phenol p-tert-butyl-phenol
		Other aromatics	1-Ethyl-2,4-dimethyl-benzene

Extractable hydrocarbons identified in CBM and shale produced and formation water			
Type	**Location**	**Compound classes**	**Representative example compounds**
CBM, cont.	Tongue River Basin Montana, cont.	Phthalates	Alkyl phthalates
		Fatty acids	Tetradecanoic acid octadecanoic acid
		Nonaromatic compounds	Pentadecane pentacosane
	Black Warrior Basin Alabama	PAHs	Methylnaphthalene dimethylnaphthalene
		Heterocyclic compounds	Benzothiazole dibenzothiophene caprolactam quinoline *isoquinoline*
		Phenols	Dimethylphenol 4-(1,1,3,3-tetramethylbutyl)-phenol 2,4-bis(1,1-dimethylethyl)-phenol
		Other aromatics	*Acetophenone* biphenyl methylbiphenyl
		Phthalates	Dioctyl phthalate dibutyl phthalate
		Fatty acids	Hexadecanoic acid
		Nonaromatic compounds	Alkyl phosphates
	Illinois Basin Illinois	PAHs	*Naphthalene* methylnaphthalene methylphenanthrene
		Heterocyclic compounds	Benzothiazole
		Phenols	2,4-Bis(1,1-dimethylethyl)-phenol
		Other aromatics	1-(3-Methylbutyl)-2,3,4-trimethyl-benzene
		Phthalates	Alkyl phthalates
		Fatty acids	Hexadecanoic acid octadecanoic acid
		Nonaromatic compounds	C23–C36 alkanes 2,6-di(tert-butyl)-4-hydroxy-4-methyl-2,5-cyclohexadien-1-on

Extractable hydrocarbons identified in CBM and shale produced and formation water

Type	Location	Compound classes	Representative example compounds
CBM, cont.	Williston Basin North Dakota	PAHs	*Naphthalene* methylnaphthalene methylphenanthrene
		Heterocyclic compounds	Benzothiazole
		Phenols	Bis(1,1-dimethylethyl)-phenol trichlorophenol 4,4'-(1-methylethylidene)bis-phenol
		Other aromatics	Benzophenone
		Phthalates	Alkyl phthalates benzyl butyl phthalate
		Fatty acids	C12, C14, C16, C18 fatty acids
		Nonaromatic compounds	C23–C35 alkanes alkyl phosphates 2,6-bis(1,1-dimthylethyl)-2,5-cyclohexadiene-1,4-dione
Shale gas	Marcellus Shale Pennsylvania	PAHs	Decahydro-4,4,8,9,10-pentamethylnaphthalene
		Heterocyclic compounds	*Hexahydro-1,3,5-trimethyl-1,3,5-triazine-2-thione* (a biocide)
		Aliphatic alcohols	*Ethylene glycol* diethylene glycol monododecyl ether triethylene glycol monodocecyl ether
		Other aromatics	(1-Methoxyethyl)-benzene
		Phthalates	Di-n-octyl phthalate
		Fatty acids	C12, C14, C16, C18 fatty acids
		Nonaromatic compounds	C11–C37 alkanes/alkenes 2,2,4-trimethyl-1,3-pentanediol tetramethylbutanedinitrile
	New Albany Shale Indiana and Kentucky	PAHs	1,2,3,4-Tetrahydro-naphthalene *naphthalene* methylphenanthrene pyrene perylene
		Heterocyclic compounds	Benzothiazole trimethyl-piperdine *quinoline* quinindoline

Extractable hydrocarbons identified in CBM and shale produced and formation water			
Type	**Location**	**Compound classes**	**Representative example compounds**
Shale gas, cont.	New Albany Shale Indiana and Kentucky, cont.	Aromatic amines	3,3'-5,5'-Tetramethyl-[1,1'-biphenyl]-4,4'-diamine
		Phenols	Bis(1,1-dimethylethyl)-phenol tert-butyl-phenol bis-(1,1-dimethylethyl)-phenol
		Other aromatics	Triphenyl phosphate methylbiphenyl octylphenyl ethoxylate
		Phthalates	Alkyl phthalates
		Fatty acids	Dodecanoic acid tetradecanoic acid octadecanoic acid
		Nonaromatic compounds	***2-(2-Butoxyethoxy)ethanol***

Table E-11. Reported concentrations (µg/L) of organic constituents in produced water for two shale formations, presented as: average (minimum-maximum) or *median* (minimum-maximum).

Both averages and medians are reported because this table summarizes published information and authors differed in their use of averages or medians.

Parameter	Barnett[a]	Marcellus[b]
States	**TX**	**MD, NY, OH, PA, VA, WY**
Acetone	145 (27 – 540)	*83* (14 – 5,800)
Carbon disulfide	-	*400* (19 – 7,300)
Chloroform	-	28
Isopropylbenzene	35 (0.8 – 69)	*120* (86 – 160)
Naphthalene	238 (4.8 – 3,100)	*195* (14 – 1,400)
Phenolic compounds	*119.65* (9.3 – 230)	-
1,2,4-Trimethylbenzene	173 (6.9 – 1,200)	*66.5* (7.7 – 4,000)
1,3,5-Trimethylbenzene	59 (6.4 – 300)	*33* (5.2 – 1,900)
1,2-Diphenylhydrazine	4.2 (0.5 – 7.8)	-
1,4-Dioxane	6.5 (3.1 – 12)	-
2-Methylnaphthalene	1,362 (5.4 – 20,000)	*3.4* (2 – 120)
2-Methylphenol	28.3 (5.8 – 76)	*13* (11 – 15)
2,4-Dichlorophenol	(ND – 15)	-
2,4-Dimethylphenol	14.5 (8.3 – 21)	12
3-Methylphenol and 4-Methylphenol	41 (7.8 – 100)	*11.5* (0.35 – 16)
Acetophenone	(ND – 4.6)	*13* (10 – 22)
Benzidine	(ND – 35)	-
Benzo(a)anthracene	(ND – 17.0)	-
Benzo(a)pyrene	(ND – 130.0)	6.7
Benzo(b)fluoranthene	42.2 (0.5 – 84.0)	10
Benzo(g,h,i)perylene	42.3 (0.7 – 84.0)	6.9
Benzo(k)fluoranthene	32.8 (0.6 – 65.0)	5.9
Benzyl alcohol	81.5 (14.0 – 200)	*41* (17 – 750)
Bis(2-Ethylhexyl) phthalate	210 (4.8 – 490)	*20* (9.6 – 870)
Butyl benzyl phthalate	34.3 (1.9 – 110)	-

Parameter	Barnett[a]	Marcellus[b]
States	TX	MD, NY, OH, PA, VA, WY
Chrysene	120 (0.57 – 240)	-
Di-n-octyl phthalate	(ND – 70)	15
Di-n-butyl phthalate	41 (1.5 – 120)	*14* (11 – 130)
Dibenz(a,h)anthracene	77 (3.2 – 150)	*3.2* (2.3 – 11)
Diphenylamine	5.3 (0.6 – 10.0)	-
Fluoranthene	(ND – 0.18)	6.1
Fluorene	0.8 (0.46 – 1.3)	8.4
Indeno(1,2,3-cd)pyrene	71 (2.9 – 140)	*3.1* (2.4 – 9.5)
N-Nitrosodiphenylamine	8.9 (7.8 – 10)	2.7
N-Nitrosomethylethylamine	(ND – 410)	-
Phenanthrene	107 (0.52 – 1,400)	*9.75* (3 – 22)
Phenol	63 (17 – 93)	*10* (2.4 – 21)
Pyrene	0.2 (ND – 0.18)	13
Pyridine	413 (100 – 670)	*250* (10 – 2,600)

-, no value available; ND, not detected.

[a] Hayes and Severin (2012a). n = 16. Data from days 1 – 23 of flowback. This data source reported concentrations without presentation of raw data.

[b] Hayes (2009). n = 1 – 35. Data from days 1 – 90 of flowback. Concentrations were calculated from Hayes' raw data. Non-detects and contaminated blanks omitted.

Table E-12. Reported concentrations of organic constituents in 65 samples of produced water from the Black Warrior CBM Basin (Alabama and Mississippi), presented as: average (minimum-maximum).

Parameter	Number of observations	Concentration (µg/L)[a]
Benzothiazole	45	0.25 (0.01 – 3.04)
Caprolactam	10	0.75 (0.02 – 2.39)
Cyclic octaatomic sulfur	29	1.06 (0.10 – 9.63)
Dimethyl-naphthalene	39	0.79 (0.01 – 9.51)
Dioctyl phthalate	57	0.21 (0.01 – 2.30)
Dodecanoic acid	30	1.13 (0.67 – 2.52)
Hexadecanoic acid	50	1.58 (1.17 – 3.02)
Hexadecenoic acid	25	1.69 (1.13 – 8.37)
Methyl-biphenyl	18	0.25 (0.01 – 2.13)
Methyl-naphthalene	52	0.77 (0.01 – 15.55)
Methyl-quinoline	31	0.96 (0.03 – 3.75)
Naphthalene	49	0.41 (0.01 – 6.57)
Octadecanoic acid	32	1.95 (1.62 – 3.73)
Octadecenoic acid	29	1.87 (1.60 – 3.47)
Phenol, 2,4-bis(1,1-dimethyl)	21	0.45 (0.01 – 4.94)
Phenol, 4-(1,1,3,3-tetramethyl)	17	1.65 (0.01 – 18.34)
Phenolic compounds	-	19.06 (ND – 192.00)
Tetradecanoic acid	53	1.51 (0.94 – 5.32)
Tributyl phosphate	23	0.26 (0.01 – 2.66)
Trimethyl-naphthalene	23	0.65 (0.01 – 4.49)
Triphenyl phosphate	6	1.18 (0.01 – 6.77)

-, no value available.

[a] DOE (2014). Concentrations were calculated based on the authors' raw data.

Table E-13. Organic chemical concentrations reported from three specific studies of produced water (Khan et al., 2016; Lester et al., 2015; Orem et al., 2007).

The complete list of chemicals which were identified in produced water are listed in Tables H-4 and H-5.

Chemical	Minimum or only value (µg/L)	Average (µg/L)	Maximum (µg/L)	Standard deviation (µg/L)	Formation type (S for shale, C for coalbed)	Reference
(Z)-9-Tricosene	0.98				C	Orem et al. (2007)
1-(2-Hydroxy-5-methylphenyl)-2-hexen-1-one	0.29				C	Orem et al. (2007)
1,1-Dimethyl-1,2,3,4-tetrahydro-7-isopropyl phenanthrene	0.19		0.68		C	Orem et al. (2007)
1,2-Di-but-2-enyl-cyclohexane	0.77				C	Orem et al. (2007)
1,2-Di-but-2-enyl-cyclohexanone	0.09				C	Orem et al. (2007)
1,4-[13C]-1,2,3,4-Tetrahydro-5-naphthaleneamine	0.33				C	Orem et al. (2007)
1,4-dioxane	60				C	Lester et al. (2015)
1,6-Dimethyl-4-(1-methylethyl)naphthalene	0.01		0.32		C	Orem et al. (2007)
1,7,11-Trimethylcyclotetradecane	1.06				C	Orem et al. (2007)
1,7,11-Trimethyl-cyclotetradecane	0.47				C	Orem et al. (2007)
1-1Methylenebis(4-methyl)-benzene	0.09		0.11		C	Orem et al. (2007)
15-Isobutyl-(13.α.H)-isocopalane	1.75				C	Orem et al. (2007)

Chemical	Minimum or only value (µg/L)	Average (µg/L)	Maximum (µg/L)	Standard deviation (µg/L)	Formation type (S for shale, C for coalbed)	Reference
17-Pentatriacontene	1				C	Orem et al. (2007)
1-Allyl-3-methylindole-2-carbaldehyde	0.49				C	Orem et al. (2007)
1-Allyl-3-methylindole-2-carbaldehyde	1.49				C	Orem et al. (2007)
1-Butyl-2-ethyloctahydro-4,7-epoxy	0.9				C	Orem et al. (2007)
1-Chloro-octadecane	2.12				C	Orem et al. (2007)
1-Docosene	2.33				C	Orem et al. (2007)
1-Ethyl-9,10-anthracenedione	0.04		0.12		C	Orem et al. (2007)
1-Hexacosene	2.04				C	Orem et al. (2007)
1-Methyl-7-(1-methylethyl)phenanthrene	0.02		3.19		C	Orem et al. (2007)
1-Methyl-9H-fluorene	0.51				C	Orem et al. (2007)
1-Nonadecene	2.15				C	Orem et al. (2007)
2-(2-Butoxyethoxy)-ethanol	0.45				C	Orem et al. (2007)
2(3H)-Benzothiazolone	0.04		3.9		C	Orem et al. (2007)
2-(Methylthio)-benzothiazole	0.05		0.54		C	Orem et al. (2007)
2,3',5-Trimethyldiphenylmethane	0.04		0.05		C	Orem et al. (2007)
2,3-Dihydro-1,1,2,3,3-pentamethyl-1H-indene	0.45				C	Orem et al. (2007)

Chemical	Minimum or only value (µg/L)	Average (µg/L)	Maximum (µg/L)	Standard deviation (µg/L)	Formation type (S for shale, C for coalbed)	Reference
2,4,6-Trimethyl-azulene	0.49				C	Orem et al. (2007)
2,4-dimethylphenol	790				C	Lester et al. (2015)
2,5-Cyclohexadiene-1,4-dione	0.01		0.08		C	Orem et al. (2007)
2,6,10,14-Tetramethyl-hexadecane	1.65				C	Orem et al. (2007)
2,6,10-Trimethyl-dodecane	0.96				C	Orem et al. (2007)
2,6-Bis(dimethylethyl)-2,5-cyclohexadiene-1,4-dione	0.04		0.28		C	Orem et al. (2007)
2,6-Bis(dimethylethyl)-phenol	0.31				C	Orem et al. (2007)
2-[2-[4-(1,1,3,3-Tetramethylbutyl)phenoxy]ethoxy]-ethanol	0.08		1.34		C	Orem et al. (2007)
22-Tricosenoic acid	0.43				C	Orem et al. (2007)
28-Nor-17.α.(H)-hopane	1.26				C	Orem et al. (2007)
28-Nor-17.α.(H)-hopane	0.84				C	Orem et al. (2007)
2a,7a-(Epoxymethano)-2H-cyclobutyl	0.33				C	Orem et al. (2007)
2-Butanone	240				C	Orem et al. (2007)
2-Dodecen-1-yl(-)succinic anhydride	1.16				C	Orem et al. (2007)
2-Ethylhexyl diphenyl phosphate (Octicizer)	0.1		0.75		C	Orem et al. (2007)
2-Mercaptobenzothiazole	0.89				C	Orem et al. (2007)

Appendix E – Produced Water Handling Supplemental Information

Chemical	Minimum or only value (µg/L)	Average (µg/L)	Maximum (µg/L)	Standard deviation (µg/L)	Formation type (S for shale, C for coalbed)	Reference
2-Methyl-8-propyl-dodecane	0.52				C	Orem et al. (2007)
2-methylnaphthalene	4				C	Lester et al. (2015)
2-Methyl-nonadecane	2.58				C	Orem et al. (2007)
2-Methyl-N-phenyl-benzenamine	0.41		3.53		C	Orem et al. (2007)
2-methylphenol	150				C	Lester et al. (2015)
2-Octadecyl-propane-1,3-diol	0.42				C	Orem et al. (2007)
3&4 methylphenol	170				C	Lester et al. (2015)
3-(4-Methoxyphenyl)-2-ethylhexylester-2-propenoic acid	0.01		2.78		C	Orem et al. (2007)
3-(4-Methoxyphenyl)-2-propenoic acid	0.06		0.16		C	Orem et al. (2007)
3-(Hexahydro-1H-azepin-1-yl)-1,1-dioxide-1,2-benzisothiazole	0.66				C	Orem et al. (2007)
3,4-Dihydro-1,9(2H,10H)acridinedione	0.02		1.35		C	Orem et al. (2007)
3,5-Di-tetra-butyl-4-hydroxybenzaldehyde	0.42				C	Orem et al. (2007)
4-(1-Methyl-phenylethyl)-phenol	1.18				C	Orem et al. (2007)
4-(4-Ethylcyclohexyl)-cyclohexene	1.66				C	Orem et al. (2007)

Chemical	Minimum or only value (µg/L)	Average (µg/L)	Maximum (µg/L)	Standard deviation (µg/L)	Formation type (S for shale, C for coalbed)	Reference
4,40-(1-Methylethylidene)bis-phenol	<=16.17				C	Orem et al. (2007)
4,4-Diacetyldiphenylmethane	0.37				C	Orem et al. (2007)
4,6,8-Trimethyl-2-propylazulene	0.4				C	Orem et al. (2007)
4-Hydroxy-3-methoxy-benzaldehyde	4.31				C	Orem et al. (2007)
4-Propyl-xanthen-9-one	0.03		0.07		C	Orem et al. (2007)
5-(1,1-Dimethylethyl)-1H-indene	0.03		0.1		C	Orem et al. (2007)
5,6-Azulenedimethanol,1,2,3,3a,8,	0.4				C	Orem et al. (2007)
7-Bromomethyl-pentadec-7-ene	2.77				C	Orem et al. (2007)
7-Bromomethyl-pentadec-7-ene	0.92				C	Orem et al. (2007)
7-Ethenylphenanthrene	0.04		0.22		C	Orem et al. (2007)
7-Tetradecyne	0.38				C	Orem et al. (2007)
8-Hexadecyne	0.28				C	Orem et al. (2007)
8-Isopropyl-2,5-dimethyl-terralin	0.36				C	Orem et al. (2007)
9,10-Dimethoxy-2,3-dihydroanthracene	0.04		0.34		C	Orem et al. (2007)
9H-Fluoren-9-ol	0.07		0.32		C	Orem et al. (2007)

Chemical	Minimum or only value (µg/L)	Average (µg/L)	Maximum (µg/L)	Standard deviation (µg/L)	Formation type (S for shale, C for coalbed)	Reference
9-Methoxyfluorene	0.06		0.18		C	Orem et al. (2007)
9-Methoxyfluorene	0.54				C	Orem et al. (2007)
9-Phenyl-tetrahydro-1H-benz[f]isoindol-1-one	0.24				C	Orem et al. (2007)
9-Phenyl-tetrahydro-1H-benz[f]isoindol-1-one	0.24				C	Orem et al. (2007)
Acetone	16,000				S	Lester et al. (2015)
Alkyl benzene	74,630	1,119,350	5,092,600	1,698,910	S	Khan et al. (2016)
Alkyl naphthalene	380	1,460	4,200	1,180	S	Khan et al. (2016)
Alkyl propo-benzene	9,340	61,900	209,150	67,220	S	Khan et al. (2016)
Benzene	1,500	107,320	778,510	271,570	S	Khan et al. (2016)
Benzenemethanol	0.33				C	Orem et al. (2007)
Benzisothiazole derivative	0.06		0.32		C	Orem et al. (2007)
Benzothiazole	0.51		14.27		C	Orem et al. (2007)
Benzyl butyl phthalate	0.04		0.33		C	Orem et al. (2007)
Biphenyl	0.16		0.3		C	Orem et al. (2007)
Bis(2-ethylhexyl) phthalate	29				S	Lester et al. (2015)
Bis(2-ethylhexyl)-hexanedioic acid	0.13		0.7		C	Orem et al. (2007)
Bis-(octylphenyl)-amine	0.05		0.19		C	Orem et al. (2007)
Butanoic acid, butyl ester	0.44				C	Orem et al. (2007)

Chemical	Minimum or only value (µg/L)	Average (µg/L)	Maximum (µg/L)	Standard deviation (µg/L)	Formation type (S for shale, C for coalbed)	Reference
butyl benzyl phthalate	4.2				S	Lester et al. (2015)
Caffeine	0.09		0.5		C	Orem et al. (2007)
Chloro-benzene	20	100	350	110	S	Khan et al. (2016)
Cholesterol	0.26				C	Orem et al. (2007)
Cyclotriacontane	1.08				C	Orem et al. (2007)
Dibutyl phthalate	<=1.27				C	Orem et al. (2007)
Didecyl phthalate	<=7.23				C	Orem et al. (2007)
Diethyl phthalate	<=14.9				C	Orem et al. (2007)
Dihydro-(-)-neocloven-(II)	0.1		1.04		C	Orem et al. (2007)
Dihydro-1-methylphenanthrene	1.06				C	Orem et al. (2007)
Dihydrophenanthrene	0.03		0.48		C	Orem et al. (2007)
Dimethyl phthalate	0.11		0.28		C	Orem et al. (2007)
	15				S	Lester et al. (2015)
Dimethyl-biphenyl	0.07		2.01		C	Orem et al. (2007)
Dimethyl-ethylindene	0.02		0.07		C	Orem et al. (2007)
Dimethylnaphthalene	0.01		1.44		C	Orem et al. (2007)
Dimethylphenanthrene	0.62		1.49		C	Orem et al. (2007)
Dimethylphenol	1.38				C	Orem et al. (2007)
Dimethyl-tetracyclo[5.2.1.0(2,6)-0(3,5)]decane	0.27				C	Orem et al. (2007)

Appendix E – Produced Water Handling Supplemental Information

Chemical	Minimum or only value (μg/L)	Average (μg/L)	Maximum (μg/L)	Standard deviation (μg/L)	Formation type (S for shale, C for coalbed)	Reference
Di-n-octyl phthalate	0.58		4.63		C	Orem et al. (2007)
Dioctyldiphenylamine	0.03		0.18		C	Orem et al. (2007)
Diphenylamine	0.04		3.73		C	Orem et al. (2007)
Diphenylmethane	0.01		0.43		C	Orem et al. (2007)
Di-tetra-butyl-4-hydroxbenzaldehyde	0.16		0.53		C	Orem et al. (2007)
Docosane	1.94				C	Orem et al. (2007)
Dodecanoic acid	1.33		1.7		C	Orem et al. (2007)
Drometrizole	0.91				C	Orem et al. (2007)
Ethylbenzene	2,010	72,610	399,840	134,630	S	Khan et al. (2016)
Ethyl dimethyl azulene	0.46				C	Orem et al. (2007)
Ethyl phenylmethyl benzene	0.1				C	Orem et al. (2007)
Ethyl-cyclodocosane	1.54				C	Orem et al. (2007)
Ethyl-cyclodocosane	0.65				C	Orem et al. (2007)
Ethyl-tetrahydronaphthalene	0.46				C	Orem et al. (2007)
Fluorene	0.05		0.24		C	Orem et al. (2007)
Heptacosane	0.95				C	Orem et al. (2007)
Hexacosane	1.73				C	Orem et al. (2007)
Isopropyl myristate	1.79				C	Orem et al. (2007)
Kaur-16-ene	0.06		1.36		C	Orem et al. (2007)

Chemical	Minimum or only value (µg/L)	Average (µg/L)	Maximum (µg/L)	Standard deviation (µg/L)	Formation type (S for shale, C for coalbed)	Reference
Methoxyanthracene	0.04		0.22		C	Orem et al. (2007)
Methoxynaphthalene derivative	0.04		0.25		C	Orem et al. (2007)
Methyl-(2,5-dimethoxyphenol)-methanoate	0.31				C	Orem et al. (2007)
Methyl(Z)-3,3-diphenyl-4-hexenoate	2				C	Orem et al. (2007)
Methyl-2-octylcyclopropene-1-octane	0.38				C	Orem et al. (2007)
Methyl-2-quinolinecarboxylic acid	6.65				C	Orem et al. (2007)
Methyl-9H-fluorene	0.52		1.16		C	Orem et al. (2007)
Methylanthracene	0.07		0.48		C	Orem et al. (2007)
Methyl-biphenyl	0.15		1		C	Orem et al. (2007)
Methylethylnaphthalene	0.55				C	Orem et al. (2007)
Methylnaphthalene	0.14		0.48		C	Orem et al. (2007)
Methylphenanthrene	0.03		1.37		C	Orem et al. (2007)
Methylpyrene	0.01		0.02		C	Orem et al. (2007)
Naphthalene	0.26		0.66		C	Orem et al. (2007)
Naphthalenone derivative	0.11		1.38		C	Orem et al. (2007)
n-Hexadecanoic acid	0.63		2.56		C	Orem et al. (2007)
Nonyl-phenol	0.09		7.91		C	Orem et al. (2007)

Chemical	Minimum or only value (µg/L)	Average (µg/L)	Maximum (µg/L)	Standard deviation (µg/L)	Formation type (S for shale, C for coalbed)	Reference
Octahydroanthracene	0.54				C	Orem et al. (2007)
Other alkyl phenols	<=5.89				C	Orem et al. (2007)
Other aromatic compounds	0.01		0.42		C	Orem et al. (2007)
Other benzenamines	0.06		0.25		C	Orem et al. (2007)
Other benzene alkyl compounds	0.02		0.62		C	Orem et al. (2007)
Other heterocyclics	<=17.87				C	Orem et al. (2007)
Other indene derivatives	0.09		0.16		C	Orem et al. (2007)
Other naphthalene alkyl compounds	0.04		0.82		C	Orem et al. (2007)
Other phthalates	<=18.68				C	Orem et al. (2007)
Other terpenoid compounds	0.12		0.37		C	Orem et al. (2007)
Pentacosane	1.54				C	Orem et al. (2007)
Pentadecanoic acid	0.84				C	Orem et al. (2007)
Phenanthrene	0.06		0.52		C	Orem et al. (2007)
Phenanthrene derivative	3				S	Lester et al. (2015)
Phenanthrene-1-carboxlic acid	0.07				C	Orem et al. (2007)
Phenanthrenone	0.02		0.12		C	Orem et al. (2007)
Phenol	0.05		0.09		C	Orem et al. (2007)
Phosphoric acid, tributyl ester	830				S	Lester et al. (2015)
	0.1		18.96		C	Orem et al. (2007)

Chemical	Minimum or only value (µg/L)	Average (µg/L)	Maximum (µg/L)	Standard deviation (µg/L)	Formation type (S for shale, C for coalbed)	Reference
Propane-diphenyl	0.03		0.22		C	Orem et al. (2007)
p-Tert-butylphenol	0.07		0.19		C	Orem et al. (2007)
p-Xylene	10	150	460	160	S	Khan et al. (2016)
Pyrene	0.01		0.04		C	Orem et al. (2007)
	0.9				S	Lester et al. (2015)
Pyreno[4,5-c]furan	1.83				C	Orem et al. (2007)
Quinolo-furazan derivative	0.82				C	Orem et al. (2007)
Squalene	<=0.24				C	Orem et al. (2007)
Sterane	0.51				C	Orem et al. (2007)
Tetracosane	1.86				C	Orem et al. (2007)
Tetradecane	0.54				C	Orem et al. (2007)
Tetradecanoic acid	0.15		0.54		C	Orem et al. (2007)
Tetrahydro-dimethylnaphthalene	0.19		3.25		C	Orem et al. (2007)
Tetrahydromethylnaphthalene	0.01		0.69		C	Orem et al. (2007)
Tetrahydronaphthalene	0.06		0.82		C	Orem et al. (2007)
Tetrahydrophenanthrene	0.03		0.42		C	Orem et al. (2007)
Tetrahydro-trimethylnaphthalene	0.5				C	Orem et al. (2007)
Tetramethylacenaphthylene	0.03		0.07		C	Orem et al. (2007)
Tetramethylnaphthalene	0.43		0.79		C	Orem et al. (2007)

Appendix E – Produced Water Handling Supplemental Information

Chemical	Minimum or only value (µg/L)	Average (µg/L)	Maximum (µg/L)	Standard deviation (µg/L)	Formation type (S for shale, C for coalbed)	Reference
Tetramethylphenanthrene	0.01		0.68		C	Orem et al. (2007)
Toluene	100	1,560	5,610	1,940	S	Khan et al. (2016)
Total xylenes	30				S	Lester et al. (2015)
Tricosane	1.7				C	Orem et al. (2007)
Tricyclo[4.4.0.0(3,9]decane	0.26				C	Orem et al. (2007)
Tridecanedial	0.86				C	Orem et al. (2007)
Trimethoxy-benzaldehyde	0.39				C	Orem et al. (2007)
Trimethylnaphthalene	0.04		2.6		C	Orem et al. (2007)
Trimethylphenanthrene	0.04		0.12		C	Orem et al. (2007)
Triphenyl phosphate	0.07		0.21		C	Orem et al. (2007)

E.3.6. Chemical Reactions

Section 7.3.4.9 describes general aspects of subsurface chemical reactions that might occur during hydraulic fracturing operations. Here we augment the discussion by describing subsurface chemical processes.

E.3.6.1. Injected Chemical Processes

Hydraulic fracturing injects relatively oxygenated fluids into a reducing environment, which may mobilize trace or major constituents into solution. Injection of oxygenated fluids may lead to short-term changes in the subsurface redox state, as conditions may shift from reducing to oxidizing. The chemical environment in hydrocarbon-rich unconventional reservoirs, such as black shales, is generally reducing, as evidenced by the presence of pyrite and methane (Engle and Rowan, 2014; Dresel and Rose, 2010). For black shales, reducing conditions are a product of original accumulations of organic matter whose decay depleted oxygen to create rich organic sediments within oil- and gas-producing formations (Tourtelot, 1979; Vine and Tourtelot, 1970). Yet reactions resulting from temporary redox shifts are likely to be less important than those resulting from other longer-term physical and geochemical processes. Temporary subsurface redox shifts may be due to the short timeframe for fluid injection (a few days to a few weeks).

Hydraulic fracturing fluid injection introduces novel chemicals into the subsurface.[1] As such, the geochemistry of injected and native fluids will not be in equilibrium. Over the course of days to months, a complex series of reactions will equilibrate disparate fluid chemistries. The evolution of flowback and produced water geochemistry are dependent upon the exposure of formation solids and fluids to novel chemicals within hydraulic fracturing fluid. Additives interact with reservoir solids and either mobilize constituents or themselves become adsorbed to solids. Such additives include metallic salts, elemental complexes, salts of organic acids, organometallics, and other metal compounds (Montgomery, 2013; House of Representatives, 2011).

The salts, elemental complexes, organic acids, organometallics, and other metal-containing compounds may interact with metals and metalloids in the target formation through processes such as ion exchange, adsorption, desorption, chelation, and complexation. For instance, natural organic ligands (e.g., citrate) are molecules that can form coordination compounds with heavy metals such as cadmium, copper, and lead (Martinez and McBride, 2001; Stumm and Morgan, 1981; Bloomfield et al., 1976). Citrate-bearing compounds are used in hydraulic fracturing fluids as surfactants, iron control agents, and biocides. Studies of the additives' interactions with formation solids at concentrations representative of hydraulic fracturing fluids are lacking.

Furthermore, pH will likely play a role in the nature and extent of these processes, as the low pH of hydraulic fracturing fluids may mobilize trace constituents. The pH of hydraulic fracturing fluids may differ from existing subsurface conditions due to the use of dilute acids (e.g., hydrochloric or acetic) used for cleaning perforations and fractures during hydraulic fracturing treatments (Montgomery, 2013; GWPC and ALL Consulting, 2009). Metals within formation solids may be

---

[1] For more information on additive usage, refer to Chapter 5 (Chemical Mixing).

released through the dissolution of acid-soluble phases such as iron and manganese oxides or hydroxides (Yang et al., 2009; Kashem et al., 2007; Filgueiras et al., 2002). Thus, the pH of hydraulic fracturing fluids, or changes in system pH that may occur as fluid recovery begins, may influence which metals and metalloids are likely to be retained within the formation and which may be recovered in flowback. Ultimately, more research is needed to fully understand how the injection of hydraulic fracturing fluids affects subsurface geochemistry and resultant flowback and produced water chemistry.

E.3.7. Microbial Community Processes and Content

By design, hydraulic fracturing releases hydrocarbons and other reduced mineral species from freshly fractured shale, sandstone, and coal, resulting in saltier in situ fluids, the release of formation solids, and increased interconnected fracture networks with rich colonization surfaces that are ideal for microbial growth (Wuchter et al., 2013; Curtis, 2002). The use of biocides, in contrast, is intended to inhibit microbial growth. Recent work by Kahrilas et al. (In Press) performed laboratory experiments to simulate downhole chemistry of the biocide glutaraldehyde at 200 °C temperature, 10 MPa pressure, and high salinity. The laboratory results suggested that in hot, alkaline shales, the effectiveness of glutaraldehyde as a biocide is limited by contact time; and is not so limited in cooler, more acidic, saline formations like the Marcellus.

Depending upon the formation, microorganisms may be native to the subsurface and/or introduced from non-sterile equipment and fracturing fluids. Additionally, microorganisms compete for novel organics in the form of additives (Wuchter et al., 2013; Arthur et al., 2009). Since large portions of hydraulic fracturing fluid can remain emplaced in the targeted formation, long-term microbial activity is supported through these novel carbon and energy resources (Orem et al., 2014; Murali Mohan et al., 2013a; Struchtemeyer and Elshahed, 2012; Bottero et al., 2010). Such physical and chemical changes to the environment at depth stimulate microbial activity and influence flowback and produced water content in important ways.

Several studies characterizing produced water from unconventional reservoirs (i.e., the Barnett, Marcellus, Utica, and Antrim Shales) indicate that taxa with recurring physiologies compose shale flowback and produced water microbial communities (Murali Mohan et al., 2013b; Wuchter et al., 2013). Such physiologies include sulfur cyclers (e.g., sulfidogens: sulfur , sulfate , and thiosulfate reducers); fermenters; acetogens; hydrocarbon oxidizers; methanogens; and iron, manganese, and nitrate reducers (Davis et al., 2012).

Based on their physiologies, microorganisms cycle substrates at depth by mobilizing or sequestering constituents in and out of solution. Mobilization can occur through biomethylation, complexation, and leaching. Sequestration can occur through intracellular sequestration, precipitation, and sorption to biomass.

The extent to which constituents are mobilized or sequestered depends upon the prevailing geochemical environment after hydraulic fracturing and through production. Significant environmental factors that influence the extent of microbially mediated reactions are increases in ionic content (i.e., salinity, conductivity, total nitrogen, bromide, iron, and potassium); decreases in acidity, and organic and inorganic carbon; the availability of diverse electron acceptors and donors; and the availability of sulfur-containing compounds (Cluff et al., 2014; Murali Mohan et al., 2013b; Davis et al., 2012). Examples follow that illustrate how subsurface microbial activity influences the content of produced water.

Under prevailing anaerobic and reducing conditions, microorganisms can mobilize or sequester metals found in produced water from unconventional reservoirs (Gadd, 2004). Microbial enzymatic reduction carried out by chromium-, iron-, manganese-, and uranium-reducing bacteria can both mobilize and sequester metals (Vanengelen et al., 2008; García et al., 2004; Mata et al., 2002; Gauthier et al., 1992; Myers and Nealson, 1988; Lovley and Phillips, 1986). For instance, iron and manganese species go into solution when reduced, while chromium and uranium species precipitate when reduced (Gadd, 2004; Newman, 2001; Ahmann et al., 1994).

Metals can also be microbially solubilized by complexing with extracellular metabolites, siderophores (metal-chelating compounds), and microbially generated bioligands (e.g., organic acids) (Glorius et al., 2008; Francis, 2007; Gadd, 2004; Hernlem et al., 1999). For example, Pseudomonas spp. secrete acids that act as bioligands to form complexes with uranium(VI) (Glorius et al., 2008).

Many sulfur-cycling taxa have been found in hydraulic fracturing flowback and produced water communities (Murali Mohan et al., 2013b; Mohan et al., 2011). Immediately following injection, microbial sulfate reduction is stimulated by diluting high-salinity formation waters with fresh water (high salinities inhibit sulfate reduction). Microbial sulfate reduction oxidizes organic matter and decreases aqueous sulfate concentrations, thereby increasing the solubility of barium (Cheung et al., 2010; Lovley and Chapelle, 1995).

Sulfidogens also reduce sulfate, as well as elemental sulfur and other sulfur species (e.g., thiosulfate) prevalent in the subsurface, contributing to biogenic sulfide or hydrogen sulfide gas in produced water (Alain et al., 2002; Ravot et al., 1997). Sulfide can also sequester metals in sulfide phases (Ravot et al., 1997; Lovley and Chapelle, 1995). Sources of sulfide also include formation solids (e.g., pyrite in shale) and remnants of drilling muds (e.g., barite and sulfonates), or other electron donor sources (Davis et al., 2012; Kim et al., 2010; Collado et al., 2009; Grabowski et al., 2005).

Additionally, anaerobic hydrocarbon oxidizers associated with shale produced water can readily degrade simple and complex carbon compounds across a considerable salinity and redox range (Murali Mohan et al., 2013b; Fichter et al., 2012; Timmis, 2010; Lalucat et al., 2006; Yakimov et al., 2005; McGowan et al., 2004; Hedlund et al., 2001; Cayol et al., 1994; Gauthier et al., 1992; Zeikus et al., 1983).

Lastly, microbial fermentation produces organic acids, alcohols, and gases under anaerobic conditions, as is the case during methanogenesis. Some nitrogen-cycling genera have been identified in shale gas systems. These include genera involved in nitrate reduction and denitrification (Kim et al., 2010; Yoshizawa et al., 2010; Yoshizawa et al., 2009; Lalucat et al., 2006).

These genera likely couple sugar, organic carbon, and sulfur species oxidation to nitrate reduction and denitrification processes.

Consequently, using a variety of recurring physiologies, microorganisms mobilize and sequester constituents in and out of solution to influence the content of produced water.

E.4.   Produced Water Content Spatial Trends

E.4.1. Variability between Plays of the Same Rock Type

E.4.1.1. Shale Formation Variability

The content of shale produced water varies geographically, as shown by data from four formations (the Bakken, Barnett, Fayetteville, and Marcellus Shales; see Table E-2, Table E-4, Table E-6, Table E-8, Table E-9, and Table E-11). For several constituents, variability between shale formations is common. The average/median TDS concentrations in the Marcellus (87,800 to 106,390 mg/L) and Bakken (196,000 mg/L) Shales are one order of magnitude greater than the average TDS concentrations reported for the Barnett and Fayetteville Shales (Table E-2). As Fayetteville produced water contains the lowest reported average TDS concentration (13,290 mg/L), average concentrations for many inorganics (i.e., bromide, calcium, chloride, magnesium, sodium, and strontium) that contribute to dissolved solids loads are the lowest compared to average concentrations for the same inorganics in Bakken, Barnett, and Marcellus produced water (Table E-4 and Table E-6). Average concentrations for metals reported within Bakken and Marcellus produced water are also higher than those within the Barnett or Fayetteville formations (Table E-6).

Additionally, Marcellus produced water is enriched in barium (average concentration of 2,224 mg/l in Barbot et al. (2013) or median calculated from Hayes (2009) of 542.5 mg/L) and strontium (average concentration of 1,695 mg/L (Barbot et al., 2013) or median calculated from Hayes (2009) of 1,240 mg/L) by one to three orders of magnitude compared to Bakken, Barnett, and Fayetteville produced water (Table E-6). Subsequently, radionuclide variability expressed as isotopic ratios (e.g., radium-228/radium-226, strontium-87/strontium-86) are being used to determine the reservoir source for produced water (Chapman et al., 2012; Rowan et al., 2011; Blauch et al., 2009). Lastly, Barnett and Bakken produced waters are enriched in sulfate.

Although organic data are limited, average BTEX concentrations are higher in Marcellus compared to Barnett produced water by one order of magnitude, whereas concentrations of benzene alone are marginally higher in Barnett compared to Marcellus produced water (Table E-9 and Table E-11).

E.4.1.2. Tight Formation Variability

The average concentrations for various constituents in tight formation produced water vary geographically between sandstone formations (the Cotton Valley Group, Devonian sandstone, and the Mesaverde and Oswego), as shown in Table E-2, Table E-4, Table E-6, Table E-8, and Table E-9. The average TDS concentrations in the Devonian sandstone (235,125 mg/L) and Cotton Valley

Group (164,683 mg/L) are one to two orders of magnitude greater than the average TDS concentrations reported for the Mesaverde (15,802 mg/L) and Oswego Formations (73,082 mg/L) (Table E-2). Mesaverde produced water also contained the lowest average concentrations for many of the inorganic components of TDS (i.e., calcium, chloride, iron, magnesium, and sodium) (Table E-4 and Table E-6).

Little variability was reported in pH between these four tight formations (E-2). Mesaverde produced water was enriched in sulfate, with an average concentration of 837 mg/L (Table E-4), whereas Devonian produced water was enriched in barium, which had an average concentration of 1,488 mg/L (Table E-6).

E.4.1.3. Coalbed Variability

Geochemical analysis showed that the Powder River Basin is predominately characterized by bicarbonate water types with a large intrusion of sodium-type waters across a large range of magnesium and calcium concentrations (Dahm et al., 2011).[1] In contrast, the Raton Basin is typified by sodium-type waters with low calcium and magnesium concentrations. A combination of Powder River and Raton produced water compositional characteristics typifies the San Juan Basin (Dahm et al., 2011). Lastly, Black Warrior Basin produced water is differentiated based upon its sodium bicarbonate- or sodium chloride-type waters (DOE, 2014; Pashin et al., 2014).

Regional variability is observed in average produced water concentrations for various constituents of four CBM basins (Powder River, Raton, San Juan, and Black Warrior (Table E-3, Table E-5, Table E-7, Table E-9, and Table E-12), but particularly between produced water of the Black Warrior Basin and the others. As the average TDS concentration in Black Warrior Basin produced water (14,319 mg/L) is one to two orders of magnitude higher than that of the other three presented in Table E-5, average concentrations for TDS contributing ions (i.e., calcium, chloride, and sodium) were also higher than in the Powder River, Raton, and San Juan Basins. These high levels follow from the marine depositional environment of the Black Warrior Basin (Horsey, 1981).

Powder River Basin produced water has the lowest average TDS concentration (997 mg/L), which is consistent with Dahm et al. (2011) reporting that nearly a quarter of all the produced water sampled from the Powder River Basin meets the U.S. drinking water secondary standard for TDS (less than 500 mg/L).[2] In addition, the Black Warrior Basin appears to be slightly enriched in barium, compared to the other three CBM basins (Table E-5). Lastly, the three western CBM basins

---

[1] Water is classified as a "type" if the dominant dissolved ion is greater than 50% of the total. A sodium-type water contains more that 50% of the cation milliequivalents (mEq) as sodium. Similarly, a sodium-bicarbonate water contains 50% of the cation mEq as sodium, and 50% of the anion mEq as bicarbonate (USGS, 2002).

[2] MCL refers to the highest level of a contaminant that is allowed in drinking water. MCLs are enforceable standards. These include primary MCLs for barium, cadmium, chromium, lead, mercury, and selenium. National Secondary Drinking Water Regulations (NSDWRs or secondary standards) are non-enforceable guidelines regulating contaminants that may cause cosmetic effects (such as skin or tooth discoloration) or aesthetic effects (such as taste, odor, or color) in drinking water. Secondary MCLs are recommended for aluminum, chloride, copper, iron, manganese, pH, silver, sulfate, TDS, and others. See http://water.epa.gov/drink/contaminants/index.cfm#Primary for more information.

(Powder River, Raton, and San Juan) are much more alkaline and enriched in bicarbonate than their eastern counterpart (the Black Warrior Basin; Table E-3).

Average concentrations of benzene, ethylbenzene, and xylenes are higher in San Juan compared to Raton produced water by two orders of magnitude, whereas concentrations of toluene are marginally higher in Raton compared to San Juan produced water (Table E-9).

E.4.2. Local Variability

Spatial variability of produced water content frequently exists within a single producing formation. For instance, Marcellus Shale barium levels increase along a southwest to northeast transect (Barbot et al., 2013). Additionally, produced water from the northern and southern portions of the San Juan Basin differ in TDS, due to groundwater recharge in the northern basin leading to higher chloride concentrations than in the southern portion (Dahm et al., 2011; Van Voast, 2003).

Spatial variability of produced water content also exists at a local level due to the stratigraphy surrounding the producing formation. For example, deep saline aquifers, if present in the over- or underlying strata, may over geologic time encroach upon shales, coals, and sandstones via fluid intrusion processes (Blauch et al., 2009). Evidence of deep brine migration from adjacent strata into shallow aquifers via natural faults and fractures has been noted previously in the Michigan Basin and the Marcellus Shale (Vengosh et al., 2014; Warner et al., 2012; Weaver et al., 1995). By extension, in situ hydraulic connectivity, which is stimulated by design during hydraulic fracturing, may lead to the migration of brine-associated constituents in under- and overlying strata into producing formations, as discussed in Chapter 6.

E.5.   North Dakota Spill Analysis

E.5.1. Materials and Methods

Incidents were reported to the North Dakota Department of Health from across the Bakken Gas Shale, Late Devonian to Early Mississippian in age. We reviewed incidents occurring during the years 2001-2015, and categorized them by release type: salt water (SW), oil, and other.[1] First, two years (2014 and 2015) of Oil Field dataset was retrieved from the North Dakota Spills Database Website operated by the North Dakota Department of Health, Division of Water Quality (http://www.ndhealth.gov/EHS/Spills/). The entire public dataset to date was later (March 15, 2016) obtained directly from the ND Department of Health for our analysis of the years 2001 to 2015. The data from 2014 and 2105 were used to summarize causes of spills.

Our method of data-cleaning involved eliminating data with empty cells (NA), or reports of "0" values. If data were presented as "0" bbl or gal for SW, oil, and other spills, we omitted those values from the dataset (n= 434). Additionally, cells containing "0" or "NA" for SW, oil, and other reported spills were omitted from the dataset (n= 98). A single spill with unit "lbs", referring to dust used in

---

[1] The "other" category also includes spills categorized as: freshwater, condensate, drilling mud, injection fluid, emulsion, injection chemical, petroleum, product, misc, uncharacterized, oil and water, freshwater and brine, and drill cuttings. Some incidents did not release a liquid as, for example, the release could have only been gas.

processing of drill cuttings, was omitted (n=1). All values were converted to barrels (bbl when necessary).

The dataset containing SW, oil, and other, was further divided into three datasets based on spill type. The compiled statistics only included releases with volumes above (SW: n=6238, oil: n=4882, and other: n=9863). Unlike the Oklahoma study reported in the main text (Fisher and Sublette, 2005), we are not able to identify salt water spills whose volume was not estimated.

The spill rates were determined by dividing the spill counts and volumes by the number of active production wells. The latter data were obtained from the North Dakota Oil and Gas Division web site (https://www.dmr.nd.gov/oilgas/stats/statisticsvw.asp). Monthly well counts are available for the years of interest, and we used the active well count for December of each year in our calculations. Alternatively different months or the average for the entire year could be used. Through testing, we found no meaningful differences in the estimates. The median (or middle) volume of produced water (SW) spills was consistently about 340 gal (1,300 L) for the period 2001 to 2015 (Figure 7-13).[1] The data are represented by box plots in the main text (Figure E-6).

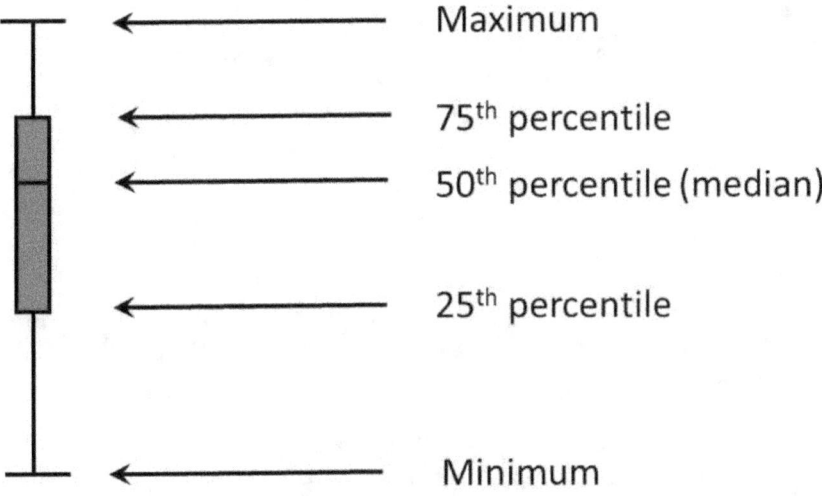

Figure E-6. Illustration of a "box" or "box and whisker" plot.

---

[1] 50% of spill volumes were below and 50% above the median value. Medians are less sensitive to extreme values than means (averages). Means above the median indicate that the distribution is skewed by a relatively small number of incidents with high spill volumes.

E.5.2. Results

For comparison with the other types of spilled liquids, after 2009 the median volume for oil spills tended toward 130 gal (480 L) for oil (Figure E-7) and 210 gal (790 L) for all other spills (Figure E-8). In each case, however, the mean numbers of spills were higher than the medians, indicating that although the majority of SW spills were 340 gal (1,300 L) or less, larger volume spills occurred and increased the mean value. For SW spills, the largest spill recorded was 2,900,000 gal (11,000,000 L) occurring in January 2015. Although most of the SW spills contained 340 gal (1,300 L) or less, large spills (400,000 gal (11,000,000 L) or more) occurred in 2013, 2014, and 2015 (Figure 7-13).

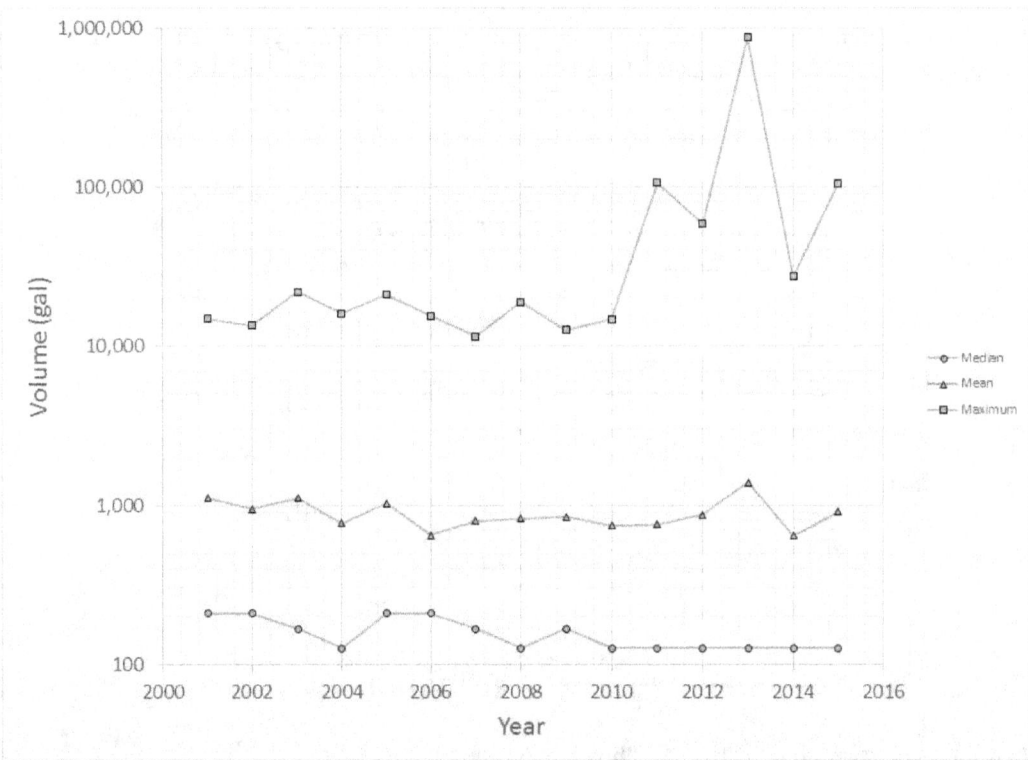

Figure E-7. Median, mean, and maximum volume of oil spills in North Dakota for 2001 to 2015.

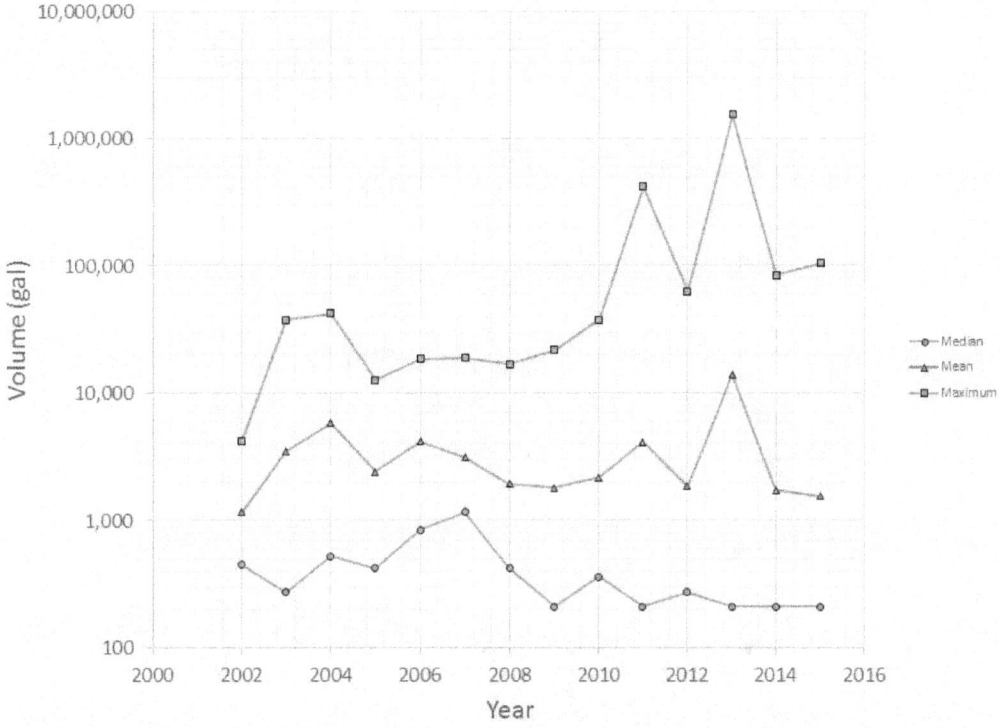

Figure E-8. Median, mean, and maximum volume of "other" spills in North Dakota for 2002 to 2015.

The number of spills increased with increasing numbers of active wells (Figure E-9). Each type of spill decreased from 2014 to 2015 (Figure E-9). From 2001 to 2007 the rate of oil and produced water spills were roughly the same (Figure E-9), afterwards there were fewer produced water spills. From 2010 to 2015, the rate of produced water spills ranged from 4.7 to 7.2 per hundred active wells; oil spills from 6.1 to 10.0 per hundred active wells and other spills from 1.7 to 3.7 per hundred active wells. By the end of 2015 there were over 13,000 active production wells in North Dakota, and these fractions corresponded to 613 produced water, 825 oil, and 369 other spills (Figure E-9). Although there were more oil than produced water spills, the median and maximum produced water spills were larger than the median oil spills.

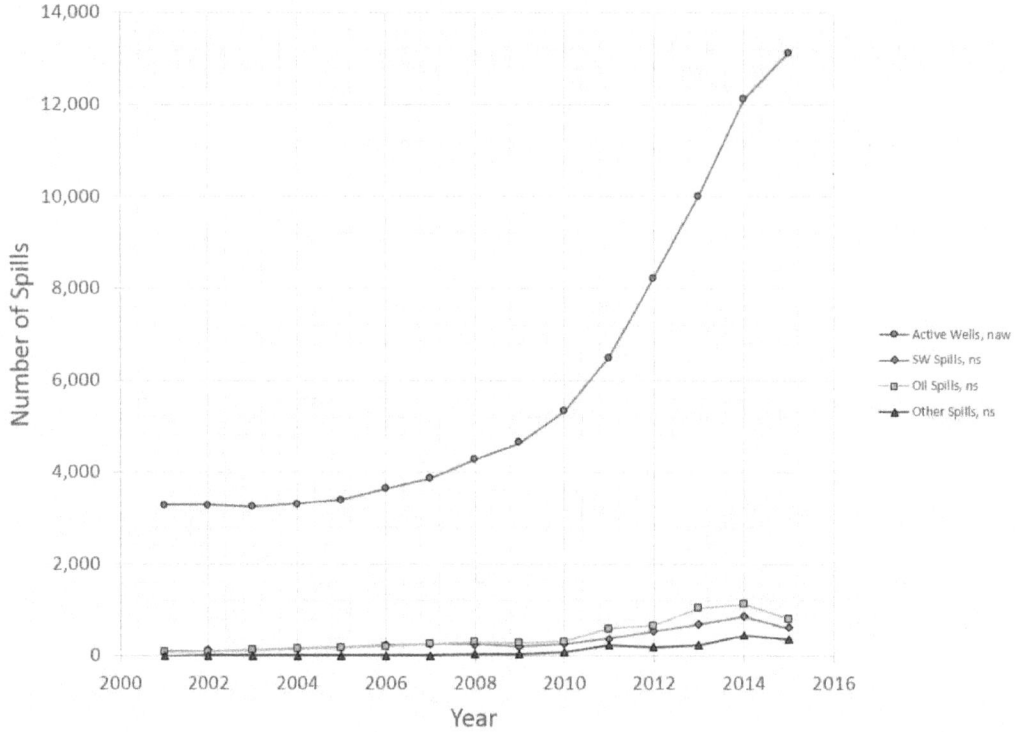

Figure E-9. Count of spills and active wells in North Dakota for the years 2001 to 2015.

North Dakota distinguishes between spills that are and are not contained within the boundaries of the production or exploration facility (http://www.ndhealth.gov/ehs/spills/). For each type of spill, more were contained than not contained (Figure E-10). The maximum spill sizes were generally higher in the not contained category (Figure E-12).

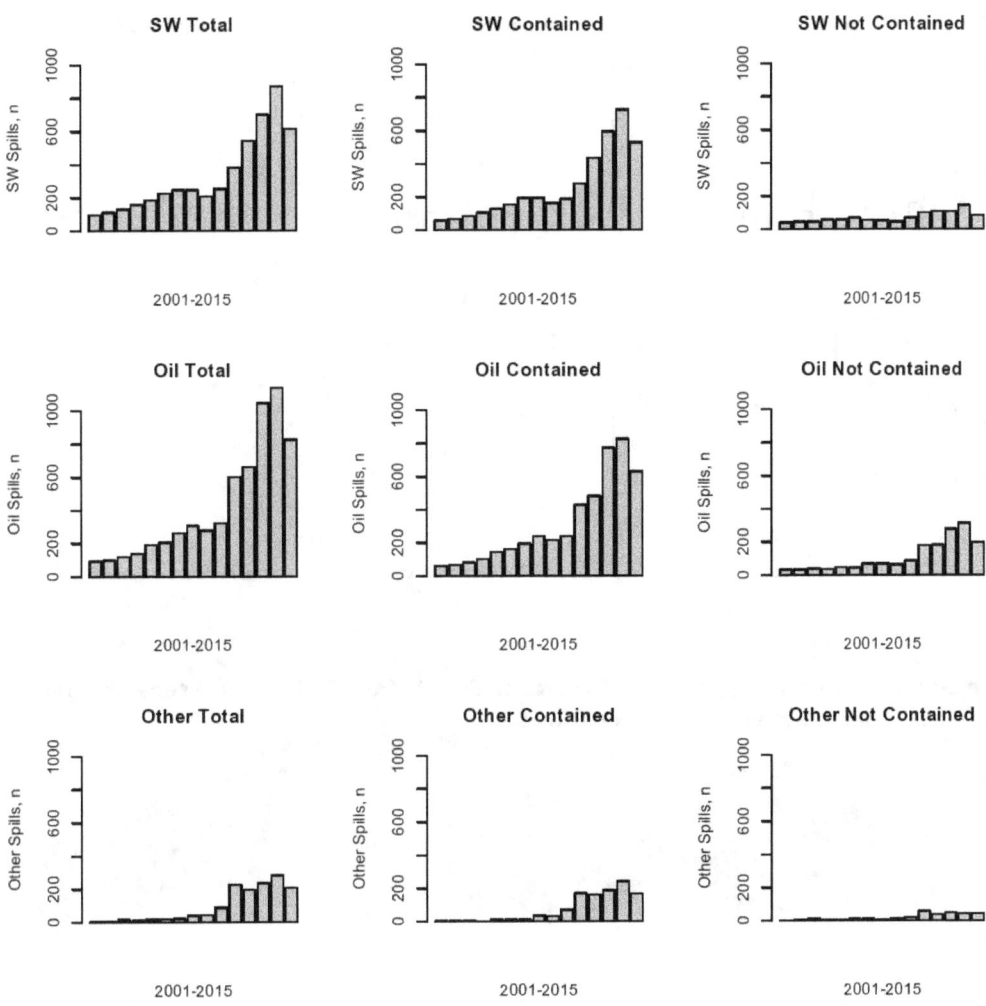

Figure E-10. Number of spills in North Dakota from 2001 to 2015 separated by type and by "contained" versus "not contained."

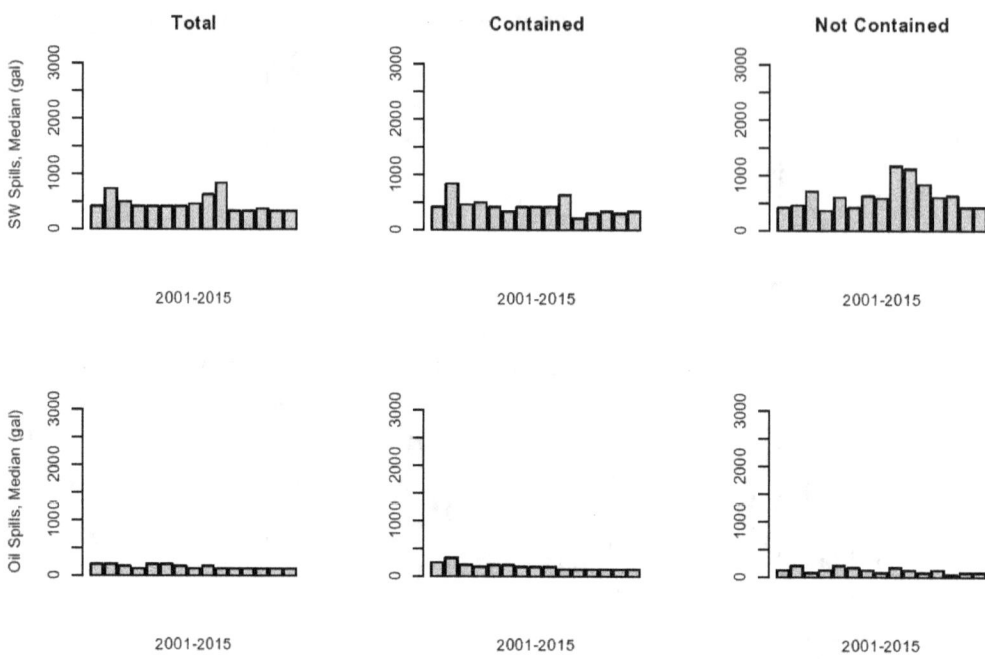

Figure E-11. Median volume (gal) of spills in North Dakota from 2001 to 2015 separated by type and by "contained" versus "not contained."

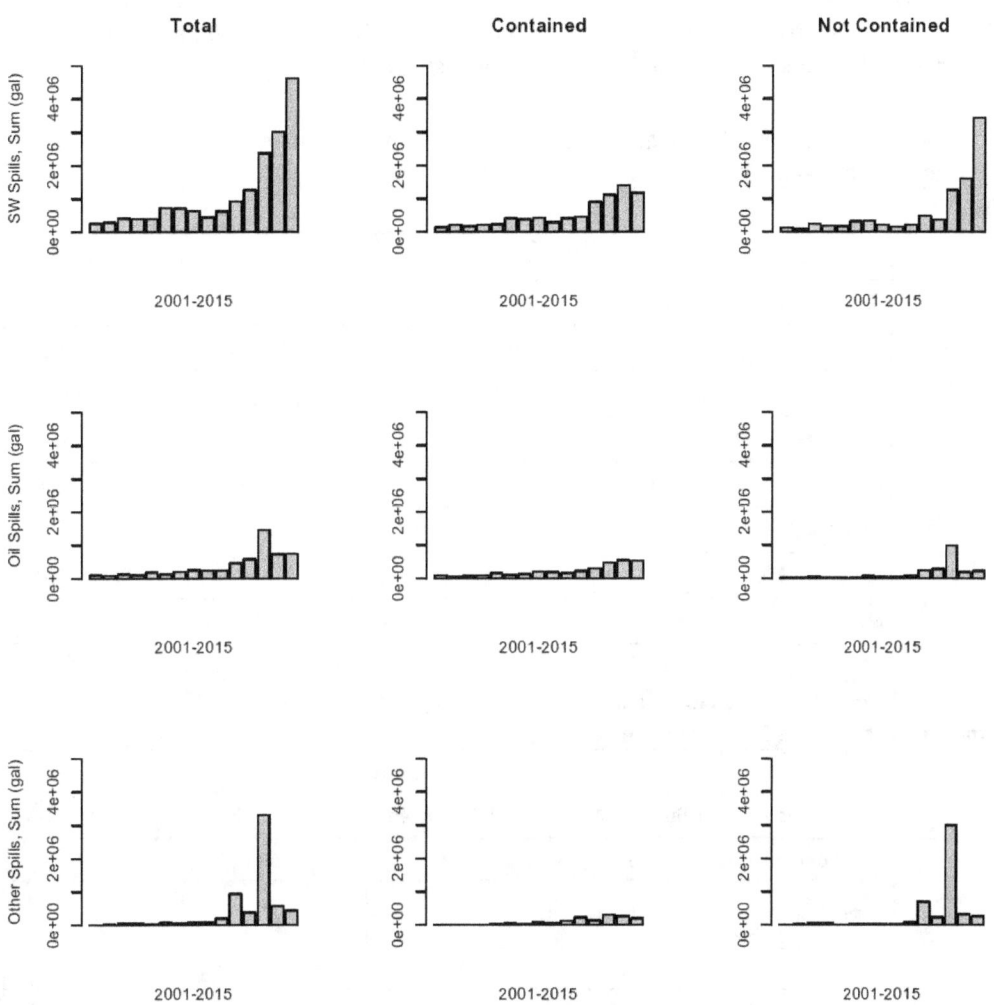

Figure E-12. Yearly sum of spill volume (gal) of spills in North Dakota from 2001 to 2015 separated by type and by "contained" versus "not contained."

The distribution of spills of each type is skewed. For 2015, the medians range from 8 to 80 gal (300 to 3,000 L) (considering contained and not contained of each type) but the maximums are much higher ranging from 50,000 to 2,900,000 gal (190,000 to 11,000,000 L) (Table E-14 and Table E-15). Further, the maximums are much higher than the 75th percentiles, indicating a relatively small number of large spills. Only a very few spills occur that are greater than 20,000 gal (80,000 L) (Table E-16). In the case of produced water, there were 12 spills over 20,000 gal (80,000 L), five over 40,000 gal (160,000 L) , and one greater than 400,000 gal (1,600,000 L) (Table E-16).

Spill causes were discussed for the composited produced water spills in the main text. Although small in absolute numbers, proportionately more pipeline leaks, "other," and stuffing box leaks caused produced water spills in 2015 (Figure E-14 and Figure E-15).

Table E-14. Volume distribution in gallons (minimum, 25th percentile, median, 75th percentile and maximum) for each type of spill in North Dakota for 2015.

Type	Spills	Min	25th	Med	75th	Max
Oil	Contained	1	40	130	420	94,000
	Not Contained	1	40	80	290	105,000
	All	1	40	130	340	105,000
Other	Contained	0.04	80	210	840	50,000
	Not Contained	2	80	840	840	105,000
	All	0.04	80	210	840	105,000
SW	Contained	1	80	340	1,300	340,000
	Not Contained	1	130	420	2,100	2,900,000
	All	1	80	340	1,300	2,900,000

Table E-15. Numbers of 2015 North Dakota spills in ranges defined by the spill volume statistics (Table E-14) for each type.

Type	Status	Count			
		Min ≤ x < 25th	25th ≤ x < Med	Med ≤ x < 75th	75th ≤ x ≤ Max
Oil	Contained	71	212	188	158
	Not Contained	59	29	57	51
	All	130	257	217	221
Other	Contained	55	32	44	35
	Not Contained	12	9	0	21
	All	67	35	50	56
SW	Contained	77	184	127	141
	Not Contained	21	18	21	24
	All	94	202	163	154

Table E-16. Number of 2015 North Dakota spills which exceed thresholds (20,000 gal, 40,000 gal, and 400,000 gal) for each type of spill.

Type	Status	Size ≥ 20,000 gal	≥ 40,000 gal	≥ 400,000 gal
Oil	Contained	3	3	0
	Not Contained	2	1	0
	All	5	4	0
Other	Contained	1	1	0
	Not Contained	2	2	0
	All	3	3	0
SW	Contained	6	2	0
	Not Contained	6	3	1
	All	12	5	1

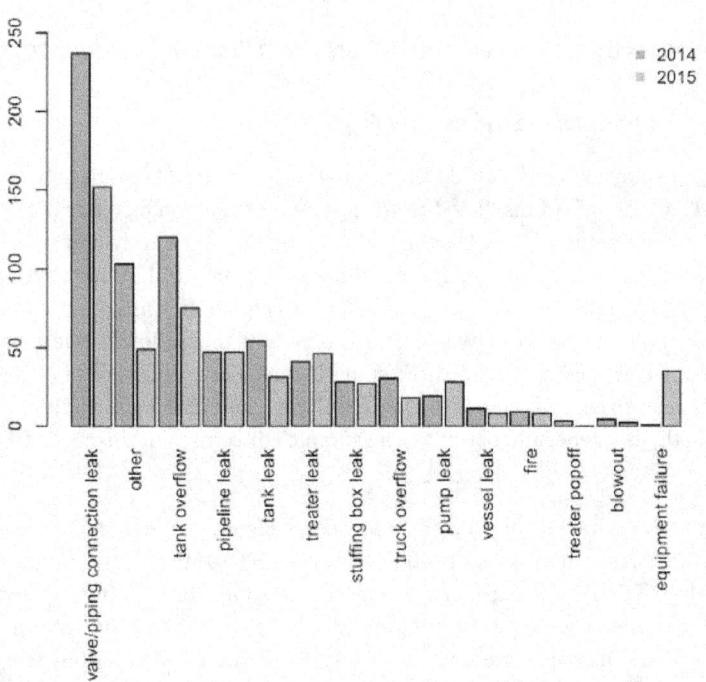

Figure E-13. Numbers of contained spills in North Dakota by cause for 2014 and 2015.

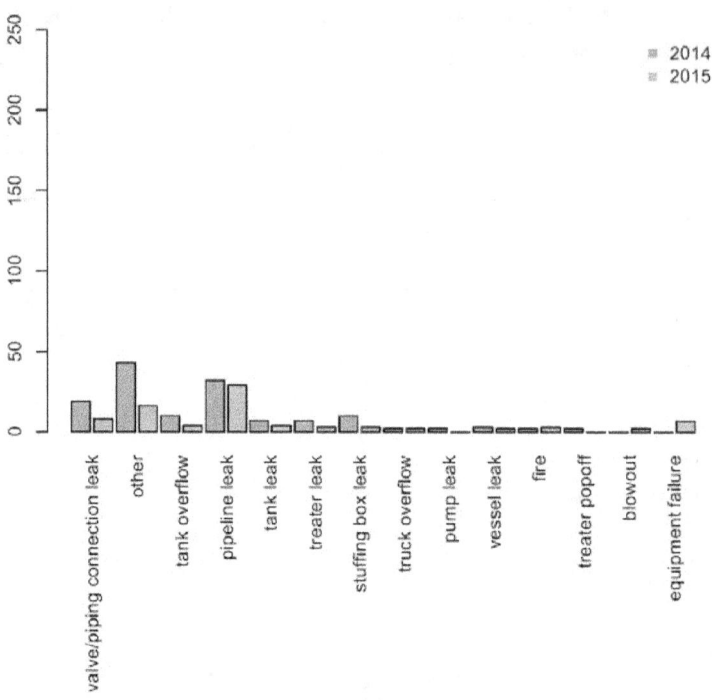

Figure E-14. Numbers of not contained spills in North Dakota by cause for 2014 and 2015.

E.5.3. Summary of Additional Studies on Spills

Gross et al. (2013) analyzed the Colorado Oil and Gas Conservation Commission's database for groundwater BTEX concentrations linked to storage and production facilities between July 2010 and July 2011 in Weld County, CO. Only spills with an impact on groundwater were included in the study. The 77 reported spills accounted for less than 0.5% of nearly 18,000 active wells. Forty-six of the 77 spills consisted of produced water and oil. Of the remaining spills, 23 consisted of only oil and eight consisted of only produced water. Thus the results that follow include cases with no produced water spill. From these composited spills, benzene concentrations in 90% of the groundwater samples exceeded 5 µg/L, the U.S. drinking water standard. Additionally, 30% of toluene, 12% of ethylbenzene, and 8% of xylene sample concentrations exceeded 1 mg/L, 0.7 mg/L and 10 mg/L, respectively (Gross et al., 2013).

Based on five spills for which volumes were reported, the average volume of a produced water spill was 294 gal (1,110 L), ranging from 42 (160 L) to 1,176 gal (4,450 L) (Gross et al., 2013). Spill areas averaged 2,120 ft² (197 m²) with an average depth of 7 ft (2 m). Tank battery systems and production facilities were the biggest volume sources of spills with groundwater impacts. Equipment failure was the most common cause of spills with groundwater impacts. Of the 77 reported spills, secondary containment was absent from 51 of them (Gross et al., 2013).

As noted from the Colorado (Gross et al., 2013) and Oklahoma (Fisher and Sublette, 2005) studies, oil releases may occur alongside produced water spills. Review of recent oil field incidents in North Dakota (from information on the state's website at http://www.ndhealth.gov/EHS/Spills/) also shows incidents with both produced water and oil releases. Oil releases are characterized by a number of features including their unique hydrocarbon composition and physical properties. Impacts can include: surface runoff, infiltration into soils, formation of sheens and oil slicks on surface waters, evaporation, oxidation, biodegradation, emulsion formation, and particle deposition (U.S. EPA, 1999b).

Brantley et al. (2014) reviewed PA DEP's online oil and gas compliance database for notices of violation issued to companies developing gas resources in unconventional reservoirs. Between May 2009 and April 2013, eight spills of flowback and produced water ranging from more than 4,000 gal (15,000 L) to more than 57,000 gal (220,000 L) reached surface water resources. The spills typically resulted in local impacts to environmental receptors and required remediation and monitoring. However, the study indicated the likelihood of a leak or spill of hydraulic fracturing-related fluids was low (less than 1%, based on 32 large spills out of more than 4,000 complete wells). Due to lack of data, specific impacts to the eight receiving surface waters were not discussed, other than noting the produced water had contacted the surface water.

E.6.   Evaluation of Impacts

As an example of set of criteria for assessing sites potentially contaminated by hydraulic fracturing activities, the U.S. EPA (2012e) developed an approach to study sites with suspected impacts from hydraulic fracturing activities. The approach was based on a tiered scheme where results from each tier are used to refine activities in higher tiers. The four tiers, with some modification, are as follows:

Verify potential issue:

- Evaluate existing data and information from operators, private citizens, federal, state and local agencies, and tribes (as appropriate). Including studies of local groundwater quality that might have been conducted by USGS.
- Conduct site visits.
- Interview stakeholders and interested parties.

Determine approach for detailed investigations:

- Establish sampling locations
- Conduct initial sampling of water wells, taps, surface water, and soils.
- Identify potential evidence of drinking water contamination.
- Develop conceptual site model describing possible sources and pathways of the reported or potential contamination.
- Develop, calibrate, and test fate and transport model(s).

Conduct detailed investigations to detect and evaluate potential sources of contamination:

- Conduct additional sampling of soils, aquifer, surface water, and produced water pits/tanks where present.

- Conduct additional testing, including further water testing with new monitoring points, soil gas surveys, geophysical testing, well mechanical integrity testing, and stable isotope analyses.

- Refine conceptual site model and further test exposure scenarios.

- Refine fate and transport model(s) based on new data.

Determine the source(s) of any impacts to drinking water resources:

- Develop multiple lines of evidence to determine the source(s) of impacts to drinking water resources.

- Exclude possible sources and pathways of the reported contamination.

- Assess uncertainties associated with conclusions regarding the source(s) of impacts.

This tiered assessment strategy provides an outline for collecting data and evaluating lines of evidence to determine whether impacts have occurred. An outline of the quality assurance project plan (QAPP) for the EPA's *Retrospective case study in northeastern Pennsylvania: Study of the potential impacts of hydraulic fracturing on drinking water resources* (U.S. EPA, 2014d, 2012d) is given in Table E-17, and a graphical presentation of the relationships among quality assurance blanks is shown in Figure E-15. Table E-18 summarizes the lines of evidence used in the EPA's *Retrospective case study in Wise County, Texas: Study of the potential impacts of hydraulic fracturing on drinking water resources* (U.S. EPA, 2015i).

Table E-17. Outline of Northeastern Pennsylvania Retrospective Case Study QAPP.

Topic	Elements
Sampling Process Design	Background information on geology, hydrology, and geochemistry
	Groundwater and surface water monitoring
Sampling Methods	Domestic wells
	Surface waters: springs, ponds, and streams
Sampling Handling and Custody	Water sample labeling
	Water sample packing, shipping, and receipt at laboratories
Analytical Methods	Groundwater and surface water
Quality Control	Quality metrics for aqueous analysis
	Measured and Calculated solute concentration data evaluation
	Detection limits
	QA/QC calculations

Topic	Elements
Instrumentation	Testing, inspection, and maintenance
	Equipment calibration and frequency
	Acceptance of supplies and consumables
Non-direct Data	Assurance of Quality of 3RD party data (i.e., USGS background water quality data, university research publications)
Data Management	Recording
	Storage
	Analysis
Assessment and Oversight	Assessments
	Assessment Reporting
Data Validation and Usability	Data review, verification, and validation
	Verification and validation methods
	Reconciliation with user requirements

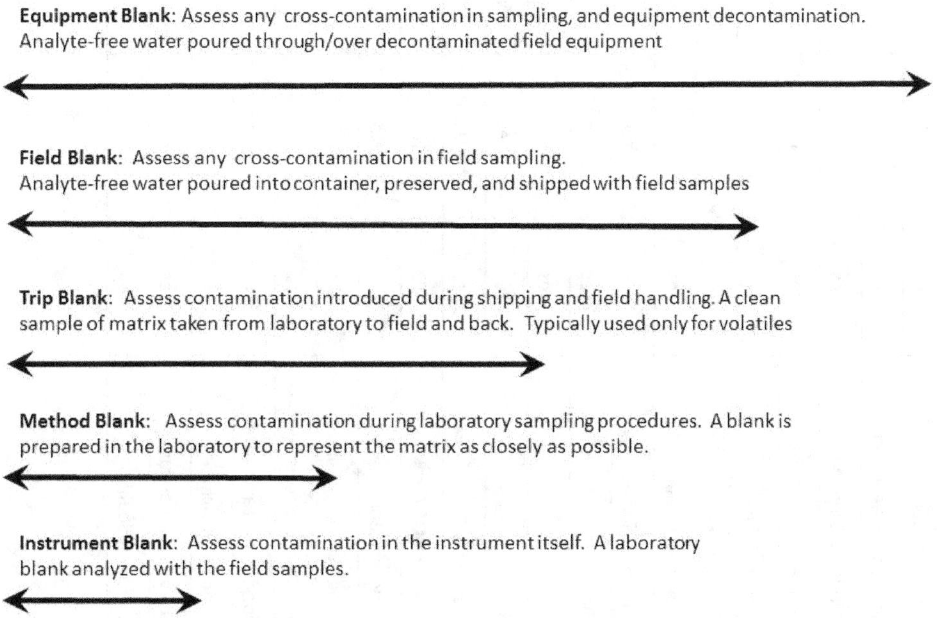

Equipment Blank: Assess any cross-contamination in sampling, and equipment decontamination. Analyte-free water poured through/over decontaminated field equipment

Field Blank: Assess any cross-contamination in field sampling. Analyte-free water poured into container, preserved, and shipped with field samples

Trip Blank: Assess contamination introduced during shipping and field handling. A clean sample of matrix taken from laboratory to field and back. Typically used only for volatiles

Method Blank: Assess contamination during laboratory sampling procedures. A blank is prepared in the laboratory to represent the matrix as closely as possible.

Instrument Blank: Assess contamination in the instrument itself. A laboratory blank analyzed with the field samples.

Figure E-15. Quality assurance blanks illustrating giving their purpose, brief procedure, and the span of their scope (modified from US EPA Region 3 Quality Control Tools: Blanks, April 27, 2009).
For example, the equipment blank spans all aspects of sampling and analysis from field to laboratory, while the instrument blank only assess contamination in the instrument itself. Reviewing results from all of these blanks could narrow down the source of sample cross-contamination.

Table E-18. Source delineation analysis table from the EPA retrospective case study in Wise County, Texas.

Well	Technique	Brine	Sea Water	Halite/ Road Salt	Landfill Leachate	Sewage/ Septic Tank	Animal Waste
WISETXGW01	Bromide vs. Boron	Yes	Yes	Yes	No	No	No
	Chloride vs. Magnesium	Yes	Yes	No	No	No	No
	Chloride vs. Bromide	Yes	Yes	Yes	No	No	Yes
	Chloride vs. Bicarbonate	Yes	Yes	Yes	No	No	No
	Chloride vs. Calcium	Yes	Yes	Yes	No	No	No
	Chloride vs. Potassium	Yes	Yes	Yes	No	No	No
	Chloride vs. Sodium	Yes	Yes	Yes	No	No	No
	Chloride vs. Sulfate	Yes	Yes	Yes	No	No	No
	Cl/Br	Yes	Yes	Yes	Yes	No	No
	Cl/I	Yes	Yes	Yes	Yes	No	Yes
	K/Rb	Yes	Yes	No Data[a]	No Data[a]	No Data[a]	No Data[a]
	Sr Isotope	Yes	No Data[a]	No Data[a]	No Data[a]	No Data[a]	No Data[a]
	Percentage Of Yes[b]	100	100	90	20	0	20
WISETXGW05	Bromide vs. Boron	No	No	No	No	No	No
	Chloride vs. Magnesium	No	No	No	No	Yes	Yes
	Chloride vs Bromide	No	No	No	No	No	No
	Chloride vs. Bicarbonate	No	No	No	Yes	No	Yes
	Chloride vs. Calcium	No	No	No	No	Yes	Yes
	Chloride vs. Potassium	Yes	Yes	Yes	Yes	Yes	Yes
	Chloride vs. Sodium	Yes	No	No	Yes	Yes	Yes

Well	Technique	Brine	Sea Water	Halite/Road Salt	Landfill Leachate	Sewage/Septic Tank	Animal Waste
WISETXGW05, cont.	Chloride vs. Sulfate	Yes	Yes	Yes	No	No	No
	Cl/Br	No	No	No	No	No	No
	Cl/I	No Data[a]	No Data[a]	No Data[a]	No Data[a]	No Data[a]	No Data[a]
	K/Rb	Yes	Yes	No Data[a]	No Data[a]	No Data[a]	No Data[a]
	Sr Isotope	Yes	No Data[a]	No Data[a]	No Data[a]	No Data[a]	No Data[a]
	Percentage Of Yes[b]	**45**	**30**	**22**	**33**	**44**	**46**
WISETXGW08	Bromide vs. Boron	Yes	Yes	Yes	No	No	No
	Chloride vs. Magnesium	Yes	Yes	No	No	No	No
	Chloride vs. Bromide	Yes	Yes	Yes	No	No	Yes
	Chloride vs. Bicarbonate	Yes	Yes	Yes	No	No	No
	Chloride vs. Calcium	Yes	Yes	Yes	No	No	No
	Chloride vs. Potassium	Yes	Yes	Yes	No	No	No
	Chloride vs. Sodium	Yes	Yes	Yes	No	No	No
	Chloride vs. Sulfate	Yes	Yes	Yes	Yes	No	No
	Cl/Br	Yes	Yes	Yes	Yes	No	No
	Cl/I	Yes	Yes	Yes	Yes	No	Yes
	K/Rb	Yes	No Data[b]	No Data[b]	No Data[b]	No Data[b]	No Data[b]
	Sr Isotope[b]	Yes	No Data[b]	No Data[b]	No Data[b]	No Data[b]	No Data[b]
	Percentage Of Yes[b]	**100**	**100**	**90**	**20**	**0**	**20**

[a] Although there was no data for the other sources, the analysis done for brine sources is consistent with brines as a source of the observed impacts (see Figure 50 and the discussion in the "Source Identification" section of this report).

[b] K/Rb and Sr isotope data were not found in the literature for these sources.

E.7. Transport Properties

The identified constituents of flowback and produced water include inorganic chemicals in the form of cations and anions (including various types of metals, metalloids, and non-metals, and radioactive materials, among others) and organic chemicals, including identified compounds in various classes, and unidentified materials measured as TOC and DOC. Environmental transport of these chemicals depends on the properties of the chemical and properties of the environment, and is extensively discussed in Section 5.8.3.

Transport of inorganic chemicals depends on the nature of groundwater and vadose zone flow, and potential reactions among the inorganic chemical, solid surfaces, and geochemistry of the water. Some inorganic anions (i.e., chloride and bromide) move with their carrier liquid and are mostly impacted by physical transport mechanisms: flow of water and dispersion. In addition to the flow-related processes, transport of most inorganics depends upon three mechanisms related to partitioning to the solid phase: adsorption, absorption, and precipitation. The effects of these mechanisms depend on both chemical and site-specific environmental characteristics, including the surface reactivity, solubility, and redox sensitivity of the contaminant; the type and abundance of reactive mineral phases, and the ground-water chemistry (U.S. EPA, 2007). Generalized characterization of inorganic transport is not possible, but through the use of transport models, the effects of physical transport mechanisms and chemical processes can be integrated. Examples of transport models for reactive metals include the Geochemist's Workbench (Bethke, 2014) and Hydrus (Šimunek et al., 1998).

Properties of organic chemicals which tend to affect the likelihood that a chemical will reach and impact drinking water resources if spilled include high chemical mobility in water and low volatility. Biodegradation, which depends on properties of the chemical, subsurface microorganisms, and the environment, governs the fate of these contaminants.

Using the EPA chemical database EPI Suite™, we were able to obtain actual or estimated physicochemical properties for 521 of the individual organic chemicals identified in produced water and listed in Appendix H. A portion of these, 59, are used in the chemical mixing stage (Table C-9). The EPI Suite™ results are constrained by their applicability to one temperature (25 °C), and salinity (low). Temperature changes impact Henry's law constant, K_{ow}, and solubility, and depend on the characteristics of the chemical and ions present (Borrirukwisitsak et al., 2012; Schwarzenbach et al., 2002). In some cases, the effect changes exponentially with salinity (Schwarzenbach et al., 2002). Therefore, property values that depart from the EPI Suite™ values are expected for produced water at elevated temperature and salinity. Although little is known concerning attenuation of hydraulic fracturing fluid constituents, Kekacs et al. (2015) report that salinity above 40,000 mg/L initially inhibited aerobic degradation of the organic constituents of a synthetic fracturing fluid (for 6.5 days), even though the bacterial communities were pre-acclimated to the salts.

E.8. Example Calculation for Roadway Transport

This section provides background information for the roadway transport calculation appearing in Chapter 7.

An estimate of releases from truck transport of produced water could be made as follows:

$$Total\ number\ of\ truckloads = \frac{Produced\ water\ volume\ per\ well}{Produced\ water\ volume\ per\ truck}$$

Then the total distance traveled by all trucks is given by:

$$Total\ distance\ traveled = Total\ number\ of\ truckloads\ \times Distance\ per\ truck$$

The number of crashes impacting drinking water resources can be estimated from:

$$Fraction\ of\ crashes\ impacting\ drinking\ water\ resources$$
$$= Fraction\ of\ crashes\ releasing\ waste\ that\ impacts\ drinking\ water\ resources$$
$$\times\ Fraction\ of\ all\ crashes\ releasing\ waste\ \times Crash\ rate$$
$$\times\ Total\ distance\ traveled$$

Because the chances of a crash is low, the results are expressed as one truck trip with a crash to total truck trips without a crash (Table E-21). Estimates of all but one of the quantities in these calculations can be made from various literature sources, which are described in the subsequent sections. A key parameter is the number of crashes of trucks per distance traveled. In 2012, the U.S. Department of Transportation (DOT) estimated that the number of crashes per 100 million highway miles driven of a type of large truck was 110, which is a relatively small number. A key parameter that is unknown is the number of crashes which impact drinking water resources, so definitive estimates of impacts to drinking water resources cannot be made. Alternatively, as an upper bound on drinking water resource impacts, the fraction of crashes which release waste can be estimated.

E.8.1. Estimation of Transport Distance

In a study of wastewater management for the Marcellus Shale, Rahm et al. (2013) used data reported to the Pennsylvania Department of Environmental Protection (PA DEP) to estimate the average distance wastewater was transported. For the period from 2008 to 2010, the distance transported was approximately 100 km, but it was reduced by 30% for 2011. The reduction was attributed to increased treatment infrastructure in Lycoming County, an area of intensive hydraulic fracturing operations in northeastern Pennsylvania. For the part of Pennsylvania within the Susquehanna River Basin, Gilmore et al. (2013) estimated the likely transport distances for drilling waste to landfills (256 km or 159 mi); produced water to disposal wells (388 km or 241 mi); and commercial wastewater treatment plants (CWTPs) (158 km or 98 mi). These distances are longer than the values from Rahm et al. (2013), in part, because wells in the Susquehanna Basin are further to the east of Ohio disposal wells and some CWTPs.

E.8.2. Estimation of Wastewater Volumes

In an example water balance calculation, Gilmore et al. (2013) used 380,000 gal (1.4 million L) of flowback as the volume transported to CWTPs, 450,000 gal (1.7 million L) of flowback transported to injection wells, and 130,000 gal (490,000 L) of un-reusable treated water also transported to injection wells for a total estimated wastewater volume of 960,000 gal (3.6 million L) per well.

E.8.3. Estimation of Roadway Accidents

The U.S. Department of Transportation (DOT) published statistics on roadway accidents (U.S. Department of Transportation, 2012) which indicate that the combined total of combination truck crashes in 2012 was 179,736, or 110 per 100 million vehicle miles (1.77 million km) (Table E-19). As an indicator of the uncertainty of these data, DOT reported 122,240 large truck crashes from a differing set of databases (Table E-20), with a rate of 75 per 100 million vehicle miles, which is 68% of the number of combination truck crashes.

Table E-19. Combination truck crashes in 2012 for the 2,469,094 registered combination trucks, which traveled 163,358 million miles.
Source: U.S. Department of Transportation (2012). A combination truck is defined as a truck tractor pulling any number of trailers.

Type of crash	Combination trucks involved in crashes	Rates per 100 million vehicle miles traveled by combination trucks
Property damage only	135,000	82.8
Injury	42,000	25.5
Fatal	2,736	1.74
Total	179,736	110

Table E-20. Large truck crashes in 2012.
Source: U.S. Department of Transportation (2012). A large truck is defined as a truck with a gross vehicle weight rating greater than 10,000 pounds.

Type of crash	Total crashes	Large trucks with cargo tanks	
		Number	Percentage
Towaway crashes	72,644	4,364	6.0%
Injury	45,794	3,245	7.1%
Fatal	3,802	360	9.5%
Totals	122,240	7,969	6.5%

E.8.4. Estimation of Material Release Rates in Crashes

Estimates ranging from 5.6% to 36% have been made for the probability of material releases from crashed trucks. Craft (2004) used data from three databases to estimate the probability of spills in fatality accidents at 36%, which may overestimate the probability for all types of accidents (Rozell and Reaven, 2012).[1] The U.S. Department of Transportation (2012) provides estimates of hazardous materials releases from large truck crashes. For all types of hazardous materials carried, 408 of 2,903 crashes, or 14%, were known to have hazardous materials releases. The occurrence of a release was unknown for 18% of the crashes. These crashes were not distinguished by truck type, so they likely overestimated the number of tanker crashes. Harwood et al. (1993) used accident data from three states (California, Illinois, and Michigan) to develop hazardous materials release rate estimates for different types of roadways, accidents, and settings (urban or rural). For roadways in rural settings the probability of release ranged from 8.1% to 9.0%, while in urban settings the probability ranged from 5.6% to 6.9%.

E.8.5. Estimation of Volume Released in Accidents

Based on the estimated volumes, disposal distances, truck sizes, and accident rates used by Gilmore et al. (2013), Rahm et al. (2013), and King (2012), the total travel distance by trucks ranges from 9,620 mi (14,900 km) to 22,875 mi (36,814 km) per well (Table E-21).

Based on the varying assumptions of each author (Gilmore et al., 2013; King, 2012; Rahm and Riha, 2012) the chances of an accident which releases produced water over the lifetime of a well ranges from 1:110 to 1:13,000 (Table E-21).[2] These estimates are dependent on the volumes, transport distances, and crash rates chosen for analysis. The results show, however, that the expected number of releases is relatively low.

Several limitations are inherent in this analysis, including differing rural road accident rates and highway rates, differing wastewater endpoints, and differing amounts of produced water transport. Further, the estimates present an upper bound on impacts, because not all releases of wastewater would reach or impact drinking water resources.

Impacts to groundwater might occur following a spill on land. When the liquid is highly saline, its migration is affected by its high density and viscosity compared with that of fresh water. When spilled flowback or produced water flows over land, a fraction of the liquid is subject to infiltration. The fraction depends on the rate of release, surface cover (i.e., pavement, cracked pavement, vegetation, bare soil, etc.), slope of the land surface, subsurface permeability, and the moisture content in the subsurface.

---

[1] The three databases were the Trucks Involved in Fatal Accidents developed by the Center for National Truck Statistics at the University of Michigan, the National Automotive Sampling System's General Estimates System (GES) produced by the National Highway Transportation Safety Agency, and the Motor Carrier Management Information System (MCMIS) Crash File produced by the Federal Motor Carrier Safety Administration.

[2] The chances of a crash releasing produced water are calculated from the material release rate times the crash rate times the total miles traveled. The results are expressed as 1 to the reciprocal of this number (i.e., 1:5,900).

Table E-21. Chances of a crash releasing produced water based on the total produced water volume per well, transport distances, crash rates, and material release rates.

Action	Waste per well (million gal)	Trucks (20 m³/truck)[a]	Miles traveled per truck	Total miles traveled (per well)	Chances of a Crash Releasing Produced Water				
					Material release rate bounds[b]				
					3.4%	5.6%		36%	
					Crash rate (per 100 million miles)				
					28	75	110	75	110
Gilmore et al. (2013) distance estimates									
Produced water to CWTP	0.38	72	29.6	2,131	n/a	n/a	n/a	n/a	n/a
Produced water to disposal well	0.45	85	147	12,495	n/a	n/a	n/a	n/a	n/a
CWTP effluent to disposal well	0.13	25	133	3,325	n/a	n/a	n/a	n/a	n/a
Total	0.96	182		17,951	1:5,900	1:1,300	1:210	1:910	1:140
Rahm et al. (2013) distance estimates									
Transport 100 km	0.96	182	62.1	11,300	1:9,300	1:2,100	1:330	1:1,400	1:220
Transport 70 km	0.96	182	43.5	9,620	1:13,000	1:3,000	1:470	1:2,100	1:320
King (2012) distance estimates									
Assumptions of King (2012)	5.0	915	25	22,875	1:4,600	1:1,000	1:162	1:710	1:110

[a] King (2012) assumed a truck volume of 5,440 gal (20,600 L) versus the assumption of 5,300 gal (20,100 L) for the other rows of the table.

[b] King (2012) assumed a release rate of 3.4% from truck crashes and an accident rate of 28 crashes per 100 million mi (160 million km).

Appendix F. Wastewater Disposal and Reuse Supplemental Information

This page is intentionally left blank.

Appendix F. Wastewater Disposal and Reuse Supplemental Information

This appendix provides additional information for context and background to support the discussions of hydraulic fracturing wastewater management and treatment in Chapter 8. Information in this appendix includes: estimates of volumes of wastewater generated compiled for several states in regions where hydraulic fracturing is occurring; an overview of the technologies that can be used to treat hydraulic fracturing wastewater; reported and estimated removal efficiencies for specific treatment technologies and contaminants of concern; a description of common technologies currently in use at centralized waste treatment plants (CWTs) and their discharge options; considerations for water reuse in hydraulic fracturing and the necessary water quality; and legacy impacts of hydraulic fracturing on publicly owned treatment works (POTWs). Discussion is also provided on disinfection byproduct (DBP) formation concerns related to hydraulic fracturing.

F.1. Estimates of Wastewater Production in Regions where Hydraulic Fracturing is Occurring

Table F-1 presents estimated wastewater volumes for several states in areas with hydraulic fracturing activity. These data were compiled from production data available in state databases and were tabulated by year. For California, data were compiled for Kern County, where about 95% of California's hydraulic fracturing takes place (CCST, 2015b). Production records from Colorado, Utah, and Wyoming include the producing formation for each well reported. Data presented for these three states include statewide estimates as well as estimates for selected basins that were identified in the literature as targets for hydraulic fracturing. Data from New Mexico are available in files for three basins (the Permian, Raton, and San Juan) as well as for the state as a whole.

Results in Table F-1 illustrate some of the challenges associated with obtaining estimates of hydraulic fracturing wastewater volumes, especially using publicly available data. Some of the estimates likely include volumes from conventional wells that are not hydraulically fractured. For example, the well counts for California, Colorado, Utah, and Wyoming were in the thousands or tens of thousands at least as early as 2000, several years before the surge of modern hydraulic fracturing began in the mid-2000s. The data used for California were from Kern County where hydraulic fracturing is conducted, but are not specific to hydraulic fracturing activity. Where producing formations are listed, but there is no indication of whether the well was hydraulically fractured, the accuracy of the estimates depends on whether hydraulically fractured formations were correctly identified based on other information. If formations (and the associated wells) were inadvertently omitted, the volumes will be underestimates.

Table F-1. Estimated volumes (millions of gallons) of wastewater based on state data for selected years and numbers of wells producing fluid. The wastewater is likely associated with an unknown combination of wells not hydraulically fractured and some hydraulically fractured.

State	Basin	Principal lithologies	Data type	2000	2004	2008	2010	2011	2012	2013	2014	Comments
California	San Joaquin[a]	Shale, unconsoli-dated sands	Produced water	46,000	48,000	58,000	65,000	71,000	75,000	74,000	-	Data from CA Department of Conservation, Oil and Gas Division.[a] Produced water data compiled for Kern County. Data may also represent contributions from production without hydraulic fracturing. Not specified whether flowback is included.
			Wells	33,695	39,088	46,519	49,201	51,031	51,567	52,763	-	
Colorado	All basins with hy-draulically fractured formations	-	Produced water	7,300	11,000	21,000	14,000	12,000	12,000	7,700	-	Data from CO Oil and Gas Conservation Commission.[b] Produced water includes flowback. Data filtered for formations indicated in literature as undergoing hydraulic fracturing and matched to corresponding basins. Example basins selected for presentation as well as estimated state total.
			Wells	11,264	14,934	28,282	33,929	35,999	38,371	37,618	-	
	Denver	Sandstone, shale	Produced water	140	160	170	160	160	150	110	-	
			Wells	1,829	1,511	1,277	1,204	1,193	1,131	1,072	-	

State	Basin	Principal lithologies	Data type	2000	2004	2008	2010	2011	2012	2013	2014	Comments
Colorado, cont.	Piceance	Sandstone	Produced water	3,500	5,800	9,300	6,900	6,500	6,800	4,300	-	
			Wells	1,134	2,478	6,486	9,105	10,057	10,868	10,954	-	
	Raton	Coalbed methane	Produced water	2,400	4,100	8,900	4,300	3,200	2,700	2,100	-	
			Wells	681	1,634	2,795	2,734	2,778	2,710	2,545	-	
	San Juan	Coalbed methane	Produced water	1,000	1,100	1,300	2,000	1,200	1,100	650	-	
			Wells	1,183	1,605	1,975	2,220	2,308	2,328	2,333	-	
New Mexico	Permian	Shale, sandstone	Produced water	-	-	-	-	-	31,000	31,000	20,000	Data from New Mexico Oil Conservation Division.[c] Data provided by the state broken out by basin. Unclear how much contribution from production without hydraulic fracturing. Produced water includes flowback.
			Wells	-	-	-	-	-	29,839	30,386	30,287	
	Raton	Coalbed methane	Produced water	-	-	-	-	-	510	540	310	
			Wells	-	-	-	-	-	1,495	1,502	1,526	
	San Juan	Coalbed methane	Produced water	-	-	-	-	-	1,700	2,000	1,100	
			Wells	-	-	-	-	-	22,492	22,349	22,076	

State	Basin	Principal lithologies	Data type	2000	2004	2008	2010	2011	2012	2013	2014	Comments
New Mexico, cont.	Total	-	Produced water	-	-	-	-	-	33,000	34,000	22,000	
			Wells	-	-	-	-	-	53,826	54,237	53,889	
Utah	All basins with hydraulically fractured formations	-	Produced water	1,200	1,200	2,300	2,400	2,700	2,900	3,400	2,800	Data from State of Utah Oil and Gas Program.[d] Produced water may or may not include flowback. Data filtered by formation indicated in the literature as hydraulically fractured and matched to basins. Data presented for selected basins as well as for all formations likely to be hydraulically fractured.
			Wells	3,080	4,377	7,409	8,432	9,101	10,075	10,661	10,900	
	Kaiparowits/ Uinta	Coalbed methane	Produced water	860	740	1,300	1,400	1,800	2,000	2,400	1,900	
			Wells	1,718	2,517	3,761	4,329	4,838	5,538	6,046	6,334	
	San Juan/ Uinta	Coalbed methane	Produced water	2	49	350	270	240	230	190	120	
			Wells	62	223	910	933	959	951	867	870	
	Uinta	Shale/sand-stone	Produced water	350	420	560	680	700	640	830	790	
			Wells	1,067	1,396	2,282	2,745	2,888	3,115	3,257	3,223	

State	Basin	Principal lithologies	Data type	2000	2004	2008	2010	2011	2012	2013	2014	Comments
Wyoming	All basins with hydraulically fractured formations	-	Produced water	1,300	1,400	1,300	1,500	1,600	1,700	1,600	1,800	Data from Wyoming Oil and Gas Conservation Commission.[e] Produced water may include flowback. Data filtered by formation indicated in the literature as hydraulically fractured and matched to basins. Data presented for selected basins as well as for all formations likely to be hydraulically fractured.
			Wells	3,470	3,378	3,585	3,620	3,728	3,843	4,030	4,213	
	Big Horn	Sandstone	Produced water	380	350	350	380	430	440	420	440	
			Wells	365	359	387	397	412	414	407	403	
	Denver	Sandstone	Produced water	54	44	49	59	76	90	97	170	
			Wells	142	118	124	140	167	204	230	278	
	Green River	Sandstone/ shale	Produced water	0	1	2	8	5	5	9	15	
			Wells	44	44	60	67	67	59	64	67	
	Powder River	Coalbed methane	Produced water	690	630	620	660	700	840	970	1,100	
			Wells	1,953	1,900	2,001	2,028	2,119	2,207	2,352	2,565	
	Wind River/ Powder River	Sandstone/ shale	Produced water	130	330	330	400	420	290	110	41	

Appendix F – Wastewater Disposal and Reuse Supplemental Information

State	Basin	Principal lithologies	Data type	2000	2004	2008	2010	2011	2012	2013	2014	Comments
Wyoming, cont.	Wind River/ Powder River, cont.	Sandstone/ shale, cont.	Wells	966	957	1,013	988	963	959	977	900	

[a] California Department of Conservation, Oil and Gas Division. Oil & Gas – Online Data. Monthly Production and Injection Databases: ftp://ftp.consrv.ca.gov/pub/oil/new database format/.

[b] Colorado Oil and Gas Conservation Commission. Data: Downloads: Production Data: http://cogcc.state.co.us/data2.html#/downloads.

[c] New Mexico Oil Conservation Division. Production Data. Production Summaries: All Wells Data: http://gotech.nmt.edu/gotech/Petroleum_Data/allwells.aspx.

[d] Utah Department of Natural Resources. Division of Oil, Gas, and Mining. Data Research Center. Database Download Files: http://oilgas.ogm.utah.gov/Data_Center/DataCenter.cfm#production.

[e] Wyoming Oil and Gas Conservation Commission. Production files by county and year: http://wogcc.state.wy.us/productioncountyyear.cfm?Oops=#oops#&RequestTimeOut=6500.

F.2. Overview of Treatment Processes for Treating Hydraulic Fracturing Wastewater

Treatment technologies discussed in this appendix are classified as basic or advanced. Basic treatment technologies are ineffective for reducing total dissolved solids (TDS) and are typically not labor intensive. Advanced treatment technologies can remove TDS and/or are complex in nature (e.g., energy- and labor-intensive).

F.2.1. Basic Treatment

Basic treatment technologies include physical separation, coagulation/oxidation, electrocoagulation, sedimentation, and disinfection. These technologies are effective at removing total suspended solids (TSS), oil and grease, scale-forming compounds, and metals, and they can minimize microbial activity. Basic treatment is typically incorporated in a permanent treatment facility (i.e., fixed location), but can also be part of a mobile unit for on-site treatment applications.

F.2.1.1. Physical Separation

The most basic treatment needed for oil and gas wastewaters, including those from hydraulic fracturing operations, is separation to remove suspended solids and oil and grease. The separation method largely depends on the type(s) of resource(s) targeted by the hydraulic fracturing operation. Down-hole separation techniques, including mechanical blocking devices and water shut-off chemicals (e.g., specialized polymers) to prevent or minimize water flow to the well, may be used during production in shale plays containing greater amounts of liquid hydrocarbons. To treat water at the surface, separation technologies such as hydrocyclones, dissolved air or induced gas flotation systems, media (sand) filtration, and biological aerated filters can remove suspended solids and some organics from hydraulic fracturing wastewater.

Media filtration can also remove hardness and some metals if chemical precipitation (i.e., coagulation, lime softening) is also employed (Boschee, 2014). An example of a CWT that uses chemical precipitation and media filtration to treat hydraulic fracturing waste is the Water Tower Square Gas Well Wastewater Processing Facility in Pennsylvania (Table F-6). One or more of these technologies is typically used prior to advanced treatment such as reverse osmosis (RO) because advanced treatment processes foul, scale, or otherwise do not operate effectively in the presence of TSS, certain organics, and/or some metals and metalloid compounds (Boschee, 2014; Drewes et al., 2009). The biggest challenge associated with use of these separation technologies is solids disposal from the resulting sludge (Igunnu and Chen, 2014).

F.2.1.2. Coagulation/Oxidation

Coagulation is the process of agglomerating small, unsettleable particles into larger particles to promote settling. Chemical coagulants such as alum, iron chloride, and polymers can be used to precipitate TSS, some dissolved solids (except monovalent ions such as sodium and chloride), and metals from hydraulic fracturing wastewater. Adjusting the pH using chemicals such as lime or caustic soda can increase the potential for some constituents, including dissolved metals, to form precipitates. Chemical precipitation is often used in industrial wastewater treatment as a pretreatment step to decrease the pollutant loading on subsequent advanced treatment technologies; this strategy can save time, money, energy consumption, and the lifetime of the infrastructure.

Processes using advanced oxidation and precipitation have been applied to hydraulic fracturing wastewaters in on-site and mobile systems. Hydroxyl radicals generated by cavitation processes and the addition of ozone can degrade organic compounds and inactivate micro-organisms. The process can also aid in the precipitation of ions that cause hardness and scaling in the treated water (e.g., calcium, magnesium). The process can also reduce sulfate and carbonate concentrations in the treated water. With the removal of constituents that contribute to scaling, this type of treatment can be very effective for on-site reuse of wastewater (Ely et al., 2011).

The produced solid residuals from coagulation/oxidation processes typically require further treatment, such as de-watering (Duraisamy et al., 2013; Hammer and VanBriesen, 2012).

F.2.1.3. Electrocoagulation

Electrocoagulation (EC) (Figure F-1) combines the principles of coagulation and electrochemistry into one process (Gomes et al., 2009). An electrical current added to the wastewater produces coagulants that then neutralize the charged particles, causing them to destabilize, precipitate, and settle. EC may be used in place of, or in addition to, chemical coagulation. EC can be effective for removal of organics, TSS, and metals, but it is not effective at removing TDS and sulfate (Halliburton, 2014). Although it is still considered an emerging technology for unconventional oil and gas wastewater treatment, EC has been used in mobile treatment systems to treat hydraulic fracturing wastewaters (Halliburton, 2014; Igunnu and Chen, 2014). This technology has the potential to cause scaling, corrosion, and bacterial growth (Gomes et al., 2009).

Figure F-1. Electrocoagulation unit.
Source: Dunkel (2013). Photo courtesy of Pioneer Natural Resources.

Testing of EC has been performed in the Green River Basin (Halliburton, 2014) and the Eagle Ford Shale (Gomes et al., 2009). While showing promising results in some trials, results of these early studies have illustrated challenges, with removal efficiencies impacted by factors such as pH and salt content.

F.2.1.4. Sedimentation

Treatment plants may include sedimentation tanks, clarifiers, or some other form of settling basin to allow larger particles to settle out of the water where they can eventually be collected, dewatered, and disposed of at a landfill or other approved location. These types of tanks/basins all serve the same purpose – to reduce the amount of solids going to subsequent processes (i.e., to prevent overload of the media filters).

F.2.1.5. Disinfection

Some hydraulic fracturing applications may require disinfection to kill bacteria after treatment and prior to reuse or discharge. Chlorine is a common disinfectant. Chlorine dioxide, ozone, or ultraviolet light can also be used. This is an important step for reused water because bacteria can cause problems for further hydraulic fracturing operations by multiplying rapidly and causing build-up in the wellbore, which decreases gas extraction efficiency.

F.2.2. Advanced Treatment

Advanced treatment technologies consist of membranes (RO, nanofiltration, ultrafiltration, microfiltration, electrodialysis, forward osmosis, and membrane distillation), thermal distillation technologies, crystallizers, ion exchange, and adsorption. These technologies are effective for removing TDS and/or targeted compounds. They typically require pretreatment to remove solids and other constituents that may damage or otherwise impede the technology from operating as designed. Advanced treatment technologies can be energy-intensive and are typically employed when a purified water effluent is necessary for direct discharge, indirect discharge, or reuse. In some instances, these water treatment technologies can use methane generated by the gas well as an energy source. Some advanced treatment technologies can be made mobile for on-site treatment.

F.2.2.1. Membranes

Pressure-Driven Membrane Processes

Pressure-driven membrane processes, including microfiltration, ultrafiltration, nanofiltration, and RO (Figure F-2), are being used in some settings to treat oil and gas wastewater. These processes use hydraulic pressure to overcome the osmotic pressure of the influent waste stream, forcing clean water through the membrane (Drewes et al., 2009). Microfiltration and ultrafiltration processes are advanced processes that do not reduce TDS but can remove TSS and some metals and organics (Drewes et al., 2009). RO and nanofiltration are capable of removing TDS, including anions and radionuclides. RO, however, may be limited to treating TDS levels of less than 35,000 mg/L (Shaffer et al., 2013; Younos and Tulou, 2005). Boron is not easily removed by RO, achieving less than 50% removal at neutral pH (Drewes et al., 2009).

Figure F-2. Photograph of reverse osmosis system.
Source: U.S. DOI (2016).

Osmotic-Driven Membrane Processes

Forward osmosis, an emerging technology for treating hydraulic fracturing wastewater, uses an osmotic pressure gradient across a membrane to draw water from a low osmotic solution (the feed water) to a high osmotic solution (the draw solution) (Drewes et al., 2009). The draw solution (typically composed of sodium chloride) becomes diluted as more water passes through the membrane while the feed side becomes more concentrated. For the diluted draw solution, a separation process is employed to further treat the product water and concentrate the sodium chloride for reuse.

Thermally-Drive Membrane Processes

Another emerging technology, membrane distillation, relies on a thermal gradient across a membrane surface to volatilize pure water and capture it in the distillate (Drewes et al., 2009). Membrane distillation has shown promise in removing heavy metals and boron from wastewaters (Camacho et al., 2013).

F.2.2.2. Electrodialysis

Electrodialysis relies on electrodes (anode and cathode) and ion exchange membranes to separate positively- and negatively-charged contaminants from the feed water (Drewes et al., 2009) (Figure

F-3). Electrodialysis has been considered for use by the shale gas industry, but is currently not widely utilized (ALL Consulting, 2013). TDS concentrations above 15,000 mg/L are difficult to treat by electrodialysis (ALL Consulting, 2013), and oil and divalent cations (e.g., calcium, iron, magnesium) can foul/scale the membranes (Hayes and Severin, 2012b; Guolin et al., 2008). Pretreatment is necessary to avoid membrane scaling (ALL Consulting, 2013; Drewes et al., 2009).

Figure F-3. Picture of mobile electrodialysis units in Wyoming.
Source: DOE (2006). Reproduced with permission from ALL Consulting.

F.2.2.3. Thermal Distillation

Thermal distillation technologies, such as mechanical vapor recompression (MVR) (Figure F-4) and dewvaporation, use liquid-vapor separation by applying heat to the waste stream, vaporizing the water to separate out impurities, and condensing the vapor into distilled water (Drewes et al., 2009; LEau LLC, 2008; Hamieh and Beckman, 2006). MVR and dewvaporation can treat high-TDS waters and have been proven in the field as effective for treating oil and gas wastewater (Hayes and Severin, 2012b; Drewes et al., 2009). Like RO, these processes are energy-intensive and are used when the objective is very clean water (i.e., TDS less than 500 mg/L) for direct/indirect discharge or if clean water is needed for reuse. As with membrane processes, scaling is an issue with these technologies, and scale inhibitors may be needed for them to operate effectively (Igunnu and Chen, 2014).

Figure F-4. Picture of a mechanical vapor recompression unit near Decatur, Texas.
Source: Drewes et al. (2009). Reproduced with permission.

CWTs such as the Judsonia Central Water Treatment Facility (Arkansas), Casella-Altela Regional Environmental Services (Pennsylvania), and Clarion Altela Environmental Services (Pennsylvania) have National Pollutant Discharge Elimination System (NPDES) permits and use MVR or thermal distillation for TDS removal. Figure F-5 shows a diagram of the treatment train at another facility, the Maggie Spain facility in Texas, which used MVR in its treatment of Barnett Shale wastewater (Hayes and Severin, 2012a).

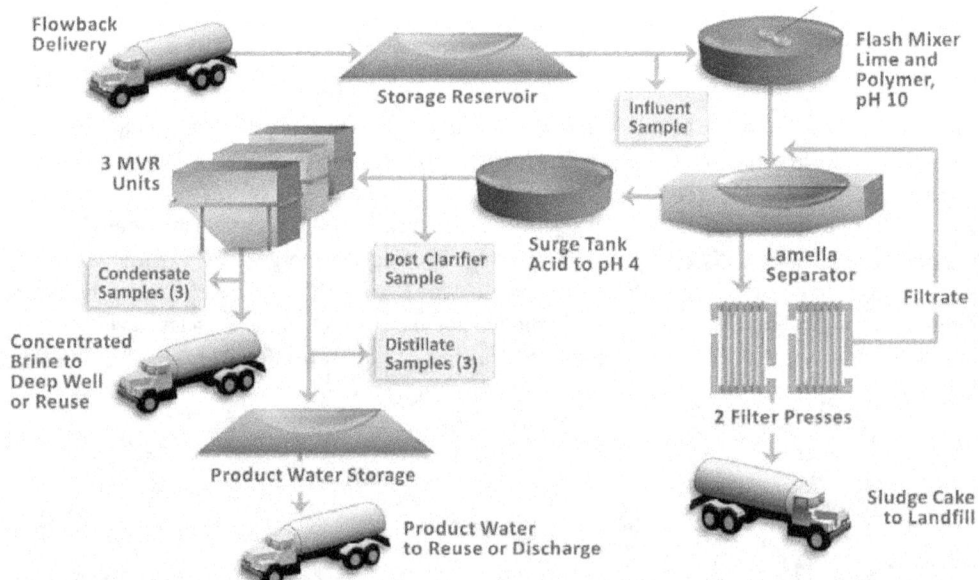

Figure F-5. Mechanical vapor recompression process design – Maggie Spain Facility.
Adapted from: Hayes and Severin (2012a). Reproduced with permission.

Crystallizers can be employed at CWTs to treat high-TDS waters or to further concentrate the waste stream from a distillation process, reducing residual waste disposal volumes. The crystallized salt can be landfilled, deep-well injected, or used to produce pure salt products that may be salable (Ertel et al., 2013).

Another thermal method, freeze-thaw evaporation, involves spraying wastewater onto a freezing pad, allowing ice crystals to form, and the brine mixture that remains in solution to drain from the ice (Drewes et al., 2009). In warmer weather, the ice thaws and the purified water is collected. This technology cannot treat waters with high methanol concentrations and is only suitable for areas where the temperature is below freezing in the winter months (Igunnu and Chen, 2014). In addition, freeze-thaw evaporation can only reduce TDS concentrations to approximately 1,000 mg/L, which is higher than the 500 mg/L TDS surface water discharge limit required by most permits (Igunnu and Chen, 2014).

F.2.2.4. Ion Exchange and Adsorption

Ion exchange (Figure F-6) is the process of exchanging ions on a media referred to as resin for unwanted ions in the water. Ion exchange is used to treat for target ions that may be difficult to remove by other treatment technologies or that may interfere with the effectiveness of advanced treatment processes.

Figure F-6. Picture of a compressed bed ion exchange unit.
Source: Drewes et al. (2009). Reproduced with permission.

Adsorption is the process of adsorbing contaminants onto a charged granular media surface. Adsorption technologies can effectively remove organics, heavy metals, and some anions (Igunnu and Chen, 2014). With ion exchange and adsorption processes, the type of resin or adsorptive media used (e.g., activated carbon, organoclay, zeolites) dictates the specific contaminants that will be removed from the water (Drewes et al., 2009; Fakhru'l-Razi et al., 2009).

Because they can be easily overloaded by contaminants, ion exchange and adsorption treatment processes are generally used as a polishing step following other treatment processes or as a unit process in a treatment train rather than as stand-alone treatment (Drewes et al., 2009). Stand-alone units require more frequent regeneration and/or replacement of the spent media making these technologies more costly to operate (Igunnu and Chen, 2014). The Pinedale Anticline Water Reclamation Facility located in Wyoming uses an ion exchange unit with boron-selective resin as a polishing step to treat hydraulic fracturing wastewater specifically for boron (Boschee, 2012) (Figure F-7).

F.3. Treatment Technology Removal Capabilities

Table F-2 provides removal efficiencies for common hydraulic fracturing wastewater constituents by treatment technology. With the exception of TSS and TDS, the studies cited demonstrate removal for a subset of constituents in a category (e.g., Gomes et al. (2009) reported that electrodialysis was an effective treatment for oil and grease, not all organics). The removal efficiencies include ranges of 1 to 33% (denoted by +), 34% to 66% (denoted by ++), and greater than 66% removal (denoted by +++). Cells denoted with "--" indicate that the treatment technology is not suitable for removal of that constituent or group of constituents. If a particular treatment technology only lists removal efficiencies for TDS, it can be assumed that, in some cases, cations and anions would also be removed by that technology; therefore, where specific results were not provided in literature, cells denoted with "Assumed" refer to cations and anions that comprise TDS.

Table F-2. Removal efficiency of different hydraulic fracturing wastewater constituents using various wastewater treatment technologies.[a]

Treatment technology	Hydraulic fracturing wastewater constituents					
	TSS	TDS	Anions	Metals	Radio-nuclides	Organics
Hydrocyclones	+++ (Duraisamy et al., 2013)	--	--	--	--	++ (Duraisamy et al., 2013)
Evaporation (freeze-thaw evaporation)	+++ (Igunnu and Chen, 2014; Drewes et al., 2009)	+++ (Igunnu and Chen, 2014; Drewes et al., 2009; Arthur et al., 2005)	Assumed	+++ (Igunnu and Chen, 2014; Drewes et al., 2009; Arthur et al., 2005)	--	+++ (Igunnu and Chen, 2014; Duraisamy et al., 2013; Drewes et al., 2009)
Filtration (granular media)	+++ (Barrett, 2010)	--	--	+++[b] (Duraisamy et al., 2013)	--	+++ (Shafer, 2011; Drewes et al., 2009)
Chemical precipitation	+++ (Fakhru'l-Razi et al., 2009)	--	--	+++ (Fakhru'l-Razi et al., 2009; AWWA, 1999)	+++[c] (Zhang et al., 2014)	+++ (Fakhru'l-Razi et al., 2009)

Treatment technology	Hydraulic fracturing wastewater constituents					
	TSS	TDS	Anions	Metals	Radio-nuclides	Organics
Sedimentation (clarifier)	++ (NMSU DACC WUTAP, 2007)	--	--	--	--	--
Dissolved air flotation	+++ (Shammas, 2010)	--	--	--	--	++/+++ (Duraisamy et al., 2013; Fakhru'l-Razi et al., 2009)
Electro-coagulation	+++ (Igunnu and Chen, 2014; Bukhari, 2008)	--	--	+ (Igunnu and Chen, 2014)	--	+++ (Igunnu and Chen, 2014; Duraisamy et al., 2013; Fakhru'l-Razi et al., 2009)
Advanced oxidation and precipitation	--	+ (Abrams, 2013)	--	+/+++ (Abrams, 2013)	--	+++[d] (Duraisamy et al., 2013; Fakhru'l-Razi et al., 2009)
Reverse osmosis	--	++/+++[e] (Alzahrani et al., 2013; Drewes et al., 2009)	+++ (Alzahrani et al., 2013; Arthur et al., 2005)	++/+++[f] (Alzahrani et al., 2013; Drewes et al., 2009; AWWA, 1999)	+++ (Drewes et al., 2009)	+/++/+++[g] (Drewes et al., 2009; Munter, 2000)
Membrane filtration (UF/MF)	+++ (Arthur et al., 2005)	--	--	+++ (Fakhru'l-Razi et al., 2009)	--	++/+++ (Duraisamy et al., 2013; Fakhru'l-Razi et al., 2009; Hayes and Arthur, 2004; AWWA, 1999)[h]
Forward osmosis	--	+++ (Drewes et al., 2009)	Assumed	Assumed	--	--
Distillation, including thermal distillation (e.g., mechanical vapor recompression (MVR))		+++[i] (Hayes et al., 2014; Bruff and Jikich, 2011; Drewes et al., 2009)	+++ (Bruff and Jikich, 2011; Drewes et al., 2009)	+++ (Hayes et al., 2014; Bruff and Jikich, 2011; Drewes et al., 2009)	+++ (Bruff and Jikich, 2011; Drewes et al., 2009)	+/++/+++ (Hayes et al., 2014; Duraisamy et al., 2013; Drewes et al., 2009; Fakhru'l-Razi et al., 2009)
Ion exchange	--	--	+++ (Drewes et al., 2009)	+++ (Drewes et al., 2009; Arthur et al., 2005)	+++ (Drewes et al., 2009)	+/++/+++ (Fakhru'l-Razi et al., 2009; Munter, 2000)[j]

Treatment technology	Hydraulic fracturing wastewater constituents					
	TSS	TDS	Anions	Metals	Radio-nuclides	Organics
Crystallization	--	+++ (ER, 2014)	Assumed	Assumed	--	--
Electrodialysis	--	+++[k] (Drewes et al., 2009; Gomes et al., 2009; Arthur et al., 2005)	++/+++ (Banasiak and Schäfer, 2009)	+/++/+++ (Banasiak and Schäfer, 2009)	--	--
Capacitive deionization (emerging technology)	--	+++[l] (Drewes et al., 2009)	--	--	--	--
Adsorption[m]	--	--	+/++/+++[n] (Habuda-Stanic et al., 2014)	+++ (Igunnu and Chen, 2014; Drewes et al., 2009)	--	+/++/+++ (Arthur et al., 2005; Hayes and Arthur, 2004; Munter, 2000)
Biological treatment	+++ (Igunnu and Chen, 2014; Drewes et al., 2009)	--	--	--	--	+/++/+++ (Igunnu and Chen, 2014; Drewes et al., 2009; Fakhru'l-Razi et al., 2009)
Constructed wetland/reed beds	++/+++ (Manios et al., 2003)	+ (Arthur et al., 2005)	--	++/+++ (Fakhru'l-Razi et al., 2009)	--	+/ +++ (Fakhru'l-Razi et al., 2009; Arthur et al., 2005)

[a] To the extent possible, removal efficiencies are based on an individual treatment technology that does not assume extensive pretreatment or combined treatment processes. However, it should be noted that some processes such as RO, media filtration, and sedimentation cannot effectively operate without pretreatment.

[b] Pretreatment (pH adjustment, aeration, solids separation) required.

[c] Radium co-precipitation with barium sulfate.

[d] The Fenton process.

[e] Typically requires pretreatment. Not a viable technology if TDS influent >50,000 mg/L.

[f] Iron and manganese oxides will foul the membranes.

[g] Some organics will foul the membranes (e.g., organic acids).

[h] Ultrafiltration membrane was modified with nanoparticles.

[i] Can typically handle high TDS concentrations.

[j] Resin consisted of modified zeolites that targeted removal of BTEX.

[k] Influent TDS for this technology should be <8,000 mg/L.

[l] Specific technology was an electronic water purifier which is a hybrid of capacitive deionization. Influent TDS for this technology should be <3,000 mg/L.

[m] Typically polishing step, otherwise can overload bed quickly with organics.

[n] Removal efficiency is dependent on the type of adsorbent used and the water quality characteristics (e.g., pH).

F.3.1. Estimated Treatment Removal Efficiencies

There are relatively few studies that have evaluated the ability of individual treatment processes to remove constituents from hydraulic fracturing wastewater and reported the resulting water quality. Furthermore, although a specific technology may demonstrate a high removal percentage for a particular constituent, if the influent concentration of that constituent is extremely high, the constituent concentration in the treated water may still exceed permit limits and/or disposal requirements. Table F-3 presents estimated effluent concentrations that could be produced by a variety of unit treatment processes for several example constituents and for various influent concentrations. This analysis uses simple calculations pairing average hydraulic fracturing wastewater concentrations from Chapter 7 and Appendix E with treatment process removal efficiencies reported in the literature in Table F-2. This analysis is intended to highlight the potential impacts of influent concentration on treatment outcome and to illustrate the relative capabilities of various treatment processes for an example set of constituents. The removal efficiencies represent a variety of studies (primarily at bench and pilot scale) that have been conducted using either conventional or hydraulic fracturing wastewater. Removal efficiency for a given treatment process can vary due to a number of factors, and constituent removal may be different in a full-scale facility that uses several processes. Thus, the calculations shown in Table F-3 are intended to be rough approximations for illustrative purposes.

As an example, radium in wastewater from the Marcellus Shale and Upper Devonian sandstones can be in the thousands of pCi/L. With a 95% removal rate, chemical precipitation may result in effluent that still exceeds 100 pCi/L. Distillation and RO might produce effluent with concentrations in the tens of pCi/L. A radium concentration of 120 pCi/L, however, could be reduced to less than 5 pCi/L by RO or distillation. Wastewater with barium concentrations in the range of 140 – 160 mg/L (e.g., the Cotton Valley and Mesaverde tight sands) might be reduced to concentrations under 5 mg/L by distillation and roughly 11 – 13 mg/L by RO. Barium concentrations in the thousands of mg/L would be substantially reduced by any of several processes, but might still be relatively high, potentially exceeding 100 mg/L. Table F-3 also illustrates the potential for achieving low concentrations of organic compounds in wastewater treated with freeze-thaw evaporation or advanced oxidation and precipitation.

Table F-3. Estimated effluent concentrations for example constituents based on treatment process removal efficiencies.

Shale/ sandstone play	Contaminant	Units (for all entries)	MCL	Avg. influent conc.	Freeze-thaw evaporation	Media filtration	Chemical precipitation	Flotation (DAF)	Electro-coagulation	Advanced oxidation and precipitation	Reverse osmosis	Membrane filtration (UF/MF)	Distillation	Ion exchange	Electrodialysis	Adsorption	Biological treatment (biodisks, BAFs)	Constructed wetland
Bakken	Barium	mg/L	2	10		1				0.44	0.8		0.1 – 0.03	ND – 0.7				2.2
Barnett	Barium	mg/L	2	3.6		0.4				0.16	0.29		0.0036 – -0.11	ND – 0.3				0.8
Fayetteville	Barium	mg/L	2	4		0.4				0.18	0.32		0.04 – 0.12	ND – 0.3				0.9
Marcellus	Barium	mg/L	2	2200		220				98	180		22 – 67	ND – 160				490
Cotton Valley	Barium	mg/L	2	160		16				7	13		1.6 – 4.8	ND – 11				35
Mesa Verde	Barium	mg/L	2	140		14				6.1	11		1.4 – 4.2	ND – 9.7				31
Marcellus	Cadmium	µg/L	5	25	2.5	2.5					13					5		15
Bakken	Strontium	mg/L	--	760		76							7.6 – 23	53				
Barnett	Strontium	mg/L	--	530		53							5.3 – 16	37				
Fayetteville	Strontium	mg/L	--	27		2.7							0.27 – 0.81	1.9				
Marcellus	Strontium	mg/L	--	1700		170							17 – 51	120				
Cotton Valley	Strontium	mg/L	--	2300		230							23 – 69	160				

Appendix F – Wastewater Disposal and Reuse Supplemental Information

Shale/sandstone play	Contaminant	Units (for all entries)	MCL	Avg. influent conc.	Freeze-thaw evaporation	Media filtration	Chemical precipitation	Flotation (DAF)	Electro-coagulation	Advanced oxidation and precipitation	Reverse osmosis	Membrane filtration (UF/MF)	Distillation	Ion exchange	Electrodialysis	Adsorption	Biological treatment (biodisks, BAFs)	Constructed wetland
Devonian Sandstone	Strontium	mg/L	--	3900		390							39 – 120	270				
Marcellus	Radium 226	pCi/L	--	620			32 – 440				6.2		6.2 – 19	44				
Devonian Sandstone	Radium 226	pCi/L	--	2400			120 – 1700				24		24 – 71	170				
Marcellus	Radium 228	pCi/L	--	120			6.2 – 85				1.2		1.2 – 3.6	8.4				
Marcellus	Total radium	pCi/L	5	2500			130 – 1800				25		25 – 76	180				
Barnett	TOC	pCi/L	--	9.8					58			0.2				0.9 – 2.9	2.1 – 4	1
Marcellus	TOC	pCi/L	--	160								3.2				16 – 48	35 – 58	16
Cotton Valley	TOC	mg/L	--	200					4			4				20 – 59	44 – 71	20
Barnett	BOD	mg/L	--	580										290 – 440			29 – 87	47
Marcellus	BOD	mg/L	--	40					4					20 – 30			2 – 6	3.2
Barnett	Oil and grease	mg/L	--	160		16						16			8	1.6	43	9.8
Marcellus	Oil and grease	mg/L	--	74		7.4						7.4			3.7	0.74	19	4.4
Barnett	Benzene	µg/L	5	680	68							310	6.8			110		ND
Marcellus	Benzene	µg/L	5	360	36							170	3.6			58		ND

Appendix F – Wastewater Disposal and Reuse Supplemental Information

Shale/sandstone play	Contaminant	Units (for all entries)	MCL	Avg. influent conc.	Freeze-thaw evaporation	Media filtration	Chemical precipitation	Flotation (DAF)	Electro-coagulation	Advanced oxidation and precipitation	Reverse osmosis	Membrane filtration (UF/MF)	Distillation	Ion exchange	Electrodialysis	Adsorption	Biological treatment (biodisks, BAFs)	Constructed wetland
Barnett	Toluene	µg/L	1,000	760	76							350				84		ND
Marcellus	Toluene	µg/L	1,000	1100	110							510				120		ND
Barnett	Ethylbenzene	µg/L	700	29	2.9			17								3.2		ND
Marcellus	Ethylbenzene	µg/L	700	150	15			90								17		ND
Barnett	Xylenes	µg/L	10,000	360	36							170				14		ND
Marcellus	Xylenes	µg/L	10,000	1300	130							600				52		ND
Barnett	BTEX	µg/L	--	1800	180					7.3			91	270 – 550		3.7 – 91		
Marcellus	BTEX	µg/L	--	2900	290					12			150	440 – 870		5.8 – 150		
Barnett	Naphthalene	µg/L	--	240						0.95								
Marcellus	Naphthalene	µg/L	--	360						1.4								
Barnett	1,2,4-Trimethyl-benzene	µg/L	--	170						0.69								
Marcellus	1,2,4-Trimethyl-benzene	µg/L	--	430						1.7								
Barnett	1,2,4-Trimethyl-benzene	µg/L	--	59						0.24								
Marcellus	1,2,4-Trimethyl-benzene	µg/L	--	310						1.2								

ND = Non-detect

F.4. Treatment for Constituents of Concern

Constituents of concern in hydraulic fracturing wastewater include TSS, TDS, anions (e.g., chloride, bromide, and sulfate), metals, radionuclides, and organic compounds (see Section 8.3 and Chapter 7). If the end use of the wastewater necessitates treatment, a variety of technologies can be employed to remove or reduce the constituent concentrations. Table F-4 provides an overview of influent and effluent results and removal percentages for constituents of concern at oil and gas treatment facilities reported in literature (both conventional and unconventional) and the specific technology(ies) used to remove them.

Table F-4. Studies of removal efficiencies and influent/effluent data for various processes and facilities.

Constituents of concern	Location and results				
	Pinedale Anticline Water Reclamation Facility, Wyoming (Shafer, 2011)	Maggie Spain Water-Recycling Facility, Barnett Shale, Texas (Hayes et al., 2014)	Judsonia, Sunnydale, Arkansas (U.S. EPA, 2015e)	9-month study treating Marcellus Shale waste using thermal distillation (Boschee, 2014; Bruff and Jikich, 2011)	San Ardo Water Reclamation Facility, San Ardo, California (conventional oil and gas) (Dahm and Chapman, 2014; Webb et al., 2009)
TSS	Results not reported.	90% Inf. = 1,272 mg/L Eff. = 9 mg/L Chemical oxidation, coagulation, and clarification	No influent data. Eff.: <4 mg/L Meets NPDES Permit Settling, biological treatment, and induced gas flotation	>90% Inf.: 35 to 114 mg/L Eff.: <3 to 3 mg/L 100 micron mesh bag filter	Results not reported.
TDS	>99% Inf. = 8,000 to 15,000 mg/L Eff. = 41 mg/L RO	99.7% Inf. = 49,550 mg/L Eff. = 171 mg/L MVR (3 units in parallel)	Results not reported. MVR	98% Inf.: 22,350 to 37,600 mg/L Eff.: 9 to 400 mg/L Thermal distillation	97% Inf. = 7,000 mg/L Eff. = 180 mg/L Ion exchange softening and double-pass RO

Constituents of concern	Location and results				
	Pinedale Anticline Water Reclamation Facility, Wyoming (Shafer, 2011)	Maggie Spain Water-Recycling Facility, Barnett Shale, Texas (Hayes et al., 2014)	Judsonia, Sunnydale, Arkansas (U.S. EPA, 2015e)	9-month study treating Marcellus Shale waste using thermal distillation (Boschee, 2014; Bruff and Jikich, 2011)	San Ardo Water Reclamation Facility, San Ardo, California (conventional oil and gas) (Dahm and Chapman, 2014; Webb et al., 2009)
Anions	Chloride: >99% Inf. = 3,600 to 6,750 mg/L Eff. = 18 mg/L RO Sulfate: 99% Inf. = 10 to 100 mg/L Eff. = non-detect Clarification and filtration	Sulfate: 98% Inf. = 309 mg/L Eff. = 6 mg/L Chemical oxidation, coagulation, clarification, and MVR	Sulfate: No influent data. Eff.: 12 mg/L Meets NPDES Permit MVR	Bromide: >99% Inf.: 101 to 162.5 mg/L Eff.: <0.1 to 1.6 mg/L Chloride: 98% Inf.: 9,760 to 16,240 mg/L Eff.: 2.9 to 184.2 mg/L Sulfate: 93% Inf.: 20.4 to <100 mg/L Eff.: <1 to 2.2 mg/L Fluoride: 96% Inf.: <2 to <20 mg/L Eff.: <0.2 to 0.42 mg/L Thermal distillation	Chloride: >99% Inf. = 3,400 mg/L Eff. = 11 mg/L Double-pass RO Sulfate: 6% Inf. = 133 mg/L Eff. = 125 mg/L Sulfuric acid is added after RO to neutralize the pH so no sulfate removal is expected.

Constituents of concern	Pinedale Anticline Water Reclamation Facility, Wyoming (Shafer, 2011)	Maggie Spain Water-Recycling Facility, Barnett Shale, Texas (Hayes et al., 2014)	Location and results Judsonia, Sunnydale, Arkansas (U.S. EPA, 2015e)	9-month study treating Marcellus Shale waste using thermal distillation (Boschee, 2014; Bruff and Jikich, 2011)	San Ardo Water Reclamation Facility, San Ardo, California (conventional oil and gas) (Dahm and Chapman, 2014; Webb et al., 2009)
Metals	Boron: 99% Inf. = 15 to 30 mg/L Eff. = non-detect Ion exchange	Iron: >99% Inf. = 28 mg/L Eff. = 0.1 mg/L For iron, 90% attributed to chemical oxidation, coagulation, and clarification Boron: 98% Inf. = 17 mg/L Eff. = 0.4 mg/L Barium: >99% Inf. = 15 mg/L Eff. = 0.1 mg/L Calcium: >99% Inf. = 2,916 mg/L Eff. = 3.2 mg/L Magnesium: >99% Inf. = 316 mg/L Eff. = 0.4 mg/L	Cobalt: No influent data. Eff.: <0.007 mg/L Arsenic: No influent data. Eff.: <0.001 mg/L Cadmium: No influent data. Eff.: <0.0001 mg/L Chromium: No influent data. Eff.: <0.007 mg/L Copper: No influent data. Eff.: <0029 mg/L Lead: No influent data. Eff.: <0.001 mg/L	Copper: >99% Inf. = <0.2 to <1.0 mg/L Eff. = <0.02 to <0.08 mg/L Zinc: inf below detect Inf. = <0.2 to <1.0 mg/L Eff. = <0.02 to 0.05 mg/L Barium: >99% Inf. = 260.5 to 405.5 mg/L Eff. = <0.1 to 4.54 mg/L Strontium: 98% Inf. = 233 to 379 mg/L Eff. = 0.026 to 3.93 mg/L Iron: Inf. = 13.9 to 22.9 mg/L Eff. = <0.02 to 0.06 mg/L Boron: 97% Inf. = <1 to 3.12 mg/L Eff. = 0.02 to 0.06 mg/L	Sodium: 98% Inf. = 2,300 mg/L Eff. = 50 mg/L Boron: >99% Inf. = 26 mg/L Eff. = 0.1 mg/L RO with elevated influent pH

Constituents of concern	Pinedale Anticline Water Reclamation Facility, Wyoming (Shafer, 2011)	Maggie Spain Water-Recycling Facility, Barnett Shale, Texas (Hayes et al., 2014)	Location and results Judsonia, Sunnydale, Arkansas (U.S. EPA, 2015e)	9-month study treating Marcellus Shale waste using thermal distillation (Boschee, 2014; Bruff and Jikich, 2011)	San Ardo Water Reclamation Facility, San Ardo, California (conventional oil and gas) (Dahm and Chapman, 2014; Webb et al., 2009)
Metals, cont.		Sodium: >99% Inf. = 10,741 mg/L Eff. = 14.3 mg/L Strontium: >99% Inf. = 505 mg/L Eff. = 0.5 mg/L MVR	Mercury: No influent data. Eff.: <0.005 mg/L Zinc: No influent data. Eff.: 0.02 mg/L Meets NPDES permit except for TMDLs for hexavalent chromium and mercury Settling, biological treatment, induced gas flotation, and MVR	Calcium: 98% Inf. = 1,175 to 1,933 mg/L Eff. = 0.36 to 22.2 mg/L Sodium: 98% Inf. = 4,712 to 7,781 mg/L Eff. = 0.37 to 87.9 mg/L Arsenic: 82% Inf. = <0.01 to 0.028 mg/L Eff. = <0.005 mg/L Thermal distillation	

	Location and results			
Constituents of concern	Pinedale Anticline Water Reclamation Facility, Wyoming (Shafer, 2011)	Maggie Spain Water-Recycling Facility, Barnett Shale, Texas (Hayes et al., 2014)	Judsonia, Sunnydale, Arkansas (U.S. EPA, 2015e)	9-month study treating Marcellus Shale waste using thermal distillation (Boschee, 2014; Bruff and Jikich, 2011)	San Ardo Water Reclamation Facility, San Ardo, California (conventional oil and gas) (Dahm and Chapman, 2014; Webb et al., 2009)
Radionuclides	Results not reported.	Results not reported.	Not regulated under permit – believed to be absent.	Radium-226: 97% - 99% Inf. = 130 to 162 pCi/L Eff. = 0.224 to 2.87 pCi/L Radium-228: 97% - 99% Inf. = 45 to 85.5 pCi/L Eff. = 0.259 to 1.32 pCi/L Gross Alpha: 97% - >99% Inf. = 161 to 664 pCi/L Eff. = 0.841 to 6.49 pCi/L Gross Beta: 98% - >99% Inf. = 79.7 to 847 pCi/L Eff. = 0.259 to 1.57 pCi/L Thermal distillation	Results not reported.

Constituents of concern	Pinedale Anticline Water Reclamation Facility, Wyoming (Shafer, 2011)	Maggie Spain Water-Recycling Facility, Barnett Shale, Texas (Hayes et al., 2014)	Judsonia, Sunnydale, Arkansas (U.S. EPA, 2015e)	9-month study treating Marcellus Shale waste using thermal distillation (Boschee, 2014; Bruff and Jikich, 2011)	San Ardo Water Reclamation Facility, San Ardo, California (conventional oil and gas) (Dahm and Chapman, 2014; Webb et al., 2009)
			Location and results		
Organics	Oil & Grease: 99% Inf. = 50 to 2,400 mg/L Eff. = non-detect BTEX: 99% Inf. = 28 to 80 mg/L Eff. = non-detect GRO: 99% Inf. = 88 to 420 mg/L Eff. = non-detect DRO: 99% Inf. = 77 to 1,100 mg/L Eff. = non-detect Methanol: 99% Inf. = 40 to 1,500 mg/L Eff. = non-detect Oil-water separator, anaerobic and aerobic biological treatment, coagulation, flocculation, flotation, sand filtration, membrane bioreactor, and ultrafiltration	TPH: >80% Inf. = 388 mg/L Eff. = 4.6 mg/L BTEX: 94% Inf. = 3.3 mg/L Eff. = 0.2 mg/L TOC: 48% Inf. = 42 mg/L Eff. = 22 mg/L Coagulation, sedimentation, MVR	Biochemical oxygen demand: No influent data. Eff.: <2 mg/L Oil & Grease: No influent data. Eff.: <5 mg/L Benzo (k) fluoranthene: No influent data. Eff.: <0.005 mg/L Bis (2-Ethylhexyl) Phthalate: No influent data. Eff.: <0.001 mg/L Butyl benzyl phthalate: No influent data. Eff.: <0.001 mg/L Meets NPDES permit Settling, biological treatment, induced gas flotation, and MVR	Acetone: 93% Inf. = 8.71 to 13.8 mg/L Eff. = 0.524 to 0.949 mg/L Toluene: >80% Inf. = 0.0083 to 0.0015 mg/L Eff. = non-detect to 0.0013 mg/L Methane: >99% Inf. = 0.748 to 5.49 mg/L Eff. = non-detect to 0.0013 mg/L DRO: 0 to 82% Inf. = 4 to 7.1 mg/L Eff. = 0.99 to 4.9 mg/L Oil & Grease: No removal Thermal distillation	Results not reported.

F.4.1. Total Suspended Solids

The reduction of TSS is typically required before wastewater can be reused for subsequent hydraulic fracturing jobs. Hydraulic fracturing wastewaters containing suspended solids can plug the well and damage equipment if reused for other fracturing operations (Tiemann et al., 2014; Hammer and VanBriesen, 2012). For treated water that is discharged to a surface water body, the EPA has a secondary treatment standard for POTWs that limits TSS in the effluent to 30 mg/L (30-day average). In addition, most advanced treatment technologies require the removal of TSS prior to treatment to avoid operational problems, such as membrane fouling/scaling, and to extend the life of the treatment unit.

TSS removal efficiencies shown in Table F-4 (90% and over 90%) were achieved with chemical oxidation, coagulation, and clarification, as well as filtration. Technologies that remove TSS have also been employed in another Marcellus Shale study (sedimentation and filtration) (Mantell, 2013); Utica Shale (chemical precipitation and filtration) (Mantell, 2013); Barnett Shale (chemical precipitation and inclined plate clarifier, >90% removal) (Hayes et al., 2014); and Utah (EC, 90% removal) (Halliburton, 2014).

F.4.2. Total Dissolved Solids

The TDS concentration of hydraulic fracturing wastewater is a key treatment consideration, with the required level of TDS removal dependent upon the intended use of the treatment effluent. POTW treatment and basic treatment processes at a CWT (i.e., chemical precipitation, sedimentation, and filtration) are typically not reliable methods for removing TDS. Reduction requires more advanced treatment processes such as RO, nanofiltration, thermal distillation (including MVR), evaporation, and/or crystallization (Olsson et al., 2013; Boschee, 2012; Drewes et al., 2009). Pretreatment (e.g., chemical precipitation, flotation, etc.) is typically needed to remove constituents that may cause fouling or scaling with the advanced treatment processes or to remove specific constituents not removed by a particular advanced process. TDS removal efficiencies reported in Table F-4 ranged from 97% to >99% with RO, thermal distillation, MVR, and ion exchange softening with a double pass RO.

RO and thermal distillation processes can treat waste streams with TDS concentrations up to 35,000 mg/L and more than 100,000 mg/L, respectively (Tiemann et al., 2014). Extremely high TDS waters may require a series of advanced treatment processes to remove TDS to desired levels. However, the cost of treating high-TDS waters may preclude facilities from choosing treatment if other options, such as deep well injection, are available and more cost-effective (Tiemann et al., 2014).

F.4.3. Anions

Although chemical precipitation processes can reduce concentrations of multivalent anions such as sulfate, monovalent anions (e.g., bromide and chloride) are not removed by basic treatment processes and require more advanced treatment such as RO, thermal distillation (including MVR), evaporation, and/or crystallization (Hammer and VanBriesen, 2012). As shown in Table F-4, anion removal efficiencies in the four studies where sulfate removal was measured ranged from 93% to >99%.

F.4.4. Metals and Metalloids

Removal of dissolved and precipitated metals and metalloids is commonly needed prior to discharge to a waterbody or reuse. The facilities in Table F-4 report removals of 98%–99% for a number of metals. Other work demonstrating effective removal includes a 99% reduction in barium using chemical precipitation (Marcellus Shale region) (Warner et al., 2013a) and over 90% boron removal with RO (at pH of 10.8) at two California facilities (Webb et al., 2009; Kennedy/Jenks Consultants, 2002). However, influent concentration must be considered together with removal efficiency to determine whether effluent quality meets the requirements dictated by end use or by regulations. In the case of the facility described by Kennedy/Jenks Consultants (2002), the boron effluent concentration of 1.9 mg/L (average influent concentration of 16.5 mg/L) was not low enough to meet California's action level of 1 mg/L.

F.4.5. Radionuclides

Data on radionuclide removals achieved in active treatment plants are scarce. The literature does provide some data from the Marcellus Shale region on use of distillation and chemical precipitation (co-precipitation of radium with barium sulfate). As shown in Table F-4, one nine-month pilot scale study conducted by Bruff and Jikich (2011) reported that distillation treatment produced removal efficiencies between 97% and >99% for radium, gross alpha, and gross beta, and 71% to 90% for thorium. In a separate study, Warner et al. (2013b) reported that a CWT was estimated to have achieved over 99% removal of radium via co-precipitation of radium with barium sulfate (radium 226 influent of 3231 pCi/L and effluent of 4 pCi/L; radium 228 influent of 452 pCi/L and effluent of 2 pCi/L). However, in both studies, radionuclides were detected in effluent samples, and the CWT was discharging to a surface water body during this time (Warner et al., 2013b; Bruff and Jikich, 2011) (Section 8.5.2). Between 2010 and 2012, samples of wastewater effluent from a western Pennsylvania CWT contained a mean radium level of 4 pCi/L (Warner et al., 2013a).

F.4.6. Organics

Facilities have demonstrated the capability to treat for organic compounds in hydraulic fracturing wastewaters. Table F-4 shows that one facility achieved 99% removal of oil and grease, BTEX (benzene, toluene, ethylbenxene, xylenes), gasoline range organics (GRO), diesel range organics (DRO), and methanol while another facility reported >80% removal of total petroleum hydrocarbons (TPH), 94% removal of BTEX, and 48% removal of total organic carbon (TOC).

Given the variety of properties among classes of organic constituents, different treatment processes may be required depending upon the types of organic compounds needing removal. Table F-5 lists treatment processes and the classes of organic compounds they can treat. It should be noted that in many studies, rather than testing for several organic constituents, researchers often measure organics in terms of biochemical oxygen demand and/or chemical oxygen demand, which are indirect measures of the amount of organic compounds in the water. Organic compounds may also be measured and/or reported in groupings such as TPH (which includes GRO, DRO, oil and grease), volatile organic compounds (VOCs) (which include BTEX), and semi-volatile organic compounds (SVOCs).

Table F-5. Treatment processes for hydraulic fracturing wastewater organic constituents.

Treatment processes	Organic compounds removed	References
Adsorption with activated carbon	Soluble organic compounds	Fakhru'l-Razi et al. (2009)
Adsorption with organoclay media	Insoluble organic compounds	Fakhru'l-Razi et al. (2009)
Air stripping	Volatile organic compounds	Tchobanoglous et al. (2013)
Dissolved air flotation	Volatile organic compounds, dispersed oil	Drewes et al. (2009)
Freeze/thaw evaporation[a]	TPH, volatile organic compounds, semi-volatile organic compounds	Duraisamy et al. (2013); Drewes et al. (2009)
Ion exchange (with modified zeolites)	BTEX, chemical oxygen demand, biochemical oxygen demand	Hayes et al. (2014); Duraisamy et al. (2013); Drewes et al. (2009); Fakhru'l-Razi et al. (2009); Munter (2000)
Distillation	BTEX, polycyclic aromatic hydrocarbons (PAHs)	Hayes et al. (2014); Duraisamy et al. (2013); Drewes et al. (2009); Fakhru'l-Razi et al. (2009).
Chemical precipitation	Oil & grease	Drewes et al. (2009); Fakhru'l-Razi et al. (2009)
Chemical Oxidation	Oil & grease	Drewes et al. (2009); Fakhru'l-Razi et al. (2009)
Media filtration (walnut shell media or sand)	Oil & grease	Drewes et al. (2009); Fakhru'l-Razi et al. (2009)
Microfiltration	Oil & grease	Drewes et al. (2009); Fakhru'l-Razi et al. (2009)
Ultrafiltration	Oil & grease, BTEX	Drewes et al. (2009); Fakhru'l-Razi et al. (2009)
Reverse osmosis[b]	Dissolved organics	Drewes et al. (2009); U.S. EPA (2005)
Electrocoagulation	Chemical oxygen demand, Biochemical oxygen demand	Fakhru'l-Razi et al. (2009)
Biologically aerated filters	Oil & grease, TPH, BTEX	Fakhru'l-Razi et al. (2009)
Reed bed technologies	Oil & grease, TPH, BTEX	Fakhru'l-Razi et al. (2009)
Hydrocyclone separators	Dispersed oil	Drewes et al. (2009)

[a] Technology cannot be used if the methanol concentration in the hydraulic fracturing wastewater exceeds 5%.

[b] RO will remove specific classes of organic compounds with removal efficiencies dependent on the compound's structure and the physical and chemical properties of the hydraulically fractured wastewater. Organoacids will foul membranes.

F.5. Centralized Waste Treatment Facilities and Waste Management Options

CWTs are designed to treat for site-specific wastewater constituents so that the effluent meets the requirements of the designated disposal option(s) (i.e., reuse, direct/indirect discharge). The most basic treatment processes that a CWT might use include (Easton, 2014; Duhon, 2012):

- Physical treatment technologies such as dissolved air or induced gas flotation systems, media filtration, hydrocyclones, and settling including sedimentation/clarification;

- Chemical treatment technologies such as chemical precipitation (coagulation) and chemical oxidation; and

- Biological treatment technologies such as biological aerated filter systems and reed beds.

Although these technologies are effective at removing oil and grease, suspended solids, scale-forming compounds, and some heavy metals, advanced processes such as RO, thermal distillation, or evaporation are necessary if TDS should be reduced as required by the intended disposal option.

This section provides an overview of treatment technologies employed at CWTs treating for oil and gas wastewaters and their discharge options.

F.5.1. Design of Treatment Trains for CWTs

Based on the chemical composition of the hydraulic fracturing wastewater and the desired effluent water quality, a series of treatment technologies will most likely be necessary. The possible combinations of unit processes combined into treatment trains are extensive. One report identified 41 different treatment unit processes that have been used in the treatment of oil and gas wastewater and 19 unique treatment trains (Drewes et al., 2009). Fakhru'l-Razi et al. (2009) also provide examples of process flow diagrams that have been used in pilot-scale and commercial applications for treating oil and gas wastewater. Figure F-7 shows the treatment train for the Pinedale Anticline Facility as of 2012, which includes pretreatment for dispersed oil, VOCs, and heavy metals and advanced treatment for removal of TDS, dissolved organics, and boron. This CWT can either discharge to surface water or provide the treated wastewater to operators for reuse.

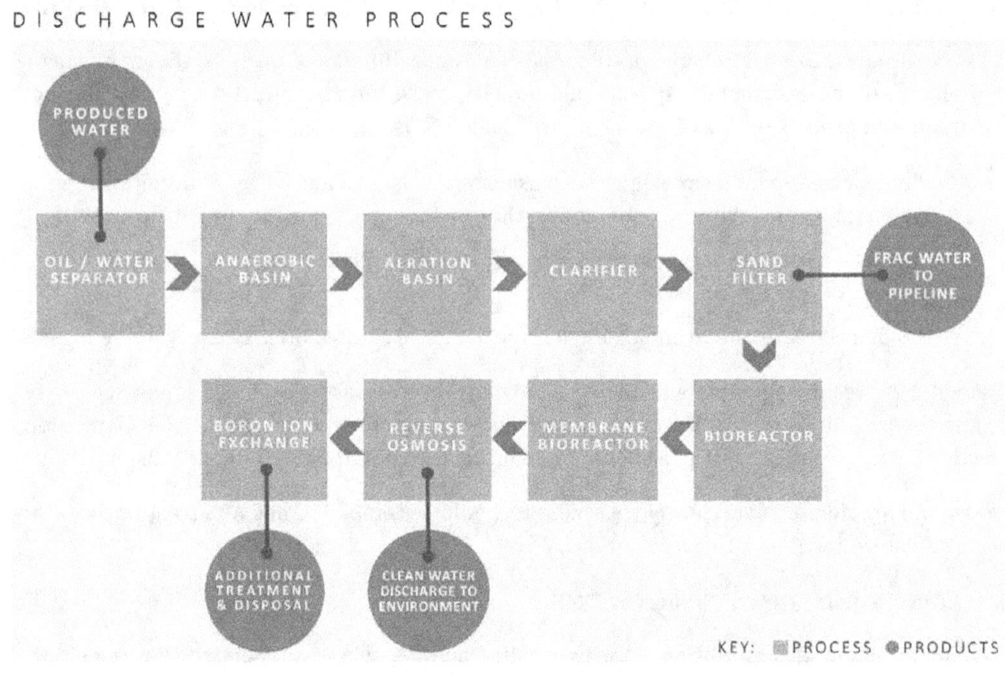

DISCHARGE WATER PROCESS

Figure F-7. Full discharge water process used in the Pinedale Anticline field.
Source: Redrawn and adapted from a figure in Boschee (2012).

Table F-6 provides information on some CWTs in locations across the country and the processes they employ. The table also notes for each facility whether data on effluent quality are readily available. Comprehensive and systematic data on influent and effluent quality from CWTs that treat to a variety of water quality levels are difficult to procure. This makes it challenging to understand removal efficiencies and resulting effluent quality, especially when a facility offers varying degrees of treatment to meet the water quality needs for different end uses (e.g., reuse vs. discharge). For those facilities with NPDES permits, discharge monitoring report (DMR) data may be available for some constituents, although if the facility does not discharge regularly, these data will be sporadic.

As of July 2016, the Pinedale Anticline Facility, the Judsonia Facility, the Eureka Resources Standing Stone Facility, and Wellington Operating Company's facility appear to be the only CWTs in Table F-6 discharging to surface water or a groundwater aquifer.[1]

---

[1] For Pinedale Anticline Water Reclamation Facility, surface water discharges are permitted under 40 CFR 435 Subpart E (beneficial use subcategory agricultural and wildlife water) not 40 CFR 437 (the discharge permit for CWTs). For the purposes of this assessment, this facility is included with CWTs.

Table F-6. Examples of centralized waste treatment facilities.

Facility	Locality	Description of unit processes	Does CWT have a NPDES permit for discharge?	Does CWT provide effluent for reuse?	Does CWT have advanced process for TDS removal?	What is the status of the facility as of July, 2016?	Are effluent quality data available through literature search?
Pinedale Anticline Water Reclamation Facility	WY	Oil/water separation, biological treatment, aeration, clarification, sand filtration, bioreactor, membrane bioreactor, RO, ion exchange, and desalinization	No – However, facility is permitted to discharge under 40 CFR 435 Subpart E (WY0054224). Facility is permitted to discharge up to 25% of its effluent stream	Yes	Yes, RO (Boschee, 2014, 2012)	The treatment plant produces treated water for reuse and for discharge to outfalls located at the New Fork River and at Sand Draw.	Yes – DMR data available on Wyoming DEQ website. Some information can also be obtained from Shafer (2011).
SEECO – Judsonia Water Reuse Recycling Facility	AR	Settling, biological treatment, induced gas flotation, and MVR	Yes - AR0052051	Yes	Yes, MVR	The treatment plant provides treated water for reuse and for discharge to surface water. Based on DMR data from late 2015- early 2016, the system is discharging treated water to a surface water body, though intermittently.	DMR data available
Eureka Resources – Williamsport 2nd Street Facility	PA	Settling, oil/water separation, chemical precipitation, clarification, MVR. Can treat with or without TDS removal.	No – However, future plans to install RO for direct discharge capability	Yes	Yes, MVR	Per Ertel et al. (2013), the facility provides treatment wastewater for reuse and indirect discharge. The facility treats entirely or almost entirely hydraulic fracturing wastewater.	No

Appendix F – Wastewater Disposal and Reuse Supplemental Information

Facility	Locality	Description of unit processes	Does CWT have a NPDES permit for discharge?	Does CWT provide effluent for reuse?	Does CWT have advanced process for TDS removal?	What is the status of the facility as of July, 2016?	Are effluent quality data available through literature search?
Eureka Resources – Standing Stone Facility, Bradford County	PA	Settling, oil/water separation, chemical precipitation, clarification, MVR, crystallization	Yes - PA0232351	Yes	Yes, MVR, crystallizer	The facility can provide treated wastewater for reuse and also has received an NPDES permit for direct discharge. The facility treats hydraulic fracturing wastewater.	No
Wellington Operating Company, LLC – 3W Production Water Treatment Facility	CO	Dissolved air flotation, pre-filtration, microfiltration with ceramic membranes, activated carbon adsorption. Water is pumped to rapid-infiltration pit which then percolates to a tributary aquifer. The aquifer supplies water to an RO plant (Alzahrani et al., 2013).	Shallow groundwater percolation pit permits issued by COGCC – 281818 and 281824	Yes	Yes, RO but only after the water is sent to an aquifer storage and recovery well	Per Stewart (2013), the facility is providing treated wastewater for reuse, for agricultural use, to a shallow well to augment the municipal drinking water supply, and for discharge to the Colorado River.	No

Facility	Locality	Description of unit processes	Does CWT have a NPDES permit for discharge?	Does CWT provide effluent for reuse?	Does CWT have advanced process for TDS removal?	What is the status of the facility as of July, 2016?	Are effluent quality data available through literature search?
Casella Altela Regional Environmental Services (CARES) McKean Facility	McKean County, PA	Pretreatment system (not defined in literature) and thermal distillation	Yes – PA0102288	Yes	Yes – thermal distillation	The treatment plant is capable of reuse and recycle for fracturing operations and surface water discharge of excess water. However, the vendor has indicated that the facility is only treating water for reuse/recycle as of early 2015.	No - just NPDES discharge requirements
Clarion Altela Environmental Services (CAES) Facility	Clarion County, PA	Pretreatment system (not defined in literature) and thermal distillation	Yes – PA0103632	Yes	Yes – thermal distillation	The treatment plant is capable of reuse and recycle for fracturing operations and surface water discharge of excess water. However, the facility has indicated that it is only treating water for reuse/recycle as of early 2015.	No – just NPDES discharge requirements
Terraqua Resource Management (aka. Water Tower Square Gas Well Wastewater Processing Facility)	Lycoming County, PA	Equalization tanks, oil-water separation via chemical addition (sulfuric acid, emulsion breaker), pH adjustment, coagulation, flocculation, inclined plate clarifier, sand filtration	No	Yes	No – However, TARM recognizes that they can't discharge, until they install TDS treatment	Listed as proposed CWT, Part I NPDES permit issued, awaiting Part II WQM application per PA DEP website visited August 25, 2016. (See DEP's list of Waste Water Treatment Facilities and http://mshaletaskforce. org/Site_Locations.html) .	No

Facility	Locality	Description of unit processes	Does CWT have a NPDES permit for discharge?	Does CWT provide effluent for reuse?	Does CWT have advanced process for TDS removal?	What is the status of the facility as of July, 2016?	Are effluent quality data available through literature search?
Maggie Spain Water-Recycling Facility	Decatur, TX	Settling, flash mixer with lime and polymer addition, inclined plate clarifier, surge tank, MVR	No	Yes	Yes – MVR	The facility reuses/recycles treated water for fracturing operations. It is unclear if the MVR mobile unit is still at this facility. A pilot study is in progress at the facility that began in 2015 looking at the addition of a hollow fiber air stripping membrane unit for CO2 removal prior to an UF/RO unit.	Yes – Some information can be obtained from Hayes et al. (2014).
Fountain Quail/NAC Services - Kenedy	Kenedy, TX	Oil-water separator, coagulation, flocculation, sedimentation, filtration, MVR.	No	Yes	Yes – MVR	According to its website, the facility reuses/recycles treated water for fracturing operations (http://www.fountainquail.com/water-recycling-solutions/clean-brine/case-studies/eagle-ford-shale-texas).	No
Purestream - Gonzales facility	Gonzales, TX	Induced gas flotation and MVR	No	Yes	Yes - MVR	Per Dahm and Chapman (2014) commercial operations deployed March 2014 for reuse/recycle for fracturing operations.	No

Facility	Locality	Description of unit processes	Does CWT have a NPDES permit for discharge?	Does CWT provide effluent for reuse?	Does CWT have advanced process for TDS removal?	What is the status of the facility as of July, 2016?	Are effluent quality data available through literature search?
FourPoint Energy, LLC (formerly owned by Linn Energy - Granite Wash	Wheeler County, TX	Induced gas flotation and MVR	No	Yes	Yes - MVR	AVARA system installed for reuse/recycle in June 2014, according to http://purestream.com/index.php/water-management/vapor-recompression/photos-and-videos. Assume still operational but status unclear as new private company acquired Linn Energy's oil and gas assets in 2014.	No
Fluid Recovery Service Josephine Facility[a]	PA	Oil-water separator, aeration, chemical precipitation with sodium sulfate, lime, and a polymer, inclined plate clarifier	PA0095273 Permit renewal application submitted and under review by PA DEP.	No	No	The facility stopped accepting Marcellus wastewater September 30, 2011 (Ferrar et al., 2013). It treats conventional oil and gas wastewater. The facility plans to upgrade to include evaporative technology to attain monthly average TDS levels of 500 mg/L or less. Not upgraded as of July, 2016.	Yes – Some effluent results obtained from Ferrar et al. (2013) and Warner et al. (2013a). Also minimal DMR data from the EPA.

Appendix F – Wastewater Disposal and Reuse Supplemental Information

Facility	Locality	Description of unit processes	Does CWT have a NPDES permit for discharge?	Does CWT provide effluent for reuse?	Does CWT have advanced process for TDS removal?	What is the status of the facility as of July, 2016?	Are effluent quality data available through literature search?
Fluid Recovery Service Franklin Facility[a]	PA	Oil-water separator, aeration, chemical precipitation with sodium sulfate, lime, and a polymer, inclined plate clarifier	PA0101508 Permit renewal application submitted and under review by PA DEP.	No	No	This facility is not accepting wastewater from hydraulic fracturing operations as of July 2016. It treats conventional oil and gas wastewater. The facility plans to upgrade to include evaporative technology to attain monthly average TDS levels of 500 mg/L or less. Not upgraded as of July, 2016.	Minimal DMR data from the EPA.
Hart Resources-Creekside Facility[a]	PA	Oil-water separator, aeration, chemical precipitation with sodium sulfate, lime, and a polymer, inclined plate clarifier	PA0095443 Permit renewal application submitted and under review by PA DEP.	No	No	This facility is not accepting wastewater from hydraulic fracturing operations as of July 2016. It treats conventional oil and gas wastewater. The facility plans to upgrade to include evaporative technology to attain monthly average TDS levels of 500 mg/L or less. Not upgraded as of July, 2016.	Minimal DMR data from the EPA.

[a] As of May 15, 2013, these facilities are under an administrative order (AO). According to the AO, these facilities must comply with a monthly effluent limit for TDS not to exceed 500 mg/L. This will allow them to treat high-saline wastewaters typical of unconventional oil and gas operations. To meet the requirements of the AO, they have applied to the Pennsylvania Department of Environmental Protection (PA DEP) for a NPDES permit and are planning to install treatment for TDS.

F.5.2. Discharge Options for CWTs

Direct discharge CWTs are allowed to discharge treated wastewater directly to surface waters under the NPDES permit program. Discharge limitations may be based on water quality standards in the NPDES and technology-based effluent limitation guidelines under 40 CFR Part 437. In addition, permitting authorities have permitted facilities for discharge under 40 CFR 435, Subpart E. Judsonia Central Water Treatment Facility in Sunnydale, Arkansas is permitted to directly discharge treated effluent from produced water from the Fayetteville Shale play to Byrd pond located on the property. Pinedale Anticline Field Wastewater Treatment Facility in Wyoming, WY, originally designed to treat produced water from tight gas plays in the Pinedale Anticline Field to levels suitable for reuse, was upgraded to include desalinization and RO treatment for discharge to a local river. CWTs with NPDES discharge permits may also opt to treat oil and gas wastewater for reuse as shown in Table F-6. Some facilities have the ability to treat wastewater to different qualities (e.g., with or without TDS removal), which they might do to target various reuse water quality criteria. Both the Judsonia and Pinedale facilities discussed above have the ability to employ either TDS- or non-TDS-removal treatments depending on the customers' needs.

Indirect discharge CWTs may treat hydraulic fracturing wastewater and then discharge the treated wastewater effluent to a POTW. Discharge to the POTW is controlled by an Industrial User mechanism, which incorporates pretreatment standards established in 40 CFR Part 437. Two facilities, one located in Pennsylvania (Eureka Resources) and the other in Ohio (Patriot Water Treatment), include indirect discharge as an option in wastewater treatment. The Eureka-Williamsport facility accepts wastewater (primarily from the Marcellus Shale play) and either treats it for reuse or discharges it to the local POTW. The Patriot facility offers services to hydraulic fracturing operators in the Marcellus and Utica Shale plays for removal of solids and metals using chemical treatment. As of March 2015, however, the Patriot facility is limited by the Ohio Environmental Protection Agency to accepting only "low salinity" (<50,000 mg/L TDS) produced water and may only discharge 100,000 gal (380,000 L) per day to the Warren Ohio POTW.

Zero-discharge CWTs do not discharge treated wastewater; instead, the wastewater is treated and reused in subsequent hydraulic fracturing operations. WVWRI (2012) state that this practice reduces potential effects on surface drinking water resources by reducing both direct and indirect discharges. Zero-discharge facilities may offer varying levels of treatment, including minimal treatment (for example, filtration), low-level treatment (chemical precipitation), and/or advanced treatment (evaporation, crystallization). Reserved Environmental Services (RES) in Mt. Pleasant, Pennsylvania, is a zero liquid discharge facility permitted by PA DEP to treat wastewater from the Marcellus Shale play for reuse. Residual solids are dewatered and sent to a landfill. Treated wastewater effluent is stored, monitored, and chlorinated for reuse (ONG Services, 2015).

F.6. Water Reuse

With the scarcity of freshwater supplies and limited access to disposal wells in some areas of the country, reuse of hydraulic fracturing wastewaters for subsequent hydraulic fracturing activity has become more prevalent (Section 8.4.4). This section discusses factors to consider in adopting reuse and the recommended or otherwise observed water quality needed.

F.6.1. Factors in Considering Reuse

In making the decision whether to manage wastewater via reuse, operators have several factors to consider (Slutz et al., 2012; NPC, 2011):

- Wastewater generation rates compared to water demand for future fracturing operations,
- Wastewater quality and treatment requirements for use in future operations,
- The costs and benefits of wastewater management for reuse compared with other management strategies,
- Available infrastructure and treatment technologies, and
- Regulatory considerations.

Among these factors, costs may be the most significant driver, weighing the costs of transportation from the generating well to the treatment facility and to the new well against the costs for transport to alternative locations (a disposal well or CWT). Trucking large quantities of water can be relatively expensive (from $0.50 to $8.00 per barrel), rendering on-site treatment technologies and reuse potentially economically competitive in some settings (Dahm and Chapman, 2014; Guerra et al., 2011). Also, logistics, including proximity of the water sources for aggregation, may be a factor in implementing reuse. For example, Boschee (2014) notes that in the Permian Basin, older conventional wells are linked by pipelines to a centralized transfer facility, enabling movement of treated water to areas where it is needed for reuse.

Regulatory factors may facilitate reuse. In 2013, the Texas Railroad Commission adopted rules intended to encourage statewide water conservation. These rules facilitate reuse by eliminating the need for a permit when operators reuse on their own lease or transfer the fluids to another operator for use in hydraulic fracturing (Rushton and Castaneda, 2014). Data for the years after 2013 will allow evaluation of whether reuse increased after this regulatory change.

Recommended compositional ranges for the base fluid used to formulate hydraulic fracturing fluid may shift in the future as fracturing fluid technology continues to develop. Development of fracturing mixture additives that are brine-tolerant have allowed for the use of high TDS wastewaters (up to tens of thousands of mg/L) for reuse in fracturing (Tiemann et al., 2014; GTI, 2012; Minnich, 2011). Some new fracturing fluid systems are claimed to be able to tolerate salt concentrations exceeding 300,000 mg/L (Boschee, 2014). This greater flexibility in acceptable water chemistry can facilitate reuse both logistically and economically by reducing treatment needs.

Reuse rates may also fluctuate with changes in the supply and demand of treated wastewater and the availability of fresh water. Flowback may be preferable to later-stage produced water for reuse because it is typically generated in larger quantities from a single location as opposed to water produced later on, which is generated in smaller volumes over time from many different locations. Flowback also tends to have lower TDS concentrations than later-stage produced water. In the Marcellus, TDS has been shown to increase from tens of thousands to about 100,000 mg/L during the first 30 days (Barbot et al., 2013; Maloney and Yoxtheimer, 2012); see Chapter 7. As more wells go into production, the changing production rate and quality of wastewaters generated in a region need to be taken into account, as well as a possible reduction in the demand for reused water as plays mature (Lutz et al., 2013; Hayes and Severin, 2012b; Slutz et al., 2012).

F.6.1.1. On-Site Treatment for Reuse

On-site systems that treat produced water for reuse can reduce potential impacts on drinking water resources associated with transportation and disposal, and they can facilitate the logistics of reuse by preparing the water close to well sites. These systems sometimes consist of mobile units containing one or more treatment processes that can be moved from site to site to treat waters in newly developed sites that are not yet producing at full-scale. Semi-permanent facilities that serve specific areas also exist (Halldorson, 2013; Boschee, 2012).

Treatment systems are typically tailored for site-specific produced water chemical concentrations and desired water quality treatment goals, including whether significant TDS removal is needed. If low TDS water is needed, more advanced treatment will be required (as discussed in Section F.2). This more extensive treatment can increase the treatment costs by three to four times compared to treatment systems that do not remove TDS (Halldorson, 2013). On-site facilities may be warranted where truck hauling or seasonal accessibility to and from a central facility is an issue (Boschee, 2014; Tiemann et al., 2014). Operators may also consider on-site facilities if they have not fully committed to an area and the well counts are initially low. In those instances, they can later decide to add or remove units based on changing production volumes (Boschee, 2014).

F.6.2. Water Quality for Reuse

As of 2016, there is no consensus on the water quality requirements for reuse of wastewater for hydraulic fracturing, and operator opinions vary on the minimum standards for the water quality needed for fracturing fluids (Vidic et al., 2013; Acharya et al., 2011). Table F-7 provides a list of constituents and the recommended or observed target concentrations for reuse applications. The wide concentration ranges for many constituents (e.g., TDS ranging from 500 to 70,000 mg/L) suggest that water quality requirements for reuse are dictated by operation-specific requirements, including operator preference and selection of fracturing fluid chemistry.

Table F-7. Water quality requirements for reuse.
Source: U.S. EPA (2015m).

Constituent	Reasons for limiting concentrations	Recommended or observed base fluid target concentrations (mg/L, after blending)[b]
TDS	Fluid stability	500 – 70,000
Chloride	Fluid stability	2,000 – 90,000
Sodium	Fluid stability	2,000 – 5,000

Constituent	Reasons for limiting concentrations	Recommended or observed base fluid target concentrations (mg/L, after blending)[b]
Metals		
Iron	Scaling	1 – 15
Strontium	Scaling	1
Barium	Scaling	2 – 38
Silica	Scaling	20
Calcium	Scaling	50 – 4,200
Magnesium	Scaling	10 – 1,000
Sulfate	Scaling	124 – 1,000
Potassium	Scaling	100 – 500
Scale formers[a]	Scaling	2,500
Other		
Phosphate	Not Reported	10
TSS	Plugging	50 – 1,500
Oil	Fluid stability	5 – 25
Boron	Fluid stability	0 – 10
pH (S.U.)	Fluid stability	6.5 – 8.1
Bacteria (counts/mL)	Bacterial growth	0 – 10,000

[a] Includes total of barium, calcium, manganese, and strontium.

[b] Unless otherwise noted.

Wastewater quality can be managed for reuse either by blending it with freshwater and allowing dilution to bring the concentrations of problematic constituents to an acceptable range or through treatment (Veil, 2010). Treatment, if needed, can be conducted at facilities that are mobile, semi-permanent modular systems, or fully permanent CWTs (Nicot et al., 2012). At a minimum, hydraulic fracturing service providers generally prefer that the wastewater be treated to remove TSS, microorganisms, and constituents that form scale or inhibit crosslinking in gelled fluid systems (Boschee, 2014). Figure F-8 shows a schematic of a treatment system to treat wastewater for reuse that can remove suspended solids, hardness, and organic constituents.

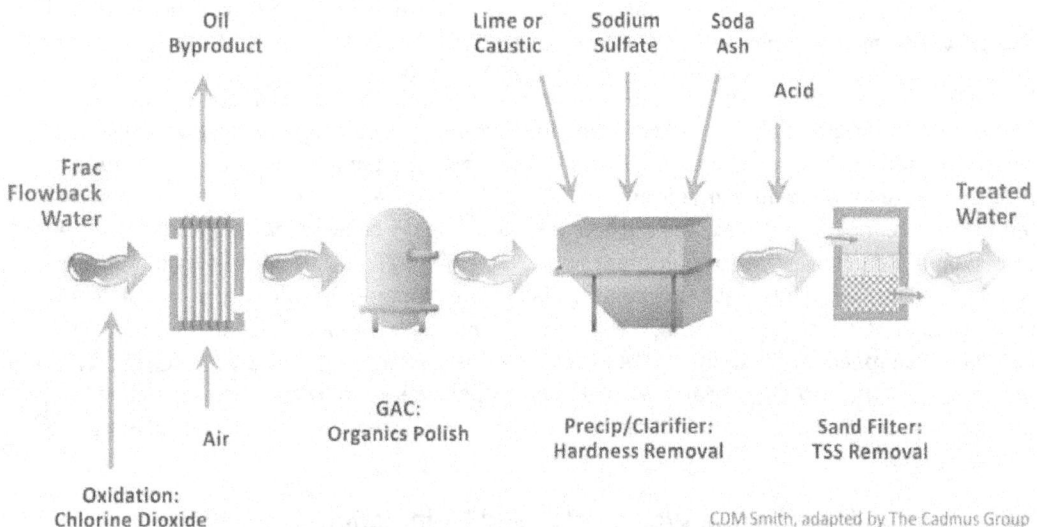

Figure F-8. Diagram of treatment for reuse of flowback and produced water.
Source: Kimball (2010). Reprinted with permission from CDM Smith.

In the Marcellus, the wastewater to be reused is generally treated with oil/gas-water separation, filtration, and dilution (Ma et al., 2014). Although many Marcellus treatment facilities only supply basic reuse treatment that removes oil and solids, advanced treatment facilities that use techniques such as RO or distillation methods are also in operation (Veil, 2010).

Reuse concerns can vary with the type of hydraulic fracturing fluid used (e.g., slickwater, linear gel, crosslinked gel, foam) (Wasylishen and Fulton, 2012) and the anticipated changes in water chemistry over time during the transition from flowback to produced water (Hammer and VanBriesen, 2012). Elevated TDS is a concern, but residual constituents from previous fluid mixtures (e.g., breakers) may also cause difficulties when reusing water for subsequent fracturing operations (Montgomery, 2013; Walsh, 2013).

F.7. Hydraulic Fracturing Wastewater Impacts on POTWs

Wastewater treatment processes used by POTWs are generally not designed or operated to treat wastewater containing high salt concentrations (>0.1-5% salt), and sudden increases in chloride concentration above 5 – 8 g/L may cause problems for wastewater treatment (Ludzack and Noran, 1965). Four basic problems for biological treatment of saline water have been described (Woolard and Irvine, 1995): (1) microbes in POTW treatment systems tend to be sensitive to changes in ionic strength; (2) microbial metabolic functions are disrupted, leading to decreased degradation of carbon compounds; (3) effluent suspended solids are increased due to cell lysis and/or a reduction in organisms that promote flocculation; and (4) the extent to which biomass at a POTW can acclimate to a salty environment is limited. To address concerns with high salinity and other contaminants that are either not removed by or can adversely impact the POTW treatment system,

EPA has promulgated pretreatment standards intended to prevent pollutants in unconventional oil and gas wastewaters from reaching POTWs (Chapter 8).

If indirect discharge to a POTW is being considered or is employed, some adaptations can be useful for wastewater treatment trains at CWTs handling hydraulic fracturing wastewater to meet the established federal limits. For example, biological pre-treatment may be beneficial as an added process prior to indirect discharge from a CWT to a POTW for removal of organic contaminants. Specialized treatment systems using salt-tolerant bacteria may be beneficial as an additional level of treatment for pre-treating (or polishing) wastewaters at CWTs. (These processes differ from conventional biological processes in standard wastewater treatment, which are not suitable for large volumes of hydraulic fracturing wastewater). In particular, membrane bioreactors (MBRs) have been examined for the treatment of oil and gas wastewater (Dao et al., 2013; Kose et al., 2012; Miller, 2011). MBRs provide advantages over conventional aeration basin processes as they can be incorporated into existing treatment trains more easily and have a much smaller areal footprint than aeration basins.

F.8. Hydraulic Fracturing Wastewater and Disinfection Byproducts

F.8.1. Disinfection Byproducts

This section provides background information on disinfection byproducts (DBPs) and their formation to support the discussion in Section 8.5.1 of Chapter 8 regarding impacts on surface waters and downstream drinking water utilities due to elevated bromide and iodide in hydraulic fracturing wastewaters.

Regulated DBPs are a small subset of the full spectrum of DBPs that include other chlorinated, brominated, iodated, and nitrogenous DBPs. Some of the emerging unregulated DBPs may be more toxic than their regulated counterparts (Harkness et al., 2015; McGuire et al., 2014; Parker et al., 2014). Of the many types of DBPs that can form when drinking water is disinfected, Safe Drinking Water Act (SDWA) Stage 1 and Stage 2 DBP Rules regulate four total trihalomethanes (TTHM), five haloacetic acids (HAA5), bromate, and chlorite (U.S. EPA, 2006).

Most brominated DBPs form when water containing organic material and bromide reacts with a disinfectant such as chlorine or chloramines during drinking water treatment. Parameters that affect DBP formation include concentration and type of organic material, disinfectant type, disinfectant concentration, pH, water temperature, and disinfectant contact time. In addition, many studies have found that elevated bromide levels correlate with increased DBP formation (AWWA, 2010; Obolensky and Singer, 2008; Matamoros et al., 2007; Hua et al., 2006; Yang and Shang, 2004). Some studies found similar results for iodide as well (McGuire et al., 2014; Parker et al., 2014). Pope et al. (2007) reported that increased bromide levels are the second best indicator of DBP formation, with pH being the best.

In addition, research finds that higher levels of bromide and iodide contribute to increased concentrations of the brominated and iodated forms of DBPs (both regulated and unregulated), which tend to be more cytotoxic, genotoxic, and carcinogenic than chlorinated species (McGuire et al., 2014; Parker et al., 2014; States et al., 2013; Krasner, 2009; Richardson et al., 2007). Studies generally report that the ratios of halogen incorporation into DBPs reflect the ratio of halogen concentrations in the source water (Criquet et al., 2012; Jones et al., 2012; Obolensky and Singer, 2008) but that bromide is preferentially incorporated into halogenated DBPs (McGuire et al., 2014; Parker et al., 2014; States et al., 2013; Krasner, 2009; Obolensky and Singer, 2008; Richardson et al., 2007; Hua et al., 2006).

From a regulatory perspective, elevated bromide levels create difficulties in meeting drinking water maximum contaminant levels (MCLs). When the TTHM are predominately in the form of brominated DBPs, the higher molecular weight of bromide (79.9 g/mol) relative to chloride (35.5 g/mol) causes the overall mass of the TTHM sum to increase. This can lead to elevated concentrations of TTHM, in turn potentially leading to violations of the TTHM MCL for the drinking water utility (Francis et al., 2009).

High bromide levels are also cited as causing formation of nitrogenous DBP N-nitrosodimethylamine (NDMA) in water disinfected with chloramines (Luh and Mariñas, 2012). Although NDMA is not regulated by the EPA as of 2016, it is listed as a priority toxic pollutant, and the EPA is planning to evaluate NDMA and other nitrosamines as candidates for regulation during the six-year review of the Microbial and Disinfection Byproducts (MDBP) rules (U.S. EPA, 2014a).

F.8.2. Studies Modeling Bromide in Receiving Waters from CWT Effluents

Contaminant modeling by Weaver et al. (2016) found that reducing effluent concentrations (e.g., discharging flowback versus produced water), discharging during higher stream flow periods, and using a pulsing or intermittent discharge can reduce bromide levels in receiving streams. Input data for the model came from several sources. Effluent bromide concentrations and permitted discharge flows came from eight commercial wastewater treatment plants in western Pennsylvania. Receiving stream flows were based on U.S. Geological Survey gage data. Data on flow accretion based on an analysis of tracer data from literature and EPA studies. The model assessed both steady-state and transient scenarios. The steady-state model assumed fixed discharges and flows and calculated mass and volume flow balance in a river network. The transient (i.e., pulsed or intermittent discharge) model simulations were based on a model developed by Weaver et al. (2016) assuming discharges of 12, 8, and 4 hours per day as well as a 24-hour simulation for comparison to a steady-state scenario. For steady-state scenarios, bromide concentrations were lowest under high flow conditions in the source water and with lower concentrations of bromide in the effluent. Bromide concentrations were generally lower for the pulsed scenarios than for the steady-state discharge scenarios.

In a separate study, U.S. EPA (2015m) evaluated the relative contributions of bromide, chloride, nitrate, and sulfate from CWTs primarily treating hydraulic fracturing wastewater to the Allegheny River Basin and to two downstream public water system intakes. The Allegheny River and its tributaries receive runoff and discharges containing an array of contaminants. Contaminant sources include discharges from CWTs for oil and gas wastewater, runoff from acid mine drainage and mining operations, discharges from coal-fired electric power stations, industrial wastewater treatment plant effluents, and POTW discharges. The Allegheny River is the water supply for thirteen public water systems that serve over 500,000 people in western Pennsylvania underscoring the importance of a full understanding of upstream contaminant contributions.

In Pennsylvania, wastewater produced from hydraulic fracturing of the Marcellus formation has been mostly diverted from CWTs and POTWs that discharge to public waters in the state to other management practices such as reuse (Hammer and VanBriesen, 2012). Wastewater produced from hydraulic fracturing of non-Marcellus formations, however, continues to be sent to CWTs and POTWs on the Allegheny River.

In order to quantify relative contributions of anions as a contaminant source at public drinking water system intakes, an EPA source apportionment study determined relative contributions of bromide from several upstream activities (U.S. EPA, 2015m). The study developed chemical source profiles for discharges upstream of the drinking water system intakes, characterized water quality in the river upstream and downstream of the CWTs and other facilities, characterized the water quality at the drinking water system intakes, and analyzed the sampling data collected with the EPA Positive Matrix Factorization (PMF) receptor model. The study focused on low-flow conditions. Researchers found that CWTs and coal-fired power plants with flue gas desulfurization were responsible for the majority of bromide at the two public water supply intakes. CWTs accounted for a substantial contribution of bromide, with 88-89% at one intake and 37% at the other. Coal-fired power plants with flue gas desulfurization were the other substantial contributors, with 50% at the second intake but less than 1% at the first. Sediment and acid mine drainage were also minor contributors in the range of 1 to 11% (U.S. EPA, 2015m).

Appendix G. Identification and Hazard Evaluation of Chemicals across the Hydraulic Fracturing Water Cycle Supplemental Information

This page is intentionally left blank.

Appendix G. Identification and Hazard Evaluation of Chemicals across the Hydraulic Fracturing Water Cycle Supplemental Information

G.1. Introduction

Appendix G provides detail and supporting information on the oral reference values (RfVs) and oral slope factors (OSFs) that were identified in Chapter 9 of this assessment.[1] Section G.2 provides detail on the criteria used to select sources of RfVs, OSFs, and qualitative cancer classifications for chemicals used or detected in hydraulic fracturing processes, and lists all sources that were considered for this study. Section G.3 provides a glossary of the toxicity terminology that is used by these various sources. Section G.4 provides a brief description of other potential tools and approaches that could be used by stakeholders to prioritize and estimate toxicity of chemicals that have a limited toxicity database. Lastly, all of the toxicity data collected from the sources that met the criteria for inclusion in this study are provided. Table G-1a through G-1e show the available RfVs, OSFs, and qualitative cancer classifications for chemicals used in hydraulic fracturing fluids, and Table G-2a through Table G-2e show the available RfVs, OSFs, and qualitative cancer classifications for chemicals detected in produced water from hydraulically fractured wells. These tables also indicate whether each chemical has available data on physicochemical properties or occurrence.

G.2. Criteria for Selection and Inclusion of Reference Value (RfV), Oral Slope Factor (OSF), and Qualitative Cancer Classification Data Sources

The criteria listed below were used to evaluate the quality of RfVs, OSFs, and qualitative cancer classifications considered for use in the hazard analyses conducted in Chapter 9. These criteria were originally outlined in the hydraulic fracturing research plan (U.S. EPA, 2011a) and interim progress report (U.S. EPA, 2012e). Only data sources that met these criteria were considered of sufficient quality to be included in the analyses.

The following criteria had to be met for a source to be deemed of sufficient quality:

1. The body or organization generating or producing the peer-reviewed RfVs, peer-reviewed OSFs, or peer reviewed qualitative assessment must be a governmental or intergovernmental body.

   a. Governmental bodies include sovereign states, and federated states/units.

   b. Intergovernmental bodies are those whose members are sovereign states, and the subdivisions or agencies of such intergovernmental bodies. The United Nations is an example of an intergovernmental body. The International Agency for Research

---

[1] As defined in Chapter 9, the term RfV refers to reference values for noncancer effects occurring via the oral route of exposure and for chronic durations, except where noted.

on Cancer (IARC) is an agency of the World Health Organization (WHO), which is itself an agency of the United Nations. Thus, IARC is considered a subdivision of the United Nations.

2. The data source must include peer-reviewed RfVs, peer-reviewed OSFs, or peer reviewed qualitative assessments.

   a. A committee that is established to derive the RfVs, OSFs, or qualitative assessments can have members of that same committee provide the peer review, so long as either the entire committee, or members of the committee who did not participate in the derivation of a specific section of a work product, conduct the review.

   b. Peer reviewers who work for grantees of the organization deriving the RfVs, OSFs, or qualitative assessments are generally allowed, and this will not be considered to constitute a conflict/duality of interest.

   c. Peer reviewers may work in the same or different office, so long as they did not participate in any way in the development of the product, and these individuals must be free of conflicts/duality of interest with respect to the chemical(s) assigned.

      i. For instance, peer reviewers for Program X, conducted by Office A, may also be employed by Office A so long as they did not participate in the creation of the Program X product they are reviewing.

3. The RfVs, OSFs, or qualitative assessments must be based on peer-reviewed scientific data.

   a. There are cases where industry reports that were not published in a peer-reviewed, scholarly journal may be used, if the industry report has been adequately peer-reviewed by an external body (external to the group generating the report, and external to the group generating the peer-reviewed RfVs, peer-reviewed OSFs, or peer-reviewed qualitative assessment) that is free of conflicts/dualities of interest.

4. The RfVs, OSFs, or qualitative assessments must be focused on protection of the general public.

   a. Sources that are focused on workers are not appropriate as workers are assumed to accommodate additional risk than the general public due to their status as workers.

5. The body generating the values or qualitative assessments must be free of conflicts of interest with respect to the chemicals for which it derives RfVs, OSFs, or qualitative assessments.

   a. If a body generating the RfVs, OSFs, or qualitative assessments accepts funding from an interested party (i.e., a company or organization that may be impacted by past, present, or future values or qualitative assessments), then the body has a conflict of interest.

      b. For instance, if a non-profit organization is funded by an industry trade group, and the non-profit generates RfVs, OSFs, or qualitative assessments for chemicals that trade group is interested in, then the non-profit is considered to have a conflict of interest with respect to those chemicals.

It is important to note that having a conflict/duality of interest for one chemical is sufficient to disqualify the entire database, as it is assumed that conflicts/dualities of interest may exist for other chemicals as well.

G.2.1. Included Sources

We applied our criteria to 16 different sources of RfVs and/or OSFs. After application of our criteria, we were left with eight sources. For those sources which did not meet our criteria, we provide an explanation of why they were excluded.

The following sources were evaluated, met our criteria, and were selected as sources of reference doses or cancer slope factors for this analysis:

- U.S. EPA Integrated Risk Information System (IRIS).
- U.S. EPA Human Health Benchmarks for Pesticides (HHBP).
- U.S. EPA Provisional Peer-Reviewed Toxicity Values (PPRTVs).
- U.S. Agency for Toxic Substances and Disease Registry (ATSDR) Minimal Risk Levels (MRLs).
- California EPA (CalEPA) Toxicity Criteria Database.
- International Programme on Chemical Safety (IPCS) Concise International Chemical Assessment Documents (CICAD).

The following sources were evaluated, met our criteria, and were selected as sources of qualitative cancer classifications:

- International Agency for Research on Cancer (IARC).
- US National Toxicology Program (NTP) Report on Carcinogens (RoC).

RfVs, OSFs, and qualitative cancer characterizations from these data sources are listed in Tables G-1a through G-1e for chemicals used in hydraulic fracturing fluid formulation, and Tables G-2a through G-2e for chemicals reported in hydraulic fracturing produced water.

In addition, Table G-1a and Table G-2a also list the EPA's drinking water maximum contaminant levels (MCLs) and maximum contaminant goal levels (MCLG) when available. These values are generally based on IRIS values, and MCLs are treatment-based.

G.2.2. Excluded Sources

The following sources were excluded:

- **American Conference of Governmental Industrial Hygienists:** The assessments derived by this body are specific to workers and are not generalizable to the general public. In addition, this body is not a governmental or intergovernmental body. Thus, these values were excluded based on criteria 1 and 4.

- **European Chemicals Bureau, Classification and Labeling Annex I of Directive 67/548/EEC:** These assessments are not based on peer-reviewed values, but are based on data supplied by manufacturers. Further, the enabling legislation states that "Manufacturers, importers, and downstream users shall examine the information...to ascertain whether it is adequate, reliable and scientifically valid for the purpose of the evaluation..." This clearly demonstrates that the data and the evaluation are not required to be peer-reviewed. Thus, these values were excluded based on criterion 2.

- **Toxicology Excellence for Risk Assessment's (TERA's) International Toxicity Estimates for Risk Assessment (ITER):** The ITER database is developed by TERA a 501(c)(3) non-profit. TERA accepts funding from various sources, including interested parties that may be impacted by their assessment work. Thus, ITER is excluded based on criteria 1 and 5.

- **Other U.S. states:** The EPA evaluated values from all states that had values reported on their websites. If a state's values were determined to be largely duplicative of the EPA's values (e.g., the state adopts EPA values, such as the regional screening levels, and does not typically generate its own peer-reviewed values), that state's values were no longer considered. The EPA contacted those states whose values were determined to not be duplicative of the EPA's values, and confirmed whether or not a peer review process was used to develop the state's values. The EPA determined that of the states with values not duplicative of the EPA's values, only California's values met all of the EPA's criteria for this report. Other states with publicly accessible RfVs and/or OSFs include: Alabama, Florida, Hawaii, and Texas.

- **WHO Guidelines for Drinking-Water Quality:** The WHO Guidelines' values are not RfVs, but rather drinking water values.

G.3. Glossary of Toxicity Value Terminology

This section defines the toxicity values and qualitative cancer classifications that are frequently found in the sources identified above.

Lowest-observed-adverse-effect level (LOAEL): The lowest exposure level at which there are biologically significant increases in frequency or severity of adverse effects between the exposed population and its appropriate control group. Source: U.S. EPA (2011c).

Maximum allowable daily level (MADL): The maximum allowable daily level of a reproductive toxicant at which the chemical would have no observable adverse reproductive effect, assuming exposure at 1,000 times that level. Source: OEHHA (2012).

Maximum contaminant level (MCL): The highest level of a contaminant that is allowed in drinking water. MCLs are set as close to the MCLG as feasible using the best available analytical and treatment technologies and taking cost into consideration. MCLs are enforceable standards. Source: U.S. EPA (2012a).

Maximum contaminant level goal (MCLG): A non-enforceable health benchmark goal which is set at a level at which no known or anticipated adverse effect on the health of persons is expected to occur and which allows an adequate margin of safety. Source: U.S. EPA (2012a).

Minimal risk level (MRL): An ATSDR estimate of daily human exposure to a hazardous substance at or below which the substance is unlikely to pose a measurable risk of harmful (adverse), noncancerous effects. MRLs are calculated for a route of exposure (inhalation or oral) over a specified time period (acute, intermediate, or chronic). MRLs should not be used as predictors of harmful (adverse) health effects.

- **Chronic MRL:** Duration of exposure is 365 days or longer.
- **Intermediate MRL:** Duration of exposure is >14 to 364 days.
- **Acute MRL:** Duration of exposure is 1 to 14 days.

Source: ATSDR (2009).

No-observed-adverse-effect level (NOAEL): The highest exposure level at which there are no biologically significant increases in the frequency or severity of adverse effect between the exposed population and its appropriate control; some effects may be produced at this level, but they are not considered adverse or precursors of adverse effects. Source: U.S. EPA (2011c).

Oral slope factor (OSF): An upper-bound, approximating a 95% confidence limit, on the increased cancer risk from a lifetime oral exposure to an agent. This estimate, usually expressed in units of proportion (of a population) affected per mg/kg-day, is generally reserved for use in the low-dose region of the dose-response relationship, that is, for exposures corresponding to risks less than 1 in 100. Source: U.S. EPA (2011c).

Reference dose (RfD): An estimate (with uncertainty spanning perhaps an order of magnitude) of a daily oral exposure to the human population (including sensitive subgroups) that is likely to be without an appreciable risk of deleterious effects during a lifetime. It can be derived from a NOAEL, LOAEL, or benchmark dose, with uncertainty factors generally applied to reflect limitations of the data used. Generally used in the EPA's noncancer health assessments.

- **Chronic RfD:** Duration of exposure is up to a lifetime.
- **Subchronic RfD (sRFD):** Duration of exposure is up to 10% of an average lifespan.
- **Acute RfD:** Duration of exposure is 24 hours or less.

Source: U.S. EPA (2011c).

Reference value (RfV): An estimate of an exposure for a given duration to the human population (including susceptible subgroups) that is likely to be without an appreciable risk of adverse health effects over a lifetime. RfV is a generic term not specific to a given route of exposure. In the context of this report, the term RfV refers to reference values for noncancer effects occurring via the oral route of exposure and for chronic durations, except where noted. Source: U.S. EPA (2011c).

Tolerable daily intake (TDI): An estimate of the intake of a substance, expressed on a body mass basis, to which an individual in a (sub) population may be exposed daily over its lifetime without appreciable health risk. Source: WHO (2015).

Qualitative cancer classifications: A system used for the hazard identification of potential carcinogens, in which human data, animal data, and other supporting evidence are combined to characterize the weight of evidence (WOE) regarding the potential of an agent to cause cancer in humans.

- **EPA 1986 guidelines:** Under the EPA's 1986 risk assessment guidelines, the WOE was described by categories "A through E," with Group A for known human carcinogens through Group E for agents with evidence of noncarcinogenicity. Five standard WOE descriptors were used:
    - A: Human carcinogen.
    - B1: Probable human carcinogen—based on limited evidence of carcinogenicity in humans and sufficient evidence of carcinogenicity in animals.
    - B2: Probable human carcinogen—based on sufficient evidence of carcinogenicity in animals.
    - C: Possible human carcinogen.
    - D: Not classifiable as to human carcinogenicity.
    - E: Evidence of noncarcinogenicity for humans.

    Source: U.S. EPA (2011c).

- **EPA 1996 proposed guidelines:** The EPA's 1996 proposed guidelines outlined a major change in the way hazard evidence was weighted in reaching conclusions about the human carcinogenic potential of agents. These guidelines replaced the WOE letter categories with the use of standard descriptors of conclusions incorporated into a brief narrative. Three categories of descriptors with the narrative were used:
    - Known/likely.
    - Cannot be determined.
    - Not likely.

    Source: U.S. EPA (1996).

- **EPA 1999 guidelines:** The 1999 guidelines adopted a framework incorporating hazard identification, dose-response assessment, exposure assessment, and risk characterization with an emphasis on characterization of evidence and conclusions in each part of the assessment. Five descriptors summarizing the WOE in the narrative were used:

- o Carcinogenic to humans.
- o Likely to be carcinogenic to humans.
- o Suggestive evidence of carcinogenicity, but not sufficient to assess human carcinogenic potential.
- o Data are inadequate for an assessment of human carcinogenic potential.
- o Not likely to be carcinogenic to humans.

Source: U.S. EPA (1999a).

- **EPA 2005 guidelines:** The approach outlined in the EPA's 2005 guidelines for carcinogen risk assessment considers all scientific information in determining whether and under what conditions an agent may cause cancer in humans and provides a narrative approach to characterize carcinogenicity rather than categories. Five standard WOE descriptors are used as part of the narrative:

  - o Carcinogenic to humans.
  - o Likely to be carcinogenic to humans.
  - o Suggestive evidence of carcinogenic potential.
  - o Inadequate information to assess carcinogenic potential.
  - o Not likely to be carcinogenic to humans.

  Source: U.S. EPA (2011c).

- **IARC Monographs on the evaluation of carcinogenic risks to humans:** The IARC classifies carcinogen risk as a matter of scientific judgement that reflects the strength of the evidence derived from studies in humans, in experimental animals, from mechanistic data, and from other relevant data. Five WOE classifications are used:

  - o Group 1: Carcinogenic to humans.
  - o Group 2A: Probably carcinogenic to humans.
  - o Group 2B: Possibly carcinogenic to humans.
  - o Group 3: Not classifiable as to its carcinogenicity to humans.
  - o Group 4: Probably not carcinogenic to humans.

  Source: IARC (2015).

- **NTP:** The NTP describes the results of individual experiments on a chemical agent and notes the strength of the evidence for conclusions regarding each study. Negative results, in which the study animals do not have a greater incidence of neoplasia than control animals, do not necessarily mean that a chemical is not a carcinogen, inasmuch as the experiments are conducted under a limited set of conditions. Positive results demonstrate that a chemical is carcinogenic for laboratory animals under the conditions of the study and indicate that exposure to the chemical has the potential for hazard to humans. For each separate experiment, one of the following five categories is selected to describe the findings. These categories refer to the strength of the experimental evidence and not to potency or mechanism.

- o Clear evidence of carcinogenic activity.
- o Some evidence of carcinogenic activity.
- o Equivocal evidence of carcinogenic activity.
- o No evidence of carcinogenic activity.
- o Inadequate study of carcinogenic activity.

Source: NTP (2014a).

- **The RoC** is a congressionally mandated, science-based, public health report that identifies agents, substances, mixtures, or exposures (collectively called "substances") in our environment that may potentially put people in the United States at increased risk for cancer. NTP prepares the RoC on behalf of the Secretary of the Health and Human Services. The listing criteria in the RoC Document are:

  - o Known to be a human carcinogen.
  - o Reasonably anticipated to be a human carcinogen.

Source: NTP (2014b).

G.4. Additional Tools for Hazard Evaluation

In addition to the methods and approaches utilized in this chapter, there are other potential tools that could be used by stakeholders to prioritize and estimate toxicity of chemicals that have a limited toxicity database. We describe three such approaches here. This list is not intended to be exhaustive, but provides examples of tools that stakeholders may find useful when faced with many data-poor chemicals at a field site. Toxicity predictions from these additional data sources can be either quantitative or qualitative, and may be used to fill and address gaps related to risk assessment.

G.4.1. Threshold of Toxicological Concern (TTC)

The TTC approach is a risk assessment tool based on the concept that there is an exposure threshold value for all chemicals below which there is a very low probability of risk to human health (Kroes et al., 2005; Kroes et al., 2004). The TTC approach proposes that such a de minimis value can be identified for many chemicals based on knowledge of chemical structure (Lapenna and Worth, 2011; Kroes et al., 2005; Kroes et al., 2004). The estimated TTC is integrated with an estimate of human exposure to that chemical, and used by the model to determine if there is potential for concern or if more detailed chemical specific data are necessary (Kroes et al., 2005; Kroes et al., 2004). As a preliminary step in risk assessment, this approach can be applied as a screening tool, for ranking and prioritization, and as an indicator of data needs (Lapenna and Worth, 2011; Kroes et al., 2005; Kroes et al., 2004).

The various TTC approaches are based on a decision tree proposed by Cramer et al. (1978), which classifies chemicals into categories of high (Class III), medium (Class II), or low (Class I) level of concern, based on structure and reactivity. Based on the analysis of chronic oral toxicity data within each of these structural classes, Munro et al. (1996) proposed oral intake TTC values of 1.5, 9.0, and 30 µg/kg body weight/day for Class III, II, and I, respectively. A tiered decision tree proposed by Kroes et al. (Kroes et al., 2005; Kroes et al., 2004) expanded these approaches by including structural alerts for possible genotoxic and/or high potency carcinogens as well as a TTC value for organophosphates. Recently, in order to help facilitate the consistent and transparent application of the TTC approach, a freely available software tool – Toxtree (http://toxtree.sourceforge.net/) – was developed by the European Commission Joint Research Centre (JRC) for predicting toxicological effects and mechanism of action (Lapenna and Worth, 2011). Toxtree implements the approaches relevant to TTC assessment, including the original Cramer decision tree and the expanded TTC decision tree by (Kroes et al., 2004), and includes improvements to the original Cramer scheme to overcome to potential for chemical misclassification.

G.4.2. Organisation for Economic Co-operation and Development (OECD) Quantitative Structure-Activity Relationship (QSAR) Toolbox

The OECD QSAR Toolbox is another available QSAR-based software tool developed to fill in toxicity data gaps for assessing the hazards of chemicals (OECD, 2016), and serves as a platform that incorporates various modules, databases, and structure-activity relationship models from a wide range of sources. This approach also implements read-across concepts by grouping chemicals into categories based on profiles related to physicochemical properties, human health, ecotoxicity, and environmental fate. The main features of the OECD QSAR Toolbox are: identification of relevant structural characteristics and potential mechanism or mode of action of a target chemical; identification of other chemicals that have the same structural characteristics and/or mechanism or mode of action; and use of existing experimental data to fill the data gap(s). The Toolbox's key strengths are for screening environmental fate endpoints, physicochemical properties, acute ecotoxicity endpoints and toxicity endpoints such as skin/eye irritation, sensitization and mutagenicity.

G.4.3. Application of Data from High Throughput Screening Assays

In addition to the tools outlined above, there have been recent advances in emerging technologies such as high throughput screening (HTS) assays that may aid in prioritizing chemical inventories for potential hazard (Wambaugh et al., 2013). HTS assays are in vitro assays that allow rapid screening of chemicals for potential toxicity and biological activity across multiple cellular pathways and targets (Wetmore et al., 2012; Rotroff et al., 2010). Recent advances have been made in dosimetry methods that extrapolate in vitro concentration data to a human oral equivalent dose, providing a quantitative estimate of the dose of a chemical that would result in an adverse effect (Wetmore et al., 2015; Wetmore et al., 2012; Judson et al., 2011; Rotroff et al., 2010). HTS data may also be combined with emerging methods to estimate exposure potential, providing a method to refine risk-based prioritization for chemicals with limited toxicity information (Wambaugh et al., 2013). Consequently, the integration of data from emerging technologies with estimates of human oral dose and exposure may provide another potential approach to address risk management needs when in vivo toxicology data are not available.

Table G-1a. Chemicals reported to be used in hydraulic fracturing fluids, with available chronic oral RfVs, OSFs, and qualitative cancer classifications from United States federal sources.

Chemicals from the FracFocus database are listed first, ranked by IRIS reference dose (RfD). The "--" symbol indicates that no value was available from the sources consulted. Additionally, an "x" indicates the availability of usage data from FracFocus (U.S. EPA, 2015a) and physicochemical properties data from EPI Suite™ (see Appendix C). Italicized chemicals are found in both hydraulic fracturing fluids and produced water.

Chemical name	CASRN	Frac Focus data available	Physico-chemical data available	IRIS Chronic RfD[a] (mg/kg-day)	IRIS OSF[b] (per mg/kg-day)	IRIS Cancer WOE characterization[c]	PPRTV Chronic RfD[a] (mg/kg-day)	PPRTV OSF[b] (per mg/kg-day)	PPRTV Cancer WOE characterization[c]	ATSDR Chronic oral MRL[d] (mg/kg-day)	HHBP Chronic RfD[a] (mg/kg-day)	NPDWRs Public health goal[e] (MCLG) (mg/L)	NPDWRs MCL[f] (mg/L)
Acrylamide	79-06-1	x	x	0.002	0.5	"Likely to be carcinogenic to humans"	--	--	--	0.001	--	0	TT[g]
Propargyl alcohol	107-19-7	x	x	0.002	--	--	--	--	--	--	--	--	--
Furfural	98-01-1	x	x	0.003	--	--	--	--	--	--	0.01	--	--
Benzene	71-43-2	x	x	0.004	0.015-0.055	A (Human carcinogen)	--	--	--	0.0005	--	0	0.005
Dichloromethane	75-09-2	x	x	0.006	0.002	"Likely to be carcinogenic in humans"	--	--	--	0.06	--	0	0.005
1,2,3-Trimethyl-benzene	526-73-8	x	x	0.01	--		--	--	"Data are inadequate for an assessment of human carcinogenic potential"	--	--	--	--
1,2,4-Trimethyl-benzene	95-63-6	x	x	0.01	--	--	--	--	--	--	--	--	--

Appendix G – Identification and Hazard Evaluation of Chemicals across the Hydraulic Fracturing Water Cycle Supplemental Information

Chemical name	CASRN	Frac Focus data available	Physico-chemical data available	IRIS Chronic RfD[a] (mg/kg-day)	IRIS OSF[b] (per mg/kg-day)	IRIS Cancer WOE characterization[c]	PPRTV Chronic RfD[a] (mg/kg-day)	PPRTV OSF[b] (per mg/kg-day)	PPRTV Cancer WOE characterization[c]	ATSDR Chronic oral MRL[d] (mg/kg-day)	HHBP Chronic RfD[a] (mg/kg-day)	NPDWRs Public health goal[e] (MCLG) (mg/L)	MCL[f] (mg/L)
1,3,5-Trimethyl-benzene	108-67-8	x	x	0.01	--	--	--	--	"Data are inadequate for an assessment of human carcinogenic potential"	--	--	--	--
Trimethylbenzene	25551-13-7	x		0.01	--	--	--	--	--	--	--	--	--
Chlorobenzene	108-90-7	x	x	0.02	--	D (Not classifiable as to human carcino-genicity)	--	--	--	--	--	0.1	0.1
Naphthalene	91-20-3	x	x	0.02	--	"Data are inadequate to assess human carcinogenic potential"	--	--	--	--	--	--	--
1,3-Dichloropro-pene	542-75-6	x	x	0.03	0.05	"Likely to be a human carcinogen"	--	--	--	0.03	--	--	--
1,4-Dioxane	123-91-1	x	x	0.03	0.1	"Likely to be carcinogenic to humans"	--	--	--	0.1	--	--	--

Appendix G – Identification and Hazard Evaluation of Chemicals across the Hydraulic Fracturing Water Cycle Supplemental Information

Chemical name	CASRN	Frac Focus data available	Physico-chemical data available	IRIS Chronic RfD[a] (mg/kg-day)	IRIS OSF[b] (per mg/kg-day)	IRIS Cancer WOE characterization[c]	PPRTV Chronic RfD[a] (mg/kg-day)	PPRTV OSF[b] (per mg/kg-day)	PPRTV Cancer WOE characterization[c]	ATSDR Chronic oral MRL[d] (mg/kg-day)	HHBP Chronic RfD[a] (mg/kg-day)	NPDWRs Public health goal[e] (MCLG) (mg/L)	NPDWRs MCL[f] (mg/L)
Chlorine dioxide	10049-04-4	x		0.03	--	"Data are inadequate to assess human carcinogenicity"	--	--	--	--	--	--	--
Sodium chlorite	7758-19-2	x		0.03	--	"Data are inadequate to assess human carcinogenicity"	--	--	--	--	--	1	0.8
Bisphenol A	80-05-7	x	x	0.05	--	--	--	--	--	--	--	--	--
Bisphenol A	80-05-7	x	x	0.05	--	--	--	--	--	--	--	--	--
Toluene	108-88-3	x	x	0.08	--	"Inadequate information to assess the carcinogenic potential"	--	--	--	--	--	1	1
1-Butanol	71-36-3	x	x	0.1	--	"D (Not classifiable as to human carcinogenicity)"	--	--	--	--	--	--	--
2-Butoxyethanol	111-76-2	x	x	0.1	--	"Not likely to be carcinogenic to humans"	--	--	--	--	--	--	--

Appendix G – Identification and Hazard Evaluation of Chemicals across
the Hydraulic Fracturing Water Cycle Supplemental Information

Chemical name	CASRN	Frac Focus data available	Physico-chemical data available	IRIS Chronic RfD[a] (mg/kg-day)	IRIS OSF[b] (per mg/kg-day)	IRIS Cancer WOE characterization[c]	PPRTV Chronic RfD[a] (mg/kg-day)	PPRTV OSF[b] (per mg/kg-day)	PPRTV Cancer WOE characterization[c]	ATSDR Chronic oral MRL[d] (mg/kg-day)	HHBP Chronic RfD[a] (mg/kg-day)	NPDWRs Public health goal[e] (MCLG) (mg/L)	NPDWRs MCL[f] (mg/L)
Acetophenone	98-86-2	x	x	0.1	--	D (Not classifiable as to human carcinogenicity)	--	--	--	--	--	--	--
Cumene	98-82-8	x	x	0.1	--	D (Not classifiable as to human carcinogenicity)	--	--	--	--	--	--	--
Ethylbenzene	100-41-4	x	x	0.1	--	D (Not classifiable as to human carcinogenicity)	--	--	--	--	--	0.7	0.7
Boron	7440-42-8	x		0.2	--	"Data are inadequate to assess the carcinogenic potential"	--	--	--	--	--	--	--
Formaldehyde	50-00-0	x	x	0.2	--	B1 (Probable human carcinogen)	--	--	--	0.2	--	--	--
Xylenes	1330-20-7	x	x	0.2	--	"Data are inadequate to assess the carcinogenic potential"	--	--	--	0.2	--	10	10
2-Methyl-1-propanol	78-83-1	x	x	0.3	--	--	--	--	--	--	--	--	--

Appendix G – Identification and Hazard Evaluation of Chemicals across the Hydraulic Fracturing Water Cycle Supplemental Information

Chemical name	CASRN	Frac Focus data available	Physico-chemical data available	IRIS			PPRTV			ATSDR	HHBP	NPDWRs	
				Chronic RfD[a] (mg/kg-day)	OSF[b] (per mg/kg-day)	Cancer WOE characterization[c]	Chronic RfD[a] (mg/kg-day)	OSF[b] (per mg/kg-day)	Cancer WOE characterization[c]	Chronic oral MRL[d] (mg/kg-day)	Chronic RfD[a] (mg/kg-day)	Public health goal[e] (MCLG) (mg/L)	MCL[f] (mg/L)
Phenol	108-95-2	x	x	0.3	--	"Data are inadequate for an assessment of human carcinogenic potential"	--	--		--	--	--	--
Acetone	67-64-1	x	x	0.9	--	"Data are inadequate for an assessment of human carcinogenic potential"	--	--		--	--	--	--
Ethyl acetate	141-78-6	x	x	0.9	--	--	--	--	IN	--	--	--	--
Ethylene glycol	107-21-1	x	x	2	--	--	--	--	--	--	--	--	--
Methanol	67-56-1	x	x	2	--	--	--	--	--	--	--	--	--
Benzoic acid	65-85-0	x	x	4	--	D (Not classifiable as to human carcinogenicity)	--	--	--	--	--	--	--
(E)-Crotonaldehyde	123-73-9	x	x	--	--	C (Possible human carcinogen)	0.001	--	--	--	--	--	--
1,2-Propylene glycol	57-55-6	x	x	--	--	--	20	--	NL	--	--	--	--

Appendix G – Identification and Hazard Evaluation of Chemicals across the Hydraulic Fracturing Water Cycle Supplemental Information

Chemical name	CASRN	Frac Focus data available	Physico-chemical data available	IRIS Chronic RfD[a] (mg/kg-day)	IRIS OSF[b] (per mg/kg-day)	IRIS Cancer WOE characterization[c]	PPRTV Chronic RfD[a] (mg/kg-day)	PPRTV OSF[b] (per mg/kg-day)	PPRTV Cancer WOE characterization[c]	ATSDR Chronic oral MRL[d] (mg/kg-day)	HHBP Chronic RfD[a] (mg/kg-day)	NPDWRs Public health goal[e] (MCLG) (mg/L)	NPDWRs MCL[f] (mg/L)
2-(2-Butoxyethoxy) ethanol	112-34-5	x	x	--	--	--	0.03	--	*IN*	--	--	--	--
2-(Thiocyanomethylthio)benzothiazole	21564-17-0	x	x	--	--	--	--	--	--	--	0.01	--	--
Aluminum	7429-90-5	x		--	--	--	1	--	*IN*	1	--	--	--
Ammonium phosphate	7722-76-1	x		--	--	--	49	--	*IN*	--	--	--	--
Aniline	62-53-3	x	x	--	0.0057	B2 (Probable human carcinogen)	0.007	--	--	--	--	--	--
Benzenesulfonic acid, C10-16-alkyl derivs.	68584-22-5	x		--	--	--	--	--	--	--	0.5	--	--
Benzyl chloride	100-44-7	x	x	--	0.17	*B2 (Probable human carcinogen)*	0.002	--	--	--	--	--	--
Bis(2-chloroethyl) ether	111-44-4	x	x	--	1.1	*B2 (Probable human carcinogen)*	--	--	--	--	--	--	--
Didecyldimethyl-ammonium chloride	7173-51-5	x	x	--	--		--	--	--	--	0.1	--	--
Dodecylbenzene-sulfonic acid	27176-87-0	x	x	--	--	--	--	--	--	--	0.5	--	--
Epichlorohydrin	106-89-8	x	x	--	0.0099	B2 (Probable human carcinogen)	0.006	--	--	--	--	0	--

Appendix G – Identification and Hazard Evaluation of Chemicals across the Hydraulic Fracturing Water Cycle Supplemental Information

Chemical name	CASRN	Frac Focus data available	Physico-chemical data available	IRIS Chronic RfD[a] (mg/kg-day)	IRIS OSF[b] (per mg/kg-day)	IRIS Cancer WOE characterization[c]	PPRTV Chronic RfD[a] (mg/kg-day)	PPRTV OSF[b] (per mg/kg-day)	PPRTV Cancer WOE characterization[c]	ATSDR Chronic oral MRL[d] (mg/kg-day)	HHBP Chronic RfD[a] (mg/kg-day)	NPDWRs Public health goal[e] (MCLG) (mg/L)	NPDWRs MCL[f] (mg/L)
Ethylenediamine	107-15-3	x	x	--	--	D (Not classifiable as to human carcinogenicity)	0.09	--	IN	--	--	--	--
Formic acid	64-18-6	x	x	--	--	--	0.9	--	IN	--	--	--	--
Hexanedioic acid	124-04-9	x	x	--	--	--	2	--	--	--	--	--	--
Hydrazine	302-01-2	x		--	3	B2 (Probable human carcinogen)	--	--	--	--	--	--	--
Iron	7439-89-6	x		--	--	--	0.7	--	IN	--	--	--	--
Mineral oil - includes paraffin oil	8012-95-1	x		--	--	--	3	--	IN	--	--	--	--
N,N-Dimethylformamide	68-12-2	x	x	--	--	--	0.1	--	IN	--	--	--	--
o-Xylene	95-47-6	x	x	--	--	--	--	--	--	0.2	--	10	10
Phosphoric acid	7664-38-2	x		--	--	--	48.6	--	IN	--	--	--	--
Potassium phosphate, tribasic	7778-53-2	x		--	--	--	49	--	IN	--	--	--	--
Quaternary ammonium compounds, benzyl-C12-16-alkyldimethyl, chlorides	68424-85-1	x		--	--	--	--	--	--	--	0.44	--	--

Appendix G – Identification and Hazard Evaluation of Chemicals across the Hydraulic Fracturing Water Cycle Supplemental Information

Chemical name	CASRN	Frac Focus data available	Physico-chemical data available	IRIS Chronic RfD[a] (mg/kg-day)	IRIS OSF[b] (per mg/kg-day)	IRIS Cancer WOE characterization[c]	PPRTV Chronic RfD[a] (mg/kg-day)	PPRTV OSF[b] (per mg/kg-day)	PPRTV Cancer WOE characterization[c]	ATSDR Chronic oral MRL[d] (mg/kg-day)	HHBP Chronic RfD[a] (mg/kg-day)	NPDWRs Public health goal[e] (MCLG) (mg/L)	NPDWRs MCL[f] (mg/L)
Quinoline	91-22-5	x	x	--	3	"Likely to be carcinogenic in humans"	--	--	--	--	--	--	--
Sodium chlorate	7775-09-9	x		--	--	--	--	--	--	--	0.03	--	--
Sodium trimetaphosphate	7785-84-4	x		--	--	--	49	--	IN	--	--	--	--
Tetrasodium pyrophosphate	7722-88-5	x		--	--	--	49	--	IN	--	--	--	--
Tricalcium phosphate	7758-87-4	x		--	--	--	49	--	IN	--	--	--	--
Triphosphoric acid, pentasodium salt	7758-29-4	x		--	--	--	49	--	IN	--	--	--	--
Trisodium phosphate	7601-54-9	x		--	--	--	49	--	IN	--	--	--	--
Arsenic	7440-38-2			0.0003	1.5	A (Human carcinogen)	--	--	--	0.0003	--	0	0.01
Phosphine	7803-51-2			0.0003	--	D (Not classifiable as to human carcinogenicity)	--	--	--	--	--	--	--
Acrolein	107-02-8		x	0.0005	--	"Data are inadequate for an assessment of human carcinogenic potential"	--	--	--	--	--	--	--

Appendix G – Identification and Hazard Evaluation of Chemicals across the Hydraulic Fracturing Water Cycle Supplemental Information

Chemical name	CASRN	Frac Focus data available	Physico-chemical data available	IRIS Chronic RfD[a] (mg/kg-day)	IRIS OSF[b] (per mg/kg-day)	IRIS Cancer WOE characterization[c]	PPRTV Chronic RfD[a] (mg/kg-day)	PPRTV OSF[b] (per mg/kg-day)	PPRTV Cancer WOE characterization[c]	ATSDR Chronic oral MRL[d] (mg/kg-day)	HHBP Chronic RfD[a] (mg/kg-day)	NPDWRs Public health goal[e] (MCLG) (mg/L)	NPDWRs MCL[f] (mg/L)
Chromium (VI)	18540-29-9			0.003	--	Inhaled: A (Human carcinogen); Oral: D (Not classifiable as to human carcinogenicity)	--	--	--	0.0009	--		--
Di(2-ethylhexyl) phthalate	117-81-7		x	0.02	0.014	B2 (Probable human carcinogen)	--	--	--	0.06	--	0	0.006
Chlorine	7782-50-5			0.1	--	--	--	--	--	--	--	--	--
Styrene	100-42-5		x	0.2	--	--	--	--	--	--	--	0.1	0.1
Zinc	7440-66-6			0.3	--	"Inadequate information to assess carcinogenic potential"	--	--	--	0.3	--	--	--
Acrylic acid	79-10-7		x	0.5	--	--	--	--	IN	--	--	--	--
Chromium (III)	16065-83-1			1.5	--	"Data are inadequate for an assessment of human carcinogenic potential"	--	--	--	--	--	--	--
Phthalic anhydride	85-44-9		x	2	--	--	--	--	--	--	--	--	--
Cyclohexanone	108-94-1		x	5	--	--	--	--	IN	--	--	--	--

Appendix G – Identification and Hazard Evaluation of Chemicals across the Hydraulic Fracturing Water Cycle Supplemental Information

Chemical name	CASRN	Frac Focus data available	Physico-chemical data available	IRIS Chronic RfD[a] (mg/kg-day)	IRIS OSF[b] (per mg/kg-day)	IRIS Cancer WOE characterization[c]	PPRTV Chronic RfD[a] (mg/kg-day)	PPRTV OSF[b] (per mg/kg-day)	PPRTV Cancer WOE characterization[c]	ATSDR Chronic oral MRL[d] (mg/kg-day)	HHBP Chronic RfD[a] (mg/kg-day)	NPDWRs Public health goal[e] (MCLG) (mg/L)	MCL[f] (mg/L)
1,2-Propylene oxide	75-56-9		x	--	0.24	B2 (Probable human carcinogen)	--	--		--	0.001	--	--
2-(2-Ethoxyethoxy) ethanol	111-90-0		x	--	--	--	0.06	--	IN	--	--	--	--
2-Methoxyethanol	109-86-4		x	--	--	--	0.005	--	IN	--	--	--	--
Lead	7439-92-1			--	--	B2 (Probable human carcinogen)	--	--	--	--	--	0	*TT; Action Level=0. 015[h]*
Phosphoric acid, aluminium sodium salt	7785-88-8			--	--	--	49	--	IN	--	--	--	--
Phosphoric acid, diammonium salt	7783-28-0			--	--	--	49	--	IN	--	--	--	--
Polyphosphoric acids, sodium salts	68915-31-1		x	--	--	--	49	--	IN	--	--	--	--
p-Xylene	106-42-3		x	--	--	--	--	--	--	0.2	--	10	10
Sodium pyrophosphate	7758-16-9			--	--	--	49	--	IN	--	--	--	--
Tributyl phosphate	126-73-8		x	--	--	--	0.01	0.009	LI	0.08	--	--	--

CASRN = Chemical Abstract Service Registry Number; IRIS = Integrated Risk Information System; PPRTV = Provisional Peer Reviewed Toxicity Values; ATSDR = Agency for Toxic Substances and Disease Registry; HHBP = Human Health Benchmarks for Pesticides; NPDWRs = National Primary Drinking Water Regulations.

[a] Reference dose (RfD): An estimate (with uncertainty spanning perhaps an order of magnitude) of a daily oral exposure to the human population (including sensitive subgroups) that is likely to be without an appreciable risk of deleterious effects during a lifetime. It can be derived from a no observed-adverse-effect level (NOAEL), lowest observed-adverse-effect level (LOAEL), or benchmark dose (BMD), with uncertainty factors generally applied to reflect limitations of the data used. The RfD is generally used in the EPA's noncancer health assessments. Chronic RfD: Duration of exposure is up to a lifetime.

[b] Oral slope factor (OSF): An upper-bound, approximating a 95% confidence limit, on the increased cancer risk from a lifetime oral exposure to an agent. This estimate, usually expressed in units of proportion (of a population) affected per mg/kg-day, is generally reserved for use in the low dose region of the dose response relationship, that is, for exposures corresponding to risks less than 1 in 100.

[c] Weight of evidence (WOE) characterization for carcinogenicity: A system used for characterizing the extent to which the available data support the hypothesis that an agent causes cancer in humans. See glossary for details.

[d] Minimal risk level (MRL): An ATSDR estimate of daily human exposure to a hazardous substance at or below which the substance is unlikely to pose a measurable risk of harmful (adverse), noncancerous effects. MRLs are calculated for a route of exposure (inhalation or oral) over a specified time period (acute, intermediate, or chronic). MRLs should not be used as predictors of harmful (adverse) health effects. Chronic MRL: Duration of exposure is 365 days or longer.

[e] Maximum contaminant level goal (MCLG): A non-enforceable health benchmark goal which is set at a level at which no known or anticipated adverse effect on the health of persons is expected to occur and which allows an adequate margin of safety.

[f] Maximum contaminant level (MCL): The highest level of a contaminant that is allowed in drinking water. MCLs are set as close to the MCLG as feasible using the best available analytical and treatment technologies and taking cost into consideration. MCLs are enforceable standards.

[g] In public water systems, acrylamide is regulated by a Treatment Technique (TT). Public water systems must certify annually that when acrylamide is used to treat water, the combination of dose and monomer level does not exceed 0.05% dosed at 1 mg/l (or equivalent).

[h] In public water systems, lead is regulated by a Treatment Technique (TT) that requires systems to control the corrosiveness of their water. If more than 10% of tap water exceeds the action level of 0.015 mg/l, water systems must take additional steps.

Table G-1b. Chemicals reported to be used in hydraulic fracturing fluids, with available chronic oral RfVs and OSFs from state sources.

Chemicals from the FracFocus database are listed first, ranked by CalEPA maximum allowable daily level (MADL). The "--" symbol indicates that no value was available from the sources consulted. Additionally, an "x" indicates the availability of usage data from FracFocus (U.S. EPA, 2015a) and physicochemical properties data from EPI Suite™ (see Appendix C). Italicized chemicals are found in both hydraulic fracturing fluids and produced water.

Chemical name	CASRN	FracFocus data available	Physico-chemical data available	CalEPA	
				Oral MADL[a] (µg/day)	OSF[b] (per mg/kg-day)
Ethylene oxide	75-21-8	x	x	20	0.31
Benzene	*71-43-2*	*x*	*x*	*24*	*0.1*
Acrylamide	79-06-1	x	x	140	4.5
N-Methyl-2-pyrrolidone	872-50-4	x	x	17000	--
1,3-Butadiene	106-99-0	x	x	--	0.6
1,3-Dichloropropene	542-75-6	x	x	--	0.091
1,4-Dioxane	*123-91-1*	*x*	*x*	*--*	*0.027*
Aniline	62-53-3	x	x	--	0.0057
Benzyl chloride	*100-44-7*	*x*	*x*	*--*	*0.17*
Bis(2-chloroethyl) ether	*111-44-4*	*x*	*x*	*--*	*2.5*
Dichloromethane	*75-09-2*	*x*	*x*	*--*	*0.014*
Epichlorohydrin	106-89-8	x	x	--	0.08
Ethylbenzene	*100-41-4*	*x*	*x*	*--*	*0.011*
Hydrazine	*302-01-2*	*x*		*--*	*3*
Nitrilotriacetic acid	139-13-9	x	x	--	0.0053
Nitrilotriacetic acid trisodium monohydrate	18662-53-8	x	x	--	0.01
Thiourea	62-56-6	x	x	--	0.072
Lead	*7439-92-1*			*0.5*	*0.0085*
Chromium (VI)	*18540-29-9*			*8.2*	*0.5*

Chemical name	CASRN	FracFocus data available	Physico-chemical data available	CalEPA	
				Oral MADL[a] (μg/day)	OSF[b] (per mg/kg-day)
Di(2-ethylhexyl) phthalate	*117-81-7*		x	*20 (neonate male); 58 (infant male); 410 (adult)*	*0.003*
2-Methoxyethanol	109-86-4		x	63	--
2-Ethoxyethanol	110-80-5		x	750	--
1,2-Propylene oxide	75-56-9		x	--	0.24
Arsenic	*7440-38-2*			--	*9.5*

CASRN = Chemical Abstract Service Registry Number; CalEPA = California Environmental Protection Agency.

[a] Maximum allowable daily level (MADL): The maximum allowable daily level of a reproductive toxicant at which the chemical would have no observable adverse reproductive effect, assuming exposure at 1,000 times that level.

[b] Oral slope factor (OSF): An upper-bound, approximating a 95% confidence limit, on the increased cancer risk from a lifetime oral exposure to an agent. This estimate, usually expressed in units of proportion (of a population) affected per mg/kg day, is generally reserved for use in the low-dose region of the dose-response relationship, that is, for exposures corresponding to risks less than 1 in 100.

Table G-1c. Chemicals reported to be used in hydraulic fracturing fluids, with available chronic oral RfVs and OSFs from international sources.

Chemicals from the FracFocus database are listed first, ranked by CICAD reference value (TDI- tolerable daily intake). An "x" indicates the availability of usage data from FracFocus (U.S. EPA, 2015a) and physicochemical properties data from EPI Suite™ (see Appendix C). Italicized chemicals are found in both hydraulic fracturing fluids and produced water.

Chemical name	CASRN	FracFocus data available	Physicochemical data available	IPCS CICAD Chronic TDI[a] (mg/kg-day)
Potassium iodide	7681-11-0	x		0.01
Sodium iodide	7681-82-5	x		0.01
Copper(I) iodide	7681-65-4	x		0.01
Ethylene glycol	*107-21-1*	*x*	*x*	*0.05*
D-Limonene	*5989-27-5*	*x*	*x*	*0.1*
Glyoxal	107-22-2	x	x	0.2
N-Methyl-2-pyrrolidone	872-50-4	x	x	0.6
Chromium (VI)	*18540-29-9*			*0.0009*
Strontium chloride	10476-85-4			0.13

CASRN = Chemical Abstract Service Registry Number; IPCS = International Programme on Chemical Safety; CICAD = Concise International Chemical Assessment Documents.

[a] Tolerable daily intake (TDI): An estimate of the intake of a substance, expressed on a body mass basis, to which an individual in a (sub) population may be exposed daily over its lifetime without appreciable health risk.

Table G-1d. Chemicals reported to be used in hydraulic fracturing fluids, with available less-than-chronic oral RfVs and OSFs.

Chemicals from the FracFocus database are listed first, ranked by PPRTV subchronic reference dose (sRfD). The "--" symbol indicates that no value was available from the sources consulted. Additionally, an "x" indicates the availability of usage data from FracFocus (U.S. EPA, 2015a) and physicochemical properties data from EPI Suite™ (see Appendix C). Italicized chemicals are found in both hydraulic fracturing fluids and produced water.

Chemical name	CASRN	FracFocus data available	Physico-chemical data available	PPRTV sRfD[a] (mg/kg-day)	ATSDR Acute oral MRL[b] (mg/kg-day)	ATSDR Inter-mediate oral MRL[b] (mg/kg-day)
Benzyl chloride	100-44-7	x	x	0.002	--	--
Epichlorohydrin	106-89-8	x	x	0.006	--	--
(E)-Crotonaldehyde	123-73-9	x	x	0.01	--	--
Benzene	71-43-2	x	x	0.01	--	--
Ethylbenzene	100-41-4	x	x	0.05	--	0.4
Chlorobenzene	108-90-7	x	x	0.07	--	0.4
Ethylenediamine	107-15-3	x	x	0.2	--	--
2-(2-Butoxyethoxy)ethanol	112-34-5	x	x	0.3	--	--
Hexane	110-54-3	x	x	0.3	--	--
N,N-Dimethylform-amide	68-12-2	x	x	0.3	--	--
Xylenes	1330-20-7	x	x	0.4	1	0.4
Antimony trioxide	1309-64-4	x		0.5	--	--
Ethyl acetate	141-78-6	x	x	0.7	--	--
Iron	7439-89-6	x		0.7	--	--
Toluene	108-88-3	x	x	0.8	0.8	0.02
Formic acid	64-18-6	x	x	0.9	--	--
Hexanedioic acid	124-04-9	x	x	2	--	--
Benzoic acid	65-85-0	x	x	4	--	--
1,2-Propylene glycol	57-55-6	x	x	20	--	--
Mineral oil - includes paraffin oil	8012-95-1	x		30	--	--
Phosphoric acid	7664-38-2	x		48.6	--	--

Chemical name	CASRN	FracFocus data available	Physico-chemical data available	PPRTV	ATSDR	
				sRfD[a] (mg/kg-day)	Acute oral MRL[b] (mg/kg-day)	Inter-mediate oral MRL[b] (mg/kg-day)
Ammonium phosphate	7722-76-1	x		49	--	--
Potassium phosphate, tribasic	7778-53-2	x		49	--	--
Sodium trimetaphosphate	7785-84-4	x		49	--	--
Tetrasodium pyrophosphate	7722-88-5	x		49	--	--
Tricalcium phosphate	7758-87-4	x		49	--	--
Triphosphoric acid, pentasodium salt	7758-29-4	x		49	--	--
Trisodium phosphate	7601-54-9	x		49	--	--
1,3-Dichloropropene	542-75-6	x	x	--	--	0.04
1,4-Dioxane	123-91-1	x	x	--	5	0.5
2-Butoxyethanol	111-76-2	x	x	--	0.4	0.07
Acetone	67-64-1	x	x	--	--	2
Acrylamide	79-06-1	x	x	--	0.01	0.001
Aluminum	7429-90-5	x		--	--	1
Boron	7440-42-8	x		--	0.2	0.2
Dichloromethane	75-09-2	x	x	--	0.2	--
Ethylene glycol	107-21-1	x	x	--	0.8	0.8
Formaldehyde	50-00-0	x	x	--	--	0.3
Naphthalene	91-20-3	x	x	--	0.6	0.6
o-Xylene	95-47-6	x	x	--	1	0.4
Phenol	108-95-2	x	x	--	1	--
Sodium chlorite	7758-19-2	x		--	--	0.1
Antimony trichloride	10025-91-9			0.0004	--	--
2-Methoxyethanol	109-86-4		x	0.02	--	--
Tributyl phosphate	126-73-8		x	0.03	1.1	0.08
Acrylic acid	79-10-7		x	0.2	--	--

Chemical name	CASRN	FracFocus data available	Physico-chemical data available	PPRTV	ATSDR	
				sRfD[a] (mg/kg-day)	Acute oral MRL[b] (mg/kg-day)	Inter-mediate oral MRL[b] (mg/kg-day)
2-(2-Ethoxyethoxy) ethanol	111-90-0		x	0.6	--	--
Cyclohexanone	108-94-1		x	2	--	--
Phosphoric acid, aluminium sodium salt	7785-88-8			49	--	--
Phosphoric acid, diammonium salt	7783-28-0			49	--	--
Polyphosphoric acids, sodium salts	68915-31-1			49	--	--
Sodium pyrophosphate	7758-16-9			49	--	--
Acrolein	107-02-8		x	--	--	0.004
Arsenic	7440-38-2			--	0.005	--
Chromium (VI)	18540-29-9			--	--	0.005
Copper	7440-50-8			--	0.01	0.01
Di(2-ethylhexyl) phthalate	117-81-7		x	--	--	0.1
p-Xylene	106-42-3		x	--	1	0.4
Styrene	100-42-5		x	--	0.1	--
Zinc	7440-66-6			--	--	0.3

CASRN = Chemical Abstract Service Registry Number; PPRTV = Provisional Peer Reviewed Toxicity Values; ATSDR = Agency for Toxic Substances and Disease Registry; HHBP = Human Health Benchmarks for Pesticides.

[a] Reference dose (RfD): An estimate (with uncertainty spanning perhaps an order of magnitude) of a daily oral exposure to the human population (including sensitive subgroups) that is likely to be without an appreciable risk of deleterious effects during a lifetime. It can be derived from a no observed-adverse-effect level (NOAEL), lowest observed-adverse-effect level (LOAEL), or benchmark dose (BMD), with uncertainty factors generally applied to reflect limitations of the data used. The RfD is generally used in the EPA's noncancer health assessments. Subchronic RfD (sRFD): Duration of exposure is up to 10% of an average lifespan.

[b] Minimal risk level (MRL): An ATSDR estimate of daily human exposure to a hazardous substance at or below which the substance is unlikely to pose a measurable risk of harmful (adverse), noncancerous effects. MRLs are calculated for a route of exposure (inhalation or oral) over a specified time period (acute, intermediate, or chronic). MRLs should not be used as predictors of harmful (adverse) health effects. Acute MRL: Duration of exposure is 1 to 14 days. Intermediate MRL: Duration of exposure is >14 to 364 days.

Table G-1e. Available qualitative cancer classifications for chemicals reported to be used in hydraulic fracturing fluids.

Chemicals from the FracFocus database are listed first, with chemicals classified as known carcinogens by one or more sources listed first. See the Appendix G glossary (Section G.3) for details on the weight of evidence characterizations. The "--" symbol indicates that no value was available from the sources consulted. Additionally, an "x" indicates the availability of usage data from FracFocus (U.S. EPA, 2015a) and physicochemical properties data from EPI Suite™ (see Appendix C). Italicized chemicals are found in both hydraulic fracturing fluids and produced water. Cancer classifications from IRIS and PPRTV are also listed in Table G-1a.

Chemical name	CASRN	FracFocus data available	Physico-chemical data available	Qualitative cancer classification			
				IRIS[a]	PPRTV[b]	IARC[c]	RoC[d]
1,3-Butadiene	106-99-0	x	x	"Carcinogenic to humans"	--	1	Known
Arsenic	*7440-38-2*			*A (Human carcinogen)*	--	*1*	*Known*
Benzene	*71-43-2*	*x*	*x*	*A (Human carcinogen)*	--	*1*	*Known*
Chromium (VI)	*18540-29-9*			*Inhaled: A (Human carcinogen); Oral: D (Not classifiable as to human carcino-genicity)*	--	*1*	*Known*
Ethanol	*64-17-5*	*x*	*x*	--	--	*1*	--
Ethylene oxide	75-21-8	x	x	--	--	1	Known
Formaldehyde	50-00-0	x	x	B1 (Probable human carcinogen)	--	1	Known
Nickel sulfate	7786-81-4	x		--	--	1	--
Nickel(II) sulfate hexahydrate	10101-97-0			--	--	1	--
Quartz-alpha (SiO$_2$)	14808-60-7	x		--	--	1	--
Sulfuric acid	7664-93-9	x		--	--	1	Known
(E)-Crotonaldehyde	123-73-9	x	x	C (Possible human carcinogen)	--	--	--

Chemical name	CASRN	FracFocus data available	Physico-chemical data available	Qualitative cancer classification			
				IRIS[a]	PPRTV[b]	IARC[c]	RoC[d]
1,2-Propylene oxide	75-56-9		x	B2 (Probable human carcinogen)	--	2B	RAHC
1,3-Dichloropropene	542-75-6	x	x	"Likely to be a human carcinogen"	--	2B	RAHC
1,4-Dioxane	123-91-1	x	x	"Likely to be carcinogenic to humans"	--	2B	RAHC
4-Methyl-2-pentanone	108-10-1	x	x	"Data are inadequate for an assessment of human carcinogenic potential"	--	2B	--
Acetaldehyde	75-07-0	x	x	B2 (Probable human carcinogen)	--	2B	RAHC
Acrylamide	79-06-1	x	x	"Likely to be carcinogenic to humans"	--	2A	RAHC
Aniline	62-53-3	x	x	B2 (Probable human carcinogen)	--	3	--
Antimony trioxide	1309-64-4	x		--	Inhaled: "Suggestive evidence of carcinogenic potential"; Oral: "Data are inadequate for an assessment of human carcinogenic potential"	2B	--
Attapulgite	12174-11-7	x		--	--	2B or 3	--
Benzyl chloride	100-44-7	x	x	B2 (Probable human carcinogen)	--	--	--

Chemical name	CASRN	FracFocus data available	Physico-chemical data available	Qualitative cancer classification			
				IRIS[a]	PPRTV[b]	IARC[c]	RoC[d]
Bis(2-chloroethyl) ether	111-44-4	x	x	B2 (Probable human carcinogen)	--	3	--
Carbon black	1333-86-4			--	--	2B	--
Coconut oil acid/Diethanola-mine condensate (2:1)	68603-42-9	x		--	--	2B	--
Cumene	98-82-8	x	x	D (Not classifiable as to human carcino-genicity)	--	2B	RAHC
Di(2-ethylhexyl) phthalate	117-81-7		x	B2 (Probable human carcinogen)	--	2B	RAHC
Dibromoaceto-nitrile	3252-43-5	x	x	--	--	2B	--
Dichloromethane	75-09-2	x	x	"Likely to be carcinogenic in humans"	--	2A	RAHC
Diethanolamine	111-42-2	x	x	--	--	2B	--
Epichlorohydrin	106-89-8	x	x	B2 (Probable human carcinogen)	--	2A	RAHC
Ethylbenzene	100-41-4	x	x	D (Not classifiable as to human carcino-genicity)	--	2B	--
Hydrazine	302-01-2	x		B2 (Probable human carcinogen)	--	2A	RAHC
Lead	7439-92-1			B2 (Probable human carcinogen)	--	2B	RAHC

Chemical name	CASRN	FracFocus data available	Physico-chemical data available	Qualitative cancer classification			
				IRIS[a]	PPRTV[b]	IARC[c]	RoC[d]
N,N-Dimethylfor-mamide	68-12-2	x	x	--	"Data are inadequate for an assessment of human carcinogenic potential"	2A	--
Naphthalene	91-20-3	x	x	"Data are inadequate to assess human carcinogenic potential"	--	2B	RAHC
Nitrilotriacetic acid	139-13-9	x	x	--	--	2B	RAHC
Quinoline	91-22-5	x	x	"Likely to be carcinogenic in humans"	--	--	--
Styrene	100-42-5		x	--	--	2B	RAHC
Thiourea	62-56-6	x	x	--	--	3	RAHC
Titanium dioxide	13463-67-7	x		--	--	2B	
Tributyl phosphate	126-73-8		x	--	"Likely to be carcinogenic to humans"	--	--
1,2,3-Trimethylbenzene	526-73-8	x	x	--	"Data are inadequate for an assessment of human carcinogenic potential"	--	--
1,3,5-Trimethylbenzene	108-67-8	x	x	--	"Data are inadequate for an assessment of human carcinogenic potential"	--	--
1-Butanol	71-36-3	x	x	D (Not classifiable as to human carcino-genicity)	--	--	--

Chemical name	CASRN	FracFocus data available	Physico-chemical data available	Qualitative cancer classification			
				IRIS[a]	PPRTV[b]	IARC[c]	RoC[d]
1-Propene	115-07-1	x	x	--	--	3	--
1-tert-Butoxy-2-propanol	57018-52-7	x	x	--	--	3	--
2-(2-Butoxyeth-oxy)ethanol	112-34-5	x	x	--	"Data are inadequate for an assessment of human carcinogenic potential"	--	--
2-(2-Ethoxyeth-oxy) ethanol	111-90-0		x	--	"Data are inadequate for an assessment of human carcinogenic potential"	--	--
2-Butoxyethanol	111-76-2	x	x	*"Not likely to be carcinogenic to humans"*	--	3	--
2-Methoxyethanol	109-86-4		x	--	"Data are inadequate for an assessment of human carcinogenic potential"	--	--
Acetone	67-64-1	x	x	*"Data are inadequate for an assessment of human carcinogenic potential"*	--	--	--
Acetophenone	98-86-2	x	x	*D (Not classifiable as to human carcino-genicity)*	--	--	--

Chemical name	CASRN	FracFocus data available	Physico-chemical data available	Qualitative cancer classification			
				IRIS[a]	PPRTV[b]	IARC[c]	RoC[d]
Acrolein	*107-02-8*		x	*"Data are inadequate for an assessment of human carcinogenic potential"*	--	3	--
Acrylic acid	79-10-7		x	--	"Data are inadequate for an assessment of human carcinogenic potential"	3	--
Aluminum	*7429-90-5*	x		--	*"Data are inadequate for an assessment of human carcinogenic potential"*	--	--
Amaranth	915-67-3	x	x	--	--	3	--
Ammonium phosphate	7722-76-1	x		--	"Data are inadequate for an assessment of human carcinogenic potential"	--	--
Benzoic acid	65-85-0	x	x	D (Not classifiable as to human carcino-genicity)	--	--	--
Boron	*7440-42-8*	x		*"Data are inadequate to assess the carcinogenic potential"*	--	--	--
Chlorine dioxide	10049-04-4	x		*"Data are inadequate to assess human carcino-genicity"*	--	--	--

Chemical name	CASRN	FracFocus data available	Physico-chemical data available	Qualitative cancer classification			
				IRIS[a]	PPRTV[b]	IARC[c]	RoC[d]
Chlorobenzene	108-90-7	x	x	D (Not classifiable as to human carcino-genicity)			
Chloromethane	74-87-3	x	x	"Carcinogenic potential cannot be determined"	--	3	--
Chromium (III)	16065-83-1			"Data are inadequate for an assessment of human carcinogenic potential"	--	3	--
Coumarin	91-64-5		x	--	--	3	--
Cyclohexanone	108-94-1		x	--	"Data are inadequate for an assessment of human carcinogenic potential"	3	--
Dapsone	80-08-0	x	x	--	--	3	--
D-Limonene	5989-27-5	x	x	--	--	3	--
Ethyl acetate	141-78-6	x	x	--	"Data are inadequate for an assessment of human carcinogenic potential"	--	--
Ethylene	74-85-1	x	x	--	--	3	--
Ethylenediamine	107-15-3	x	x	D (Not classifiable as to human carcino-genicity)	"Data are inadequate for an assessment of human carcinogenic potential"	--	--
FD&C Blue no. 1	3844-45-9	x	x	--	--	3	--

Chemical name	CASRN	FracFocus data available	Physico-chemical data available	Qualitative cancer classification			
				IRIS[a]	PPRTV[b]	IARC[c]	RoC[d]
FD&C Yellow 6	2783-94-0		x	--	--	3	--
Formic acid	*64-18-6*	*x*	*x*	--	*"Data are inadequate for an assessment of human carcinogenic potential"*	--	--
Furfural	98-01-1	x	x	--	--	3	--
Hematite	1317-60-8	x		--	--	3	--
Hexane	*110-54-3*	*x*	*x*	*"Inadequate information to assess the carcinogenic potential"*	--	--	--
Hydrochloric acid	*7647-01-0*	*x*		--	--	3	--
Hydrogen peroxide	7722-84-1	x		--	--	3	--
Iron	*7439-89-6*	*x*		--	*"Data are inadequate for an assessment of human carcinogenic potential"*	--	--
Iron(III) oxide	1309-37-1	x		--	--	3	--
Isopropanol	*67-63-0*	*x*	*x*	--	--	*3*	--
Latex 2000 TM	9003-55-8			--	--	3	--
Ligroine	8032-32-4			--	--	3	--
Mineral oil - includes paraffin oil	8012-95-1	x		--	*"Data are inadequate for an assessment of human carcinogenic potential"*	--	--
Mineral spirits	64475-85-0			--	--	3	--
Morpholine	110-91-8	x	x	--	--	3	--

Chemical name	CASRN	FracFocus data available	Physico-chemical data available	Qualitative cancer classification			
				IRIS[a]	PPRTV[b]	IARC[c]	RoC[d]
Pentane	*109-66-0*	x	x	--	"Data are inadequate for an assessment of human carcinogenic potential"	--	--
Petroleum	8002-05-9	x		--	--	3	--
Phenanthrene	*85-01-8*		x	--	--	3	--
Phenol	*108-95-2*	x	x	"Data are inadequate for an assessment of human carcinogenic potential"	--	3	--
Phosphine	7803-51-2			D (Not classifiable as to human carcino-genicity)	--	--	--
Phosphoric acid	7664-38-2	x		--	"Data are inadequate for an assessment of human carcinogenic potential"	--	--
Phosphoric acid, aluminium sodium salt	7785-88-8			--	"Data are inadequate for an assessment of human carcinogenic potential"	--	--
Phosphoric acid, diammonium salt	7783-28-0			--	"Data are inadequate for an assessment of human carcinogenic potential"	--	--
Policapram (Nylon 6)	25038-54-4	x		--	--	3	--

Chemical name	CASRN	FracFocus data available	Physico-chemical data available	Qualitative cancer classification			
				IRIS[a]	PPRTV[b]	IARC[c]	RoC[d]
Poly(tetrafluoro-ethylene)	9002-84-0	x		--	--	3	--
Polyphosphoric acids, sodium salts	68915-31-1			--	"Data are inadequate for an assessment of human carcinogenic potential"	--	--
Polyvinyl acetate copolymer	9003-20-7			--	--	3	--
Polyvinyl alcohol	9002-89-5			--	--	3	--
Polyvinylpyrroli-done	9003-39-8	x		--	--	3	--
Potassium phosphate, tribasic	7778-53-2	x		--	"Data are inadequate for an assessment of human carcinogenic potential"	--	--
Rhodamine B	81-88-9		x	--	--	3	--
Silica	*7631-86-9*	x		--	--	*3*	--
Sodium bisulfite	7631-90-5	x		--	--	3	--
Sodium chlorite	7758-19-2	x		"Data are inadequate to assess human carcino-genicity"	--	3	--
Sodium metabisulfite	7681-57-4	x		--	--	3	--
Sodium pyrophosphate	7758-16-9			--	"Data are inadequate for an assessment of human carcinogenic potential"	--	--
Sodium sulfite	7757-83-7	x		--	--	3	--

Chemical name	CASRN	FracFocus data available	Physico-chemical data available	Qualitative cancer classification			
				IRIS[a]	PPRTV[b]	IARC[c]	RoC[d]
Sodium trimetaphos-phate	7785-84-4	x		--	"Data are inadequate for an assessment of human carcinogenic potential"	--	--
Stoddard solvent	8052-41-3	x		--	--	3	--
Sulfan blue	129-17-9	x	x	--	--	3	--
Sulfur dioxide	7446-09-5	x		--	--	3	--
Talc	14807-96-6	x		--	--	3	--
Tetrasodium pyrophosphate	7722-88-5	x		--	"Data are inadequate for an assessment of human carcinogenic potential"	--	--
Toluene	*108-88-3*	*x*	*x*	*"Inadequate information to assess the carcinogenic potential"*	--	*3*	--
Tricalcium phosphate	7758-87-4	x		--	"Data are inadequate for an assessment of human carcinogenic potential"	--	--
Triethanolamine	102-71-6	x	x	--	--	3	--
Triphosphoric acid, penta-sodium salt	7758-29-4	x		--	"Data are inadequate for an assessment of human carcinogenic potential"	--	--
Trisodium phosphate	7601-54-9	x		--	"Data are inadequate for an assessment of human carcinogenic potential"	--	--

Chemical name	CASRN	FracFocus data available	Physico-chemical data available	Qualitative cancer classification			
				IRIS[a]	PPRTV[b]	IARC[c]	RoC[d]
Xylenes	1330-20-7	x	x	*"Data are inadequate to assess the carcinogenic potential"*	--	3	--
Zeolites	1318-02-1			--	--	3	--
Zinc	7440-66-6			*"Inadequate information to assess carcinogenic potential"*	--	--	--
1,2-Propylene glycol	57-55-6	x	x	--	*"Not likely to be carcinogenic to humans"*	--	--

CASRN = Chemical Abstract Service Registry Number; IRIS = Integrated Risk Information System; PPRTV = Provisional Peer Reviewed Toxicity Values; IARC = International Agency for Research on Cancer Monographs; RoC = National Toxicology Program 13th Report on Carcinogens.

[a] IRIS assessments use the EPA's 1986, 1996, 1999, or 2005 guidelines to establish descriptors for summarizing the weight of evidence as to whether a contaminant is or may be carcinogenic. See glossary in Appendix G for details.

[b] PPRTV assessments use the EPA's 1999 guidelines to establish descriptors for summarizing the weight of evidence as to whether a contaminant is or may be carcinogenic. See glossary in Appendix G for details.

[c] The IARC summarizes the weight of evidence as to whether a contaminant is or may be carcinogenic using five weight of evidence classifications: Group 1: Carcinogenic to humans; Group 2A: Probably carcinogenic to humans; Group 2B: Possibly carcinogenic to humans; Group 3: Not classifiable as to its carcinogenicity to humans; Group 4: Probably not carcinogenic to humans. See glossary in Appendix G for details.

[d] The listing criteria in the 13th RoC Document are: Known = Known to be a human carcinogen; RAHC = Reasonably anticipated to be a human carcinogen.

Table G-2a. Chemicals reported to be detected in produced water, with available chronic oral RfVs, OSFs, and qualitative cancer classifications from United States federal sources.

Chemicals are ranked by IRIS reference dose (RfD). The "--" symbol indicates that no value was available from the sources consulted. Additionally, an "X" indicates the availability of measured concentration data in produced water (see Appendix E) and physicochemical properties data from EPI Suite™ (see Appendix C). Italicized chemicals are found in both hydraulic fracturing fluids and produced water.

Chemical name	CASRN	Concentration data available	Physico-chemical data available	IRIS Chronic RfD[a] (mg/kg-day)	IRIS OSF[b] (per mg/kg-day)	IRIS Cancer WOE characterization[c]	PPRTV Chronic RfD[a] (mg/kg-day)	PPRTV OSF[b] (per mg/kg-day)	PPRTV Cancer WOE characterization[c]	ATSDR Chronic oral MRL[d] (mg/kg-day)	HHBP Chronic RfD[a] (mg/kg-day)	NPDWRs Public health goal[e] (MCLG) (mg/L)	NPDWRs MCL[f] (mg/L)
Heptachlor epoxide	1024-57-3		x	0.000013	9.1	B2 (Probable human carcinogen)	--	--	--	--	--	0	0.0002
Phosphorus	7723-14-0	x		0.00002	--	D (Not classifiable as to human carcinogenicity)	--	--	--	--	--	--	--
Aldrin	309-00-2		x	0.00003	17	B2 (Probable human carcinogen)	--	--	--	0.00003	--	--	--
Dieldrin	60-57-1		x	0.00005	16	B2 (Probable human carcinogen)	--	--	--	0.00005	--	--	--
Arsenic	7440-38-2	x		*0.0003*	*1.5*	*A (Human carcinogen)*	--	--	--	*0.0003*	--	*0*	*0.01*
Lindane	58-89-9		x	0.0003	--	--	--	--	--	--	--	0.0002	0.0002

Appendix G – Identification and Hazard Evaluation of Chemicals across the Hydraulic Fracturing Water Cycle Supplemental Information

Chemical name	CASRN	Concentration data available	Physico-chemical data available	IRIS			PPRTV			ATSDR	HHBP	NPDWRs	
				Chronic RfD^a (mg/kg-day)	OSP^b (per mg/kg-day)	Cancer WOE characterization^c	Chronic RfD^a (mg/kg-day)	OSP^b (per mg/kg-day)	Cancer WOE characterization^c	Chronic oral MRL^d (mg/kg-day)	Chronic RfD^a (mg/kg-day)	Public health goal^e (MCLG) (mg/L)	MCL^f (mg/L)
Antimony	7440-36-0	x		0.0004	–	–	–	–	"Data are inadequate for an assessment of human carcinogenic potential"	–	–	0.006	0.006
Acrolein	107-02-8		x	0.0005	–	"Data are inadequate for an assessment of human carcinogenic potential"	–	–	–	–	–	–	–
Heptachlor	76-44-8		x	0.0005	4.5	B2 (Probable human carcinogen)	–	–	–	–	–	0	0.0004
Cyanide	57-12-5		x	0.0006	–	"Inadequate information to assess the carcinogenic potential"	–	–	–	–	–	0.2	0.2
Pyridine	110-86-1	x	x	0.001	–	D (Not classifiable as to human carcinogenicity)	–	–	–	–	–	–	–
Methyl bromide	74-83-9		x	0.0014	–	–	–	–	"Data are inadequate for an assessment of human carcinogenic potential"	–	0.02	–	–

Appendix G – Identification and Hazard Evaluation of Chemicals across the Hydraulic Fracturing Water Cycle Supplemental Information

Chemical name	CASRN	Concentration data available	Physicochemical data available	IRIS Chronic RfD[a] (mg/kg-day)	IRIS OSF[b] (per mg/kg-day)	IRIS Cancer WOE characterization[c]	PPRTV Chronic RfD[a] (mg/kg-day)	PPRTV OSF[b] (per mg/kg-day)	PPRTV Cancer WOE characterization[c]	ATSDR Chronic oral MRL[d] (mg/kg-day)	HHBP Chronic RfD[a] (mg/kg-day)	NPDWRs Public health goal[e] (MCLG) (mg/L)	NPDWRs MCL[f] (mg/L)
Beryllium	7440-41-7	x		0.002	--	B1 (Probable human carcinogen)	--	--	--	0.002	--	0.004	0.004
Propargyl alcohol	*107-19-7*		x	*0.002*	--	--	--	--	--	--	--	--	--
2,4-Dichlorophenol	120-83-2	x	x	0.003	--		--	--	"Data are inadequate for an assessment of human carcinogenic potential"	--	--	--	--
Benzidine	92-87-5	x	x	0.003	230	A (Human carcinogen)	--	--	--	--	--	--	--
Chromium (VI)	18540-29-9	x		0.003	--	*Inhaled: A (Human carcinogen; Oral: D (Not classifiable as to human carcinogenicity)*	--	--	--	0.0009	--	--	--
2-Methylnaphthalene	91-57-6	x	x	0.004	--	*"Data are inadequate to assess human carcinogenic potential"*	--	--	--	0.04	--	--	--
Benzene	*71-43-2*	x	x	*0.004*	*0.015-0.055*	*A (Human carcinogen)*	--	--	--	*0.0005*	--	0	0.005

Appendix G – Identification and Hazard Evaluation of Chemicals across the Hydraulic Fracturing Water Cycle Supplemental Information

Chemical name	CASRN	Concentration data available	Physico-chemical data available	IRIS Chronic RfD[a] (mg/kg-day)	IRIS OSF[b] (per mg/kg-day)	IRIS Cancer WOE characterization[c]	PPRTV Chronic RfD[a] (mg/kg-day)	PPRTV OSF[b] (per mg/kg-day)	PPRTV Cancer WOE characterization[c]	ATSDR Chronic oral MRL[d] (mg/kg-day)	HHBP Chronic RfD[a] (mg/kg-day)	NPDWRs Public health goal[e] (MCLG) (mg/L)	MCL[f] (mg/L)
Molybdenum	7439-98-7	x		0.005	--	--	--	--	--	--	--	--	--
Selenium	7782-49-2	x		0.005	--	D (Not classifiable as to human carcinogenicity)			--	0.005	--	0.05	0.05
Silver	7440-22-4	x		0.005	--	D (Not classifiable as to human carcinogenicity)				--	--	--	--
Dichloromethane	75-09-2		x	0.006	0.002	"Likely to be carcinogenic in humans"	--	--	--	0.06	--	0	0.005
Tetrachloroethene	127-18-4		x	0.006	0.0021	"Likely to be carcinogenic in humans"	--	--	--	0.008	--	0	0.005
1,2,3-Trimethylbenzene	526-73-8		x	0.01	--		--	--	"Data are inadequate for an assessment of human carcinogenic potential"	--	--	--	--
1,2,4-Trichlorobenzene	120-82-1		x	0.01	--	D (Not classifiable as to human carcinogenicity)	--	0.029	"Likely to be carcinogenic to humans"	0.1	--	0.07	0.07

Appendix G – Identification and Hazard Evaluation of Chemicals across
the Hydraulic Fracturing Water Cycle Supplemental Information

Chemical name	CASRN	Concentration data available	Physico-chemical data available	IRIS Chronic RfD[a] (mg/kg-day)	IRIS OSP[b] (per mg/kg-day)	IRIS Cancer WOE characterization[c]	PPRTV Chronic RfD[a] (mg/kg-day)	PPRTV OSP[b] (per mg/kg-day)	PPRTV Cancer WOE characterization[c]	ATSDR Chronic oral MRL[d] (mg/kg-day)	HHBP Chronic RfD[a] (mg/kg-day)	NPDWRs Public health goal[e] (MCLG) (mg/L)	NPDWRs MCL[f] (mg/L)
1,2,4-Trimethyl-benzene	95-63-6	x	x	0.01	--	--	--	--	--	--	--	--	--
1,3,5-Trimethyl-benzene	108-67-8	x	x	0.01	--	--	--	--	"Data are inadequate for an assessment of human carcinogenic potential"	--	--	--	--
Chloroform	67-66-3	x	x	0.01	--	B2 (Probable human carcinogen)	--	--	--	0.01	--	--	--
Trimethylben-zene	25551-13-7			0.01	--	--	--	--	--	--	--	--	--
2,4-Dimethylphenol	105-67-9	x	x	0.02	--	--	--	--	"Data are inadequate for an assessment of human carcinogenic potential"	--	--	--	--
Bromodichloro-methane	75-27-4	x	x	0.02	0.062	B2 (Probable human carcinogen)	--	--	--	0.02	--	--	--
Bromoform	75-25-2	x	x	0.02	0.0079	B2 (Probable human carcinogen)	--	--	--	0.02	--	--	--

Appendix G – Identification and Hazard Evaluation of Chemicals across the Hydraulic Fracturing Water Cycle Supplemental Information

Chemical name	CASRN	Concentration data available	Physico-chemical data available	IRIS Chronic RfD[a] (mg/kg-day)	IRIS OSP[b] (per mg/kg-day)	IRIS Cancer WOE characterization[c]	PPRTV Chronic RfD[a] (mg/kg-day)	PPRTV OSP[b] (per mg/kg-day)	PPRTV Cancer WOE characterization[c]	ATSDR Chronic oral MRL[d] (mg/kg-day)	HHBP Chronic RfD[a] (mg/kg-day)	NPDWRs Public health goal[e] (MCLG) (mg/L)	MCL[f] (mg/L)
Chlorobenzene	108-90-7		x	0.02	--	D (Not classifiable as to human carcinogenicity)	--	--	--	--	--	0.1	0.1
Chlorodibromo-methane	124-48-1		x	0.02	0.084	C (Possible human carcinogen)	--	--	--	0.09	--	--	--
Di(2-ethylhexyl) phthalate	117-81-7	x	x	0.02	0.014	B2 (Probable human carcinogen)	--	--	--	0.06	--	0	0.006
Naphthalene	91-20-3	x	x	0.02	--	"Data are inadequate to assess human carcinogenic potential"	--	--	--	--	--	--	--
Diphenylamine	122-39-4	x	x	0.025	--	--	--	--	"Data are inadequate for an assessment of human carcinogenic potential"	--	0.1	--	--
1,4-Dioxane	123-91-1	x	x	0.03	0.1	"Likely to be carcinogenic to humans"	--	--	--	0.1	--	--	--

Appendix G – Identification and Hazard Evaluation of Chemicals across the Hydraulic Fracturing Water Cycle Supplemental Information

Chemical name	CASRN	Concentration data available	Physicochemical data available	IRIS			PPRTV			ATSDR	HHBP	NPDWRs	
				Chronic RfD[a] (mg/ kg-day)	OSF[b] (per mg/ kg-day)	Cancer WOE character-ization[c]	Chronic RfD[a] (mg/ kg-day)	OSF[b] (per mg/ kg-day)	Cancer WOE character-ization[c]	Chronic oral MRL[d] (mg/ kg-day)	Chronic RfD[a] (mg/kg-day)	Public health goal[e] (MCLG) (mg/L)	MCL[f] (mg/L)
Pyrene	129-00-0	x	x	0.03	--	D (Not classifiable as to human carcino-genicity)	--	--	--	--	--	--	--
Fluoranthene	206-44-0	x	x	0.04	--	D (Not classifiable as to human carcino-genicity)	--	--	"Data are inadequate for an assessment of human carcinogenic potential"	--	--	--	--
Fluorene	86-73-7	x	x	0.04	--	D (Not classifiable as to human carcino-genicity)	--	--	--	--	--	--	--
Bisphenol A	80-05-7	x	x	0.05	--	--	--	--	--	--	--	--	--
m-Cresol	108-39-4	x	x	0.05	--	C (Possible human carcinogen)	--	--	--	--	--	--	--
o-Cresol	95-48-7	x	x	0.05	--	C (Possible human carcinogen)	--	--	"Data are inadequate for an assessment of human carcinogenic potential"	--	--	--	--

Chemical name	CASRN	Concentration data available	Physico-chemical data available	IRIS Chronic RfD[a] (mg/ kg-day)	IRIS OSP[b] (per mg/ kg-day)	IRIS Cancer WOE characterization[c]	PPRTV Chronic RfD[a] (mg/ kg-day)	PPRTV OSP[b] (per mg/ kg-day)	PPRTV Cancer WOE characterization[c]	ATSDR Chronic oral MRL[d] (mg/ kg-day)	HHBP Chronic RfD[a] (mg/kg-day)	NPDWRs Public health goal[e] (MCLG) (mg/L)	NPDWRs MCL[f] (mg/L)
Toluene	108-88-3	x	x	0.08	--	"Inadequate information to assess the carcinogenic potential"	--	--	--	--	--	1	1
1-Butanol	71-36-3		x	0.1	--	D (Not classifiable as to human carcinogenicity)	--	--	--	--	--	--	--
2-Butoxyethanol	111-76-2		x	0.1	--	"Not likely to be carcinogenic to humans"	--	--	--	--	--	--	--
Acetophenone	98-86-2	x	x	0.1	--	D (Not classifiable as to human carcinogenicity)	--	--	--	--	--	--	--
Carbon disulfide	75-15-0	x	x	0.1	--	--	--	--	--	--	--	--	--
Chlorine	7782-50-5			0.1	--	--	--	--	--	--	--	--	--
Cumene	98-82-8	x	x	0.1	--	D (Not classifiable as to human carcinogenicity)	--	--	--	--	--	--	--

Appendix G – Identification and Hazard Evaluation of Chemicals across the Hydraulic Fracturing Water Cycle Supplemental Information

Chemical name	CASRN	Concentration data available	Physico-chemical data available	IRIS Chronic RfD[a] (mg/kg-day)	IRIS OSP[b] (per mg/kg-day)	IRIS Cancer WOE characterization[c]	PPRTV Chronic RfD[a] (mg/kg-day)	PPRTV OSP[b] (per mg/kg-day)	PPRTV Cancer WOE characterization[c]	ATSDR Chronic oral MRL[d] (mg/kg-day)	HHBP Chronic RfD[a] (mg/kg-day)	NPDWRs Public health goal[e] (MCLG) (mg/L)	MCL[f] (mg/L)
Dibutyl phthalate	84-74-2	x	x	0.1	--	D (Not classifiable as to human carcinogenicity)	--	--	--	--	--	--	--
Ethylbenzene	100-41-4	x	x	0.1	--	D (Not classifiable as to human carcinogenicity)	--	--	--	--	--	0.7	0.7
Nitrite	14797-65-0	x		0.1	--	--	--	--	--	--	--	1	1
Manganese	7439-96-5	x		0.14	--	D (Not classifiable as to human carcinogenicity)	--	--	--	--	--	--	--
Barium	7440-39-3	x		0.2	--	"Not likely to be carcinogenic to humans"	--	--	--	0.2	--	2	2
Benzyl butyl phthalate	85-68-7	x	x	0.2	--	C (Possible human carcinogen)	--	--	--	--	--	--	--
Boron	7440-42-8	x		0.2	--	*"Data are inadequate to assess the carcinogenic potential"*	--	--	--	--	--	--	--

Appendix G – Identification and Hazard Evaluation of Chemicals across
the Hydraulic Fracturing Water Cycle Supplemental Information

Chemical name	CASRN	Concentration data available	Physicochemical data available	IRIS Chronic RfD[a] (mg/kg-day)	IRIS OSP[b] (per mg/kg-day)	IRIS Cancer WOE characterization[c]	PPRTV Chronic RfD[a] (mg/kg-day)	PPRTV OSP[b] (per mg/kg-day)	PPRTV Cancer WOE characterization[c]	ATSDR Chronic oral MRL[d] (mg/kg-day)	HHBP Chronic RfD[a] (mg/kg-day)	NPDWRs Public health goal[e] (MCLG) (mg/L)	NPDWRs MCL[f] (mg/L)
Xylenes	1330-20-7	x	x	0.2	--	"Data are inadequate to assess the carcinogenic potential"	--	--	--	0.2	--	10	10
Phenol	108-95-2	x	x	0.3	--	"Data are inadequate to assess human carcinogenicity"	--	--	--	--	--	--	--
Zinc	7440-66-6	x		0.3	--	"Inadequate information to assess carcinogenic potential"	--	--	--	0.3	--	--	--
Biphenyl	92-52-4	x	x	0.5	0.008	"Suggestive evidence of carcinogenic potential"	--	--	--	--	--	--	--
Caprolactam	105-60-2	x	x	0.5	--	--	--	--	--	--	--	--	--
Methyl ethyl ketone	78-93-3		x	0.6	--	"Data are inadequate to assess carcinogenic potential"	--	--	--	--	--	--	--
Strontium	7440-24-6	x		0.6	--	--	--	--	--	--	--	--	--

Chemical name	CASRN	Concentration data available	Physicochemical data available	IRIS Chronic RfDª (mg/kg-day)	IRIS OSPᵇ (per mg/kg-day)	IRIS Cancer WOE characterizationᶜ	PPRTV Chronic RfDª (mg/kg-day)	PPRTV OSPᵇ (per mg/kg-day)	PPRTV Cancer WOE characterizationᶜ	ATSDR Chronic oral MRLᵈ (mg/kg-day)	HHBP Chronic RfDª (mg/kg-day)	NPDWRs Public health goalᵉ (MCLG) (mg/L)	MCLᶠ (mg/L)
Diethyl phthalate	84-66-2		x	0.8	--	D (Not classifiable as to human carcinogenicity)	--	--	--	--	--	--	--
Acetone	67-64-1	x	x	0.9	--	"Data are inadequate to assess human carcinogenicity"	--	--	--	--	--	--	--
Chromium (III)	16065-83-1	x		1.5	--	"Data are inadequate to assess human carcinogenicity"	--	--	--	--	--	--	--
Nitrate	14797-55-8	x		1.6	--	--	--	--	--	--	--	10	10
Ethylene glycol	107-21-1		x	2	--	--	--	--	--	--	--	--	--
Methanol	67-56-1		x	2	--	--	--	--	--	--	--	--	--
Cadmium	7440-43-9	x		0.0005 (water)	--	B1 (Probable human carcinogen)	--	--	--	0.0001	--	0.005	0.005
1,1-Dichloro-ethane	75-34-3		x	--	--	C (Possible human carcinogen)	0.2	--	--	--	--	--	--
1,2-Diphenyl-hydrazine	122-66-7	x	x	--	0.8	B2 (Probable human carcinogen)	--	--	--	--	--	--	--

Appendix G – Identification and Hazard Evaluation of Chemicals across
the Hydraulic Fracturing Water Cycle Supplemental Information

Chemical name	CASRN	Concentration data available	Physicochemical data available	IRIS Chronic RfD[a] (mg/kg-day)	IRIS OSF[b] (per mg/kg-day)	IRIS Cancer WOE characterization[c]	PPRTV Chronic RfD[a] (mg/kg-day)	PPRTV OSF[b] (per mg/kg-day)	PPRTV Cancer WOE characterization[c]	ATSDR Chronic oral MRL[d] (mg/kg-day)	HHBP Chronic RfD[a] (mg/kg-day)	NPDWRs Public health goal[e] (MCLG) (mg/L)	MCL[f] (mg/L)
1,2-Propylene glycol	57-55-6		x	--	--	--	20	--	"Not likely to be carcinogenic to humans"	--	--	--	--
1-Methylnaphthalene	90-12-0		x	--	--	--	0.007	0.029	--	0.07	--	--	--
2-(2-Butoxyethoxy)ethanol	112-34-5		x	--	--	--	0.03	--	"Data are inadequate for an assessment of human carcinogenic potential"	--	--	--	--
2-Chloroethanol	107-07-3		x	--	--	--	0.02	--	"Data are inadequate for an assessment of human carcinogenic potential"	--	0.045	--	--
Acrylonitrile	107-13-1		x	--	0.54	B1 (Probable human carcinogen)	--	--	--	0.04	--	--	--
Alpha particle	12587-46-1	x		--	--	--	--	--	--	--	--	--	15
Aluminum	7429-90-5	x		--	--	--	1	--	"Data are inadequate for an assessment of human carcinogenic potential"	1	--	--	--

Chemical name	CASRN	Concentration data available	Physico-chemical data available	IRIS			PPRTV			ATSDR	HHBP	NPDWRs	
				Chronic RfD[a] (mg/ kg-day)	OSF[b] (per mg/ kg-day)	Cancer WOE character-ization[c]	Chronic RfD[a] (mg/ kg-day)	OSF[b] (per mg/ kg-day)	Cancer WOE character-ization[c]	Chronic oral MRL[d] (mg/ kg-day)	Chronic RfD[a] (mg/kg-day)	Public health goal[e] (MCLG) (mg/L)	MCL[f] (mg/L)
Benz(a)anthra-cene	56-55-3	x	x	–	–	B2 (Probable human carcinogen)	–	0.7	–	–	–	–	–
Benzo(a)pyrene	50-32-8	x	x	–	7.3	B2 (Probable human carcinogen)	–	–	–	–	–	0	0.0002
Benzyl alcohol	100-51-6	x	x	–	–	–	0.1	–	"Data are inadequate for an assessment of human carcinogenic potential"	–	–	–	–
Benzyl chloride	100-44-7		x	–	0.17	B2 (Probable human carcinogen)	0.002	–	–	–	–	–	–
Beta particle	12587-47-2	x		–	–	–	–	–	–	–	–	–	4
beta-Hexachloro cyclohexane	319-85-7		x	–	1.8	C (Possible human carcinogen)	–	–	–	–	–	–	–
Bis(2-chloro-ethyl) ether	111-44-4		x	–	1.1	B2 (Probable human carcinogen)	–	–	–	–	–	–	–
Butylbenzene	104-51-8		x	–	–	–	0.05	–	"Data are inadequate for an assessment of human carcinogenic potential"	–	–	–	–

Appendix G – Identification and Hazard Evaluation of Chemicals across the Hydraulic Fracturing Water Cycle Supplemental Information

Chemical name	CASRN	Concentration data available	Physico-chemical data available	IRIS Chronic RfD[a] (mg/kg-day)	IRIS OSF[b] (per mg/kg-day)	IRIS Cancer WOE character-ization[c]	PPRTV Chronic RfD[a] (mg/kg-day)	PPRTV OSF[b] (per mg/kg-day)	PPRTV Cancer WOE character-ization[c]	ATSDR Chronic oral MRL[d] (mg/kg-day)	HHBP Chronic RfD[a] (mg/kg-day)	NPDWRs Public health goal[e] (MCLG) (mg/L)	MCL[f] (mg/L)
Cobalt	7440-48-4	x		--	--		0.0003	--	"Likely to be carcinogenic to humans"	--	--	--	--
Copper	7440-50-8	x		--	--	--	--	--	--	--	--	1.3	TT; Action Level=1.3 g
Dibenzothio-phene	132-65-0		x	--	--	--	0.01	--	--	--	--	--	--
Fluoride	16984-48-8	x		--	--	--	--	--	--	--	--	4	4
Formic acid	64-18-6		x	--	--	--	0.9	--	"Data are inadequate for an assessment of human carcinogenic potential"	--	--	--	--
Hydrazine	302-01-2			--	3	B2 (Probable human carcinogen)	--	--	--	--	--	--	--
Iron	7439-89-6	x		--	--	--	0.7	--	"Data are inadequate for an assessment of human carcinogenic potential"	--	--	--	--

Appendix G – Identification and Hazard Evaluation of Chemicals across the Hydraulic Fracturing Water Cycle Supplemental Information

Chemical name	CASRN	Concentration data available	Physico-chemical data available	IRIS Chronic RfD[a] (mg/kg-day)	IRIS OSF[b] (per mg/kg-day)	IRIS Cancer WOE characterization[c]	PPRTV Chronic RfD[a] (mg/kg-day)	PPRTV OSF[b] (per mg/kg-day)	PPRTV Cancer WOE characterization[c]	ATSDR Chronic oral MRL[d] (mg/kg-day)	HHBP Chronic RfD[a] (mg/kg-day)	NPDWRs Public health goal[e] (MCLG) (mg/L)	NPDWRs MCL[f] (mg/L)
Lead	7439-92-1	x		--	--	B2 (Probable human carcinogen)	--	--	--	--	--	0	TT; Action Level=0.0 15 [g]
Lithium	7439-93-2	x		--	--	--	0.002	--	"Data are inadequate for an assessment of human carcinogenic potential"	--	--	--	--
m,p-Cresol mixture	NOCAS_ 24858			--	--	--	--	--	--	0.1	--	--	--
m-Xylene	108-38-3		x	--	--	--	--	--	--	0.2	--	10	10
N,N-Dimethyl-formamide	68-12-2		x	--	--	--	0.1	--	"Data are inadequate for an assessment of human carcinogenic potential"	--	--	--	--
N-Nitrosodi-phenylamine	86-30-6	x	x	--	0.0049	B2 (Probable human carcinogen)	--	--	--	--	--	--	--
N-Nitroso-N-methylethyl-amine	10595-95-6	x	x	--	22	B2 (Probable human carcinogen)	--	--	--	--	--	--	--

Appendix G – Identification and Hazard Evaluation of Chemicals across the Hydraulic Fracturing Water Cycle Supplemental Information

Chemical name	CASRN	Concentration data available	Physico-chemical data available	IRIS Chronic RfD[a] (mg/kg-day)	IRIS OSF[b] (per mg/kg-day)	IRIS Cancer WOE characterization[c]	PPRTV Chronic RfD[a] (mg/kg-day)	PPRTV OSF[b] (per mg/kg-day)	PPRTV Cancer WOE characterization[c]	ATSDR Chronic oral MRL[d] (mg/kg-day)	HHBP Chronic RfD[a] (mg/kg-day)	NPDWRs Public health goal[e] (MCLG) (mg/L)	NPDWRs MCL[f] (mg/L)
Nonane	111-84-2		x	--	--	--	0.0003	--	"Data are inadequate for an assessment of human carcinogenic potential"	--	--	--	--
o-Xylene	95-47-6		x	--	--	--	--	--	--	0.2	--	10	10
p,p'-DDE	72-55-9		x	--	0.34	B2 (Probable human carcinogen)	--	--	--	--	--	--	--
Phorate	298-02-2		x	--	--	--	--	--	--	--	0.0005	--	--
p-Xylene	106-42-3		x	--	--	--	--	--	--	0.2	--	10	10
Quinoline	91-22-5		x	--	3	"Likely to be carcinogenic in humans"	--	--	--	--	--	--	--
Radium	7440-14-4	x		--	--	--	--	--	--	--	--	--	5 pCi/L
Radium-226	13982-63-3	x		--	--	--	--	--	--	--	--	--	5 pCi/L
Radium-228	15262-20-1	x		--	--	--	--	--	--	--	--	--	5 pCi/L
Thallium	7440-28-0	x		--	--	--	--	--	--	--	--	0.0005	0.002
Tributyl phosphate	126-73-8	x	x	--	--	--	0.01	0.009	"Likely to be carcinogenic to humans"	0.08	--	--	--
Uranium-235	15117-96-1	x		--	--	--	--	--	--	--	--	--	30
Uranium-238	7440-61-1	x		--	--	--	--	--	--	--	--	--	30

Chemical name	CASRN	Concentration data available	Physico-chemical data available	IRIS			PPRTV			ATSDR	HHBP	NPDWRs	
				Chronic RfD[a] (mg/kg-day)	OSF[b] (per mg/kg-day)	Cancer WOE character-ization[c]	Chronic RfD[a] (mg/kg-day)	OSF[b] (per mg/kg-day)	Cancer WOE character-ization[c]	Chronic oral MRL[d] (mg/kg-day)	Chronic RfD[a] (mg/kg-day)	Public health goal[e] (MCLG) (mg/L)	MCL[f] (mg/L)
Vanadium	7440-62-2	x		--	--	"Data are inadequate to assess carcinogenic potential"	0.00007	--	"Data are inadequate for an assessment of human carcinogenic potential"	--	--	--	--

CASRN = Chemical Abstract Service Registry Number; IRIS = Integrated Risk Information System; PPRTV = Provisional Peer Reviewed Toxicity Values; ATSDR = Agency for Toxic Substances and Disease Registry; HHBP = Human Health Benchmarks for Pesticides; NPDWRs = National Primary Drinking Water Regulations.

[a] Reference dose (RfD): An estimate (with uncertainty spanning perhaps an order of magnitude) of a daily oral exposure to the human population (including sensitive subgroups) that is likely to be without an appreciable risk of deleterious effects during a lifetime. It can be derived from a no observed-adverse-effect level (NOAEL), lowest observed-adverse-effect level (LOAEL), or benchmark dose (BMD), with uncertainty factors generally applied to reflect limitations of the data used. The RfD is generally used in the EPA's noncancer health assessments. Chronic RfD: Duration of exposure is up to a lifetime.

[b] Oral slope factor (OSF): An upper-bound, approximating a 95% confidence limit, on the increased cancer risk from a lifetime oral exposure to an agent. This estimate, usually expressed in units of proportion (of a population) affected per mg/kg day, is generally reserved for use in the low dose region of the dose response relationship, that is, for exposures corresponding to risks less than 1 in 100.

[c] Weight of evidence (WOE) characterization for carcinogenicity: A system used for characterizing the extent to which the available data support the hypothesis that an agent causes cancer in humans. See glossary for details.

[d] Minimal risk level (MRL): An ATSDR estimate of daily human exposure to a hazardous substance at or below which the substance is unlikely to pose a measurable risk of harmful (adverse), noncancerous effects. MRLs are calculated for a route of exposure (inhalation or oral) over a specified time period (acute, intermediate, or chronic). MRLs should not be used as predictors of harmful (adverse) health effects. Chronic MRL: Duration of exposure is 365 days or longer.

[e] Maximum contaminant level goal (MCLG): A non-enforceable health benchmark goal which is set at a level at which no known or anticipated adverse effect on the health of persons is expected to occur and which allows for an adequate margin of safety.

[f] Maximum contaminant level (MCL): The highest level of a contaminant that is allowed in drinking water. MCLs are set as close to the MCLG as feasible using the best available analytical and treatment technologies and taking cost into consideration. MCLs are enforceable standards.

[g] In public water systems, lead and copper are regulated by a Treatment Technique (TT) that requires systems to control the corrosiveness of their water. If more than 10% of tap water exceeds the action level, water systems must take additional steps. For copper, the action level is 1.3 mg/l, and for lead is 0.015 mg/l.

Table G-2b. Chemicals reported to be detected in produced water, with available chronic oral RfVs and OSFs from state sources.

Chemicals are ranked by CalEPA maximum allowable daily level (MADL). The "--" symbol indicates that no value was available from the sources consulted. Additionally, an "x" indicates the availability of measured concentration data in produced water (see Appendix E) and physicochemical properties data from EPI Suite™ (see Appendix C). Italicized chemicals are found in both hydraulic fracturing fluids and produced water.

Chemical name	CASRN	Concen-tration data available	Physico-chemical data available	CalEPA Oral MADL[a] (µg/day)	CalEPA OSF[b] (per mg/kg-day)
Lead	7439-92-1	x		0.5	0.0085
Cadmium	7440-43-9	x		4.1	15
Chromium (VI)	18540-29-9	x		8.2	0.5
Dibutyl phthalate	84-74-2	x	x	8.7	--
Benzene	71-43-2	x	x	24	0.1
Benzyl butyl phthalate	85-68-7	x	x	1200	--
1,2-Benzenedicarboxylic acid, 1,2-bis(8-methylnonyl) ester	89-16-7		x	2200	--
Diisodecyl phthalate	26761-40-0		x	2,200	--
Di(2-ethylhexyl) phthalate	117-81-7	x	x	20 (neonate male); 58 (infant male); 410 (adult)	0.003
1,2,4-Trichlorobenzene	120-82-1		x	--	0.0036
1,4-Dioxane	123-91-1	x	x	--	0.027
7,12-Dimethylbenz(a)anthracene	57-97-6		x	--	250
Acrylonitrile	107-13-1		x	--	1
Aldrin	309-00-2		x	--	17
Arsenic	7440-38-2	x		--	9.5
Benz(a)anthracene	56-55-3	x	x	--	1.2
Benzidine	92-87-5	x	x	--	500
Benzo(a)pyrene	50-32-8	x	x	--	2.9
Benzo(b)fluoranthene	205-99-2	x	x	--	1.2
Benzo(k)fluoranthene	207-08-9	x	x	--	1.2
Benzyl chloride	100-44-7		x	--	0.17

Chemical name	CASRN	Concen-tration data available	Physico-chemical data available	CalEPA	
				Oral MADL[a] (μg/day)	OSF[b] (per mg/kg-day)
beta-Hexachlorocyclohexane	319-85-7		x	--	1.5
Bis(2-chloroethyl) ether	111-44-4		x	--	2.5
Bromodichloromethane	75-27-4		x	--	0.13
Bromoform	75-25-2		x	--	0.011
Chloroform	67-66-3	x	x	--	0.019
Chrysene	218-01-9	x	x	--	0.12
Dibenz(a,h)anthracene	53-70-3	x	x	--	4.1
Dichloromethane	75-09-2		x	--	0.014
Dieldrin	60-57-1		x	--	16
Ethylbenzene	100-41-4	x	x	--	0.011
Heptachlor	76-44-8		x	--	4.1
Heptachlor epoxide	1024-57-3		x	--	5.5
Hydrazine	302-01-2			--	3
Indeno(1,2,3-cd)pyrene	193-39-5	x	x	--	1.2
Lindane	58-89-9		x	--	1.1
N-Nitrosodiphenylamine	86-30-6	x	x	--	0.009
N-Nitroso-N-methylethylamine	10595-95-6	x	x	--	22
p,p'-DDE	72-55-9		x	--	0.34
Safrole	94-59-7		x	--	0.22
Tetrachloroethene	127-18-4		x	--	0.051

CASRN = Chemical Abstract Service Registry Number; CalEPA = California Environmental Protection Agency.

[a] Maximum allowable daily level (MADL): The maximum allowable daily level of a reproductive toxicant at which the chemical would have no observable adverse reproductive effect, assuming exposure at 1,000 times that level.

[b] Oral slope factor (OSF): An upper-bound, approximating a 95% confidence limit, on the increased cancer risk from a lifetime oral exposure to an agent. This estimate, usually expressed in units of proportion (of a population) affected per mg/kg day, is generally reserved for use in the low-dose region of the dose-response relationship, that is, for exposures corresponding to risks less than 1 in 100.

Table G-2c. Chemicals reported to be detected in produced water, with available chronic oral RfVs and OSFs from international sources.

Chemicals are ranked by CICAD reference value (TDI- tolerable daily intake). An "x" indicates the availability of measured concentration data in produced water (see Appendix E) and physicochemical properties data from EPI Suite™ (see Appendix C). Italicized chemicals are found in both hydraulic fracturing fluids and produced water.

Chemical name	CASRN	Concentration data available	Physicochemical data available	IPCS CICAD Chronic TDI[a] (mg/kg-day)
Heptachlor	76-44-8		x	0.0001
Chromium (VI)	*18540-29-9*	*x*		*0.0009*
Mercury	7439-97-6	x		0.002
Beryllium	7440-41-7	x		0.002
Iodine	7553-56-2	x		0.01
Chloroform	67-66-3	x	x	0.015
Barium	7440-39-3	x		0.02
Ethylene glycol	*107-21-1*		*x*	*0.05*
Tetrachloroethene	127-18-4		x	0.05
D-Limonene	*5989-27-5*		*x*	*0.1*
Strontium	7440-24-6	x		0.13
Benzyl butyl phthalate	85-68-7	x	x	1.3
Diethyl phthalate	84-66-2		x	5

CASRN = Chemical Abstract Service Registry Number; IPCS = International Programme on Chemical Safety; CICAD = Concise International Chemical Assessment Documents.

[a] Tolerable Daily Intake (TDI): An estimate of the intake of a substance, expressed on a body mass basis, to which an individual in a (sub) population may be exposed daily over its lifetime without appreciable health risk.

Table G-2d. Chemicals reported to be detected in produced water, with available less-than-chronic oral RfVs and OSFs.
Chemicals are ranked by PPRTV subchronic reference dose (sRfD). The "–" symbol indicates that no value was available from the sources consulted. Additionally, an "x" indicates the availability of measured concentration data in produced water (see Appendix E) and physicochemical properties data from EPI Suite™ (see Appendix C). Italicized chemicals are found in both hydraulic fracturing fluids and produced water.

Chemical name	CASRN	Concentration data available	Physico-chemical data available	PPRTV sRfD[a] (mg/kg-day)	ATSDR Acute oral MRL[b] (mg/kg-day)	ATSDR Intermediate oral MRL[b] (mg/kg-day)	HHBP Acute RfD[a] (mg/kg-day)
Aldrin	309-00-2		x	0.00004	0.002	–	–
Antimony	7440-36-0	x		0.0004	–	–	–
Vanadium	7440-62-2	x		0.0007	–	0.01	–
Benzyl chloride	100-44-7		x	0.002	–	–	–
Lithium	7439-93-2	x		0.002	–	–	–
Cobalt	7440-48-4	x		0.003	–	0.01	–
Nonane	111-84-2		x	0.003	–	–	–
2-Methylnaphthalene	91-57-6	x	x	0.004	–	–	–
Methyl bromide	74-83-9		x	0.005	–	0.003	0.02
1,2,3-Trichlorobenzene	87-61-6		x	0.008	–	–	–
Bromodichloromethane	75-27-4		x	0.008	0.04	–	–
Benzene	71-43-2	x	x	0.01	–	–	–
2,4-Dichlorophenol	120-83-2	x	x	0.02	–	0.003	–
p-Cresol	106-44-5	x	x	0.02	–	–	–
Bromoform	75-25-2		x	0.03	0.7	0.2	–
Tributyl phosphate	126-73-8	x	x	0.03	1.1	0.08	–
2,4-Dimethylphenol	105-67-9	x	x	0.05	–	–	–

Appendix G – Identification and Hazard Evaluation of Chemicals across
the Hydraulic Fracturing Water Cycle Supplemental Information

Chemical name	CASRN	Concentration data available	Physico-chemical data available	PPRTV sRfD[a] (mg/kg-day)	ATSDR Acute oral MRL[b] (mg/kg-day)	ATSDR Intermediate oral MRL[b] (mg/kg-day)	HHBP Acute RfD[a] (mg/kg-day)
Ethylbenzene	100-41-4	x	x	0.05	--	0.4	--
Chlorobenzene	108-90-7		x	0.07	--	0.4	--
Chlorodibromomethane	124-48-1		x	0.07	0.1	--	--
1,2,4-Trichlorobenzene	120-82-1		x	0.09	--	0.1	--
Butylbenzene	104-51-8		x	0.1	--	--	--
Fluoranthene	206-44-0	x	x	0.1	--	0.4	--
2-Chloroethanol	107-07-3		x	0.2	--	--	0.045
o-Cresol	95-48-7	x	x	0.2	--	--	--
2-(2-Butoxyethoxy)ethanol	112-34-5		x	0.3	--	--	--
Benzyl alcohol	100-51-6	x	x	0.3	--	--	--
Hexane	110-54-3		x	0.3	--	--	--
N,N-Dimethylformamide	68-12-2		x	0.3	--	--	--
Pyrene	129-00-0	x	x	0.3	--	--	--
Xylenes	1330-20-7	x	x	0.4	1	0.4	--
Iron	7439-89-6	x	x	0.7	--	--	--
Toluene	108-88-3	x	x	0.8	0.8	0.02	--
Formic acid	64-18-6		x	0.9	--	--	--
1,1-Dichloroethane	75-34-3		x	2	--	--	--
1,2-Propylene glycol	57-55-6		x	20	--	--	--
1,4-Dioxane	123-91-1	x	x	--	5	0.5	--
2-Butoxyethanol	111-76-2		x	--	0.4	0.07	--

Appendix G – Identification and Hazard Evaluation of Chemicals across the Hydraulic Fracturing Water Cycle Supplemental Information

Chemical name	CASRN	Concentration data available	Physico-chemical data available	PPRTV sRfD[a] (mg/kg-day)	ATSDR Acute oral MRL[b] (mg/kg-day)	Intermediate oral MRL[b] (mg/kg-day)	HHBP Acute RfD[a] (mg/kg-day)
Acetone	67-64-1	x	x	--	--	2	--
Acrolein	107-02-8		x	--	--	0.004	--
Acrylonitrile	107-13-1		x	--	0.1	0.01	--
Aluminum	7429-90-5	x		--	--	1	--
Arsenic	7440-38-2	x		--	0.005	--	--
Barium	7440-39-3	x		--	--	0.2	--
beta-Hexachlorocyclohexane	319-85-7		x	--	0.05	0.0006	--
Boron	7440-42-8	x		--	0.2	0.2	--
Cadmium	7440-43-9	x		--	--	0.0005	--
Carbon disulfide	75-15-0	x	x	--	0.01	--	--
Chloroform	67-66-3	x	x	--	0.3	0.1	--
Chromium (VI)	18540-29-9	x		--	--	0.005	--
Copper	7440-50-8	x		--	0.01	0.01	--
Di(2-ethylhexyl) phthalate	117-81-7	x	x	--	--	0.1	--
Dibutyl phthalate	84-74-2	x	x	--	0.5	--	--
Dichloromethane	75-09-2		x	--	0.2	--	--
Dieldrin	60-57-1		x	--	--	0.0001	--
Diethyl phthalate	84-66-2		x	--	7	6	--
Diethyltoluamide	134-62-3		x	--	--	1	--
Dioctyl phthalate	117-84-0	x	x	--	3	0.4	--
Ethylene glycol	107-21-1		x	--	0.8	0.8	--

Appendix G – Identification and Hazard Evaluation of Chemicals across the Hydraulic Fracturing Water Cycle Supplemental Information

Chemical name	CASRN	Concentration data available	Physico-chemical data available	PPRTV sRfD[a] (mg/kg-day)	ATSDR Acute oral MRL[b] (mg/kg-day)	ATSDR Intermediate oral MRL[b] (mg/kg-day)	HHBP Acute RfD[a] (mg/kg-day)
Fluorene	86-73-7	x	x	--	--	0.4	--
Heptachlor	76-44-8		x	--	0.0006	0.0001	--
Lindane	58-89-9		x	--	0.003	0.00001	--
m,p-Cresol mixture	NOCAS_24858			--	--	0.1	--
m-Xylene	108-38-3		x	--	1	0.4	--
Naphthalene	91-20-3	x	x	--	0.6	0.6	--
o-Xylene	95-47-6		x	--	1	0.4	--
Phenol	108-95-2	x	x	--	1	--	--
Phosphorus	7723-14-0	x		--	--	0.0002	--
p-Xylene	106-42-3		x	--	1	0.4	--
Strontium	7440-24-6	x		--	--	2	--
Tetrachloroethene	127-18-4		x	--	0.008	0.008	--
Tin	7440-31-5	x		--	--	0.3	--
Zinc	7440-66-6	x		--	--	0.3	--

CASRN = Chemical Abstract Service Registry Number; PPRTV = Provisional Peer Reviewed Toxicity Values; ATSDR = Agency for Toxic Substances and Disease Registry; HHBP = Human Health Benchmarks for Pesticides.

[a] Reference dose (RfD): An estimate (with uncertainty spanning perhaps an order of magnitude) of a daily oral exposure to the human population (including sensitive subgroups) that is likely to be without an appreciable risk of deleterious effects during a lifetime. It can be derived from a no-observed-adverse-effect level (NOAEL), lowest observed-adverse-effect level (LOAEL), or benchmark dose (BMD), with uncertainty factors generally applied to reflect limitations of the data used. The RfD is generally used in the EPA's noncancer health assessments. Subchronic RfD (sRfD): Duration of exposure is up to 10% of an average lifespan. Acute RfD: Duration of exposure is 24 hours or less.

[b] Minimal risk level (MRL): An ATSDR estimate of daily human exposure to a hazardous substance at or below which the substance is unlikely to pose a measurable risk of harmful (adverse), noncancerous effects. MRLs are calculated for a route of exposure (inhalation or oral) over a specified time period (acute, intermediate, or chronic). MRLs should not be used as predictors of harmful (adverse) health effects. Acute MRL: Duration of exposure is 1 to 14 days. Intermediate MRL: Duration of exposure is >14 to 364 days.

Table G-2e. Available qualitative cancer classifications for chemicals reported to be detected in produced water.

Chemicals classified as known carcinogens by one or more sources are listed first. The "--" symbol indicates that no value was available from the sources consulted. Additionally, an "x" indicates the availability of measured concentration data in produced water (see Appendix E) and physicochemical properties data from EPI Suite™ (see Appendix C). Italicized chemicals are found in both hydraulic fracturing fluids and produced water. Cancer classifications from IRIS and PPRTV are also listed in Table G-2a.

Chemical name	CASRN	Concen- tration data available	Physico- chemical data available	Qualitative cancer classification			
				IRIS[a]	PPRTV[b]	IARC[c]	RoC[d]
Alpha particle	12587-46-1	x		--	--	1	--
Arsenic	*7440-38-2*	x		A (Human carcinogen)	--	1	*Known*
Benzene	*71-43-2*	x	x	A (Human carcinogen)	--	1	*Known*
Benzidine	92-87-5	x	x	A (Human carcinogen)	--	1	Known
Benzo(a)pyrene	50-32-8	x	x	B2 (Probable human carcinogen)	--	1	RAHC
Beryllium	7440-41-7	x		B1 (Probable human carcinogen)	--	1	Known
Beta particle	12587-47-2	x		--	--	1	--
Cadmium	7440-43-9	x		B1 (Probable human carcinogen)	--	1	Known
Chromium (VI)	*18540-29-9*	x		*Inhaled: A (Human carcinogen); Oral: D (Not classifiable as to human carcinogenicity)*	--	1	*Known*
Lindane	58-89-9		x	--	--	1	RAHC
Radium-226	13982-63-3	x		--	--	1	--
Radium-228	15262-20-1	x		--	--	1	--
Ethanol	*64-17-5*		x	--	--	*1*	--
Radium	7440-14-4	x		--	--	1	--
1,2,4-Trichlorobenzene	120-82-1		x	D (Not classifiable as to human carcinogenicity)	"Likely to be carcinogenic to humans"	--	--

Chemical name	CASRN	Concen-tration data available	Physico-chemical data available	Qualitative cancer classification			
				IRIS[a]	PPRTV[b]	IARC[c]	RoC[d]
1,2-Diphenyl-hydrazine	122-66-7	x	x	B2 (Probable human carcinogen)	--	--	RAHC
1,4-Dioxane	123-91-1	x	x	"Likely to be carcinogenic to humans"	--	2B	RAHC
2-Mercapto-benzothiazole	149-30-4	x	x	--	--	2A	--
Acrylonitrile	107-13-1		x	B1 (Probable human carcinogen)	--	2B	RAHC
Aldrin	309-00-2		x	B2 (Probable human carcinogen)	--	3	--
Benz(a)anthra-cene	56-55-3	x	x	B2 (Probable human carcinogen)	--	2B	RAHC
Benzo(b)fluoran-thene	205-99-2	x	x	--	--	2B	RAHC
Benzo(k)fluoran-thene	207-08-9	x	x	--	--	2B	RAHC
Benzophenone	119-61-9		x	--	--	2B	--
Benzyl butyl phthalate	85-68-7	x	x	C (Possible human carcinogen)	--	3	--
beta-Hexachloro-cyclohexane	319-85-7		x	C (Possible human carcinogen)	--	--	--
Biphenyl	92-52-4	x	x	"Suggestive evidence of carcinogenic potential"	--	--	--
Bis(2-chloroethyl) ether	111-44-4		x	B2 (Probable human carcinogen)	--	3	--
Bromodichloro-methane	75-27-4		x	B2 (Probable human carcinogen)	--	2B	RAHC
Bromoform	75-25-2		x	B2 (Probable human carcinogen)	--	3	--
Chlorodibromo-methane	124-48-1		x	C (Possible human carcinogen)	--	3	--

Chemical name	CASRN	Concen-tration data available	Physico-chemical data available	Qualitative cancer classification			
				IRIS[a]	PPRTV[b]	IARC[c]	RoC[d]
Chloroform	67-66-3	x	x	B2 (Probable human carcinogen)	--	2B	RAHC
Chrysene	218-01-9	x	x	B2 (Probable human carcinogen)	--	2B	--
Cobalt	7440-48-4	x		--	"Likely to be carcinogenic to humans"	2B	--
Cumene	*98-82-8*	*x*	*x*	*D (Not classifiable as to human carcinogenicity)*	--	*2B*	*RAHC*
Di(2-ethylhexyl) phthalate	*117-81-7*	*x*	*x*	*B2 (Probable human carcinogen)*	--	*2B*	*RAHC*
Dibenz(a,h)an-thracene	53-70-3	x	x	--	--	2A	RAHC
Dichloromethane	*75-09-2*		*x*	*"Likely to be carcinogenic in humans"*	--	*2A*	*RAHC*
Dieldrin	60-57-1		x	B2 (Probable human carcinogen)	--	3	--
Ethylbenzene	*100-41-4*	*x*	*x*	*D (Not classifiable as to human carcinogenicity)*	--	*2B*	--
Heptachlor	76-44-8		x	B2 (Probable human carcinogen)	--	2B	--
Heptachlor epoxide	1024-57-3		x	B2 (Probable human carcinogen)	--	--	--
Indeno(1,2,3-cd)pyrene	193-39-5	x	x	--	--	2B	RAHC
Lead	*7439-92-1*	*x*		*B2 (Probable human carcinogen)*	--	*2B*	*RAHC*
m-Cresol	108-39-4	x	x	C (Possible human carcinogen)	--	--	--

Chemical name	CASRN	Concen-tration data available	Physico-chemical data available	Qualitative cancer classification			
				IRIS[a]	PPRTV[b]	IARC[c]	RoC[d]
Naphthalene	*91-20-3*	x	x	"Data are inadequate to assess human carcinogenic potential"	--	2B	RAHC
Nickel	7440-02-0	x		--	--	2B	RAHC
Nitrate	14797-55-8	x		--	--	2A	--
Nitrite	14797-65-0	x		--	--	2A	--
N-Nitrosodiphen-ylamine	86-30-6	x	x	B2 (Probable human carcinogen)	--	3	--
N-Nitroso-N-methylethylamine	10595-95-6	x	x	B2 (Probable human carcinogen)	--	2B	--
o-Cresol	95-48-7	x	x	C (Possible human carcinogen)	"Data are inadequate for an assessment of human carcinogenic potential"	--	--
p-Cresol	106-44-5	x	x	C (Possible human carcinogen)	--	--	--
p,p'-DDE	72-55-9		x	B2 (Probable human carcinogen)	--	--	--
Safrole	94-59-7		x	--	--	2B	RAHC
Tetrachloroeth-ene	127-18-4		x	"Likely to be carcinogenic in humans"	--	2A	RAHC
1,1-Dichloroethane	75-34-3		x	C (Possible human carcinogen)	--	--	--
Benzyl chloride	*100-44-7*		x	*B2 (Probable human carcinogen)*	--	--	--
Hydrazine	*302-01-2*			*B2 (Probable human carcinogen)*	--	2A	RAHC

Chemical name	CASRN	Concentration data available	Physico-chemical data available	Qualitative cancer classification			
				IRIS[a]	PPRTV[b]	IARC[c]	RoC[d]
N,N-Dimethylfor-mamide	68-12-2		x	--	"Data are inadequate for an assessment of human carcinogenic potential"	2A	--
Quinoline	91-22-5		x	"Likely to be carcinogenic in humans"	--	--	--
Tributyl phosphate	126-73-8	x	x	--	"Likely to be carcinogenic to humans"	--	--
Dibromoaceto-nitrile	3252-43-5		x	--	--	2B	--
Acetaldehyde	75-07-0		x	B2 (Probable human carcinogen)	--	2B	RAHC
1,2,3-Trichloro-benzene	87-61-6		x	--	"Data are inadequate for an assessment of human carcinogenic potential"	--	--
1,3,5-Trimethylbenzene	108-67-8	x	x	--	"Data are inadequate for an assessment of human carcinogenic potential"	--	--
2,4-Dichlorophenol	120-83-2	x	x	--	"Data are inadequate for an assessment of human carcinogenic potential"	--	--
2,4-Dimethylphenol	105-67-9	x	x	--	"Data are inadequate for an assessment of human carcinogenic potential"	--	--

Chemical name	CASRN	Concentration data available	Physicochemical data available	Qualitative cancer classification			
				IRIS[a]	PPRTV[b]	IARC[c]	RoC[d]
2,5-Cyclohexa-diene-1,4-dione	106-51-4	x	x	--	--	3	--
2-Methylnaph-thalene	91-57-6	x	x	"Data are inadequate to assess human carcinogenic potential"	--	--	--
Acetone	67-64-1	x	x	"Data are inadequate for an assessment of human carcinogenic potential"	--	--	--
Acetophenone	98-86-2	x	x	D (Not classifiable as to human carcinogenicity)	--	--	--
Acrolein	107-02-8		x	"Data are inadequate for an assessment of human carcinogenic potential"	--	3	--
Aluminum	7429-90-5	x		--	"Data are inadequate for an assessment of human carcinogenic potential"	--	--
Antimony	7440-36-0	x		--	"Data are inadequate for an assessment of human carcinogenic potential"	--	--
Benzo(g,h,i)peryl-ene	191-24-2	x	x	--	--	3	--

Chemical name	CASRN	Concen- tration data available	Physico- chemical data available	Qualitative cancer classification			
				IRIS[a]	PPRTV[b]	IARC[c]	RoC[d]
Benzyl alcohol	100-51-6	x	x	--	"Data are inadequate for an assessment of human carcinogenic potential"	--	--
Boron	*7440-42-8*	*x*		*"Data are inadequate to assess the carcinogenic potential"*	--	--	--
Butylbenzene	104-51-8		x	--	"Data are inadequate for an assessment of human carcinogenic potential"	--	--
Caffeine	58-08-2	x	x	--	--	3	--
Chloromethane	*74-87-3*		x	*"Carcinogenic potential cannot be determined"*	--	*3*	--
Cholesterol	57-88-5	x	x	--	--	3	--
Chromium	7440-47-3			--	--	3	--
Chromium (III)	*16065-83-1*	*x*		*"Data are inadequate for an assessment of human carcinogenic potential"*	--	*3*	--
Cyanide	57-12-5		x	*"Inadequate information to assess the carcinogenic potential"*	--	--	--
delta- Hexachlorocyclo- hexane	319-86-8		x	D (Not classifiable as to human carcinogenicity)	--	--	--

Chemical name	CASRN	Concentration data available	Physico-chemical data available	Qualitative cancer classification			
				IRIS[a]	PPRTV[b]	IARC[c]	RoC[d]
Dibenzothiophene	132-65-0		x	--	"Data are inadequate for an assessment of human carcinogenic potential"	3	--
Dibutyl phthalate	84-74-2	x	x	D (Not classifiable as to human carcinogenicity)	--	--	--
Diethyl phthalate	84-66-2		x	D (Not classifiable as to human carcinogenicity)	--	--	--
Dimethyl phthalate	131-11-3	x	x	D (Not classifiable as to human carcinogenicity)	--	--	--
Diphenylamine	122-39-4	x	x	--	"Data are inadequate for an assessment of human carcinogenic potential"	--	--
Fluoranthene	206-44-0	x	x	D (Not classifiable as to human carcinogenicity)	"Data are inadequate for an assessment of human carcinogenic potential"	3	--
Fluorene	86-73-7	x	x	D (Not classifiable as to human carcinogenicity)	--	3	--
Fluoride	16984-48-8	x		--	--	3	--
Formic acid	*64-18-6*		*x*	--	*"Data are inadequate for an assessment of human carcinogenic potential"*	--	--

Chemical name	CASRN	Concen-tration data available	Physico-chemical data available	Qualitative cancer classification			
				IRIS[a]	PPRTV[b]	IARC[c]	RoC[d]
Iron	7439-89-6	x		--	"Data are inadequate for an assessment of human carcinogenic potential"	--	--
Isopropanol	67-63-0		x	--	--	3	--
Lithium	7439-93-2	x		--	"Data are inadequate for an assessment of human carcinogenic potential"	--	--
Manganese	7439-96-5	x		D (Not classifiable as to human carcinogenicity)	--	--	--
Mercury	7439-97-6	x		D (Not classifiable as to human carcinogenicity)	--	3	--
Methyl bromide	74-83-9		x	D (Not classifiable as to human carcinogenicity)	"Data are inadequate for an assessment of human carcinogenic potential"	3	--
Methyl ethyl ketone	78-93-3		x	"Data are inadequate to assess carcinogenic potential"	--	--	--
Perylene	198-55-0		x	--	--	3	--
Phenanthrene	85-01-8	x	x	--	--	3	--
Phenol	108-95-2	x	x	"Data are inadequate for an assessment of human carcinogenic potential"	--	3	--

Chemical name	CASRN	Concen-tration data available	Physico-chemical data available	Qualitative cancer classification			
				IRIS[a]	PPRTV[b]	IARC[c]	RoC[d]
Phosphorus	7723-14-0	x		D (Not classifiable as to human carcinogenicity)	--	--	--
Pyrene	129-00-0	x	x	D (Not classifiable as to human carcinogenicity)	--	3	--
Pyridine	110-86-1	x	x	--	--	3	--
Selenium	7782-49-2	x		D (Not classifiable as to human carcinogenicity)	--	3	--
Silica	*7631-86-9*			--	--	*3*	--
Silver	7440-22-4	x		D (Not classifiable as to human carcinogenicity)	--	--	--
Toluene	*108-88-3*	*x*	*x*	*"Inadequate information to assess the carcinogenic potential"*	--	*3*	--
Vanadium	7440-62-2	x		*"Data are inadequate to assess the carcinogenic potential"*	"Data are inadequate for an assessment of human carcinogenic potential"	--	--
Xylenes	*1330-20-7*	*x*	*x*	*"Data are inadequate to assess the carcinogenic potential"*	--	*3*	--
Zinc	*7440-66-6*	*x*		*"Inadequate information to assess carcinogenic potential"*	--	--	--

Chemical name	CASRN	Concen-tration data available	Physico-chemical data available	Qualitative cancer classification			
				IRIS[a]	PPRTV[b]	IARC[c]	RoC[d]
2-Chloroethanol	107-07-3		x	--	"Data are inadequate for an assessment of human carcinogenic potential"	--	--
Nonane	111-84-2		x	--	"Data are inadequate for an assessment of human carcinogenic potential"	--	--
2-Butoxyethanol	111-76-2		x	"Not likely to be carcinogenic to humans"	--	3	--
D-Limonene	5989-27-5		x	--	--	3	--
Chlorobenzene	108-90-7		x	D (Not classifiable as to human carcinogenicity)	--	--	--
1-Butanol	71-36-3		x	D (Not classifiable as to human carcinogenicity)	--	--	--
Hydrochloric acid	7647-01-0			--	--	3	--
2-(2-Butoxyeth-oxy)ethanol	112-34-5		x	--	"Data are inadequate for an assessment of human carcinogenic potential"	--	--
Hexane	110-54-3		x	"Inadequate information to assess the carcinogenic potential"	--	--	--

Chemical name	CASRN	Concen- tration data available	Physico- chemical data available	Qualitative cancer classification			
				IRIS[a]	PPRTV[b]	IARC[c]	RoC[d]
Pentane	109-66-0		x	--	"Data are inadequate for an assessment of human carcinogenic potential"	--	--
1,2,3-Trimethyl-benzene	526-73-8		x	--	"Data are inadequate for an assessment of human carcinogenic potential"	--	--
1,2-Propylene glycol	57-55-6		x	--	"Not likely to be carcinogenic to humans"	--	--
Barium	7440-39-3	x		"Not likely to be carcinogenic to humans"	--	--	--
Caprolactam	105-60-2	x	x	--	--	4	--

CASRN = Chemical Abstract Service Registry Number; IRIS = Integrated Risk Information System; PPRTV = Provisional Peer Reviewed Toxicity Values; IARC = International Agency for Research on Cancer Monographs; RoC = National Toxicology Program 13th Report on Carcinogens.

[a] IRIS assessments use the EPA's 1986, 1996, 1999, or 2005 guidelines to establish descriptors for summarizing the weight of evidence as to whether a contaminant is or may be carcinogenic. See glossary in Appendix G for details.

[b] PPRTV assessments use the EPA's 1999 guidelines to establish descriptors for summarizing the weight of evidence as to whether a contaminant is or may be carcinogenic. See glossary in Appendix G for details.

[c] The IARC summarizes the weight of evidence as to whether a contaminant is or may be carcinogenic using five weight of evidence classifications: Group 1: Carcinogenic to humans; Group 2A: Probably carcinogenic to humans; Group 2B: Possibly carcinogenic to humans; Group 3: Not classifiable as to its carcinogenicity to humans; Group 4: Probably not carcinogenic to humans. See glossary in Appendix G for details.

[d] The listing criteria in the 13th RoC Document are: Known = Known to be a human carcinogen; RAHC = Reasonably anticipated to be a human carcinogen.

Appendix H. Chemicals Identified in Hydraulic Fracturing Fluids and/or Produced Water

This page is intentionally left blank.

Appendix H. Chemicals Identified in Hydraulic Fracturing Fluids and/or Produced Water

H.1. Supplemental Tables and Information

The EPA identified authoritative sources for information on hydraulic fracturing-related chemicals and, to the extent possible, verified the chemicals used in hydraulic fracturing fluids and detected in produced water of hydraulically fractured wells. The EPA used 28 sources of information to identify the chemicals used in hydraulic fracturing fluids or detected in produced water of hydraulically fracturing wells. The sources include compilations of industry-provided data (all Toxic Substance Control Act (TSCA) confidential business information (CBI) chemical lists handled by the EPA were managed in accordance with TSCA CBI procedures); publications that represent collaborations between state, non-profit, academic, and/or industry groups; and peer-reviewed journal articles. Most of the listed chemicals were cited by multiple sources.

Seven of the 28 sources obtained information about the chemicals used in hydraulic fracturing fluids from Material Safety Data Sheets (MSDSs) provided by chemical manufacturers for the products they sell, as required by the Occupational Safety and Health Administration (OSHA). The MSDSs must list all hazardous ingredients if they comprise at least 1% of the product; for carcinogens, the reporting threshold is 0.1%. However, chemical manufacturers may withhold information (e.g., chemical name, concentration of the substance in a mixture) about a hazardous substance from MSDSs if it is claimed as confidential business information (CBI), provided that certain conditions are met (OSHA, 2013).

Table H-1. Sources used to create lists of chemicals used in fracturing fluids or detected in produced water.
The number next to each citation in the reference column corresponds to numbers in the reference columns found in Table H-2, Table H-3, Table H-4, and Table H-5.

Reference	Citation
House of Representatives (U.S. House of Representatives). (2011). Chemicals used in hydraulic fracturing. Washington, D.C.: U.S. House of Representatives, Committee on Energy and Commerce, Minority Staff. http://www.conservation.ca.gov/dog/general_information/Documents/Hydraulic%ic-Fracturing-Chemicals-2011-4-18.pdf.	House of Representatives (2011) [a] (1)
Colborn, T; Kwiatkowski, C; Schultz, K; Bachran, M. (2011). Natural gas operations from a public health perspective. Hum Ecol Risk Assess 17: 1039-1056. http://dx.doi.org/10.1080/10807039.2011.605662.	Colborn et al. (2011) [a] (2)
NYSDEC (New York State Department of Environmental Conservation). (2011). Revised draft supplemental generic environmental impact statement (SGEIS) on the oil, gas and solution mining regulatory program: Well permit issuance for horizontal drilling and high-volume hydraulic fracturing to develop the Marcellus shale and other low-permeability gas reservoirs. Albany, NY: NY SDEC. http://www.dec.ny.gov/energy/75370.html.	NYSDEC (2011) [a,b] (3)

Reference	Citation
U.S. EPA (U.S. Environmental Protection Agency). (2013). Data received from oil and gas exploration and production companies, including hydraulic fracturing service companies 2011 to 2013. Non-confidential business information source documents are located in Federal Docket ID: EPA-HQ-ORD2010-0674. Available at http://www.regulations.gov.	U.S. EPA (2013a) [a] (4)
Material Safety Data Sheets. (a) Encana/Halliburton Energy Services, Inc.: Duncan, Oklahoma. Provided by Halliburton Energy Services during an onsite visit by the EPA on May 10, 2010; (b) Encana Oil and Gas (USA), Inc.: Denver, Colorado. Provided to US EPA Region 8.	Material Safety Data Sheets [a] (5)
U.S. EPA (U.S. Environmental Protection Agency). (2004). Evaluation of impacts to underground sources of drinking water by hydraulic fracturing of coalbed methane reservoirs. (EPA/816-R-04/003). Washington, DC.: U.S. Environmental Protection Agency, Office of Solid Waste.	U.S. EPA (2004) [a] (6)
PA DEP (Pennsylvania Department of Environmental Protection). (2010). Chemicals used by hydraulic fracturing companies in Pennsylvania for surface and hydraulic fracturing activities. Harrisburg, PA: Pennsylvania Department of Environmental Protection (PADEP). http://files.dep.state.pa.us/OilGas/BOGM/BOGMPortalFiles/MarcellusShale/Frac%20list%206-30-2010.pdf.	PA DEP (2010) [a] (7)
U.S. EPA (U.S. Environmental Protection Agency). (2015). Analysis of hydraulic fracturing fluid data from the FracFocus chemical disclosure registry 1.0: Project database [EPA Report]. (EPA/601/R-14/003). Washington, D.C.: U.S. Environmental Protection Agency, Office of Research and Development. http://www2.epa.gov/hfstudy/epa-project-database-developed-fracfocus-1-disclosures.	U.S. EPA (2015c) [a] (8)
Hayes, T. (2009). Sampling and analysis of water streams associated with the development of Marcellus shale gas. Des Plaines, IL: Marcellus Shale Coalition http://energyindepth.org/wp-content/uploads/marcellus/2012/11/MSCommission-Report.pdf.	Hayes (2009) [b] (9)
U.S. EPA (U.S. Environmental Protection Agency). (2011). Sampling data for flowback and produced water provided to EPA by nine oil and gas well operators (non-confidential business information). US Environmental Protection Agency. http://www.regulations.gov/#!docketDetail;rpp=100;so=DESC;sb=docId;po=0;D=EPA-HQ-ORD-2010-0674.	U.S. EPA (2011b) [b] (10)
Akob, DM; Cozzarelli, IM; Dunlap, DS; Rowan, EL; Lorah, MM. (2015). Organic and inorganic composition and microbiology of produced waters from Pennsylvania shale gas wells. Appl Geochem 60: 116-125. http://dx.doi.org/10.1016/j.apgeochem.2015.04.011.	Akob et al. (2015) [b] (11)
Cluff, M; Hartsock, A; Macrae, J; Carter, K; Mouser, PJ. (2014). Temporal changes in microbial ecology and geochemistry in produced water from hydraulically fractured Marcellus Shale Gas Wells. Environ Sci Technol 48: 6508-6517. http://dx.doi.org/10.1021/es501173p.	Cluff et al. (2014) [b] (12)

Reference	Citation
Digiulio, DC; Jackson, RB. (2016). Impact to underground sources of drinking water and domestic wells from production well stimulation and completion practices in the Pavillion, Wyoming, Field. Environ Sci Technol 50: 4524-4536. http://dx.doi.org/10.1021/acs.est.5b04970.	Digiulio and Jackson (2016)[b] (13)
Geological Survey of Alabama. (2014). Water management strategies for improved coalbed methane production in the Black Warrior Basin. (DE-FE0000888). Washington, DC: U.S. Department of Energy, National Energy Technology Library. https://www.netl.doe.gov/research/oil-and-gas/natural-gas-resources/00888-geosurveyalabama.	Geological Survey of Alabama (2014)[b] (14)
Hayes, T; Severin, B. (2012). Characterization of flowback water from the the Marcellus and the Barnett shale regions. Barnett and Appalachian shale water management and reuse technologies. (08122-05.09; Contract 08122-05). Hayes, T; Severin, B. http://www.rpsea.org/media/files/project/2146b3a0/08122-05-RT-Characterization_Flowback_Waters_Marcellus_Barnett_Shale_Regions-03-20-12.pdf.	Hayes and Severin (2012a)[b] (15)
Khan, NA; Engle, M; Dungan, B; Holguin, FO; Xu, P; Carroll, KC. (2016). Volatile-organic molecular characterization of shale-oil produced water from the Permian Basin. Chemosphere 148: 126-136. http://dx.doi.org/10.1016/j.chemosphere.2015.12.116.	Khan et al. (2016)[b] (16)
Lester, Y; Ferrer, I; Thurman, EM; Sitterley, KA; Korak, JA; Aiken, G; Linden, KG. (2015). Characterization of hydraulic fracturing flowback water in Colorado: Implications for water treatment. Sci Total Environ 512-513: 637-644. http://dx.doi.org/10.1016/j.scitotenv.2015.01.043.	Lester et al. (2015)[b] (17)
Maguire-Boyle, SJ; Barron, AR. (2014). Organic compounds in produced waters from shale gas wells. Environ Sci Process Impacts 16: 2237-2248. http://dx.doi.org/10.1039/c4em00376d.	Maguire-Boyle and Barron (2014)[b] (18)
Olsson, O; Weichgrebe, D; Rosenwinkel, KH. (2013). Hydraulic fracturing wastewater in Germany: Composition, treatment, concerns. Environ Earth Sci 70: 3895-3906. http://dx.doi.org/10.1007/s12665-013-2535-4.	Olsson et al. (2013)[b] (19)
Orem, WH; Tatu, CA; Lerch, HE; Rice, CA; Bartos, TT; Bates, AL; Tewalt, S; Corum, MD. (2007). Organic compounds in produced waters from coalbed natural gas wells in the Powder River Basin, Wyoming, USA. Appl Geochem 22: 2240-2256. http://dx.doi.org/10.1016/j.apgeochem.2007.04.010.	Orem et al. (2007)[b] (20)
Orem, W; Tatu, C; Varonka, M; Lerch, H; Bates, A; Engle, M; Crosby, L; McIntosh, J. (2014). Organic substances in produced and formation water from unconventional natural gas extraction in coal and shale. Int J Coal Geol 126: 20-31. http://dx.doi.org/10.1016/j.coal.2014.01.003.	Orem et al. (2014)[b] (21)
Thacker, JB; Carlton, DD, Jr; Hildenbrand, ZL; Kadjo, AF; Schug, KA. (2015). Chemical analysis of wastewater from unconventional drilling operations. Water 7: 1568-1579. http://dx.doi.org/10.3390/w7041568.	Thacker et al. (2015)[b] (22)
Thurman, EM; Ferrer, I; Blotevogel, J; Borch, T. (2014). Analysis of hydraulic fracturing flowback and produced waters using accurate mass: Identification of ethoxylated surfactants. Anal Chem 86: 9653-9661. http://dx.doi.org/10.1021/ac502163k.	Thurman et al. (2014)[b] (23)

Reference	Citation
Rowan, EL; Engle, MA; Kirby, CS; Kraemer, TF. (2011). Radium content of oil- and gas-field produced waters in the northern Appalachian Basin (USA): Summary and discussion of data. (Scientific Investigations Report 20115135). Reston, VA: U.S. Geological Survey. http://pubs.usgs.gov/sir/2011/5135/.	Rowan et al. (2011) [b] (24)
PA DEP (Pennsylvania Department of Environmental Protection). (2015). Technologically enhanced naturally occurring radioactive materials (TENORM) study report. Harrisburg, PA.	PA DEP (2015) [b] (25)
Ziemkiewicz, PF; He, YT. (2015). Evolution of water chemistry during Marcellus Shale gas development: A case study in West Virginia. Chemosphere 134: 224-231. http://dx.doi.org/10.1016/j.chemosphere.2015.04.040.	Ziemkiewicz and He (2015) [b] (26)
Dresel, PE; Rose, AW. (2010). Chemistry and origin of oil and gas well brines in western Pennsylvania (pp. 48). (Open-File Report OFOG 1001.0). Harrisburg, PA: Pennsylvania Geological Survey, 4th ser. http://www.marcellus.psu.edu/resources/PDFs/brines.pdf.	Dresel and Rose (2010) [b] (27)
Barbot, E; Vidic, NS; Gregory, KB; Vidic, RD. (2013). Spatial and temporal correlation of water quality parameters of produced waters from Devonian-age shale following hydraulic fracturing. Environ Sci Technol 47: 2562-2569.	Barbot et al. (2013) [b] (28)

[a] Sources used to identify chemicals used in hydraulic fracturing fluids.

[b] Sources used to identify chemicals detected in produced water.

Once it had identified chemicals used in hydraulic fracturing fluids and chemicals detected in produced water, the EPA conducted an initial review of the chemicals for preliminary validation of provided chemical name and Chemical Abstracts Service Registry Number (CASRN) combinations. A CASRN is a numeric identifier assigned by the Chemical Abstracts Service (CAS) to a chemical substance when it enters the CAS Registry Database.

The EPA Office of Research and Development's National Center for Computational Toxicology (NCCT) processed and provided the final list of curated CASRN-chemical name matches with validated chemical structures from NCCT's Distributed Structure-Searchable Toxicity Database (DSSTox) (U.S. EPA, 2013b). As of late 2016, the DSSTox database exceeds 700,000 chemical substances. The highest quality, manually curated subset (~25,000 chemical substances) focuses on chemicals of relevance to environmental exposures, toxicity, and bioactivity. Additional content (~130,000 chemical substances) imported from the EPA's Substance Registry System (SRS) chemicals and the National Library of Medicine (NLM)'s ChemID library comprises a portion the DSSTox database with intermediate quality (NLM, 2014; U.S. EPA, 2014e). The remainder of the chemical substances are imported from lower quality, uncurated public resources such as PubChem (https://pubchem.ncbi.nlm.nih.gov/). The entire DSSTox database is searchable through the EPA public CompTox Dashboard (https://comptox.epa.gov/dashboard).

The DSSTox database is distinguished from other publicly available chemical databases by the manual curation workflow applied to high-priority EPA chemical lists, as well as by the enforcement of unique (1:1:1) mappings of CASRN to a single "preferred name" and unique chemical structure. Initial automated-processing of a CASRN and/or chemical name list yields various types of corrections, notes, and mappings to registered DSSTox chemical substance records. The simplest include: (1) CASRN- exact chemical name match; (2) CASRN-synonym chemical name match; (3) CASRN-match through mapping of a "deleted" CASRN that is no longer in use to an "active" CASRN; (4) Auto-repair of common errors in CASRN formatting resulting in a CASRN- exact or synonym chemical name match. These four situations are considered valid "matches" and the records are mapped to a DSSTox ID directly.

Other situations (such as when a CASRN-chemical name appears mismatched in an original source of information or when only a chemical name without a CASRN is provided in the original source) require various levels of curation review prior to final mapping of a CASRN to a single chemical name and unique structure. The general methodology for resolving conflicts between CASRN-chemical name combinations and other chemical identification issues differed slightly depending on the data provided by each source. To resolve CASRN-chemical name issues in data provided by the nine service companies, the EPA worked with each company to verify the CASRN-chemical name matches proposed by NCCT. In cases of CASRN-chemical name mismatches in data provided by FracFocus, chemical names were considered primary to the CASRN (i.e., the name overrode the CASRN). When the chemical name was non-specific and the CASRN was valid, then the CASRN was considered primary to the chemical name, and the correct specific chemical name from DSSTox was assigned to the CASRN. For all other sources of information, the CASRN was considered primary unless it was invalid or missing. In such cases, the chemical name was primary.

When no CASRN-chemical name match is possible, the chemical may undergo manual curation review and require registration of new DSSTox substance-structure records. Each registered DSSTox substance record, in turn, is assigned a Curation Quality Score that indicates the level of curation (automated vs. manual) and reliability of the CASRN-chemical name-structure association. The manual DSSTox curation process is carried out in accordance with the published DSSTox Chemical Information Quality Review Procedures (ftp://ftp.epa.gov/dsstoxftp/DSSTox_Archive_20150930/DSSTox_ChemInfQAProcedures_20150930.pdf).

Individual chemicals or chemical mixtures with valid CASRN-chemical name matches that are used in hydraulic fracturing fluids are presented in Table H-2. Generic chemicals used in hydraulic fracturing fluids (i.e., encompassing a general class of chemicals) or chemicals without a valid CASRN-chemical name match are presented in Table H-3. Chemicals with valid CASRN-chemical name matches that have been detected in produced water are presented in Table H-4. Generic chemicals or chemicals without a valid CASRN-chemical name match that have been detected in produced water are presented in Table H-5. Chemicals with valid CASRN-chemical name matches found in both fracturing fluids and produced water are also indicated in Table H-2 and Table H-4.

In total, 1,606 chemicals with valid CASRN-chemical name matches were reported to be used in hydraulic fracturing fluids and/or detected in produced water from hydraulically fractured wells. This total number comprises 1,084 chemicals reported to be used in hydraulic fracturing fluids from 2005–2013 and 599 chemicals detected in produced water according to the sources of information that we summarized. The number of chemicals reported to be used in hydraulic fracturing fluids from 2005–2013 that were also detected in produced water was 77.

Table H-2. Chemicals reported to be used in hydraulic fracturing fluids.

Chemicals were reported to be used in hydraulic fracturing fluids from 2005-2013, according to the references cited. An "X" indicates the availability of physicochemical properties from EPI Suite™ (Appendix C) and selected toxicity data (Appendix G). An empty cell indicates no information was available from the sources we consulted. Reference number corresponds to the citation in Table H-1.

Chemical name[a]	CASRN[b]	Known constituent of produced water	Physico-chemical properties	Selected toxicity data[c]	Reference
(13Z)-N,N-bis(2-hydroxyethyl)-N-methyldocos-13-en-1-aminium chloride	120086-58-0		X		1
(2,3-dihydroxypropyl)trimethylammonium chloride	34004-36-9		X		8
(E)-Crotonaldehyde	123-73-9		X	X	1, 4
[Nitrilotris(methylene)]tris-phosphonic acid pentasodium salt	2235-43-0		X		1
1-(1-Naphthylmethyl)quinolinium chloride	65322-65-8		X		1
1-(Alkyl* amino)-3-aminopropane *(42%C12, 26%C18, 15%C14, 8%C16, 5%C10, 4%C8)	68155-37-3		X		8
1-(Phenylmethyl)pyridinium Et Me derivs., chlorides	68909-18-2		X		1, 2, 3, 4, 6, 8
1,2,3-Trimethylbenzene	526-73-8	X	X	X	1, 4
1,2,4-Trimethylbenzene	95-63-6	X	X	X	1, 2, 3, 4, 5
1,2-Benzisothiazolin-3-one	2634-33-5		X		1, 3, 4
1,2-Dibromo-2,4-dicyanobutane	35691-65-7		X		1, 4
1,2-Ethanediamine, polymer with 2-methyloxirane	25214-63-5				8
1,2-Ethanediaminium, N,N'-bis[2-[bis(2-hydroxyethyl)methylammonio]ethyl]-N,N'-bis(2-hydroxyethyl)-N,N'-dimethyl-, tetrachloride	138879-94-4		X		1, 4
1,2-Propylene glycol	57-55-6	X	X	X	1, 2, 3, 4, 8
1,2-Propylene oxide	75-56-9		X	X	1, 4
1,3,5-Triazine	290-87-9		X		8

Chemical name[a]	CASRN[b]	Known constituent of produced water	Physico-chemical properties	Selected toxicity data[c]	Reference
1,3,5-Triazine-1,3,5(2H,4H,6H)-triethanol	4719-04-4		X		1, 4
1,3,5-Trimethylbenzene	108-67-8	X	X	X	1, 4
1,3-Butadiene	106-99-0		X	X	8
1,3-Dichloropropene	542-75-6		X	X	8
1,4-Dioxane	123-91-1	X	X	X	2, 3, 4
1,4-Dioxane-2,5-dione, 3,6-dimethyl-, (3R,6R)-, polymer with (3S,6S)-3,6-dimethyl-1,4-dioxane-2,5-dione and (3R,6S)-rel-3,6-dimethyl-1,4-dioxane-2,5-dione	9051-89-2				1, 4, 8
1,6-Hexanediamine	124-09-4		X		1, 2
1,6-Hexanediamine dihydrochloride	6055-52-3		X		1
1-[2-(2-Methoxy-1-methylethoxy)-1-methylethoxy]-2-propanol	20324-33-8		X		4
1-Amino-2-propanol	78-96-6		X		8
1-Benzylquinolinium chloride	15619-48-4		X		1, 3, 4
1-Butanol	71-36-3	X	X	X	1, 2, 3, 4, 7
1-Butoxy-2-propanol	5131-66-8		X		8
1-Decanol	112-30-1		X		1, 4
1-Dodecyl-2-pyrrolidinone	2687-96-9		X		1, 4
1-Eicosene	3452-07-1		X		3
1-Ethyl-2-methylbenzene	611-14-3	X	X		4
1-Hexadecene	629-73-2	X	X		3
1-Hexanol	111-27-3		X		1, 4, 8
1-Hexanol, 2-ethyl-, manuf. of, by products from, distn. residues	68609-68-7				4

Chemical name[a]	CASRN[b]	Known constituent of produced water	Physico-chemical properties	Selected toxicity data[c]	Reference
1H-Imidazole-1-ethanamine, 4,5-dihydro-, 2-nortall-oil alkyl derivs.	68442-97-7				2, 4
1-Methoxy-2-propanol	107-98-2		X		1, 2, 3, 4
1-Octadecanamine, acetate (1:1)	2190-04-7		X		8
1-Octadecanamine, N,N-dimethyl-	124-28-7		X		1, 3, 4
1-Octadecene	112-88-9	X	X		3
1-Octanol	111-87-5		X		1, 4
1-Pentanol	71-41-0		X		8
1-Propanaminium, 3-amino-N-(carboxymethyl)-N,N-dimethyl-, N-coco acyl derivs., chlorides, sodium salts	61789-39-7				1
1-Propanaminium, 3-amino-N-(carboxymethyl)-N,N-dimethyl-, N-coco acyl derivs., inner salts	61789-40-0				1, 2, 3, 4
1-Propanaminium, 3-chloro-2-hydroxy-N,N,N-trimethyl-, chloride	3327-22-8		X		8
1-Propanaminium, N-(3-aminopropyl)-2-hydroxy-N,N-dimethyl-3-sulfo-, N-coco acyl derivs., inner salts	68139-30-0				1, 3, 4
1-Propanaminium, N-(carboxymethyl)-N,N-dimethyl-3-[(1-oxooctyl)amino]-, inner salt	73772-46-0				8
1-Propanesulfonic acid	5284-66-2		X		3
1-Propanol	71-23-8	X	X		1, 2, 4, 5
1-Propanol, zirconium(4+) salt	23519-77-9				1, 4, 8
1-Propene	115-07-1		X	X	2
1-tert-Butoxy-2-propanol	57018-52-7		X	X	8
1-Tetradecene	1120-36-1		X		3
1-Tridecanol	112-70-9		X		1, 4

Chemical name[a]	CASRN[b]	Known constituent of produced water	Physico-chemical properties	Selected toxicity data[c]	Reference
1-Undecanol	112-42-5		X		2
2-(2-Butoxyethoxy)ethanol	112-34-5	X	X	X	2, 4
2-(2-Ethoxyethoxy)ethanol	111-90-0		X	X	1, 4
2-(2-Ethoxyethoxy)ethyl acetate	112-15-2		X		1, 4
2-(Dibutylamino)ethanol	102-81-8		X		1, 4
2-(Hydroxymethylamino)ethanol	34375-28-5		X		1, 4
2-(Thiocyanomethylthio)benzothiazole	21564-17-0		X	X	2
2,2'-(diazene-1,2-diyldiethane-1,1-diyl)bis-4,5-dihydro-1H-imidazole dihydrochloride	27776-21-2		X		3
2,2'-(Octadecylimino)diethanol	10213-78-2		X		1
2,2'-[Ethane-1,2-diylbis(oxy)]diethanamine	929-59-9		X		1, 4
2,2'-Azobis(2-amidinopropane) dihydrochloride	2997-92-4		X		1, 4
2,2-Dibromo-3-nitrilopropionamide	10222-01-2	X	X		1, 2, 3, 4, 6, 7, 8
2,2-Dibromopropanediamide	73003-80-2		X		3
2,4-Hexadienoic acid, potassium salt, (2E,4E)-	24634-61-5		X		3
2,6,8-Trimethyl-4-nonanol	123-17-1		X		8
2-Acrylamide - 2-propanesulfonic acid and N,N-dimethylacrylamide copolymer	NOCAS_51252				2
2-Acrylamido -2-methylpropanesulfonic acid copolymer	NOCAS_51255				8
2-Acrylamido-2-methyl-1-propanesulfonic acid	15214-89-8		X		1, 3
2-Amino-2-methylpropan-1-ol	124-68-5		X		8
2-Aminoethanol ester with boric acid (H₃BO₃) (1:1)	10377-81-8				8

Chemical name[a]	CASRN[b]	Known constituent of produced water	Physico-chemical properties	Selected toxicity data[c]	Reference
2-Aminoethanol hydrochloride	2002-24-6		X		4, 8
2-Bromo-3-nitrilopropionamide	1113-55-9		X		1, 2, 3, 4, 5
2-Butanone oxime	96-29-7		X		1
2-Butenediamide, (2E)-, N,N'-bis[2-(4,5-dihydro-2-nortall-oil alkyl-1H-imidazol-1-yl)ethyl] derivs.	68442-77-3				3, 8
2-Butoxy-1-propanol	15821-83-7		X		8
2-Butoxyethanol	111-76-2	X	X	X	1, 2, 3, 4, 6, 7, 8
2-Dodecylbenzenesulfonic acid- n-(2-aminoethyl)ethane-1,2-diamine(1:1)	40139-72-8		X		8
2-Ethoxyethanol	110-80-5		X	X	6
2-Ethoxynaphthalene	93-18-5		X		3
2-Ethyl-1-hexanol	104-76-7	X	X		1, 2, 3, 4, 5
2-Ethyl-2-hexenal	645-62-5		X		2
2-Ethylhexyl benzoate	5444-75-7		X		4
2-Hydroxyethyl acrylate	818-61-1		X		1, 4
2-Hydroxyethylammonium hydrogen sulphite	13427-63-9		X		1
2-Hydroxy-N,N-bis(2-hydroxyethyl)-N-methylethanaminium chloride	7006-59-9		X		8
2-Mercaptoethanol	60-24-2		X		1, 4
2-Methoxyethanol	109-86-4		X	X	4
2-Methyl-1-propanol	78-83-1		X	X	1, 2, 4
2-Methyl-2,4-pentanediol	107-41-5		X		1, 2, 4
2-Methyl-3(2H)-isothiazolone	2682-20-4		X		1, 2, 4

Chemical name[a]	CASRN[b]	Known constituent of produced water	Physico-chemical properties	Selected toxicity data[c]	Reference
2-Methyl-3-butyn-2-ol	115-19-5		X		3
2-Methylbutane	78-78-4		X		2
2-Methylquinoline hydrochloride	62763-89-7		X		3
2-Phosphono-1,2,4-butanetricarboxylic acid	37971-36-1		X		1, 4
2-Phosphonobutane-1,2,4-tricarboxylic acid, potassium salt (1:x)	93858-78-7		X		1
2-Propanol, aluminum salt	555-31-7				1
2-Propen-1-aminium, N,N-dimethyl-N-2-propenyl-, chloride, homopolymer	26062-79-3				3
2-Propenamide, homopolymer	25038-45-3				8
2-Propenoic acid, 2-(2-hydroxyethoxy)ethyl ester	13533-05-6		X		4
2-Propenoic acid, 2-ethylhexyl ester, polymer with 2-hydroxyethyl 2-propenoate	36089-45-9				8
2-Propenoic acid, 2-methyl-, polymer with 2-propenoic acid, sodium salt	28205-96-1				8
2-Propenoic acid, 2-methyl-, polymer with sodium 2-methyl-2-[(1-oxo-2-propen-1-yl)amino]-1-propanesulfonate (1:1)	136793-29-8				8
2-Propenoic acid, ethyl ester, polymer with ethenyl acetate and 2,5-furandione, hydrolyzed	113221-69-5				4, 8
2-Propenoic acid, ethyl ester, polymer with ethenyl acetate and 2,5-furandione, hydrolyzed, sodium salt	111560-38-4				8
2-Propenoic acid, polymer with 2-propenamide, sodium salt	25987-30-8				3, 4, 8
2-Propenoic acid, polymer with ethene, zinc salt	28208-80-2				8
2-Propenoic acid, polymer with ethenylbenzene	25085-34-1				8

Chemical name[a]	CASRN[b]	Known constituent of produced water	Physico-chemical properties	Selected toxicity data[c]	Reference
2-Propenoic acid, polymer with sodium ethanesulfonate, peroxydisulfuric acid, disodium salt- initiated, reaction products with tetrasodium ethenylidenebis (phosphonata)	397256-50-7				8
2-Propenoic acid, polymer with sodium phosphinate (1:1), sodium salt	129898-01-7				8
2-Propenoic acid, sodium salt (1:1), polymer with sodium 2-methyl-2-[(1-oxo-2-propen-1-yl)amino]-1-propanesulfonate (1:1)	37350-42-8				1
2-Propenoic acid, telomer with sodium 4-ethenylbenzenesulfonate (1:1), sodium 2-methyl-2-[(1-oxo-2-propen-1-yl)amino]-1-propanesulfonate (1:1) and sodium sulfite (1:1), sodium salt	151006-66-5				4
2-Propenoic, polymer with sodium phosphinate	71050-62-9				3, 4
3-(Dimethylamino)propylamine	109-55-7		X		8
3,4,4-Trimethyloxazolidine	75673-43-7		X		8
3,5,7-Triazatricyclo(3.3.1.1(superscript 3,7))decane, 1-(3-chloro-2-propenyl)-, chloride, (Z)-	51229-78-8		X		3
3,7-Dimethyl-2,6-octadienal	5392-40-5		X		3
3-Hydroxybutanal	107-89-1		X		1, 2, 4
3-Methoxypropylamine	5332-73-0		X		8
3-Phenylprop-2-enal	104-55-2		X		1, 2, 3, 4, 7
4,4-Dimethyloxazolidine	51200-87-4		X		8
4,6-Dimethyl-2-heptanone	19549-80-5		X		8
4-[Abieta-8,11,13-trien-18-yl(3-oxo-3-phenylpropyl)amino]butan-2-one hydrochloride	143106-84-7		X		1, 4
4-Ethyloct-1-yn-3-ol	5877-42-9		X		1, 2, 3, 4
4-Hydroxy-3-methoxybenzaldehyde	121-33-5		X		3

Chemical name[a]	CASRN[b]	Known constituent of produced water	Physico-chemical properties	Selected toxicity data[c]	Reference
4-Methoxybenzyl formate	122-91-8		X		3
4-Methoxyphenol	150-76-5		X		4
4-Methyl-2-pentanol	108-11-2		X		1, 4
4-Methyl-2-pentanone	108-10-1		X	X	5
4-Nonylphenol	104-40-5		X		8
4-Nonylphenol polyethoxylate	68412-54-4				2, 3, 4
5-Chloro-2-methyl-3(2H)-isothiazolone	26172-55-4		X		1, 2, 4
Acetaldehyde	75-07-0	X	X	X	1, 4
Acetic acid	64-19-7	X	X		1, 2, 3, 4, 5, 6, 7, 8
Acetic acid ethenyl ester, polymer with ethenol	25213-24-5				1, 4
Acetic acid, C6-8-branched alkyl esters	90438-79-2		X		4
Acetic acid, hydroxy-, reaction products with triethanolamine	68442-62-6		X		3
Acetic acid, mercapto-, monoammonium salt	5421-46-5		X		2, 8
Acetic acid, reaction products with acetophenone, cyclohexylamine, formaldehyde and methanol	224635-63-6				8
Acetic anhydride	108-24-7		X		1, 2, 3, 4, 7
Acetone	67-64-1	X	X	X	1, 3, 4, 6
Acetonitrile, 2,2',2''-nitrilotris-	7327-60-8		X		1, 4
Acetophenone	98-86-2	X	X	X	1
Acetyltriethyl citrate	77-89-4		X		1, 4
Acrolein	107-02-8	X	X	X	2
Acrylamide	79-06-1		X	X	1, 2, 3, 4

Appendix H – Chemicals Identified in Hydraulic Fracturing Fluids and/or Produced Water

Chemical name[a]	CASRN[b]	Known constituent of produced water	Physico-chemical properties	Selected toxicity data[c]	Reference
Acrylamide/ sodium acrylate copolymer	25085-02-3				1, 2, 3, 4, 8
Acrylamide-sodium-2-acrylamido-2-methlypropane sulfonate copolymer	38193-60-1				1, 2, 3, 4
Acrylic acid	79-10-7		X	X	2, 4
Acrylic acid, with sodium-2-acrylamido-2-methyl-1-propanesulfonate and sodium phosphinate	110224-99-2		X		8
Alcohols (C13-C15), ethoxylated	64425-86-1				8
Alcohols, C10-12, ethoxylated	67254-71-1		X		3
Alcohols, C10-14, ethoxylated	66455-15-0				3
Alcohols, C11-14-iso-, C13-rich	68526-86-3		X		3
Alcohols, C11-14-iso-, C13-rich, butoxylated ethoxylated	228414-35-5				1
Alcohols, C11-14-iso-, C13-rich, ethoxylated	78330-21-9		X		3, 4, 8
Alcohols, C12-13, ethoxylated	66455-14-9		X		4
Alcohols, C12-14, ethoxylated	68439-50-9				2, 3, 4, 8
Alcohols, C12-14, ethoxylated propoxylated	68439-51-0		X		1, 3, 4, 8
Alcohols, C12-14-secondary	126950-60-5		X		1, 3, 4
Alcohols, C12-14-secondary, ethoxylated	84133-50-6				3, 4, 8
Alcohols, C12-15, ethoxylated	68131-39-5				3, 4
Alcohols, C12-16, ethoxylated	68551-12-2		X		3, 4, 8
Alcohols, C14-15, ethoxylated	68951-67-7		X		3, 4, 8
Alcohols, C6-12, ethoxylated	68439-45-2		X		3, 4, 8
Alcohols, C7-9-iso-, C8-rich, ethoxylated	78330-19-5		X		2, 4, 8

Chemical name[a]	CASRN[b]	Known constituent of produced water	Physico-chemical properties	Selected toxicity data[c]	Reference
Alcohols, C8-10, ethoxylated propoxylated	68603-25-8				3
Alcohols, C9-11, ethoxylated	68439-46-3		X		3, 4
Alcohols, C9-11-iso-, C10-rich, ethoxylated	78330-20-8		X		1, 2, 4, 8
Alkanes C10-16-branched and linear	90622-52-9				4
Alkanes, C10-14	93924-07-3				1
Alkanes, C12-14-iso-	68551-19-9		X		2, 4, 8
Alkanes, C13-16-iso-	68551-20-2		X		1, 4
Alkenes, C>10 .alpha.-	64743-02-8		X		1, 3, 4, 8
Alkenes, C>8	68411-00-7				1
Alkenes, C24-25 alpha-, polymers with maleic anhydride, docosyl esters	68607-07-8				8
Alkyl quaternary ammonium with bentonite	71011-24-0				4
Alkyl* dimethyl ethylbenzyl ammonium chloride *(50%C12, 30%C14, 17%C16, 3%C18)	NOCAS_34320		X		8
Alkyl* dimethyl ethylbenzyl ammonium chloride *(60%C14, 30%C16, 5%C12, 5%C18)	68956-79-6		X		8
Alkylbenzenesulfonate, linear	42615-29-2		X		1, 4, 6
Almandite and pyrope garnet	1302-62-1				1, 4
alpha-[3.5-dimethyl-1-(2-methylpropyl)hexyl]-omega-hydroxy-poly(oxy-1,2-ethandiyl)	60828-78-6				3
alpha-Amylase	9000-90-2				4
alpha-Lactose monohydrate	5989-81-1		X		8
alpha-Terpineol	98-55-5		X		3
Alumina	1344-28-1				1, 2, 4

Appendix H – Chemicals Identified in Hydraulic Fracturing Fluids and/or Produced Water

Chemical name[a]	CASRN[b]	Known constituent of produced water	Physico-chemical properties	Selected toxicity data[c]	Reference
Aluminatesilicate	1327-36-2				8
Aluminum	7429-90-5	X		X	1, 4, 6
Aluminum calcium oxide (Al2CaO4)	12042-68-1				2
Aluminum chloride	7446-70-0				1, 4
Aluminum chloride hydroxide sulfate	39290-78-3				8
Aluminum chloride, basic	1327-41-9				3, 4
Aluminum oxide (Al2O3)	90669-62-8				8
Aluminum oxide silicate	12068-56-3				1, 2, 4
Aluminum silicate	12141-46-7				1, 2, 4
Aluminum sulfate	10043-01-3				1, 4
Amaranth	915-67-3		X	X	4
Amides, C8-18 and C18-unsatd., N,N-bis(hydroxyethyl)	68155-07-7				3
Amides, coco, N-[3-(dimethylamino)propyl]	68140-01-2				1, 4
Amides, coco, N-[3-(dimethylamino)propyl], alkylation products with chloroacetic acid, sodium salts	70851-07-9				1, 4
Amides, coco, N-[3-(dimethylamino)propyl], alkylation products with sodium 3-chloro-2-hydroxypropanesulfonate	70851-08-0				8
Amides, coco, N-[3-(dimethylamino)propyl], N-oxides	68155-09-9				1, 3, 4
Amides, from C16-22 fatty acids and diethylenetriamine	68876-82-4				3
Amides, tall-oil fatty, N,N-bis(hydroxyethyl)	68155-20-4				3, 4
Amides, tallow, N-[3-(dimethylamino)propyl],N-oxides	68647-77-8				1, 4
Amine oxides, cocoalkyldimethyl	61788-90-7				8

Chemical name[a]	CASRN[b]	Known constituent of produced water	Physico-chemical properties	Selected toxicity data[c]	Reference
Amines, C14-18; C16-18-unsaturated, alkyl, ethoxylated	68155-39-5				1
Amines, C8-18 and C18-unsatd. alkyl	68037-94-5				5
Amines, coco alkyl	61788-46-3				4
Amines, coco alkyl, acetates	61790-57-6				1, 4
Amines, coco alkyl, ethoxylated	61791-14-8				8
Amines, coco alkyldimethyl	61788-93-0				8
Amines, dicoco alkyl	61789-76-2				8
Amines, dicoco alkylmethyl	61788-62-3				8
Amines, ditallow alkyl, acetates	71011-03-5				8
Amines, hydrogenated tallow alkyl, acetates	61790-59-8				4
Amines, N-tallow alkyltrimethylenedi-, ethoxylated	61790-85-0				8
Amines, polyethylenepoly-, ethoxylated, phosphonomethylated	68966-36-9				1, 4
Amines, polyethylenepoly-, reaction products with benzyl chloride	68603-67-8				1
Amines, tallow alkyl	61790-33-8				8
Amines, tallow alkyl, ethoxylated, acetates (salts)	68551-33-7				1, 3, 4
Amines, tallow alkyl, ethoxylated, phosphates	68308-48-5				4
Aminotrimethylene phosphonic acid	6419-19-8		X		1, 4, 8
Ammonia	7664-41-7	X			1, 2, 3, 4, 7
Ammonium (lauryloxypolyethoxy)ethyl sulfate	32612-48-9				4
Ammonium acetate	631-61-8		X		1, 3, 4, 5, 8
Ammonium acrylate	10604-69-0		X		8
Ammonium acrylate-acrylamide polymer	26100-47-0				2, 4, 8

Appendix H – Chemicals Identified in Hydraulic Fracturing Fluids and/or Produced Water

Chemical name[a]	CASRN[b]	Known constituent of produced water	Physico-chemical properties	Selected toxicity data[c]	Reference
Ammonium bisulfate	7803-63-6				2
Ammonium bisulfite	10192-30-0				1, 2, 3, 4, 7
Ammonium chloride	12125-02-9				1, 2, 3, 4, 5, 6, 8
Ammonium citrate (1:1)	7632-50-0		X		3
Ammonium citrate (2:1)	3012-65-5		X		8
Ammonium dodecyl sulfate	2235-54-3		X		1
Ammonium fluoride	12125-01-8				1, 4
Ammonium hydrogen carbonate	1066-33-7		X		1, 4
Ammonium hydrogen difluoride	1341-49-7				1, 3, 4, 7
Ammonium hydrogen phosphonate	13446-12-3				4
Ammonium hydroxide	1336-21-6				1, 3, 4
Ammonium lactate	515-98-0		X		8
Ammonium ligninsulfonate	8061-53-8				2
Ammonium nitrate	6484-52-2				1, 2, 3
Ammonium phosphate	7722-76-1			X	1, 4
Ammonium sulfate	7783-20-2				1, 2, 3, 4, 6
Ammonium thiosulfate	7783-18-8				8
Amorphous silica	99439-28-8				1, 7
Anethole	104-46-1		X		3
Aniline	62-53-3		X	X	2, 4
Antimony pentoxide	1314-60-9				1, 4
Antimony trichloride	10025-91-9			X	1, 4

Chemical name[a]	CASRN[b]	Known constituent of produced water	Physico-chemical properties	Selected toxicity data[c]	Reference
Antimony trioxide	1309-64-4			X	8
Arsenic	7440-38-2	X		X	4
Ashes, residues	68131-74-8				4
Asphalt, sulfonated, sodium salt	68201-32-1				2
Attapulgite	12174-11-7			X	2, 3
Aziridine, polymer with 2-methyloxirane	31974-35-3				4, 8
Barium sulfate	7727-43-7				1, 2, 4
Bauxite	1318-16-7				1, 2, 4
Benactyzine hydrochloride	57-37-4		X		8
Bentonite	1302-78-9				1, 2, 4, 6
Bentonite, benzyl(hydrogenated tallow alkyl) dimethylammonium stearate complex	121888-68-4				3, 4
Benzamorf	12068-08-5		X		1, 4
Benzene	71-43-2	X	X	X	1, 3, 4
Benzene, 1,1'-oxybis-, sec-hexyl derivs., sulfonated, sodium salts	147732-60-3				8
Benzene, 1,1'-oxybis-, tetrapropylene derivs., sulfonated	119345-03-8				8
Benzene, 1,1'-oxybis-, tetrapropylene derivs., sulfonated, sodium salts	119345-04-9				3, 4, 8
Benzene, C10-16-alkyl derivs.	68648-87-3		X		1
Benzene, ethenyl-, polymer with 2-methyl-1,3-butadiene, hydrogenated	68648-89-5				8
Benzenemethanaminium, N,N-dimethyl-N-(2-((1-oxo-2-propen-1-yl)oxy)ethyl)-, chloride (1:1), polymer with 2-propenamide	74153-51-8				3
Benzenesulfonic acid	98-11-3		X		2

Chemical name[a]	CASRN[b]	Known constituent of produced water	Physico-chemical properties	Selected toxicity data[c]	Reference
Benzenesulfonic acid, (1-methylethyl)-,	37953-05-2		X		4
Benzenesulfonic acid, (1-methylethyl)-, ammonium salt	37475-88-0		X		3, 4
Benzenesulfonic acid, (1-methylethyl)-, sodium salt	28348-53-0		X		8
Benzenesulfonic acid, C10-16-alkyl derivs.	68584-22-5			X	1, 4
Benzenesulfonic acid, C10-16-alkyl derivs., compds. with cyclohexylamine	255043-08-4		X		1
Benzenesulfonic acid, C10-16-alkyl derivs., compds. with triethanolamine	68584-25-8		X		8
Benzenesulfonic acid, C10-16-alkyl derivs., potassium salts	68584-27-0		X		1, 4, 8
Benzenesulfonic acid, dodecyl-, branched, compds. with 2-propanamine	90218-35-2		X		4
Benzenesulfonic acid, mono-C10-16 alkyl derivs., compds. with 2-propanamine	68648-81-7				1, 4
Benzenesulfonic acid, mono-C10-16-alkyl derivs., sodium salts	68081-81-2		X		8
Benzoic acid	65-85-0		X	X	1, 4, 7
Benzyl chloride	100-44-7	X	X	X	1, 2, 4, 8
Benzyldimethyldodecylammonium chloride	139-07-1		X		2, 8
Benzylhexadecyldimethylammonium chloride	122-18-9		X		8
Benzyltrimethylammonium chloride	56-93-9		X		8
Bicine	150-25-4		X		1, 4
Bio-Perge	55965-84-9				8
Bis(1-methylethyl)naphthalenesulfonic acid, cyclohexylamine salt	68425-61-6		X		1
Bis(2-chloroethyl) ether	111-44-4	X	X	X	8
Bisphenol A	80-05-7	X	X	X	4

Chemical name[a]	CASRN[b]	Known constituent of produced water	Physico-chemical properties	Selected toxicity data[c]	Reference
Bisphenol A/ Epichlorohydrin resin	25068-38-6				1, 2, 4
Bisphenol A/ Novolac epoxy resin	28906-96-9				1, 4
Blast furnace slag	65996-69-2				2, 3
Borax	1303-96-4				1, 2, 3, 4, 6
Boric acid	10043-35-3				1, 2, 3, 4, 6, 7
Boric acid (H₃BO₃), compd. with 2-aminoethanol (1:x)d	26038-87-9				8
Boric oxide	1303-86-2				1, 2, 3, 4
Boron	7440-42-8	X		X	8
Boron potassium oxide (B4K2O7)	1332-77-0				8
Boron potassium oxide (B4K2O7), tetrahydrate	12045-78-2				8
Boron potassium oxide (B5KO8)	11128-29-3				1
Boron sodium oxide	1330-43-4				1, 2, 4
Boron sodium oxide pentahydrate	12179-04-3				8
Bronopol	52-51-7		X		1, 2, 3, 4, 6
Butane	106-97-8		X		2, 5
Butanedioic acid, sulfo-, 1,4-bis(1,3-dimethylbutyl) ester, sodium salt	2373-38-8		X		1
Butene	25167-67-3		X		8
Butyl glycidyl ether	2426-08-6		X		1, 4
Butyl lactate	138-22-7		X		1, 4
Butyryl trihexyl citrate	82469-79-2		X		8
C.I. Acid Red 1	3734-67-6		X		4
C.I. Acid violet 12, disodium salt	6625-46-3		X		4

Appendix H – Chemicals Identified in Hydraulic Fracturing Fluids and/or Produced Water

Chemical name[a]	CASRN[b]	Known constituent of produced water	Physico-chemical properties	Selected toxicity data[c]	Reference
C.I. Pigment Red 5	6410-41-9		X		4
C.I. Solvent Red 26	4477-79-6		X		4
C10-16-Alkyldimethylamines oxides	70592-80-2		X		4
C10-C16 ethoxylated alcohol	68002-97-1		X		1, 2, 3, 4, 8
C11-15-Secondary alcohols ethoxylated	68131-40-8				1, 2, 8
C12-14 tert-alkyl ethoxylated amines	73138-27-9		X		3
C8-10 Alcohols	85566-12-7				8
Calcined bauxite	66402-68-4				2, 8
Calcium aluminate	12042-78-3				2
Calcium bromide	7789-41-5				4
Calcium carbide (CaC2)	75-20-7				8
Calcium chloride	10043-52-4				1, 2, 3, 4, 7
Calcium dichloride dihydrate	10035-04-8				1, 4
Calcium dodecylbenzene sulfonate	26264-06-2		X		4
Calcium fluoride	7789-75-5				1, 4
Calcium hydroxide	1305-62-0				1, 2, 3, 4
Calcium hypochlorite	7778-54-3				1, 2, 4
Calcium magnesium hydroxide oxide	58398-71-3				4
Calcium oxide	1305-78-8				1, 2, 4, 7
Calcium peroxide	1305-79-9				1, 3, 4, 8
Calcium sulfate	7778-18-9				1, 2, 4
Calcium sulfate dihydrate	10101-41-4				2

Chemical name[a]	CASRN[b]	Known constituent of produced water	Physico-chemical properties	Selected toxicity data[c]	Reference
Camphor	76-22-2		X		3
Canola oil	120962-03-0				8
Carbon black	1333-86-4			X	1, 2, 4
Carbon dioxide	124-38-9	X	X		1, 3, 4, 6
Carbonic acid calcium salt (1:1)	471-34-1				1, 2, 4
Carbonic acid, dipotassium salt	584-08-7		X		1, 2, 3, 4, 8
Carboxymethyl cellulose	9000-11-7				8
Carboxymethyl guar gum, sodium salt	39346-76-4				1, 2, 4
Castor oil	8001-79-4				8
Cedarwood oil	8000-27-9				3
Cellophane	9005-81-6				1, 4
Cellulose	9004-34-6				1, 2, 3, 4
Chloride	16887-00-6	X			4, 8
Chlorine	7782-50-5	X		X	2
Chlorine dioxide	10049-04-4			X	1, 2, 3, 4, 8
Chlorobenzene	108-90-7	X	X	X	8
Chloromethane	74-87-3	X	X	X	8
Choline bicarbonate	78-73-9		X		3, 8
Choline chloride	67-48-1		X		1, 3, 4, 7, 8
Chromium (III)	16065-83-1	X		X	2, 6
Chromium (VI)	18540-29-9	X		X	6
Chromium acetate, basic	39430-51-8				2

Chemical name[a]	CASRN[b]	Known constituent of produced water	Physico-chemical properties	Selected toxicity data[c]	Reference
Chromium(III) acetate	1066-30-4				1, 2
Citric acid	77-92-9		X		1, 2, 3, 4, 7
Citronella oil	8000-29-1				3
Citronellol	106-22-9		X		3
Citrus extract	94266-47-4				1, 3, 4, 8
Coal, granular	50815-10-6				1, 2, 4
Cobalt(II) acetate	71-48-7				1, 4
Coco-betaine	68424-94-2				3
Coconut oil	8001-31-8				8
Coconut oil acid/Diethanolamine condensate (2:1)	68603-42-9			X	1
Coconut trimethylammonium chloride	61789-18-2		X		1, 8
Copper	7440-50-8	X		X	1, 4
Copper sulfate	7758-98-7				1, 4, 8
Copper(I) chloride	7758-89-6				1, 4
Copper(I) iodide	7681-65-4			X	1, 2, 4, 6
Copper(II) chloride	7447-39-4				1, 3, 4
Copper(II) sulfate, pentahydrate	7758-99-8				8
Corn flour	68525-86-0				4
Corn sugar gum	11138-66-2				1, 2, 4
Corundum (Aluminum oxide)	1302-74-5				4, 8
Cottonseed, flour	68308-87-2				2, 4
Coumarin	91-64-5		X	X	3

Chemical name[a]	CASRN[b]	Known constituent of produced water	Physico-chemical properties	Selected toxicity data[c]	Reference
Cremophor(R) EL	61791-12-6				1, 3
Cristobalite	14464-46-1				1, 2, 4
Crystalline silica, tridymite	15468-32-3				1, 2, 4
Cumene	98-82-8	X	X	X	1, 2, 3, 4
Cupric chloride dihydrate	10125-13-0				1, 4, 7
Cyclohexane	110-82-7		X		1, 7
Cyclohexanol	108-93-0		X		8
Cyclohexanone	108-94-1		X	X	1, 4
Cyclohexylamine sulfate	19834-02-7		X		8
D&C Red 28	18472-87-2		X		4
D&C Red No. 33	3567-66-6		X		8
Daidzein	486-66-8		X		8
Dapsone	80-08-0		X	X	1, 4
Dazomet	533-74-4		X		1, 2, 3, 4, 7, 8
Decamethylcyclopentasiloxane	541-02-6				8
Decyldimethylamine	1120-24-7		X		3, 4
Deuterium oxide	7789-20-0				8
D-Glucitol	50-70-4		X		1, 3, 4
D-Gluconic acid	526-95-4		X		1, 4
D-Glucopyranoside, methyl	3149-68-6		X		2
D-Glucose	50-99-7		X		1, 4
Di(2-ethylhexyl) phthalate	117-81-7	X	X	X	1, 4

Chemical name[a]	CASRN[b]	Known constituent of produced water	Physico-chemical properties	Selected toxicity data[c]	Reference
Diammonium peroxydisulfate	7727-54-0				1, 2, 3, 4, 6, 7, 8
Diatomaceous earth	68855-54-9				2, 4
Diatomaceous earth, calcined	91053-39-3				1, 2, 4
Dibromoacetonitrile	3252-43-5	X	X	X	1, 2, 3, 4, 8
Dicalcium silicate	10034-77-2				1, 2, 4
Dichloromethane	75-09-2	X	X	X	8
Didecyldimethylammonium chloride	7173-51-5		X	X	1, 2, 4, 8
Diethanolamine	111-42-2		X	X	1, 2, 3, 4, 6
Diethylbenzene	25340-17-4		X		1, 3, 4
Diethylene glycol	111-46-6		X		1, 2, 3, 4, 7
Diethylene glycol monomethyl ether	111-77-3		X		1, 2, 4
Diethylenetriamine	111-40-0		X		1, 2, 4, 5
Diethylenetriamine reaction product with fatty acid dimers	68647-57-4				2
Diisobutyl ketone	108-83-8		X		8
Diisopropanolamine	110-97-4		X		8
Diisopropylnaphthalene	38640-62-9		X		3, 4
Dimethyl adipate	627-93-0		X		8
Dimethyl glutarate	1119-40-0		X		1, 4
Dimethyl polysiloxane	63148-62-9				1, 2, 4
Dimethyl succinate	106-65-0		X		8
Dimethylaminoethanol	108-01-0		X		2, 4
Dimethyldiallylammonium chloride	7398-69-8		X		3, 4

Chemical name[a]	CASRN[b]	Known constituent of produced water	Physico-chemical properties	Selected toxicity data[c]	Reference
Diphenyl oxide	101-84-8		X		3
Dipotassium monohydrogen phosphate	7758-11-4				5
Dipropylene glycol	25265-71-8		X		1, 3, 4
Di-sec-butylphenol	31291-60-8		X		1
Disodium dodecyl(sulphonatophenoxy)benzenesulphonate	28519-02-0		X		1
Disodium ethylenediaminediacetate	38011-25-5		X		1, 4
Disodium ethylenediaminetetraacetate dihydrate	6381-92-6		X		1
Disodium octaborate	12008-41-2				4, 8
Disodium octaborate tetrahydrate	12280-03-4				1, 4
Disodium sulfide	1313-82-2				8
Distillates, petroleum, catalytic reformer fractionator residue, low-boiling	68477-31-6				1, 4
Distillates, petroleum, heavy arom.	67891-79-6				1, 4
Distillates, petroleum, hydrodesulfurized light catalytic cracked	68333-25-5				1
Distillates, petroleum, hydrodesulfurized middle	64742-80-9				1
Distillates, petroleum, hydrotreated heavy naphthenic	64742-52-5				1, 2, 3, 4
Distillates, petroleum, hydrotreated heavy paraffinic	64742-54-7				1, 2, 4
Distillates, petroleum, hydrotreated light	64742-47-8				1, 2, 3, 4, 5, 7, 8
Distillates, petroleum, hydrotreated light naphthenic	64742-53-6				1, 2, 8
Distillates, petroleum, hydrotreated light paraffinic	64742-55-8				8
Distillates, petroleum, hydrotreated middle	64742-46-7				1, 2, 3, 4, 8
Distillates, petroleum, light catalytic cracked	64741-59-9				1, 4

Chemical name[a]	CASRN[b]	Known constituent of produced water	Physico-chemical properties	Selected toxicity data[c]	Reference
Distillates, petroleum, light hydrocracked	64741-77-1				3
Distillates, petroleum, solvent-dewaxed heavy paraffinic	64742-65-0				1
Distillates, petroleum, solvent-refined heavy naphthenic	64741-96-4				1, 4
Distillates, petroleum, steam-cracked	64742-91-2				1, 4
Distillates, petroleum, straight-run middle	64741-44-2				1, 2, 4
Distillates, petroleum, sweetened middle	64741-86-2				1, 4
Ditallow alkyl ethoxylated amines	71011-04-6				3
D-Lactic acid	10326-41-7		X		1, 4
D-Limonene	5989-27-5	X	X	X	1, 3, 4, 5, 7, 8
Docusate sodium	577-11-7		X		1
Dodecamethylcyclohexasiloxane	540-97-6				8
Dodecane	112-40-3	X	X		8
Dodecylbenzene	123-01-3		X		3, 4
Dodecylbenzenesulfonic acid	27176-87-0		X	X	2, 3, 4, 8
Dodecylbenzenesulfonic acid, monoethanolamine salt	26836-07-7		X		1, 4
Edifas B	9004-32-4				2, 3, 4
EDTA, copper salt	12276-01-6				1, 5, 6
Endo-1,4.-beta.-mannanase	37288-54-3				3, 8
Epichlorohydrin	106-89-8		X	X	1, 4, 8
Epoxy resin	25085-99-8				1, 4, 8
Erucic amidopropyl dimethyl betaine	149879-98-1				1, 3
Ethanaminium, N,N,N-trimethyl-2-[(1-oxo-2-propenyl)oxy]-, chloride	44992-01-0		X		3

Chemical name[a]	CASRN[b]	Known constituent of produced water	Physico-chemical properties	Selected toxicity data[c]	Reference
Ethanaminium, N,N,N-trimethyl-2-[(1-oxo-2-propenyl)oxy]-,chloride, polymer with 2-propenamide	69418-26-4				1, 3, 4
Ethanaminium, N,N,N-trimethyl-2-[(2-methyl-1-oxo-2-propen-1-yl)oxy]-, chloride (1:1), polymer with 2-propenamide	35429-19-7				8
Ethanaminium, N,N,N-trimethyl-2-[(2-methyl-1-oxo-2-propenyl)oxy]-, methyl sulfate, homopolymer	27103-90-8				8
Ethane	74-84-0	X	X		2, 5
Ethanol	64-17-5	X	X	X	1, 2, 3, 4, 5, 6, 8
Ethanol, 2,2',2''-nitrilotris-, tris(dihydrogen phosphate) (ester), sodium salt	68171-29-9		X		4
Ethanol, 2,2'-iminobis-, N-coco alkyl derivs., N-oxides	61791-47-7				1
Ethanol, 2,2'-iminobis-, N-tallow alkyl derivs.	61791-44-4				1
Ethanol, 2,2'-oxybis-, reaction products with ammonia, morpholine derivs. residues	68909-77-3				4, 8
Ethanol, 2,2-oxybis-, reaction products with ammonia, morpholine derivs. residues, acetates (salts)	68877-16-7				4
Ethanol, 2,2-oxybis-, reaction products with ammonia, morpholine derivs. residues, reaction products with sulfur dioxide	102424-23-7				4
Ethanol, 2-[2-[2-(tridecyloxy)ethoxy]ethoxy]-, hydrogen sulfate, sodium salt	25446-78-0		X		1, 4
Ethanol, 2-amino-, polymer with formaldehyde	34411-42-2				4
Ethanol, 2-amino-, reaction products with ammonia, by-products from, phosphonomethylated	68649-44-5				4
Ethanolamine[d]	141-43-5		X		1, 2, 3, 4, 6, 8
Ethoxylated dodecyl alcohol	9002-92-0		X		4

Chemical name[a]	CASRN[b]	Known constituent of produced water	Physico-chemical properties	Selected toxicity data[c]	Reference
Ethoxylated hydrogenated tallow alkylamines	61790-82-7				4
Ethoxylated, propoxylated trimethylolpropane	52624-57-4				3
Ethyl acetate	141-78-6		X	X	1, 4, 7
Ethyl acetoacetate	141-97-9		X		1, 4
Ethyl benzoate	93-89-0		X		3
Ethyl lactate	97-64-3		X		3
Ethyl salicylate	118-61-6		X		3
Ethylbenzene	100-41-4	X	X	X	1, 2, 3, 4, 7
Ethylcellulose	9004-57-3				2
Ethylene	74-85-1		X	X	8
Ethylene glycol	107-21-1	X	X	X	1, 2, 3, 4, 6, 7, 8
Ethylene oxide	75-21-8		X	X	1, 2, 3, 4
Ethylenediamine	107-15-3		X	X	2, 4
Ethylenediaminetetraacetic acid	60-00-4		X		1, 2, 4
Ethylenediaminetetraacetic acid tetrasodium salt	64-02-8		X		1, 2, 3, 4
Ethylenediaminetetraacetic acid, diammonium copper salt	67989-88-2				4
Ethylenediaminetetraacetic acid, disodium salt	139-33-3		X		1, 3, 4, 8
Ethyne	74-86-2		X		7
Fats and Glyceridic oils, vegetable, hydrogenated	68334-28-1				8
Fatty acid, tall oil, hexa esters with sorbitol, ethoxylated	61790-90-7				1, 4
Fatty acids, C 8-18 and C18-unsaturated compounds with diethanolamine	68604-35-3				3

Chemical name[a]	CASRN[b]	Known constituent of produced water	Physico-chemical properties	Selected toxicity data[c]	Reference
Fatty acids, C14-18 and C16-18-unsatd., distn. residues	70321-73-2				2
Fatty acids, C18-unsatd., dimers	61788-89-4		X		2
Fatty acids, C18-unsatd., dimers, compds. with ethoxylated tall-oil fatty acid-polyethylenepolyamine reaction products	68132-59-2				8
Fatty acids, C18-unsatd., dimers, ethoxylated propoxylated	68308-89-4				8
Fatty acids, coco, ethoxylated	61791-29-5				3
Fatty acids, coco, reaction products with diethylenetriamine and soya fatty acids, ethoxylated, chloromethane-quaternized	68604-75-1				8
Fatty acids, coco, reaction products with ethanolamine, ethoxylated	61791-08-0				3
Fatty acids, tall oil, reaction products with acetophenone, formaldehyde and thiourea	68188-40-9				3
Fatty acids, tall-oil	61790-12-3				1, 2, 3, 4
Fatty acids, tall-oil, reaction products with diethylenetriamine	61790-69-0				1, 4
Fatty acids, tall-oil, reaction products with diethylenetriamine, maleic anhydride, tetraethylenepentamine and triethylenetetramine	68990-47-6				8
Fatty acids, tallow, sodium salts	8052-48-0				1, 3
Fatty acids, vegetable-oil, reaction products with diethylenetriamine	68153-72-0				3
Fatty quaternary ammonium chloride	61789-68-2				1, 4
FD&C Blue no. 1	3844-45-9		X	X	1, 4
FD&C Yellow 5	1934-21-0		X		8
FD&C Yellow 6	2783-94-0		X	X	8
Ferric chloride	7705-08-0				1, 3, 4
Ferric sulfate	10028-22-5				1, 4

Chemical name[a]	CASRN[b]	Known constituent of produced water	Physico-chemical properties	Selected toxicity data[c]	Reference
Ferrous sulfate monohydrate	17375-41-6				2
Ferumoxytol	1309-38-2				8
Fiberglass	65997-17-3				2, 3, 4
Formaldehyde	50-00-0		X	X	1, 2, 3, 4
Formaldehyde polymer with 4,1,1-(dimethylethyl)phenol and methyloxirane	29316-47-0				3
Formaldehyde polymer with methyl oxirane, 4-nonylphenol and oxirane	63428-92-2				4, 8
Formaldehyde, polymer with 4-(1,1-dimethylethyl)phenol, 2-methyloxirane and oxirane	30704-64-4				1, 2, 4, 8
Formaldehyde, polymer with 4-(1,1-dimethylethyl)phenol, 2-methyloxirane, 4-nonylphenol and oxirane	68188-99-8				8
Formaldehyde, polymer with 4-nonylphenol and oxirane	30846-35-6				1, 4
Formaldehyde, polymer with 4-nonylphenol and phenol	40404-63-5				8
Formaldehyde, polymer with ammonia and phenol	35297-54-2				1, 4
Formaldehyde, polymer with bisphenol A	25085-75-0				4
Formaldehyde, polymer with N1-(2-aminoethyl)-1,2-ethanediamine, benzylated	70750-07-1				8
Formaldehyde, polymer with nonylphenol and oxirane	55845-06-2				4
Formaldehyde, polymers with branched 4-nonylphenol, oxirane and 2-methyloxirane	153795-76-7				1, 3
Formaldehyde/amine	NOCAS_51232				1, 2, 3, 4
Formamide	75-12-7		X		1, 2, 3, 4
Formic acid	64-18-6	X	X	X	1, 2, 3, 4, 6, 7
Formic acid, potassium salt	590-29-4		X		1, 3, 4

Chemical name[a]	CASRN[b]	Known constituent of produced water	Physico-chemical properties	Selected toxicity data[c]	Reference
Frits, chemicals	65997-18-4				8
Fuel oil, no. 2	68476-30-2				1, 2
Fuels, diesel	68334-30-5				2
Fuels, diesel, no. 2	68476-34-6				2, 4, 8
Fuller's earth	8031-18-3				2
Fumaric acid	110-17-8		X		1, 2, 3, 4, 6
Fumes, silica	69012-64-2				8
Furfural	98-01-1		X	X	1, 4
Furfuryl alcohol	98-00-0		X		1, 4
Galantamine hydrobromide	69353-21-5		X		8
Gas oils, petroleum, straight-run	64741-43-1				1, 4
Gelatin	9000-70-8				1, 4
Gilsonite	12002-43-6				1, 2, 4
Gluconic acid	133-42-6		X		7
Glutaraldehyde	111-30-8	X	X		1, 2, 3, 4, 7
Glycerides, C14-18 and C16-18-unsatd. mono- and di-	67701-32-0				8
Glycerol	56-81-5		X		1, 2, 3, 4, 5
Glycine, N-(carboxymethyl)-N-(2-hydroxyethyl)-, disodium salt	135-37-5		X		1
Glycine, N-(hydroxymethyl)-, monosodium salt	70161-44-3		X		8
Glycine, N,N-bis(carboxymethyl)-, trisodium salt	5064-31-3		X		1, 2, 3, 4
Glycine, N-[2-[bis(carboxymethyl)amino]ethyl]-N-(2-hydroxyethyl)-, trisodium salt	139-89-9		X		1

Appendix H – Chemicals Identified in Hydraulic Fracturing Fluids and/or Produced Water

Chemical name[a]	CASRN[b]	Known constituent of produced water	Physico-chemical properties	Selected toxicity data[c]	Reference
Glycolic acid	79-14-1	X	X		1, 3, 4
Glycolic acid sodium salt	2836-32-0		X		1, 3, 4
Glyoxal	107-22-2		X	X	1, 2, 4
Glyoxylic acid	298-12-4		X		1
Goethite (Fe(OH)O)	1310-14-1				8
Guar gum	9000-30-0				1, 2, 3, 4, 7, 8
Guar gum, carboxymethyl 2-hydroxypropyl ether, sodium salt	68130-15-4				1, 2, 3, 4, 7
Gypsum (Ca(SO4).2H2O)	13397-24-5				2, 4
Hematite	1317-60-8			X	1, 2, 4
Hemicellulase	9012-54-8				1, 2, 3, 4, 5
Hemicellulase enzyme concentrate	9025-56-3				3, 4
Heptane	142-82-5	X	X		1, 2
Heptene, hydroformylation products, high-boiling	68526-88-5				1, 4
Hexadecyltrimethylammonium bromide	57-09-0		X		1
Hexane	110-54-3	X	X	X	5
Hexanedioic acid	124-04-9		X	X	1, 2, 4, 6
Humic acids, commercial grade	1415-93-6				2
Hydrazine	302-01-2	X		X	8
Hydrocarbons, terpene processing by-products	68956-56-9				1, 3, 4
Hydrochloric acid	7647-01-0	X		X	1, 2, 3, 4, 5, 6, 7, 8
Hydrogen fluoride	7664-39-3				1, 2, 4
Hydrogen peroxide	7722-84-1			X	1, 3, 4

Chemical name[a]	CASRN[b]	Known constituent of produced water	Physico-chemical properties	Selected toxicity data[c]	Reference
Hydrogen sulfide	7783-06-4				1, 2
Hydroxyethylcellulose	9004-62-0				1, 2, 3, 4
Hydroxylamine hydrochloride	5470-11-1				1, 3, 4
Hydroxylamine sulfate (2:1)	10039-54-0				4
Hydroxypropyl cellulose	9004-64-2				2, 4
Hydroxypropyl guar gum	39421-75-5				1, 3, 4, 5, 6, 8
Hydroxyvalerenic acid	1619-16-5		X		8
Hypochlorous acid	7790-92-3				8
Illite	12173-60-3				8
Ilmenite (FeTiO3), conc.	98072-94-7				8
Indole	120-72-9		X		2
Inulin, carboxymethyl ether, sodium salt	430439-54-6				1, 4
Iridium oxide	12030-49-8				8
Iron	7439-89-6	X		X	2, 4
Iron oxide	1332-37-2				1, 4
Iron oxide (Fe3O4)	1317-61-9				4
Iron(II) sulfate	7720-78-7				2
Iron(II) sulfate heptahydrate	7782-63-0				1, 2, 3, 4
Iron(III) oxide	1309-37-1			X	1, 2, 4
Isoascorbic acid	89-65-6		X		1, 3, 4
Isobutane	75-28-5		X		2
Isobutene	115-11-7		X		8

Appendix H – Chemicals Identified in Hydraulic Fracturing Fluids and/or Produced Water

Chemical name[a]	CASRN[b]	Known constituent of produced water	Physico-chemical properties	Selected toxicity data[c]	Reference
Isooctanol	26952-21-6		X		1, 4, 5
Isopentyl alcohol	123-51-3		X		1, 4
Isopropanol	67-63-0	X	X	X	1, 2, 3, 4, 6, 7
Isopropanolamine dodecylbenzene	42504-46-1		X		1, 3, 4
Isopropylamine	75-31-0		X		1, 4
Isoquinoline	119-65-3	X	X		8
Isoquinoline, reaction products with benzyl chloride and quinoline	68909-80-8		X		3
Isoquinolinium, 2-(phenylmethyl)-, chloride	35674-56-7		X		3
Isotridecanol, ethoxylated	9043-30-5				1, 3, 4, 8
Kaolin	1332-58-7				1, 2, 4
Kerosine, petroleum, hydrodesulfurized	64742-81-0				1, 2, 4
Kieselguhr	61790-53-2				1, 2, 4
Kyanite	1302-76-7				1, 2, 4
Lactic acid	50-21-5		X		1, 4, 8
Lactose	63-42-3		X		3
Latex 2000 TM	9003-55-8			X	2, 4
Lauryl hydroxysultaine	13197-76-7		X		1
Lavandula hybrida abrial herb oil	8022-15-9				3
L-Dilactide	4511-42-6		X		1, 4
Lead	7439-92-1	X		X	1, 4
Lecithin	8002-43-5				4
L-Glutamic acid	56-86-0		X		8

H-38

Chemical name[a]	CASRN[b]	Known constituent of produced water	Physico-chemical properties	Selected toxicity data[c]	Reference
Lignite	129521-66-0				2
Lignosulfuric acid	8062-15-5				2
Ligroine	8032-32-4			X	8
Limestone	1317-65-3				1, 2, 3, 4
Linseed oil	8001-26-1				8
L-Lactic acid	79-33-4		X		1, 4, 8
Magnesium carbonate (1:1)	7757-69-9				8
Magnesium carbonate (1:x)	546-93-0				1, 3, 4
Magnesium chloride	7786-30-3				1, 2, 4
Magnesium chloride hexahydrate	7791-18-6				4
Magnesium hydroxide	1309-42-8				1, 4
Magnesium iron silicate	19086-72-7				1, 4
Magnesium nitrate	10377-60-3				1, 2, 4
Magnesium oxide	1309-48-4				1, 2, 3, 4
Magnesium peroxide	14452-57-4				1, 4
Magnesium phosphide	12057-74-8				1
Magnesium silicate	1343-88-0				1, 4
Magnesium sulfate	7487-88-9				8
Maleic acid homopolymer	26099-09-2				8
Methanamine-N-methyl polymer with chloromethyl oxirane	25988-97-0				4
Methane	74-82-8		X		2, 5
Methanol	67-56-1	X	X	X	1, 2, 3, 4, 5, 6, 7, 8

Chemical name[a]	CASRN[b]	Known constituent of produced water	Physico-chemical properties	Selected toxicity data[c]	Reference
Methenamine	100-97-0		X		1, 2, 4
Methoxyacetic acid	625-45-6		X		8
Methyl cellulose	9004-67-5				8
Methyl salicylate	119-36-8		X		1, 2, 3, 4, 7
Methyl vinyl ketone	78-94-4		X		1, 4
Methylcyclohexane	108-87-2	X	X		1
Methylene bis(thiocyanate)	6317-18-6		X		2
Methylenebis(5-methyloxazolidine)	66204-44-2		X		2
Methyloxirane polymer with oxirane, mono (nonylphenol) ether, branched	68891-11-2				3
Mica	12001-26-2				1, 2, 4, 6
Mineral oil - includes paraffin oil	8012-95-1			X	4, 8
Mineral spirits	64475-85-0			X	2
Mono- and di- potassium salts of phosphorous acid	13492-26-7				8
Montmorillonite	1318-93-0				2
Morpholine	110-91-8		X	X	1, 2, 4
Morpholinium, 4-ethyl-4-hexadecyl-, ethyl sulfate	78-21-7		X		8
MT 6	76-31-3				8
Mullite	1302-93-8				1, 2, 4, 8
N-(2-Acryloyloxyethyl)-N-benzyl-N,N-dimethylammonium chloride	46830-22-2		X		3
N-(3-Chloroallyl)hexaminium chloride	4080-31-3		X		8

Chemical name[a]	CASRN[b]	Known constituent of produced water	Physico-chemical properties	Selected toxicity data[c]	Reference
N,N,N-Trimethyl-2[1-oxo-2-propenyl]oxy ethanaminimum chloride, homopolymer	54076-97-0				3
N,N,N-Trimethyl-3-((1-oxooctadecyl)amino)-1-propanaminium methyl sulfate	19277-88-4		X		1
N,N,N-Trimethyloctadecan-1-aminium chloride	112-03-8		X		1, 3, 4
N,N'-Dibutylthiourea	109-46-6		X		1, 4
N,N-Dimethyldecylamine oxide	2605-79-0		X		1, 3, 4
N,N-Dimethylformamide	68-12-2	X	X	X	1, 2, 4, 5, 8
N,N-Dimethylmethanamine hydrochloride	593-81-7		X		1, 4, 5, 7
N,N-Dimethyl-methanamine-N-oxide	1184-78-7		X		3
N,N-dimethyloctadecylamine hydrochloride	1613-17-8		X		1, 4
N,N'-Methylenebisacrylamide	110-26-9		X		1, 4
Naphtha, petroleum, heavy catalytic reformed	64741-68-0				1, 2, 3, 4
Naphtha, petroleum, hydrotreated heavy	64742-48-9				1, 2, 3, 4, 8
Naphthalene	91-20-3	X	X	X	1, 2, 3, 4, 5, 7
Naphthalenesulfonic acid, bis(1-methylethyl)-	28757-00-8		X		1, 3, 4
Naphthalenesulfonic acid, polymer with formaldehyde, sodium salt	9084-06-4				2
Naphthalenesulphonic acid, bis (1-methylethyl)-methyl derivatives	99811-86-6		X		1
Naphthenic acid ethoxylate	68410-62-8		X		4
Navy fuels JP-5	NOCAS_25704				1, 2, 3, 4, 8
Nickel sulfate	7786-81-4			X	2
Nickel(II) sulfate hexahydrate	10101-97-0			X	1, 4

Chemical name[a]	CASRN[b]	Known constituent of produced water	Physico-chemical properties	Selected toxicity data[c]	Reference
Nitriles, tallow, hydrogenated	61790-29-2				4
Nitrilotriacetamide	4862-18-4		X		1, 4, 7
Nitrilotriacetic acid	139-13-9		X	X	1, 4
Nitrilotriacetic acid trisodium monohydrate	18662-53-8		X	X	1, 4
Nitrogen	7727-37-9				1, 2, 3, 4, 6
N-Methyl-2-pyrrolidone	872-50-4		X	X	1, 4
N-Methyldiethanolamine	105-59-9		X		2, 4, 8
N-Methylethanolamine	109-83-1		X		4
N-Methyl-N-hydroxyethyl-N-hydroxyethoxyethylamine	68213-98-9		X		4
N-Oleyl diethanolamide	13127-82-7		X		1, 4
Nonyl nonoxynol-10	9014-93-1				4
Nonylphenol (mixed)	25154-52-3				1, 4
Octamethylcyclotetrasiloxane	556-67-2				8
Octoxynol-9	9036-19-5				1, 2, 3, 4, 8
Oil of eucalyptus	8000-48-4				3
Oil of lemongrass	8007-02-1				3
Oil of rosemary	8000-25-7				3
Oleic acid	112-80-1		X		2, 4
Olivine-group minerals	1317-71-1				4
Orange terpenes	8028-48-6				4
Oxirane, 2-methyl-, polymer with oxirane, ether with (chloromethyl) oxirane polymer with 4,4´-(1-methylidene) bis[phenol]	68036-95-3				8

Chemical name[a]	CASRN[b]	Known constituent of produced water	Physico-chemical properties	Selected toxicity data[c]	Reference
Oxirane, 2-methyl-, polymer with oxirane, mono(2-ethylhexyl) ether	64366-70-7				8
Oxirane, 2-methyl-, polymer with oxirane, monodecyl ether	37251-67-5				8
Oxirane, methyl-, polymer with oxirane, mono-C10-16-alkyl ethers, phosphates	68649-29-6				1, 4
Oxygen	7782-44-7				4
o-Xylene	95-47-6	X	X	X	4
Ozone	10028-15-6				8
Paraffin waxes and Hydrocarbon waxes	8002-74-2				1
Paraformaldehyde	30525-89-4				2
PEG-10 Hydrogenated tallow amine	61791-26-2				1, 3
Pentaethylenehexamine	4067-16-7		X		4
Pentane	109-66-0	X	X	X	2, 5
Pentyl acetate	628-63-7		X		3
Pentyl butyrate	540-18-1		X		3
Peracetic acid	79-21-0		X		8
Perboric acid, sodium salt, monohydrate	10332-33-9				1, 8
Perlite	93763-70-3				4
Petrolatum, petroleum, oxidized	64743-01-7				3
Petroleum	8002-05-9			X	1, 2
Petroleum distillate hydrotreated light	6742-47-8				8
Phenanthrene	85-01-8	X	X	X	6
Phenol	108-95-2	X	X	X	1, 2, 4

Chemical name[a]	CASRN[b]	Known constituent of produced water	Physico-chemical properties	Selected toxicity data[c]	Reference
Phenol, 4,4'-(1-methylethylidene)bis-, polymer with 2-(chloromethyl)oxirane, 2-methyloxirane and oxirane	68123-18-2				8
Phenol-formaldehyde resin	9003-35-4				1, 2, 4, 7
Phosphine	7803-51-2			X	1, 4
Phosphonic acid	13598-36-2				1, 4
Phosphonic acid (dimethylamino(methylene))	29712-30-9		X		1
Phosphonic acid, (((2-[(2-hydroxyethyl)(phosphonomethyl)amino)ethyl)imino)bis(methylene))bis-, compd. with 2-aminoethanol	129828-36-0		X		1
Phosphonic acid, (1-hydroxyethylidene)bis-, potassium salt	67953-76-8		X		4
Phosphonic acid, (1-hydroxyethylidene)bis-, tetrasodium salt	3794-83-0		X		1, 4
Phosphonic acid, [[(phosphonomethyl)imino]bis[2,1-ethanediylnitrilobis(methylene)]]tetrakis-	15827-60-8		X		1, 2, 4
Phosphonic acid, [[(phosphonomethyl)imino]bis[2,1-ethanediylnitrilobis(methylene)]]tetrakis-, ammonium salt (1:x)	70714-66-8		X		3
Phosphonic acid, [[(phosphonomethyl)imino]bis[2,1-ethanediylnitrilobis(methylene)]]tetrakis-, sodium salt	22042-96-2		X		3
Phosphonic acid, [[(phosphonomethyl)imino]bis[6,1-hexanediylnitrilobis(methylene)]]tetrakis-	34690-00-1		X		1, 4, 8
Phosphonic acid, [[(phosphonomethyl)imino]bis[6,1-hexanediylnitrilobis(methylene)]]tetrakis-, sodium salt (1:x)	35657-77-3				8
Phosphoric acid	7664-38-2			X	1, 2, 4
Phosphoric acid, aluminium sodium salt	7785-88-8			X	1, 2
Phosphoric acid, ammonium salt (1:x)	10124-31-9				8
Phosphoric acid, ammonium salt (1:3)	10361-65-6				8

Chemical name[a]	CASRN[b]	Known constituent of produced water	Physico-chemical properties	Selected toxicity data[c]	Reference
Phosphoric acid, diammonium salt	7783-28-0			X	2
Phosphoric acid, mixed decyl and Et and octyl esters	68412-60-2				1
Phosphorous acid	10294-56-1				1
Phthalic anhydride	85-44-9		X	X	1, 4
Pine oils	8002-09-3				1, 2, 4
Pluronic F-127	9003-11-6				1, 3, 4, 8
Policapram (Nylon 6)	25038-54-4			X	1, 4
Poly (acrylamide-co-acrylic acid), partial sodium salt	62649-23-4				3, 4
Poly(acrylamide-co-acrylic acid)	9003-06-9				4, 8
Poly(lactide)	26680-10-4				1
Poly(oxy-1,2-ethanediyl), .alpha.-(nonylphenyl)-.omega.-hydroxy-, phosphate	51811-79-1				1, 4
Poly(oxy-1,2-ethanediyl), .alpha.-(octylphenyl)-.omega.-hydroxy-, branched	68987-90-6		X		1, 4
Poly(oxy-1,2-ethanediyl), .alpha.,.alpha.,.alpha.'-[[(9Z)-9-octadecenylimino]di-2,1-ethanediyl]bis[.omega.-hydroxy-	26635-93-8				1, 4
Poly(oxy-1,2-ethanediyl), .alpha.-[(9Z)-1-oxo-9-octadecenyl]-.omega.-hydroxy-	9004-96-0				8
Poly(oxy-1,2-ethanediyl), .alpha.-hydro-.omega.-hydroxy-, mono-C10-14-alkyl ethers, phosphates	68585-36-4				8
Poly(oxy-1,2-ethanediyl), .alpha.-hydro-.omega.-hydroxy-, mono-C8-10-alkyl ethers, phosphates	68130-47-2				8
Poly(oxy-1,2-ethanediyl), .alpha.-isodecyl-.omega.-hydroxy-	61827-42-7				8

Chemical name[a]	CASRN[b]	Known constituent of produced water	Physico-chemical properties	Selected toxicity data[c]	Reference
Poly(oxy-1,2-ethanediyl), .alpha.-sulfo-.omega.-hydroxy-, C10-16-alkyl ethers, sodium salts	68585-34-2				8
Poly(oxy-1,2-ethanediyl), .alpha.-sulfo-.omega.-hydroxy-, C12-14-alkyl ethers, sodium salts	68891-38-3				1, 4
Poly(oxy-1,2-ethanediyl), alpha-(2,3,4,5-tetramethylnonyl)-omega-hydroxy	68015-67-8				1
Poly(oxy-1,2-ethanediyl), alpha-(nonylphenyl)-omega-hydroxy-, branched, phosphates	68412-53-3				1
Poly(oxy-1,2-ethanediyl), alpha-hexyl-omega-hydroxy	31726-34-8				3, 8
Poly(oxy-1,2-ethanediyl), alpha-hydro-omega-hydroxy-, (9Z)-9-octadecenoate	56449-46-8				3
Poly(oxy-1,2-ethanediyl), alpha-hydro-omega-hydroxy-, ether with alpha-fluoro-omega-(2-hydroxyethyl)poly(difluoromethylene) (1:1)	65545-80-4				1
Poly(oxy-1,2-ethanediyl), alpha-hydro-omega-hydroxy-, ether with D-glucitol (2:1), tetra-(9Z)-9-octadecenoate	61723-83-9				8
Poly(oxy-1,2-ethanediyl), alpha-sulfo-omega-(decyloxy)-, ammonium salt (1:1)	52286-19-8				4
Poly(oxy-1,2-ethanediyl), alpha-sulfo-omega-(hexyloxy)-, ammonium salt (1:1)	63428-86-4				1, 3, 4
Poly(oxy-1,2-ethanediyl), alpha-sulfo-omega-(hexyloxy)-, C6-10-alkyl ethers, ammonium salts	68037-05-8				3, 4
Poly(oxy-1,2-ethanediyl), alpha-sulfo-omega--(nonylphenoxy)-	9081-17-8				4
Poly(oxy-1,2-ethanediyl), alpha-sulfo-omega-(octyloxy)-, ammonium salt (1:1)	52286-18-7				4
Poly(oxy-1,2-ethanediyl), alpha-sulfo-omega-hydroxy-, C10-12-alkyl ethers, ammonium salts	68890-88-0				8

Chemical name[a]	CASRN[b]	Known constituent of produced water	Physico-chemical properties	Selected toxicity data[c]	Reference
Poly(oxy-1,2-ethanediyl), alpha-tridecyl-omega-hydroxy-	24938-91-8				1, 3, 4
Poly(oxy-1,2-ethanediyl), alpha-undecyl-omega-hydroxy-, branched and linear	127036-24-2				1
Poly-(oxy-1,2-ethanediyl)-alpha-undecyl-omega-hydroxy	34398-01-1				1, 3, 4, 8
Poly(oxy-1,2-ethanediyl)-nonylphenyl-hydroxy branched	127087-87-0				1, 2, 3, 4
Poly(sodium-p-styrenesulfonate)	25704-18-1				1, 4
Poly(tetrafluoroethylene)	9002-84-0			X	8
Poly[imino(1,6-dioxo-1,6-hexanediyl)imino-1,6-hexanediyl]	32131-17-2				2
Polyacrylamide	9003-05-8				1, 2, 4, 6
Polyacrylate/ polyacrylamide blend	NOCAS_51256				2
Polyacrylic acid, sodium bisulfite terminated	66019-18-9				3
Polyethylene glycol	25322-68-3				1, 2, 3, 4, 7, 8
Polyethylene glycol (9Z)-9-octadecenyl ether	9004-98-2				8
Polyethylene glycol ester with tall oil fatty acid	68187-85-9				1
Polyethylene glycol monobutyl ether	9004-77-7				1, 4
Polyethylene glycol mono-C8-10-alkyl ether sulfate ammonium	68891-29-2				1, 3, 4
Polyethylene glycol nonylphenyl ether	9016-45-9				1, 2, 3, 4, 8
Polyethylene glycol tridecyl ether phosphate	9046-01-9				1, 3, 4
Polyethyleneimine	9002-98-6				4
Polyglycerol	25618-55-7				2
Poly-L-aspartic acid sodium salt	34345-47-6				8
Polyoxyethylene sorbitan trioleate	9005-70-3				3

Chemical name[a]	CASRN[b]	Known constituent of produced water	Physico-chemical properties	Selected toxicity data[c]	Reference
Polyoxyethylene(10)nonylphenyl ether	26027-38-3				1, 2, 3, 4, 8
Polyoxyl 15 hydroxystearate	70142-34-6				8
Polyoxypropylenediamine	9046-10-0				1
Polyphosphoric acids, esters with triethanolamine, sodium salts	68131-72-6				1
Polyphosphoric acids, sodium salts	68915-31-1			X	1, 4
Polypropylene glycol	25322-69-4	X			1, 2, 4
Polypropylene glycol glycerol triether, epichlorohydrin, bisphenol A polymer	68683-13-6				1
Polyquaternium 5	26006-22-4				1, 4
Polysorbate 20	9005-64-5				8
Polysorbate 60	9005-67-8				3, 4
Polysorbate 80	9005-65-6				3, 4
Polyvinyl acetate copolymer	9003-20-7			X	2
Polyvinyl acetate, partially hydrolyzed	304443-60-5				8
Polyvinyl alcohol	9002-89-5			X	1, 2, 4
Polyvinyl alcohol/polyvinyl acetate copolymer	NOCAS_50147				2
Polyvinylidene chloride	9002-85-1				8
Polyvinylpyrrolidone	9003-39-8			X	8
Portland cement	65997-15-1				2, 4
Potassium acetate	127-08-2		X		1, 3, 4
Potassium aluminum silicate	1327-44-2				5
Potassium antimonate	29638-69-5				1, 4

Chemical name[a]	CASRN[b]	Known constituent of produced water	Physico-chemical properties	Selected toxicity data[c]	Reference
Potassium bisulfate	7646-93-7				8
Potassium borate	12712-38-8				3
Potassium borate (1:x)	20786-60-1				1, 3
Potassium carbonate sesquihydrate	6381-79-9				5
Potassium chloride	7447-40-7				1, 2, 3, 4, 5, 6, 7
Potassium dichromate	7778-50-9				4
Potassium hydroxide	1310-58-3				1, 2, 3, 4, 6
Potassium iodide	7681-11-0			X	1, 4
Potassium metaborate	13709-94-9				1, 2, 3, 4, 8
Potassium oleate	143-18-0		X		4
Potassium oxide	12136-45-7				1, 4
Potassium persulfate	7727-21-1				1, 2, 4
Potassium phosphate, tribasic	7778-53-2			X	8
Potassium sulfate	7778-80-5				2
Propane	74-98-6		X		2, 5
Propanol, 1(or 2)-(2-methoxymethylethoxy)-	34590-94-8		X		1, 2, 3, 4
Propargyl alcohol	107-19-7	X	X	X	1, 2, 3, 4, 5, 6, 7, 8
Propylene carbonate	108-32-7		X		1, 4
Propylene pentamer	15220-87-8		X		1
p-Xylene	106-42-3	X	X	X	1, 4
Pyridine, alkyl derivs.	68391-11-7				1, 4
Pyridinium, 1-(phenylmethyl)-, alkyl derivs., chlorides	100765-57-9				4, 8

Chemical name[a]	CASRN[b]	Known constituent of produced water	Physico-chemical properties	Selected toxicity data[c]	Reference
Pyridinium, 1-(phenylmethyl)-, C7-8-alkyl derivs., chlorides	70914-44-2				6
Pyrimidine	289-95-2		X		2
Pyrrole	109-97-7		X		2
Quartz-alpha (SiO2)	14808-60-7			X	1, 2, 3, 4, 5, 6, 8
Quaternary ammonium compounds (2-ethylhexyl) hydrogenated tallow alkyl)dimethyl, methyl sulfates	308074-31-9				8
Quaternary ammonium compounds, (oxydi-2,1-ethanediyl)bis[coco alkyldimethyl, dichlorides	68607-28-3				2, 3, 4, 8
Quaternary ammonium compounds, benzyl(hydrogenated tallow alkyl)dimethyl, bis(hydrogenated tallow alkyl)dimethylammonium salt with bentonite	71011-25-1				8
Quaternary ammonium compounds, benzylbis(hydrogenated tallow alkyl)methyl, salts with bentonite	68153-30-0				2, 5, 6
Quaternary ammonium compounds, benzyl-C10-16-alkyldimethyl, chlorides	68989-00-4				1, 4
Quaternary ammonium compounds, benzyl-C12-16-alkyldimethyl, chlorides	68424-85-1			X	1, 2, 4, 8
Quaternary ammonium compounds, benzyl-C12-18-alkyldimethyl, chlorides	68391-01-5				8
Quaternary ammonium compounds, benzylcoco alkyldimethyl, chlorides	61789-71-7				8
Quaternary ammonium compounds, bis(hydrogenated tallow alkyl)dimethyl, salts with bentonite	68953-58-2				2, 3, 4, 8
Quaternary ammonium compounds, bis(hydrogenated tallow alkyl)dimethyl, salts with hectorite	71011-27-3				2
Quaternary ammonium compounds, di-C8-10-alkyldimethyl, chlorides	68424-95-3		X		2

Chemical name[a]	CASRN[b]	Known constituent of produced water	Physico-chemical properties	Selected toxicity data[c]	Reference
Quaternary ammonium compounds, dicoco alkyldimethyl, chlorides	61789-77-3				1
Quaternary ammonium compounds, pentamethyltallow alkyltrimethylenedi-, dichlorides	68607-29-4				4
Quaternary ammonium compounds, trimethyltallow alkyl, chlorides	8030-78-2				1, 4
Quinaldine	91-63-4		X		8
Quinoline	91-22-5	X	X	X	2, 4
Raffinates (petroleum)	68514-29-4				5
Raffinates, petroleum, sorption process	64741-85-1				1, 2, 4, 8
Residual oils, petroleum, solvent-refined	64742-01-4				5
Residues, petroleum, catalytic reformer fractionator	64741-67-9				1, 4, 8
Rhodamine B	81-88-9		X	X	4
Rosin	8050-09-7				1, 4
Rutile titanium dioxide	1317-80-2				8
Sand	308075-07-2				8
Scandium oxide	12060-08-1				8
Sepiolite	63800-37-3				2
Silane, dichlorodimethyl-, reaction products with silica	68611-44-9				2, 4
Silica	7631-86-9	X		X	1, 2, 3, 4, 8
silica gel, cryst. -free	112926-00-8				3, 4
Silica, amorphous, fumed, cryst.-free	112945-52-5				1, 3, 4
Silica, vitreous	60676-86-0				1, 4, 8
Silicic acid, aluminum potassium sodium salt	12736-96-8				4

Chemical name[a]	CASRN[b]	Known constituent of produced water	Physico-chemical properties	Selected toxicity data[c]	Reference
Siloxanes (Polysiloxane)	9011-19-2				4
Siloxanes and Silicones, di-Me, 3-hydroxypropyl Me, ethoxylated propoxylated	68937-55-3				8
Siloxanes and Silicones, di-Me, Me hydrogen	68037-59-2				8
Siloxanes and silicones, di-Me, polymers with Me silsesquioxanes	68037-74-1				4
Siloxanes and Silicones, di-Me, reaction products with silica	67762-90-7				4
Siloxanes and silicones, dimethyl,	63148-52-7				4
Silwet L77	27306-78-1				1
Sodium 1-octanesulfonate	5324-84-5		X		3
Sodium 2-mercaptobenzothiolate	2492-26-4		X		2
Sodium acetate	127-09-3		X		1, 3, 4
Sodium aluminate	1302-42-7				2, 4
Sodium benzoate	532-32-1		X		3
Sodium bicarbonate	144-55-8		X		1, 2, 3, 4, 7
Sodium bis(tridecyl) sulfobutanedioate	2673-22-5		X		4
Sodium bisulfite	7631-90-5			X	1, 3, 4
Sodium borate	1333-73-9				1, 4, 6, 7
Sodium bromate	7789-38-0				1, 2, 4
Sodium bromide	7647-15-6				1, 2, 3, 4, 7
Sodium bromosulfamate	1004542-84-0				8
Sodium C14-16 alpha-olefin sulfonate	68439-57-6		X		1, 3, 4
Sodium caprylamphopropionate	68610-44-6		X		4

Chemical name[a]	CASRN[b]	Known constituent of produced water	Physico-chemical properties	Selected toxicity data[c]	Reference
Sodium carbonate	497-19-8		X		1, 2, 3, 4, 8
Sodium chlorate	7775-09-9			X	1, 4
Sodium chloride	7647-14-5				1, 2, 3, 4, 5, 8
Sodium chlorite	7758-19-2			X	1, 2, 3, 4, 5, 8
Sodium chloroacetate	3926-62-3		X		3
Sodium cocaminopropionate	68608-68-4				1
Sodium decyl sulfate	142-87-0		X		1
Sodium D-gluconate	527-07-1		X		4
Sodium diacetate	126-96-5		X		1, 4
Sodium dichloroisocyanurate	2893-78-9		X		2
Sodium dl-lactate	72-17-3		X		8
Sodium dodecyl sulfate	151-21-3		X		8
Sodium erythorbate (1:1)	6381-77-7		X		1, 3, 4, 8
Sodium ethasulfate	126-92-1		X		1
Sodium formate	141-53-7		X		2, 8
Sodium hydrogen sulfate	7681-38-1				4
Sodium hydroxide	1310-73-2				1, 2, 3, 4, 7, 8
Sodium hydroxymethanesulfonate	870-72-4		X		8
Sodium hypochlorite	7681-52-9				1, 2, 3, 4, 8
Sodium iodide	7681-82-5			X	4
Sodium ligninsulfonate	8061-51-6				2
Sodium l-lactate	867-56-1		X		8

Chemical name[a]	CASRN[b]	Known constituent of produced water	Physico-chemical properties	Selected toxicity data[c]	Reference
Sodium maleate (1:x)	18016-19-8		X		8
Sodium metabisulfite	7681-57-4			X	1
Sodium metaborate	7775-19-1				3, 4
Sodium metaborate dihydrate	16800-11-6				1, 4
Sodium metaborate tetrahydrate[d]	10555-76-7				1, 4, 8
Sodium metasilicate	6834-92-0				1, 2, 4
Sodium molybdate(VI)	7631-95-0				8
Sodium nitrate	7631-99-4				2
Sodium nitrite	7632-00-0				1, 2, 4
Sodium N-methyl-N-oleoyltaurate	137-20-2		X		4
Sodium octyl sulfate	142-31-4		X		1
Sodium oxide	1313-59-3				1
Sodium perborate	11138-47-9				4
Sodium perborate tetrahydrate	10486-00-7				1, 4, 5, 8
Sodium peroxoborate	7632-04-4				1
Sodium persulfate	7775-27-1				1, 2, 3, 4, 7, 8
Sodium phosphate	7632-05-5				1, 4
Sodium polyacrylate	9003-04-7				1, 2, 3, 4
Sodium pyrophosphate	7758-16-9			X	1, 2, 4
Sodium salicylate	54-21-7		X		1, 4
Sodium sesquicarbonate	533-96-0		X		1, 2
Sodium silicate	1344-09-8				1, 2, 4

Chemical name[a]	CASRN[b]	Known constituent of produced water	Physico-chemical properties	Selected toxicity data[c]	Reference
Sodium starch glycolate	9063-38-1				2
Sodium sulfate	7757-82-6				1, 2, 3, 4
Sodium sulfite	7757-83-7			X	2, 4, 8
Sodium thiocyanate	540-72-7		X		1, 4
Sodium thiosulfate	7772-98-7				1, 2, 3, 4
Sodium thiosulfate, pentahydrate	10102-17-7				1, 4
Sodium trichloroacetate	650-51-1		X		1, 4
Sodium trimetaphosphate	7785-84-4			X	8
Sodium xylenesulfonate	1300-72-7		X		1, 3, 4
Sodium zirconium lactate	15529-67-6				8
Sodium zirconium lactic acid (4:4:1)	10377-98-7				1, 4
Solvent naphtha, petroleum, heavy aliph.	64742-96-7				2, 4, 8
Solvent naphtha, petroleum, heavy arom.	64742-94-5				1, 2, 4, 5, 8
Solvent naphtha, petroleum, light aliph.	64742-89-8				8
Solvent naphtha, petroleum, light arom.	64742-95-6				1, 2, 4
Sorbic acid	110-44-1		X		8
Sorbitan sesquioleate	8007-43-0		X		4
Sorbitan, mono-(9Z)-9-octadecenoate	1338-43-8		X		1, 2, 3, 4
Sorbitan, monooctadecanoate	1338-41-6		X		8
Sorbitan, tri-(9Z)-9-octadecenoate	26266-58-0		X		8
Spirit of ammonia, aromatic	8013-59-0				8
Stannous chloride dihydrate	10025-69-1				1, 4

Appendix H – Chemicals Identified in Hydraulic Fracturing Fluids and/or Produced Water

Chemical name[a]	CASRN[b]	Known constituent of produced water	Physico-chemical properties	Selected toxicity data[c]	Reference
Starch	9005-25-8				1, 2, 4
Steam cracked distillate, cyclodiene dimer, dicyclopentadiene polymer	68131-87-3				1
Stoddard solvent	8052-41-3			X	1, 3, 4
Stoddard solvent IIC	64742-88-7				1, 2, 4
Strontium chloride	10476-85-4			X	4
Styrene	100-42-5		X	X	2
Subtilisin	9014-01-1				8
Sucrose	57-50-1		X		1, 2, 3, 4
Sulfamic acid	5329-14-6			X	1, 4
Sulfan blue	129-17-9		X	X	8
Sulfate	14808-79-8	X			1, 4
Sulfo NHS Biotin	119616-38-5				8
Sulfomethylated quebracho	68201-64-9				2
Sulfonic acids, C10-16-alkane, sodium salts	68608-21-9				6
Sulfonic acids, petroleum	61789-85-3				1
Sulfonic acids, petroleum, sodium salts	68608-26-4				3
Sulfur dioxide	7446-09-5			X	2, 4, 8
Sulfuric acid	7664-93-9			X	1, 2, 4, 7
Sulfuric acid, mono-C12-18-alkyl esters, sodium salts	68955-19-1		X		4
Sulfuric acid, mono-C6-10-alkyl esters, ammonium salts	68187-17-7		X		1, 4, 8
Symclosene	87-90-1		X		2
Talc	14807-96-6			X	1, 3, 4, 6, 7

Chemical name[a]	CASRN[b]	Known constituent of produced water	Physico-chemical properties	Selected toxicity data[c]	Reference
Tall oil	8002-26-4				4, 8
Tall oil imidazoline	61791-36-4				4
Tall oil, compound with diethanolamine	68092-28-4				1
Tall oil, ethoxylated	65071-95-6				4, 8
Tall-oil pitch	8016-81-7				4
Tallow alkyl amines acetate	61790-60-1				8
Tar bases, quinoline derivatives, benzyl chloride-quaternized	72480-70-7				1, 3, 4
Tegin M	8043-29-6				8
Terpenes and Terpenoids, sweet orange-oil	68647-72-3				1, 3, 4, 8
Terpineol	8000-41-7				1, 3
tert-Butyl hydroperoxide	75-91-2		X		1, 4
tert-Butyl perbenzoate	614-45-9		X		1
Tetra-calcium-alumino-ferrite	12068-35-8				1, 2, 4
Tetradecane	629-59-4	X	X		8
Tetradecyldimethylbenzylammonium chloride	139-08-2		X		1, 4, 8
Tetraethylene glycol	112-60-7		X		1, 4
Tetraethylenepentamine	112-57-2		X		1, 4
Tetrakis(hydroxymethyl)phosphonium sulfate	55566-30-8		X		1, 2, 3, 4, 7
Tetramethyl orthosilicate	681-84-5				1
Tetramethylammonium chloride	75-57-0		X		1, 2, 3, 4, 7, 8
Tetrasodium pyrophosphate	7722-88-5			X	8
Thiamine hydrochloride	67-03-8		X		8

Chemical name[a]	CASRN[b]	Known constituent of produced water	Physico-chemical properties	Selected toxicity data[c]	Reference
Thiocyanic acid, ammonium salt	1762-95-4		X		2, 3, 4
Thioglycolic acid	68-11-1		X		1, 2, 3, 4
Thiourea	62-56-6		X	X	1, 2, 3, 4, 6
Thiourea, polymer with formaldehyde and 1-phenylethanone	68527-49-1				1, 4, 8
Thuja plicata donn ex. D. don leaf oil	68917-35-1				3
Tin(II) chloride	7772-99-8				1
Titanium dioxide[d]	13463-67-7			X	1, 2, 4, 8
Titanium(4+) 2-[bis(2-hydroxyethyl)amino]ethanolate propan-2-olate (1:2:2)	36673-16-2				1
Titanium, isopropoxy (triethanolaminate)	74665-17-1				1, 4
Toluene	108-88-3	X	X	X	1, 3, 4
Tributyl phosphate	126-73-8	X	X	X	1, 2, 4
Tributyltetradecylphosphonium chloride	81741-28-8		X		1, 3, 4
Tricalcium phosphate	7758-87-4			X	1, 4
Tricalcium silicate	12168-85-3				1, 2, 4
Tridecane	629-50-5	X	X		8
Triethanolamine	102-71-6		X	X	1, 2, 4
Triethanolamine hydrochloride	637-39-8		X		8
Triethanolamine hydroxyacetate	68299-02-5		X		3
Triethanolamine polyphosphate ester	68131-71-5				1, 4, 8
Triethyl citrate	77-93-0		X		1, 4
Triethyl phosphate	78-40-0		X		1, 4

Chemical name[a]	CASRN[b]	Known constituent of produced water	Physico-chemical properties	Selected toxicity data[c]	Reference
Triethylene glycol	112-27-6		X		1, 2, 3
Triethylenetetramine	112-24-3		X		4
Triisopropanolamine	122-20-3		X		1, 4
Trimethanolamine	14002-32-5		X		3
Trimethyl borate	121-43-7				8
Trimethylamine	75-50-3		X		8
Trimethylamine quaternized polyepichlorohydrin	51838-31-4				1, 2, 3, 4, 5, 8
Trimethylbenzene	25551-13-7	X		X	1, 2, 4
Triphosphoric acid, pentasodium salt	7758-29-4			X	1, 4
Tripoli	1317-95-9				4
Tripotassium citrate monohydrate	6100-05-6		X		4
Tripropylene glycol monomethyl ether	25498-49-1		X		2
Trisodium citrate	68-04-2		X		3
Trisodium citrate dihydrate	6132-04-3		X		1, 4
Trisodium ethylenediaminetetraacetate	150-38-9		X		1, 3
Trisodium ethylenediaminetriacetate	19019-43-3		X		1, 4, 8
Trisodium phosphate	7601-54-9			X	1, 2, 4
Trisodium phosphate dodecahydrate	10101-89-0				1
Tritan R (X-100)	92046-34-9				8
Triton X-100	9002-93-1				1, 3, 4
Tromethamine	77-86-1		X		3, 4
Tryptone	73049-73-7				8

Appendix H – Chemicals Identified in Hydraulic Fracturing Fluids and/or Produced Water

Chemical name[a]	CASRN[b]	Known constituent of produced water	Physico-chemical properties	Selected toxicity data[c]	Reference
Ulexite	1319-33-1				1, 2, 3, 8
Undecane	1120-21-4	X	X		3, 8
Undecanol, branched and linear	128973-77-3				8
Urea	57-13-6		X		1, 2, 4, 8
Vermiculite	1318-00-9				2
Vinyl acetate ethylene copolymer	24937-78-8				1, 4
Vinylidene chloride/methylacrylate copolymer	25038-72-6				4
Water	7732-18-5				2, 4, 8
White mineral oil, petroleum	8042-47-5				1, 2, 4
Xylenes	1330-20-7	X	X	X	1, 2, 4
Yeast extract	8013-01-2				8
Zeolites	1318-02-1			X	8
Zinc	7440-66-6	X		X	2
Zinc carbonate	3486-35-9				2
Zinc chloride	7646-85-7				1, 2
Zinc oxide	1314-13-2				1, 4
Zinc sulfate monohydrate	7446-19-7				8
Zirconium nitrate	13746-89-9				2, 6
Zirconium oxide sulfate	62010-10-0				1, 4
Zirconium oxychloride	7699-43-6				1, 2, 4
Zirconium(IV) chloride tetrahydrofuran complex	21959-01-3				5
Zirconium(IV) sulfate	14644-61-2				2, 6

Chemical name[a]	CASRN[b]	Known constituent of produced water	Physico-chemical properties	Selected toxicity data[c]	Reference
Zirconium, 1,1'-((2-hydroxyethyl)(2-hydroxypropyl)amino)ethyl)imino)bis(2-propanol) complexes	197980-53-3				4
Zirconium, acetate lactate oxo ammonium complexes	68909-34-2				4, 8
Zirconium, chloro hydroxy lactate oxo sodium complexes	174206-15-6				4
Zirconium, hydroxylactate sodium complexes	113184-20-6				1, 4
Zirconium,tetrakis[2-[bis(2-hydroxyethyl)amino-kN]amino-kN]ethanolato-kO]-	101033-44-7				1, 2, 4, 5

[a] DSSTox chemical names assigned to the listed CASRN can be reformatted or change over time with additional curation review. In the case that a chemical name in this table no longer matches the DSSTox chemical name for a listed CASRN, the CASRN would be presumed to be the invariant identifier.

[b] Some chemicals are designated as "NOCAS_" which are DSSTox database-specific CAS-like identifiers assigned to a listed chemical name or substance.

[c] Chemicals are flagged as having selected toxicity data available if they have one or more oral reference values, oral slope factors, or qualitative cancer classifications available from the sources presented in Appendix G.

[d] Four chemicals have data in the EPA's FracFocus 1.0 project database for CASRNs that are different from those in this table: Ethanolamine, CASRN 9007-33-4; Sodium metaborate tetrahydrate, CASRN 35585-58-1; Boric acid (H₃BO₃), compd. with 2-aminoethanol (1:x), CASRN 68425-67-2; and Titanium dioxide, CASRN 98084-96-9. Three of these (9007-33-4, 68425-67-2, and 98084-96-9) are "deleted" CASRNs, and so were not included in this table; instead, the chemical name has been remapped here to the current "active" CASRNs. CASRN 35585-58-1 is listed for sodium metaborate tetrahydrate in the EPA's FracFocus 1.0 project database, but is assigned to a different chemical (disodium dioxoborate) in the EPA's Distributed Structure-Searchable Toxicity (DSSTox) Database, and so was not included in this table.

Table H-3. List of generic names of chemicals reportedly used in hydraulic fracturing fluids.

In some cases, the generic chemical name masks a specific chemical name and CASRN provided to the EPA and claimed as CBI by one or more of the nine hydraulic fracturing service companies.

Generic chemical name	Reference
2-Substituted aromatic amine salt	1, 4
Acetylenic alcohol	1
Acrylamide acrylate copolymer	4
Acrylamide copolymer	1, 4
Acrylamide modified polymer	4
Acrylamide-sodium acrylate copolymer	4
Acrylate copolymer	1
Acrylic copolymer	1
Acrylic polymer	1, 4
Acrylic resin	4
Acyclic hydrocarbon blend	1, 4
Acylbenzylpyridinium choride	8
Alcohol alkoxylate	1, 4
Alcohol and fatty acid blend	2
Alcohol ethoxylates	4
Alcohols	1, 4
Alcohols, C9-C22	1, 4
Aldehydes	1, 4, 5
Alfa-alumina	1, 4
Aliphatic acids	1, 2, 3, 4
Aliphatic alcohol	2
Aliphatic alcohol glycol ether	3, 4
Aliphatic alcohols, ethoxylated	2
Aliphatic amine derivative	1
Aliphatic carboxylic acid	4
Alkaline bromide salts	1, 4
Alkaline metal oxide	4
Alkanes/alkenes	4

Generic chemical name	Reference
Alkanolamine derivative	2
Alkanolamine/aldehyde condensate	1, 2, 4
Alkenes	1, 4
Alklaryl sulfonic acid	1, 4
Alkoxylated alcohols	1
Alkoxylated amines	1, 4
Alkyaryl sulfonate	1, 2, 3, 4
Alkyl alkoxylate	1, 4
Alkyl amide	4
Alkyl amine	1, 4
Alkyl amine blend in a metal salt solution	1, 4
Alkyl aryl amine sulfonate	4
Alkyl aryl polyethoxy ethanol	3, 4
Alkyl dimethyl benzyl ammonium chloride	4
Alkyl esters	1, 4
Alkyl ether phosphate	4
Alkyl hexanol	1, 4
Alkyl ortho phosphate ester	1, 4
Alkyl phosphate ester	1, 4
Alkyl phosphonate	4
Alkyl pyridines	2
Alkyl quaternary ammonium chlorides	1, 4
Alkyl quaternary ammonium salt	4
Alkylamine alkylaryl sulfonate	4
Alkylamine salts	2
Alkylaryl sulfonate	1, 4
Alkylated quaternary chloride	1, 2, 4
Alkylated sodium naphthalenesulphonate	2
Alkylbenzenesulfonate	2
Alkylbenzenesulfonic acid	1, 4, 5

Generic chemical name	Reference
Alkylethoammonium sulfates	1
Alkylphenol ethoxylates	1, 4
Alkylpyridinium quaternary	4
Alphatic alcohol polyglycol ether	2
Aluminum oxide	1, 4
Amide	4
Amidoamine	1, 4
Amine	1, 4
Amine compound	4
Amine oxides	1, 4
Amine phosphonate	1, 4
Amine salt	1
Amino compounds	1, 4
Amino methylene phosphonic acid salt	1, 4
Ammonium alcohol ether sulfate	1, 4
Ammonium salt	1, 4
Ammonium salt of ethoxylated alcohol sulfate	1, 4
Amorphous silica	4
Amphoteric surfactant	2
Anionic acrylic polymer	2
Anionic copolymer	1, 4
Anionic polyacrylamide	1, 2, 4
Anionic polyacrylamide copolymer	1, 4, 6
Anionic polymer	1, 3, 4
Anionic surfactants	2, 4, 6
Antifoulant	1, 4
Antimonate salt	1, 4
Aqueous emulsion of diethylpolysiloxane	2
Aromatic alcohol glycol ether	1
Aromatic aldehyde	1, 4

Generic chemical name	Reference
Aromatic hydrocarbons	3, 4
Aromatic ketones	1, 2, 3, 4
Aromatic polyglycol ether	1
Arsenic compounds	4
Ashes, residues	4
Bentone clay	4
Biocide	4
Biocide component	1, 4
Bis-quaternary methacrylamide monomer	4
Blast furnace slag	4
Borate salts	1, 2, 4
Cadmium compounds	4
Carbohydrates	1, 2, 4
Carboxylmethyl hydroxypropyl guar	4
Cationic polyacrylamide	4
Cationic polymer	2, 4
Cedar fiber, processed	2
Cellulase enzyme	1
Cellulose derivative	1, 2, 4
Cellulose ether	2
Cellulosic polymer	2
Ceramic	4
Chlorous ion solution	1
Chromates	1, 4
Chrome-free lignosulfonate compound	2
Citrus rutaceae extract	4
Common white	4
Complex alkylaryl polyo-ester	1
Complex aluminum salt	1, 4
Complex carbohydrate	2

Generic chemical name	Reference
Complex organometallic salt	1
Complex polyamine salt	7
Complex substituted keto-amine	1
Complex substituted keto-amine hydrochloride	1
Copper compounds	6
Coric oxide	4
Cotton dust (raw)	2
Cottonseed hulls	2
Cured acrylic resin	1, 4
Cured resin	1, 4, 5
Cured urethane resin	1, 4
Cyclic alkanes	1, 4
Defoamer	4
Dibasic ester	4
Dicarboxylic acid	1, 4
Diesel	1, 4, 6
Dimethyl silicone	1, 4
Dispersing agent	1
Emulsifier	4
Enzyme	4
Epoxy	4
Epoxy resin	1, 4
Essential oils	1, 4
Ester Salt	2, 4
Esters	2, 4
Ether compound	4
Ether salt	4
Ethoxylated alcohol blend	4
Ethoxylated alcohol/ester mixture	4
Ethoxylated alcohols	1, 2, 4, 5, 7

Generic chemical name	Reference
Ethoxylated alkyl amines	1, 4
Ethoxylated amine blend	4
Ethoxylated amines	1, 4
Ethoxylated fatty acid	4
Ethoxylated fatty acid ester	1, 4
Ethoxylated nonionic surfactant	1, 4
Ethoxylated nonylphenol	1, 2, 4
Ethoxylated sorbitol esters	1, 4
Ethylene oxide-nonylphenol polymer	4
Fatty acid amine salt mixture	4
Fatty acid ester	1, 2, 4
Fatty acid tall oil	1, 4
Fatty acid, ethoxylate	4
Fatty acids	1
Fatty alcohol alkoxylate	1, 4
Fatty alkyl amine salt	1, 4
Fatty amine carboxylates	1, 4
Fatty imidazoline	4
Fluoroaliphatic polymeric esters	1, 4
Formaldehyde polymer	1
Glass fiber	1, 4
Glyceride esters	2
Glycol	4
Glycol blend	2
Glycol ethers	1, 4, 7
Ground cedar	2
Ground paper	2
Guar derivative	1, 4
Guar gum	4
Haloalkyl heteropolycycle salt	1, 4

Generic chemical name	Reference
Hexanes	1
High molecular weight polymer	2
High pH conventional enzymes	2
Hydrocarbons	1
Hydrogen solvent	4
Hydrotreated and hydrocracked base oil	1, 4
Hydrotreated distillate, light C9-16	4
Hydrotreated heavy naphthalene	5
Hydrotreated light distillate	2, 4
Hydrotreated light petroleum distillate	4
Hydroxyalkyl imino carboxylic sodium salt	2
Hydroxycellulose	6
Hydroxyethyl cellulose	1, 2, 4
Imidazolium compound	4
Inner salt of alkyl amines	1, 4
Inorganic borate	1, 4
Inorganic chemical	4
Inorganic particulate	1, 4
Inorganic salt	2, 4
Iso-alkanes/n-alkanes	1, 4
Isomeric aromatic ammonium salt	1, 4
Latex	2, 4
Lead compounds	4
Low toxicity base oils	1, 4
Lubra-Beads course	4
Maghemite	1, 4
Magnetite	1, 4
Metal salt	1
Metal salt solution	1
Mineral	1, 4

Generic chemical name	Reference
Mineral fiber	2
Mineral filler	1
Mineral oil	4
Mixed titanium ortho ester complexes	1, 4
Modified acrylamide copolymer	2, 4
Modified acrylate polymer	4
Modified alkane	1, 4
Modified bentonite	4
Modified cycloaliphatic amine adduct	1, 4
Modified lignosulfonate	2, 4
Naphthalene derivatives	1, 4
Neutralized alkylated napthalene sulfonate	4
Nickel chelate catalyst	4
Nonionic surfactant	1
N-tallowalkyltrimethylenediamines	4
Nuisance particulates	1, 2, 4
Nylon	4
Olefinic sulfonate	1, 4
Olefins	1, 4
Organic acid salt	1, 4
Organic acids	1, 4
Organic alkyl amines	4
Organic chloride	4
Organic modified bentonite clay	4
Organic phosphonate	1, 4
Organic phosphonate salts	1, 4
Organic phosphonic acid salts	1, 4
Organic polymer	4
Organic polyol	4
Organic salt	1, 4

Generic chemical name	Reference
Organic sulfur compound	1, 4
Organic surfactants	1
Organic titanate	1, 4
Organo amino silane	4
Organo phosphonic acid	4
Organo phosphonic acid salt	4
Organometallic ammonium complex	1
Organophilic clay	4
Oxidized tall oil	2
Oxoaliphatic acid	2
Oxyalkylated alcohol	1, 4
Oxyalkylated alkyl alcohol	2, 4
Oxyalkylated alkylphenol	1, 2, 3, 4
Oxyalkylated fatty acid	1, 4
Oxyalkylated fatty alcohol salt	2
Oxyalkylated phenol	1, 4
Oxyalkylated phenolic resin	4
Oxyalkylated polyamine	1
Oxyalkylated tallow diamine	2
Oxyethylated alcohol	2
Oxylated alcohol	1, 4
P/F resin	4
Paraffin inhibitor	4
Paraffinic naphthenic solvent	1
Paraffinic solvent	1, 4
Paraffins	1
Pecan shell	2
Petroleum distallate blend	2, 3, 4
Petroleum gas oils	1
Petroleum hydrocarbons	4

Generic chemical name	Reference
Petroleum solvent	2
Phosphate ester	1, 4
Phosphonate	2
Phosphonic acid	1, 4
Phosphoric acid, mixed polyoxyalkylene aryl and alkyl esters	4
Plasticizer	1, 2
Polyacrylamide copolymer	4
Polyacrylamides	1
Polyacrylate	1, 4
Polyactide resin	4
Polyalkylene esters	4
Polyaminated fatty acid	2
Polyaminated fatty acid surfactants	2
Polyamine	1, 4
Polyamine polymer	4
Polyanionic cellulose	1
Polyaromatic hydrocarbons	6
Polycyclic organic matter	6
Polyelectrolyte	4
Polyether polyol	2
Polyethoxylated alkanol	2, 3, 4
Polyethylene copolymer	4
Polyethylene glycols	4
Polyethylene wax	4
Polyglycerols	2
Polyglycol	2
Polyglycol ether	6
Polylactide resin	4
Polymer	2, 4
Polymeric hydrocarbons	3, 4

Generic chemical name	Reference
Polymerized alcohol	4
Polymethacrylate polymer	4
Polyol phosphate ester	2
Polyoxyalkylene phosphate	2
Polyoxyalkylene sulfate	2
Polyoxyalkylenes	1, 4, 7
Polyphenylene ether	4
Polyphosphate	4
Polypropylene glycols	2
Polyquaternary amine	4
Polysaccaride polymers in suspension	2
Polysaccharide	4
Polysaccharide blend	4
Polyvinylalcohol/polyvinylactetate copolymer	4
Potassium chloride substitute	4
Quarternized heterocyclic amines	4
Quaternary amine	2, 4
Quaternary amine salt	4
Quaternary ammonium chloride	4
Quaternary ammonium compound	1, 2, 4
Quaternary ammonium salts	1, 2, 4
Quaternary compound	1, 4
Quaternary salt	1, 4
Quaternized alkyl nitrogenated compd	4
Red dye	4
Refined mineral oil	2
Resin	4
Salt of amine-carbonyl condensate	3, 4
Salt of fatty acid/polyamine reaction product	3, 4
Salt of phosphate ester	1

Generic chemical name	Reference
Salt of phosphono-methylated diamine	1, 4
Salts	4
Salts of oxyalkylated fatty amines	4
Sand	4
Sand, AZ silica	4
Sand, brown	4
Sand, sacked	4
Sand, white	4
Secondary alcohol	1, 4
Silica sand, 100 mesh, sacked	4
Silicone emulsion	1
Silicone ester	4
Sodium acid pyrophosphate	4
Sodium calcium magnesium polyphosphate	4
Sodium phosphate	4
Sodium salt of aliphatic amine acid	2
Sodium xylene sulfonate	4
Softwood dust	2
Starch blends	6
Substituted alcohol	1, 2, 4
Substituted alkene	1
Substituted alklyamine	1, 4
Substituted alkyne	4
Sulfate	4
Sulfomethylated tannin	2, 5
Sulfonate	4
Sulfonate acids	1
Sulfonate surfactants	1
Sulfonated asphalt	2
Sulfonic acid salts	1, 4

Generic chemical name	Reference
Sulfur compound	1, 4
Sulphonic amphoterics	4
Sulphonic amphoterics blend	4
Surfactant blend	3, 4
Surfactants	1, 2, 4
Synthetic copolymer	2
Synthetic polymer	4
Tallow soap	4
Telomer	4
Terpenes	1, 4
Titanium complex	4
Triethanolamine zirconium chelate	1 4
Triterpanes	4
Vanadium compounds	4
Wall material	1
Walnut hulls	1, 2, 4
Zirconium complex	2, 4
Zirconium salt	4

Table H-4. Chemicals detected in produced water.

An "X" indicates the availability of physicochemical properties from EPI Suite™ (Appendix C) and selected toxicity data (Appendix G). An empty cell indicates no information was available from the sources we consulted. Reference number corresponds to the citation in Table H-1. Formation type indicated by: "S" (shale), "C" (coalbed), or "U" (uncertain). This refers both to unknown formation types and chemicals in produced water that occur in other types of formations not specified.

Chemical Name[a]	CASRN[b]	Known constituent of hydraulic fracturing fluid	Physico-chemical properties	Selected toxicity data[c]	NCCT CASRN or name change[d]	Formation type	Reference
2,6-di(tert-butyl)-4-hydroxy-4-methyl-2,5-cyclohexadien-1-on	10396-80-2		X		Name	C	21
(1,3-Dimethylbutyl)cyclohexane	61142-19-6		X		Name	S	18
(1-Butylheptyl)cyclohexane	13151-80-9		X		Name	S	18
(1-Methoxyethyl)-benzene	4013-34-7				Name	S,C	21
(1-Methyl-1-buten-1-yl)benzene	53172-84-2		X		Name	S	18
(1-Pentyloctyl)cyclohexane	13151-91-2		X		Name	S	18
(1-Propylnonyl)cyclohexane	13151-84-3		X		Name	S	18
(3E)-3-Heptene	14686-14-7		X		Name	S	18
(3R)-3,7-Dimethyloct-6-enal	2385-77-5		X		Name	S	18
(4Z)-2-Methyl-4-tetradecene	866760-27-2		X		Name	S	18
(9E)-8-Methyl-9-tetradecen-1-yl acetate	912629-93-7		X		Name	S	18
(E)-5-Decene	7433-56-9		X			S	18
(E)-5-Methylspiro[3,5]nonan-1-one	65147-56-0		X		Name; CASRN	S	18
(Z)-1,2-Dimethylcyclohexane	2207-01-4		X			S	18
(Z)-1,2-Dimethylcyclopentane	1192-18-3		X		Name	S	18
(Z)-1,3-Dimethylcyclohexane	638-04-0		X		Name	S	18

Chemical Name[a]	CASRN[b]	Known constituent of hydraulic fracturing fluid	Physico-chemical properties	Selected toxicity data[c]	NCCT CASRN or name change[d]	Formation type	Reference
(Z)-1-Ethyl-2-methylcyclopentane	930-89-2		X		Name	S	18
(Z)-1-Ethyl-3-methylcyclohexane	19489-10-2		X		Name	S	18
(Z)-5-Octen-1-ol	64275-73-6		X		Name	S	18
(Z)-9-Methylundec-4-ene	74630-56-1		X		Name	S	18
(Z)-9-Tricosene	27519-02-4		X			C	18, 20
1-Heptadecene	6765-39-5		X			S	18
1-(1,5-Dimethylhexyl)-4-(4-methylpentyl)cyclohexane	56009-20-2		X		Name	S	18
1-(2,4-Dimethylphenyl)ethanone	89-74-7		X		Name	S	16
1-(2-Furanyl)-3-butene-1,2-diol	19261-13-3		X		Name	S	16
1-(3-Methylbutyl)-2,3,4-trimethyl-benzene	107997-59-1		X		Name	C	21
1-(Butan-2-yl)-4-methylbenzene	1595-16-0		X		Name	S	18
1-(Cyclohexylmethyl)-4-methylcyclohexane	66826-95-7		X			S	18
1-(Pentyloxy)hexane	32357-83-8		X		Name	S	18
1,8,10-Pentadecatriene	1227308-82-8		X			S	18
1,1,3,5-Tetramethylcyclohexane	4306-65-4		X			S	18
1,1,3-Trimethylcyclohexane	3073-66-3		X		Name	S	18
1,1,3-Trimethylcyclopentane	4516-69-2		X		Name	S	18
1,12-Dibromododecane	3344-70-5		X			S	18
1,1-Dichloroethane	75-34-3		X	X		S	18

Chemical Name[a]	CASRN[b]	Known constituent of hydraulic fracturing fluid	Physico-chemical properties	Selected toxicity data[c]	NCCT CASRN or name change[d]	Formation type	Reference
1,1-Dimethyl-1,2,3,4-tetrahydro-7-isopropyl phenanthrene	27530-79-6		X		Name	C	20
1,1-Dimethylcyclohexane	590-66-9		X			S	18
1,1-Dimethylcyclopropane	1630-94-0		X			S	18
1,1'-Methylenebis(4-methyl)-benzene	4957-14-6		X		Name	C	20
1,1'-Oxybisdecane	2456-28-2		X			S	18
1,2,3,4-Tetrahydro-2,5,7-trimethylnaphthalene	65001-61-8		X			S	18
1,2,3,4-Tetrahydro-2,5,8-trimethylnaphthalene	30316-17-7		X			S	18
1,2,3,4-Tetrahydro-naphthalene	119-64-2		X		Name	S,C	21
1,2,3,4-Tetramethylcyclohexane	3726-45-2		X			S	18
1,2,3,4-Tetramethylnaphthalene	3031-15-0		X			S	18
1,2,3-Trichlorobenzene	87-61-6		X	X		S	3, 9
1,2,3-Trimethylbenzene	526-73-8	X	X	X		S	18
1,2,3-Trimethylcyclopentane	2815-57-8		X			S	18
1,2,4,5-Tetramethylbenzene	95-93-2		X		Name	S	18
1,2,4-Trichlorobenzene	120-82-1		X	X		S	9
1,2,4-Trimethylbenzene	95-63-6	X	X	X		S,C	3, 9, 10, 13, 15, 18, 22
1,2,4-Trimethylcyclohexane	2234-75-5		X			S	18
1,2,4-Trimethylcyclopentane	2815-58-9		X			S	18

Chemical Name[a]	CASRN[b]	Known constituent of hydraulic fracturing fluid	Physico-chemical properties	Selected toxicity data[c]	NCCT CASRN or name change[d]	Formation type	Reference
1,2-Benzenedicarboxylic acid, 1,2-bis(8-methylnonyl) ester	89-16-7		X	X	Name	S	18
1,2-Benzenedicarboxylic acid, 1-butyl 2-(8-methylnonyl) ester	42343-36-2		X		Name	S	18
1,2-Di-but-2-enyl-cyclohexane	NOCAS_873054		X			C	20
1,2-Dimethyl-1-cycloheptene	20053-89-8		X		Name	S	18
1,2-Dimethyl-4-ethylbenzene	934-80-5		X		Name	S	18
1,2-Diphenylhydrazine	122-66-7		X	X		S	15
1,2-Epoxydodecane	2855-19-8		X		Name	S	18
1,2-Epoxyhexadecane	7320-37-8		X		Name	S	18
1,2-Propylene glycol	57-55-6	X	X	X		S	3, 9, 22
1,3,5-Trimethylbenzene	108-67-8	X	X	X	Name	S,C	3, 9, 10, 13, 15, 18, 22
1,3,5-Trimethylcyclohexane	1839-63-0		X			S	18
1,3-Dimethyl-4-ethylbenzene	874-41-9		X		Name	S,C	18, 21
1,3-Dimethyladamantane	702-79-4		X		Name	C	13
1,3-Dimethylcyclohexane	591-21-9		X			S	18
1,3-Dimethylcyclopentane	2453-00-1		X			S	18
1,4,5,8-Tetramethylnaphthalene	2717-39-7		X			S	16
1,4,5-Trimethylnaphthalene	2131-41-1		X			S	18
1,4,6-Trimethylnaphthalene	2131-42-2		X			S	18
1,4-Dihydro-1,4-methanonaphthalene	4453-90-1		X			S	18

Chemical Name[a]	CASRN[b]	Known constituent of hydraulic fracturing fluid	Physico-chemical properties	Selected toxicity data[c]	NCCT CASRN or name change[d]	Formation type	Reference
1,4-Dimethyl-2,3-diazabicyclo[2.2.1]hept-2-ene	71312-54-4		X		Name	S	16
1,4-Dimethylcyclohexane	589-90-2		X			S	18
1,4-Dimethylnaphthalene	571-58-4		X			S	18
1,4-Dioxane	123-91-1	X	X	X		S	9, 10, 15
1,4-Hexadecansultone	15224-88-1		X		Name	S	18
1,5,7-Trimethyl-1,2,3,4-tetrahydronaphthalene	21693-55-0		X		Name	S	18
1,54-Dibromotetrapentacontane	852228-22-9		X			S	18
1,5-Dimethyl-7-oxabicyclo[4.1.0]heptane	162239-52-3		X			S	18
1,5-Dimethylnaphthalene	571-61-9		X			S	18
1,6-Dimethyl-4(1-methylethyl)naphthalene	483-78-3		X		Name	C	20
1,6-Dimethylnaphthalene	575-43-9		X			S	18
10-Pentadecen-1-ol	129396-62-9		X			S	18
10,4-Dihydroxy-70-methoxy-2,30-dimethyl-,()-[1,20-binaphthalene]-5,50,8,80-tetrone	119736-96-8		X		Name	S	18
10-Methylicosane	54833-23-7		X		Name	S	18
10-Methylnonadecane	56862-62-5		X			S	18
11-(1-Ethylpropyl)-heneicosane	55282-11-6		X		Name	S	18
11,13-Dimethyl-12-tetradecen-1-yl acetate	400037-00-5		X		Name	S	18

Chemical Name[a]	CASRN[b]	Known constituent of hydraulic fracturing fluid	Physico-chemical properties	Selected toxicity data[c]	NCCT CASRN or name change[d]	Formation type	Reference
13-Tetradecen-1-ol	67400-04-8		X			S	18
13-Tetradecen-1-yl acetate	56221-91-1		X		Name	S	18
14-Bromo-1-tetradecene	74646-31-4		X			S	18
14-Methylhexadecanal	93815-50-0		X			S	18
15-Isobutyl-(13.α.H)-isocopalane	228729-94-0		X		Name	C	20
17-Methylpentatriacontane	56987-83-8		X			S	18
1a,9b-Dihydro-1H-cyclopropa[l]phenanthrene	949-41-7		X		Name	S	18
1-Allyl-3-methylindole-2-carbaldehyde	123731-75-9		X		Name	C	20
1-Bromo-11-iodoundecane	139123-69-6		X			S	18
1-Bromohexadecane	112-82-3		X			S	18
1-Bromooctadecane	112-89-0		X			S	18
1-Bromopentadecane	629-72-1		X			S	18
1-Butanol	71-36-3	X	X	X	Name	S	22
1-Butyl-2-ethyloctahydro-1H-4,7-epoxyinden-5-ol	62583-58-8		X		Name	C	20
1-Butyl-2-pentylcyclopentane	61142-52-7		X			S	18
1-Chloro-Heptacosane	62016-79-9		X		Name	S	18
1-Chlorohexadecane	4860-03-1		X			S	18
1-Decene	872-05-9		X			S	18
1-Docosanethiol	7773-83-3		X			S	18
1-Dodecene	112-41-4		X			S	18

Chemical Name[a]	CASRN[b]	Known constituent of hydraulic fracturing fluid	Physico-chemical properties	Selected toxicity data[c]	NCCT CASRN or name change[d]	Formation type	Reference
1-Dotriacontanol	6624-79-9		X			S	18
1-Ethyl-2,3-dimethylbenzene	933-98-2		X			S	18
1-Ethyl-2-methylbenzene	611-14-3	X	X			S	18
1-Ethyl-2-methylcyclohexane	3728-54-9		X			S	18
1-Ethyl-2-methylcyclopentane	3726-46-3		X			S	18
1-Ethyl-3-methylcyclohexane	3728-55-0		X			S	18
1-Ethyl-4-methylcyclohexane	3728-56-1		X			S	18
1-Ethyl-9,10-anthracenedione	24624-29-1		X		Name	C	20
1-Ethylidene-1H-indene	2471-83-2		X			S	18
1-Fluorododecane	334-68-9		X			S	18
1-Hentetracontanol	40710-42-7		X			S	18
1-Hexacosanol	506-52-5		X			S	18
1-Hexacosene	18835-33-1					C	20
1-Hexadecene	629-73-2	X	X			S	18
1-Iodo-2-methylundecane	73105-67-6		X			S	18
1-Isopropyl-2,3-dimethylcyclopentane	489-20-3		X		Name	S	18
1-Methyl-1,2-cyclohexanediol	6296-84-0		X			S	18
1-Methyl-2-pentylcyclohexane	54411-01-7		X			S	18
1-Methyl-3-(1-methylethyl)cyclopentane	53771-88-3		X		Name	S	18
1-Methyl-3-propylbenzene	1074-43-7		X		Name	S	16

Chemical Name[a]	CASRN[b]	Known constituent of hydraulic fracturing fluid	Physico-chemical properties	Selected toxicity data[c]	NCCT CASRN or name change[d]	Formation type	Reference
1-Methyl-7-(1-methylethyl)phenanthrene	483-65-8		X		Name	C	20, 21
1-Methyl-7-oxabicyclo[4.1.0]heptane	1713-33-3		X			S	18
1-Methylene-1H-indene	2471-84-3		X			S	18
1-Methylfluorene	1730-37-6		X		Name	C	20
1-Methylnaphthalene	90-12-0		X	X		S,C	18, 21, 22
1-Naphthol	90-15-3		X			S	22
1-Nonene	124-11-8		X			S	18
1-Octadecanethiol	2885-00-9		X			S	18
1-Octadecene	112-88-9	X	X			S	18
1-Oxopyridin-2-ylamine	14150-95-9		X		Name	S	18
1-Pentyl-2-propylcyclopentane	62199-51-3		X			S	18
1-Propanol	71-23-8	X	X			S	22
1-Propoxyhexane	53685-78-2		X			S	18
1-Propylcyclohexene	2539-75-5		X			S	18
1-Tricosene	18835-32-0		X			S	18
1-Tridecene	2437-56-1		X			S	18
2-(2-Buten-1-yl)-1,3,5-trimethylbenzene	63435-25-6		X			S	18
2-(2-Butoxyethoxy)ethanol	112-34-5	X	X	X		S,C	21
2(3H)-Benzothiazolone	934-34-9		X			C	20, 21
2-(Methylthio)-benzothiazole	615-22-5		X		Name	C	20

Chemical Name[a]	CASRN[b]	Known constituent of hydraulic fracturing fluid	Physico-chemical properties	Selected toxicity data[c]	NCCT CASRN or name change[d]	Formation type	Reference
2,10-Dimethylundecane	17301-27-8		X			S	18
2,2,3,3-Tetramethylhexane	13475-81-5		X			S	18
2,2,4-trimethyl-1,3-pentanediol	144-19-4		X			S,C	21
2,2-Dibromo-3-nitrilopropionamide	10222-01-2	X	X			S	11
2,2-Dichloro-3,6-dimethyl-1-oxa-2-silacyclohexa-3,5-diene	69586-09-0		X			S	18
2,3',5-Trimethyldiphenylmethane	61819-81-6		X			C	20
2,3,6-Trimethylnaphthalene	829-26-5		X			S	18
2,3-Dihydro-1,1,2,3,3-pentamethyl-1H-indene	1203-17-4		X			C	20
2,3-Dimethyldecahydronaphthalene	1008-80-6		X		Name	S	18
2,3-Dimethyldecane	17312-44-6		X			S	18
2,3-Dimethylheptane	3074-71-3		X			S	18
2,3-Dimethylnaphthalene	581-40-8		X			S	18
2,3-Dimethylundecane	17312-77-5		X			S	18
2,3-Heptanedione	96-04-8		X		Name	S	16
2,4,6-Trimethyl-azulene	NOCAS_873044				Name	C	20
2,4-Bis(1,1-dimethylethyl)phenol	96-76-4		X		Name	C	21
2,4-Dichloro-5-oxohex-2-enedioic acid	56771-78-9		X		Name	S	18
2,4-Dichlorophenol	120-83-2		X	X		S	15
2,4-dimethyl-1-(1-methylpropyl)-benzene	1483-60-9				Name	C	21
2,4-Dimethylheptane	2213-23-2		X			S	18

Chemical Name[a]	CASRN[b]	Known constituent of hydraulic fracturing fluid	Physico-chemical properties	Selected toxicity data[c]	NCCT CASRN or name change[d]	Formation type	Reference
2,4-Dimethylhexane	589-43-5		X			S	18
2,4-Dimethylphenol	105-67-9		X	X		S,C	3, 9, 10, 13, 15
2,4-Dimethylundecane	17312-80-0		X			S	18
2,5,9-Trimethyldecane	62108-22-9		X			S	18
2,5-Cyclohexadiene-1,4-dione	106-51-4		X	X	Name	C	20
2,5-Dimethyldodecane	56292-65-0		X			S	18
2,6,10-Trimethyl-9-undecenoic acid	97993-62-9		X			S	18
2,6,10-Trimethylpentadecane	3892-00-0		X			S	18
2,6,10-Trimethylundec-9-enal	141-13-9		X		Name	S	18
2,6,10-Trimethylundecanoic acid	1115-94-2		X			S	18
2,6,11-Trimethyldodecane	31295-56-4		X			S	18
2,6-Bis(dimethylethyl)-2,5-cyclohexadiene-1,4-dione	719-22-2		X		Name	C	20
2,6-Dichlorophenol	87-65-0		X			S	3, 9
2,6-Dimethyldecane	13150-81-7		X			S	18
2,6-Dimethylheptane	1072-05-5		X			S	18
2,6-Dimethylnaphthalene	581-42-0		X			S	18
2,6-Di-tert-butylphenol	128-39-2		X		Name	C	10, 14, 20
2,7-Dimethylnaphthalene	582-16-1		X			S	18
2-[2-[4-(1,1,3,3-tetramethylbutyl)phenoxy]ethoxy]-ethanol	2315-61-9		X		Name	C	20, 21
22-Tricosenoic acid	65119-95-1		X			C	20

Chemical Name[a]	CASRN[b]	Known constituent of hydraulic fracturing fluid	Physico-chemical properties	Selected toxicity data[c]	NCCT CASRN or name change[d]	Formation type	Reference
28-Nor-17.α.(H)-hopane	204781-73-7		X		Name	C	20
2-Aminoimidazole	7720-39-0		X		Name	S	18
2-Butoxyethanol	111-76-2	X	X	X		S	22
2-Butyloctan-1-ol	3913-02-8		X		Name	S	18
2-Chloroethanol	107-07-3		X	X		S	18
2-Dodecen-1-yl(-)succinic anhydride	25377-73-5		X		Name	C	20
2'-Dodecyl- 1,1':3',1''-tercyclopentane	55282-68-3		X		Name	S	18
2-Ethyl-1,1,3-trimethylcyclohexane	442662-72-8		X			S	18
2-Ethyl-1-decanol	21078-65-9		X			S	18
2-Ethyl-1-hexanol	104-76-7	X	X			S	22
2-Ethylhexyl diphenyl phosphate (Octicizer)	1241-94-7		X		Name	C	20
2-Hexyl-1-decanol	2425-77-6		X			S	18
2-Hydroxy-2-methylbut-3-en-1-yl 2-methylbut-2-enoate	1418543-90-4		X		Name	S	18
2-Hydroxy-4-(propan-2-yl)cyclopent-2-en-1-one	54639-82-6		X		Name	S	18
2-Imino-5,6-dihydro-2H-cyclopenta[d][1,3]thiazol-3(4H)-ol	738528-09-1		X		Name	S	18
2-Mercaptobenzothiazole	149-30-4		X	X		C	20
2-Methoxyfuran	25414-22-6		X			S	16
2-Methyl-2-butene	513-35-9		X			S	18
2-Methyl-7-octadecene	51050-50-1		X			S	18

Chemical Name[a]	CASRN[b]	Known constituent of hydraulic fracturing fluid	Physico-chemical properties	Selected toxicity data[c]	NCCT CASRN or name change[d]	Formation type	Reference
2-Methyl-8-propyl-dodecane	55045-07-3		X		Name	C	20
2-Methylbut-1-ene	563-46-2		X		Name	S	18
2-Methyldecane	6975-98-0		X			S	18
2-Methyldodecan-1-ol	22663-61-2		X		Name	S	18
2-Methyldodecane	1560-97-0		X			S	18
2-Methylheptane	592-27-8		X			S	18
2-Methylnaphthalene	91-57-6		X	X		S,C	3, 9, 10, 13, 15, 16, 18, 21, 22
2-Methyl-nonadecane	52845-07-5		X		Name	C	20
2-Methylnonane	871-83-0		X			S	18
2-Methyl-N-phenyl-benzenamine	1205-39-6		X		Name	C	21
2-Methyloctane	3221-61-2		X			S	18
2-Methylpentadecane	1560-93-6		X			S	18
2-Methylpentane	107-83-5		X			S	18
2-Methylphenanthrene	2531-84-2		X			S	18
2-Methylpropanoic acid	79-31-2		X			U	10
2-Methylpyridine	109-06-8		X			S	3, 9
2-Methyltetradecane	1560-95-8		X			S	18
2-Methyltridecane	1560-96-9		X			S	18
2-Methylundecane	7045-71-8		X			S	18
2-Naphthalenol	135-19-3		X		Name	S	22
2-Octadecyl-propane-1,3-diol	5337-61-1		X		Name	C	20

Chemical Name[a]	CASRN[b]	Known constituent of hydraulic fracturing fluid	Physico-chemical properties	Selected toxicity data[c]	NCCT CASRN or name change[d]	Formation type	Reference
2-Octene	111-67-1		X			S	18
2-Pentyl-2-nonenal	3021-89-4		X			S	18
2-Phenylpentane	2719-52-0		X		Name	S	18
3-(4-Methoxyphenyl)-2-ethylhexylester-2-propenoic acid	5466-77-3		X		Name	C	20
3-(4-Methoxyphenyl)-2-propenoic acid	830-09-1				Name	C	20
3-(Hexahydro-1H-azepin-1-yl)-1,1-dioxide-1,2-benzisothiazole	309735-29-3		X		Name	C	20
3,3,5,5-Tetramethylcyclopentene	38667-10-6		X		Name	S	18
3,3'-5,5'-Tetramethyl-[1,1'-biphenyl]-4,4'-diamine	54827-17-7		X		Name	S,C	21
3,4-Dihydro-1,9(2H,10H)acridinedione	80061-31-0		X		Name	C	21
3,5,24-Trimethyltetracontane	55162-61-3		X			S	18
3,5-Dimethyloctane	15869-93-9		X			S	18
3,5-Di-tert-butyl-4-hydroxybenzaldehyde	1620-98-0		X		Name	C	20
3,6-Dimethylundecane	17301-28-9		X			S	18
3,7-Dimethyldecane	17312-54-8		X			S	18
3,7-Dimethylnonane	17302-32-8		X			S	18
3,7-Dimethyloct-7-enal	141-26-4		X		Name	S	18
3,7-Dimethylundecane	17301-29-0		X			S	18
3,8-Dimethyldecane	17312-55-9		X			S	18
3,9-Dimethylundecane	17301-31-4		X			S	18

Chemical Name[a]	CASRN[b]	Known constituent of hydraulic fracturing fluid	Physico-chemical properties	Selected toxicity data[c]	NCCT CASRN or name change[d]	Formation type	Reference
3-Cyclohexylpropan-1-ol	1124-63-6		X		Name	S	18
3-Cyclopentyl-2-methylpropan-1-ol	264258-62-0		X		Name	S	18
3-Ethyl-2-methylheptane	14676-29-0		X			S	18
3-Ethylhexane	619-99-8		X			S	18
3-Ethyltoluene	620-14-4		X		Name	S	18
3-Methyl-1-heptene	4810-09-7		X			S	18
3-Methyl-2-(2-oxopropyl)furan	87773-62-4		X		Name	S	18
3-Methyl-3-hexene	42154-69-8		X			S	18
3-Methylcyclohexene	591-48-0		X			S	16
3-Methylcyclopentadecan-1-one	541-91-3		X		Name	S	18
3-Methyldecane	13151-34-3		X			S	18
3-Methyldodecane	17312-57-1		X			S	18
3-Methylnonane	5911-04-6		X			S	18
3-Methyloctane	2216-33-3		X			S	18
4-(1,1,3,3-Tetramethylbutyl)phenol	140-66-9		X		Name	C	14, 21
4,4-Diacetyldiphenylmethane	790-82-9		X		Name	C	20
4,4-Dimethyl-2-(1-methylethenyl)cyclopentanone	343270-53-1		X		Name	S	18
4,6,8-Trimethyl-2-propylazulene	160951-15-5		X			C	20
4,6-Dimethyldodecane	61141-72-8		X			S	18
4-[1-(2-Methylphenyl)ethyl]phenol	35770-76-4		X		Name	C	20
4-Decene	19398-89-1		X			S	18

Chemical Name[a]	CASRN[b]	Known constituent of hydraulic fracturing fluid	Physico-chemical properties	Selected toxicity data[c]	NCCT CASRN or name change[d]	Formation type	Reference
4-Ethyl-2,3-dimethylhex-2-ene	959028-24-1		X		Name	S	18
4-Ethyl-5-octyl-2,2-bis(trifluoromethyl)-1,3-dioxolane, cis-	38274-72-5		X		Name	S	18
4-Ethyloctane	15869-86-0		X			S	18
4-Methyl-2-pentene	4461-48-7		X			S	18
4-Methyl-2-phenyl-2-pentenal	26643-91-4		X		Name	S	18
4-Methyldecane	2847-72-5		X			S	18
4-Methyldocosane	25117-30-0		X			S	18
4-Methyldodec-3-en-1-ol	1372101-59-1		X		Name	S	18
4-Methylheptane	589-53-7		X			S	18
4-Methylnonane	17301-94-9		X			S	18
4-Methyloctane	2216-34-4		X			S	18
4-Methyltetradecane	25117-24-2		X			S	18
4-Methyltridecane	26730-12-1		X			S	18
4-Methylundecane	2980-69-0		X			S	18
4-Phenyl-1-buten-4-ol	936-58-3		X		Name	S	18
4-Propyl-3-heptene	4485-13-6		X			S	18
4-Propylcyclohexanone	40649-36-3		X			S	18
4-Propylheptane	3178-29-8		X			S	18
4-Propyl-xanthen-9-one	108837-05-4		X		Name	C	20
5-(1,1-Dimethylethyl)-1H-indene	NOCAS_873045					C	20
5-Butyl-6-hexyloctahydro-1H-indene	55044-36-5		X			S	18

Appendix H – Chemicals Identified in Hydraulic Fracturing Fluids and/or Produced Water

Chemical Name[a]	CASRN[b]	Known constituent of hydraulic fracturing fluid	Physico-chemical properties	Selected toxicity data[c]	NCCT CASRN or name change[d]	Formation type	Reference
5-Methyldecane	13151-35-4		X			S	18
5-Methyltetradecane	25117-32-2		X			S	18
5-Methyltridecane	25117-31-1		X			S	18
6-Methyl-6-ethylfulvene	3141-02-4		X		Name	S	18
6-Methyltridecane	13287-21-3		X			S	18
6-Methylundecane	17302-33-9		X			S	18
7,12-Dimethylbenz(a)anthracene	57-97-6		X	X		S	3, 9
7-Bromomethyl-pentadec-7-ene	941228-34-8		X		Name	C	20
7-Ethenylphenanthrene	68593-94-2		X		Name	C	20
7-Methylpentadecane	6165-40-8		X			S	18
7-Methyltridecane	26730-14-3		X			S	18
7-Tetradecyne	35216-11-6		X			C	20
8-Hexadecyne	19781-86-3		X			C	20
8-Methylundec-3-ene	876314-66-8		X		Name	S	18
9-Hexacosene	71502-22-2		X			S	18
9-Methylanthracene	779-02-2		X			S	18
9-Methylnonadecane	13287-24-6		X			S	18
Acetaldehyde	75-07-0	X	X	X		S	22
Acetate	71-50-1		X			S,C	11, 21
Acetic acid	64-19-7	X	X			S	3, 9, 10, 12
Acetone	67-64-1	X	X	X		S	3, 9, 10, 15, 18
Acetophenone	98-86-2	X	X	X		S,C	3, 9, 15, 21, 22

Chemical Name[a]	CASRN[b]	Known constituent of hydraulic fracturing fluid	Physico-chemical properties	Selected toxicity data[c]	NCCT CASRN or name change[d]	Formation type	Reference
Acetyl tributyl citrate	77-90-7		X		Name	S	18
Acrolein	107-02-8	X	X	X		S	9
Acrylonitrile	107-13-1		X	X		S	3, 9
Adamantane	281-23-2		X			C	13
Aldrin	309-00-2		X	X		S	3, 9
Alpha particle	12587-46-1			X	Name	S	24, 25, 26
alpha-Farnesene	502-61-4		X		Name	S	18
alpha-Methyl-1H-imidazole-1-ethanol	37788-55-9		X		Name	S	18
Aluminum	7429-90-5	X		X		S	3, 9, 10
Ammonia	7664-41-7	X				S	3, 9, 10, 18
Antimony	7440-36-0			X		S	3, 9, 10
Aroclor 1248	12672-29-6		X			S	3, 9
Arsenic	7440-38-2	X		X		S	3, 9, 10
Barium	7440-39-3			X		S	3, 9, 10
Benz(a)anthracene	56-55-3		X	X	Name; CASRN	S	15
Benzene	71-43-2	X	X	X		S,C	3, 9, 10, 12, 13, 16, 22
Benzene, 1,3 (or 1,4)-dimethyl-	179601-23-1				Name	C	13
Benzidine	92-87-5		X	X		S	15
Benzo(a)pyrene	50-32-8		X	X		S	3, 9, 15
Benzo(b)fluoranthene	205-99-2		X	X		S	3, 9, 15

Chemical Name[a]	CASRN[b]	Known constituent of hydraulic fracturing fluid	Physico-chemical properties	Selected toxicity data[c]	NCCT CASRN or name change[d]	Formation type	Reference
Benzo(g,h,i)perylene	191-24-2		X	X		S	3, 9, 10, 15
Benzo(k)fluoranthene	207-08-9		X	X		S	3, 9, 15
Benzophenone	119-61-9		X	X		C	21
Benzothiazole	95-16-9		X			S,C	14, 20, 21
Benzyl alcohol	100-51-6		X	X	Name	S	3, 9, 10, 15, 20
Benzyl butyl phthalate	85-68-7		X	X	Name	C	20, 21
Benzyl chloride	100-44-7	X	X	X		S	22
Beryllium	7440-41-7			X		S	3, 9, 10
Beta particle	12587-47-2			X	Name	S	24, 25, 26
beta-Hexachlorocyclohexane	319-85-7		X	X		S	3, 9
biphenyl	92-52-4		X	X		C	20, 21
Bis(1,1-dimethylethyl)-phenol	26746-38-3		X		Name	S,C	21
Bis(2-chloroethyl) ether	111-44-4	X	X	X		S	3, 9
Bis(2-ethylhexyl) isophthalate	137-89-3		X		Name	S	18
Bis(dichloromethyl) ether	20524-86-1		X		Name	S	18
Bis-(octylphenyl)-amine	26603-23-6				Name	C	20
Bisphenol A	80-05-7	X	X	X	Name	S	22
Boron	7440-42-8	X		X		S	3, 9, 10
Bromide	24959-67-9					S	3, 9, 10
Bromodichloromethane	75-27-4		X	X		S	3
Bromoform	75-25-2		X	X		S	3, 9, 10
Butanenitrile	109-74-0		X			S	16

Chemical Name[a]	CASRN[b]	Known constituent of hydraulic fracturing fluid	Physico-chemical properties	Selected toxicity data[c]	NCCT CASRN or name change[d]	Formation type	Reference
Butanoic acid	107-92-6		X			S	9, 10
Butanoic acid, butyl ester	109-21-7		X		Name	C	20
Butyl 8-methylnonyl phthalate	89-18-9		X		Name	S	18
Butylbenzene	104-51-8		X	X	Name	S,C	9, 10, 13
Butylcyclohexane	1678-93-9		X			S	18
Butyrate	461-55-2		X			S	11
Cadmium	7440-43-9			X		S	3, 9, 10
Caesium	7440-46-2				Name	C	14
Caesium-137	10045-97-3					S	3
Caffeine	58-08-2		X	X		C	20
Calcium	7440-70-2					S	3, 9, 10
Caprolactam	105-60-2		X	X		C	14, 21
Carbon dioxide	124-38-9	X	X			S	3, 9, 10
Carbon disulfide	75-15-0		X	X		S	3, 9, 22
Chloride	16887-00-6	X				S	3, 9, 10
Chlorine	7782-50-5	X		X		S	3, 10
Chlorobenzene	108-90-7	X	X	X	Name	S	16
Chlorodibromomethane	124-48-1		X	X		S	3
Chloroform	67-66-3		X	X		S	3, 9, 10, 18
Chloromethane	74-87-3	X	X	X		S	3, 10, 22
Chloromethyl 5-chloropentyl ether	145912-11-4		X		Name	S	18
Cholesterol	57-88-5		X	X		C	20

Chemical Name[a]	CASRN[b]	Known constituent of hydraulic fracturing fluid	Physico-chemical properties	Selected toxicity data[c]	NCCT CASRN or name change[d]	Formation type	Reference
Chromium	7440-47-3			X		S	3, 9, 10
Chromium (III)	16065-83-1	X		X		S	3
Chromium (VI)	18540-29-9	X		X		S	3, 10
Chrysene	218-01-9		X	X		S	15
cis-1,4-Dimethyladamantane	24145-89-9		X		Name	S	18
cis-Octahydro-4a-methyl-2(1H)-naphthalenone	938-06-7		X		Name	S	18
Cobalt	7440-48-4			X		S	3, 9, 10
Copper	7440-50-8	X		X		S	3, 9, 10
Cumene	98-82-8	X	X	X	Name	S	3, 9, 22
Cyanide	57-12-5		X	X		S	3, 9, 10
Cyclohexyl mercaptoacetate	16849-98-2		X		Name	S	18
Cyclohexylbenzene	827-52-1		X			S	18
Cyclopentadecane	295-48-7		X			S	18
Cyclotetracosane	297-03-0		X			S	18
Cyclotetradecane	295-17-0		X			S	18
Cyclotridecane	295-02-3		X			S	18
Decahydro-1-methyl-2-methylenenaphthalene	90548-09-7		X			S	18
Decahydro-2-methylnaphthalene	2958-76-1		X		Name	S	18
Decalin	91-17-8		X		Name	S	18
Decylcyclohexane	1795-16-0		X			S	18

Chemical Name[a]	CASRN[b]	Known constituent of hydraulic fracturing fluid	Physico-chemical properties	Selected toxicity data[c]	NCCT CASRN or name change[d]	Formation type	Reference
delta-Hexachlorocyclohexane	319-86-8		X	X		S	9
Di(2-ethylhexyl) phthalate	117-81-7	X	X	X	Name	S	3, 9, 10, 18
Dibenz(a,h)anthracene	53-70-3		X	X		S	3, 9, 15
Dibenzosuberol	1210-34-0		X		Name	S	18
dibenzothiophene	132-65-0		X	X		C	21
Dibromoacetonitrile	3252-43-5	X	X	X		S	11
Dibutyl hexanedioate	105-99-7		X		Name	S	18
Dibutyl phthalate	84-74-2		X	X		S,C	3, 9, 10, 20, 21
Dichloromethane	75-09-2	X	X	X	Name	S	9, 10, 18
didecyl phthalate	84-77-5		X			S	18
Dieldrin	60-57-1		X	X		S	9
Diethyl phthalate	84-66-2		X	X	Name	S,C	9, 20, 21
Diethyltoluamide	134-62-3		X	X	Name	C	21
Diisodecyl phthalate	26761-40-0		X	X	Name	S	18
Diisooctyl phthalate	27554-26-3		X		Name	S	18
Dimethyl phthalate	131-11-3		X	X		C	20
Dimethylnaphthalene	28804-88-8				Name	C	20, 21
Dimethylphenol	1300-71-6					C	20, 21
Dimethyl-tetracyclo[5.2.1.0(2,6)-0(3,5)]decane	74646-38-1		X		Name	C	20
DINP	28553-12-0		X		Name	S	18
Dioctadecyloate phosphoric acid	3037-89-6		X			S	18

Chemical Name[a]	CASRN[b]	Known constituent of hydraulic fracturing fluid	Physico-chemical properties	Selected toxicity data[c]	NCCT CASRN or name change[d]	Formation type	Reference
Dioctyl hexanedioate	123-79-5		X		Name	S	18
Dioctyl phthalate	117-84-0		X	X	Name	S	9, 10, 14, 18, 21
Diphenylamine	122-39-4		X	X		S,C	3, 9, 15, 20, 21
Diphenylmethane	101-81-5		X			C	20
Di-tert-butyl nitroxide	2406-25-9		X		Name	S	16
D-Limonene	5989-27-5	X	X	X		S	22
Dodecane	112-40-3	X	X			S	12, 18
Dodecanoic acid	143-07-7		X			S,C	14, 20, 21
Dotriacontane	544-85-4		X			S	18
Drometrizole	2440-22-4		X		Name	C	20
Endosulfan I	959-98-8		X			S	3, 9
Endosulfan II	33213-65-9		X			S	3, 9
Endrin aldehyde	7421-93-4		X			S	3, 9
Ethanol	64-17-5	X	X	X		S	22
Ethyl glycylglycinate	627-74-7		X		Name	S	18
Ethylbenzene	100-41-4	X	X	X		S,C	3, 9, 10, 13, 18, 22
Ethylcyclohexane	1678-91-7		X			S	18
Ethylcyclopentane	1640-89-7		X			S	18
Ethylene glycol	107-21-1	X	X	X		S,C	3, 9, 21, 22
Farnesol	4602-84-0		X		Name	S	18
Fluoranthene	206-44-0		X	X		S	3, 9, 15
Fluorene	86-73-7		X	X		S,C	3, 9, 10, 15, 20

Chemical Name[a]	CASRN[b]	Known constituent of hydraulic fracturing fluid	Physico-chemical properties	Selected toxicity data[c]	NCCT CASRN or name change[d]	Formation type	Reference
Fluoride	16984-48-8			X		S	3, 9, 10
Formate	71-47-6		X		Name	S	11, 19
Formic acid	64-18-6	X	X	X		U	10
Glutaraldehyde	111-30-8	X	X			S	22
Glycolic acid	79-14-1	X	X			S	18
Heptachlor	76-44-8		X	X		S	3, 9
Heptachlor epoxide	1024-57-3		X	X		S	3, 9
Heptacosane	593-49-7		X			S,C	18, 20
Heptane	142-82-5	X	X			S	18
Heptanoic acid	111-14-8		X			U	10
Heptylcyclohexane	5617-41-4		X			S	18
Hex-3-yne	928-49-4		X		Name	S	18
Hexadecahydropyrene	2435-85-0		X			S	18
Hexadecanoic acid	57-10-3		X			C	14, 21
Hexane	110-54-3	X	X	X		S	18
Hexanoic acid	142-62-1		X			U	10
Hexatriacontane	630-06-8		X			S	18
Hexylcyclohexane	4292-75-5		X			S	18
Hydratropaldehyde	93-53-8		X		Name	S	18
Hydrazine	302-01-2	X		X		S	18
Hydrochloric acid	7647-01-0	X		X		S	18
Hydroxyacetonitrile	107-16-4		X			S	18

Chemical Name[a]	CASRN[b]	Known constituent of hydraulic fracturing fluid	Physico-chemical properties	Selected toxicity data[c]	NCCT CASRN or name change[d]	Formation type	Reference
Imidazo[1,2-a]pyrimidine	274-95-3		X			S	18
Indeno(1,2,3-cd)pyrene	193-39-5		X	X		S	3, 9, 15
Iodine	7553-56-2			X		S	9, 14
Iron	7439-89-6	X		X		S	3, 9, 10
Isobutylbenzene	538-93-2		X		Name	S	18
Isobutylcyclohexane	1678-98-4		X		Name	S	18
Isopropanol	67-63-0	X	X	X		S	3, 9, 22
Isopropyl myristate	110-27-0		X		Name	C	20
Isoquinoline	119-65-3	X	X			C	21
Isovaleric acid	503-74-2		X			U	10
Kaur-16-ene	562-28-7		X		Name	C	21
Lead	7439-92-1	X		X		S	3, 9, 10
Lindane	58-89-9		X	X		S	3, 9
Lithium	7439-93-2			X		S	3, 9, 10
m,p-Cresol mixture	NOCAS_24858			X	Name	C	10
Magnesium	7439-95-4					S	3, 9, 10
Manganese	7439-96-5			X		S	3, 9, 10
m-Cresol	108-39-4		X	X	Name	S,C	3, 9, 10, 13, 15
m-Cymene	535-77-3		X		Name	S	18
Menthol	1490-04-6		X		Name	S	18
Mercury	7439-97-6			X		S	3, 9, 10
Methanol	67-56-1	X	X	X		S	3, 9, 22

Chemical Name[a]	CASRN[b]	Known constituent of hydraulic fracturing fluid	Physico-chemical properties	Selected toxicity data[c]	NCCT CASRN or name change[d]	Formation type	Reference
Methyl biphenyl, mixed isomers	28652-72-4				Name	C	14, 21
Methyl bromide	74-83-9		X	X		S	3, 9
Methyl crotonate	18707-60-3		X		Name	S	18
Methyl ethyl ketone	78-93-3		X	X	Name	S	3, 9, 10
Methyl(Z)-3,3-diphenyl-4-hexenoate	119296-91-2		X		Name	C	20
Methylcyclohexane	108-87-2	X	X			S	18
Methylenecyclohexane	1192-37-6		X		CASRN	S	18
Methylnaphthalene	1321-94-4				Name	C	20, 21
Methylquinoline	27601-00-9				Name	C	14
Molybdenum	7439-98-7			X		S	3, 9, 10
m-xylene	108-38-3		X	X	Name	S	18
N,N-Dimethylformamide	68-12-2	X	X	X		S	22
Naphthalene	91-20-3	X	X	X		S,C	3, 9, 10, 11, 12, 13, 14, 15, 20, 21, 22
Nickel	7440-02-0			X		S	3, 9, 10
Nitrate	14797-55-8			X		S,C	3, 9, 10
Nitrite	14797-65-0			X		S,C	3, 9, 10
N-Nitrosodiphenylamine	86-30-6		X	X		S	3, 9, 10, 15
N-Nitroso-N-methylethylamine	10595-95-6		X	X	Name	S	9, 15
Nonacosane	630-03-5		X			S	18
Nonahexacontanoic acid	40710-32-5		X			S	18
Nonane	111-84-2		X	X		S	18

Chemical Name[a]	CASRN[b]	Known constituent of hydraulic fracturing fluid	Physico-chemical properties	Selected toxicity data[c]	NCCT CASRN or name change[d]	Formation type	Reference
Norphytane	1921-70-6		X		Name	S	18
o-Cresol	95-48-7		X	X	Name	S,C	3, 9, 10, 13, 15
Octadecanoic acid	57-11-4		X			S,C	14, 21
Octahydro-2-methylpentalene	3868-64-2		X			S	18
Octane	111-65-9		X			S	18
Octasulfur	10544-50-0				Name	C	14
O-Decylhydroxylamine	29812-79-1		X			S	18
O-Isobutylhydroxylamine	5618-62-2		X		Name	S	16
o-Xylene	95-47-6	X	X	X		S	18
p,p'-DDE	72-55-9		X	X		S	3, 9
p-Cresol	106-44-5		X	X	Name	S,C	3, 9, 10, 13, 15
p-Cymene	99-87-6		X		Name	S	9, 10, 18
Pentadecanoic acid	1002-84-2		X			C	20
Pentane	109-66-0	X	X	X		S	16
Pentanoic acid	109-52-4		X			U	10
Pentatriacontane	630-07-9		X			S	18
Pentylcyclohexane	4292-92-6		X			S	18
Pentylhydroperoxide	74-80-6		X			S	18
perylene	198-55-0		X	X		S,C	21
Phenanthrene	85-01-8	X	X	X		S,C	3, 9, 10, 15, 16, 20
Phenanthrene-1-carboxlic acid	27875-89-4		X		Name	C	20
Phenol	108-95-2	X	X	X		S,C	3, 9, 10, 13, 15

Chemical Name[a]	CASRN[b]	Known constituent of hydraulic fracturing fluid	Physico-chemical properties	Selected toxicity data[c]	NCCT CASRN or name change[d]	Formation type	Reference
Phorate	298-02-2		X	X		S	9
Phosphorus	7723-14-0			X		S	3, 9
Polypropylene glycol	25322-69-4	X				S	23
Potassium	7440-09-7					S	3, 9, 10
Propane-diphenyl	25167-94-6				Name	C	20
Propargyl alcohol	107-19-7	X	X	X		S	22
Propionate	72-03-7		X		Name	C	21
Propionic acid	79-09-4		X			U	10
Propyl cyanate	1768-36-1		X		Name	S	16
Propylbenzene	103-65-1		X		Name	S	9, 13, 16, 18
Propylcyclohexane	1678-92-8		X			S	18
Propylcyclopentane	2040-96-2		X			S	18
p-Tert-butylphenol	98-54-4		X		Name	C	21
p-Xylene	106-42-3	X	X	X		S,C	13, 22
Pyrene	129-00-0		X	X		S,C	9, 10, 15, 20, 21
Pyreno[4,5-c]furan	15123-40-7					C	20
Pyridine	110-86-1		X	X		S	3, 9, 10, 15
Pyruvate	57-60-3		X			S	11
Quinoline	91-22-5	X	X	X		S,C	21
Radium	7440-14-4			X		S	3
Radium-226	13982-63-3			X		S	3, 10, 24, 25, 26, 27
Radium-228	15262-20-1			X		S	3, 10, 24, 25, 26

Appendix H – Chemicals Identified in Hydraulic Fracturing Fluids and/or Produced Water

Chemical Name[a]	CASRN[b]	Known constituent of hydraulic fracturing fluid	Physico-chemical properties	Selected toxicity data[c]	NCCT CASRN or name change[d]	Formation type	Reference
rel-(1R,2S)-1,2-Diethylcyclohexadecane	14113-60-1		X		Name	S	18
Rubidium	7440-17-7					C	14
Safrole	94-59-7		X	X		S	3, 9
sec-Butylbenzene	135-98-8		X			S,C	9, 13
Selenium	7782-49-2			X		S	3, 9, 10
Silica	7631-86-9	X		X		U	10
Silicon	7440-21-3					U	10
Silver	7440-22-4			X		S	3, 9, 10
Sodium	7440-23-5					S	3, 9, 10
Sterane	50-24-8		X		Name	C	20
Strontium	7440-24-6			X		S	3, 9, 10
Sulfate	14808-79-8	X				S	3, 9, 10
Sulfide	18496-25-8					S	9, 14
Sulfite	14265-45-3					S	3
syn-1,6:8,13-Bismethano[14]annulene	55821-04-0		X		Name	S	18
tert-Butylbenzene	98-06-6		X			C	13
Tetrachloroethene	127-18-4		X	X	Name	S	3, 9, 11
Tetracontane	4181-95-7		X			S	18
Tetradecanal	124-25-4		X			S	18
Tetradecane	629-59-4	X	X			S,C	18, 20
Tetradecanoic acid	544-63-8		X			S,C	14, 20, 21
Tetradecyl trifluoroacetate	6222-02-2		X		Name	S	18

Chemical Name[a]	CASRN[b]	Known constituent of hydraulic fracturing fluid	Physico-chemical properties	Selected toxicity data[c]	NCCT CASRN or name change[d]	Formation type	Reference
tetramethylbutanedinitrile	3333-52-6		X		Name	S,C	21
Thallium	7440-28-0			X		S	3, 9, 10
Tin	7440-31-5			X		S	9, 10
Titanium	7440-32-6					S	3, 9, 10
Toluene	108-88-3	X	X	X		S	3, 9, 10, 12, 13, 18, 22
trans-1,4-Dimethyladamantane	24145-88-8		X		Name	S	18
Triacontane	638-68-6		X			S	18
Tributyl citrate	77-94-1		X		Name	S	18
Tributyl phosphate	126-73-8	X	X	X	Name	C	14
Trichlorodocosylsilane	7325-84-0		X			S	18
trichlorophenol	25167-82-2					C	21
Tricyclo[4.4.0.0(3,9)]decane	NOCAS_873040		X			C	20
Tridecanal	10486-19-8		X			S	18
Tridecane	629-50-5	X	X			S	18
Tridecanedial	63521-76-6		X			C	20
Tridecyloate-2,2,3,3,4,4,4-heptafluorobutanoic acid	959088-59-6		X			S	18
Triethylene glycol monododecyl ether	3055-94-5		X			S,C	13, 21
Trimethylbenzene	25551-13-7	X		X	Name	S,C	12, 21
Triphenyl phosphate	115-86-6		X			S,C	14, 20, 21
Tritetracontane	7098-21-7		X			S	18

Chemical Name[a]	CASRN[b]	Known constituent of hydraulic fracturing fluid	Physico-chemical properties	Selected toxicity data[c]	NCCT CASRN or name change[d]	Formation type	Reference
Undecane	1120-21-4	X	X			S	12, 18
Undecyl heptafluorobutanoate	959103-74-3		X		Name	S	18
Uranium-235	15117-96-1			X		S	28
Uranium-238	7440-61-1			X		S	26, 28
Vanadium	7440-62-2			X		S	3, 10
Vellerdiol	51276-18-7		X		Name	C	20
Xylenes	1330-20-7	X	X	X	Name	S	3, 9, 10
Zinc	7440-66-6	X		X		S	3, 9, 10
Zirconium	7440-67-7			X		S	3, 9, 10

[a] The following chemicals were found in literature Reference #18 as being present in produced water, but were inadvertently not included in our chemical name/CASRN matching process. Chemical name/CASRN match was made by the authors of that study and may or may not reflect the preferred match as appears in DSSTox: 1-Chlorooctadecane, CASRN 3386-33-2; 1-Nonadecene, CASRN 18435-45-5; 17-Pentatriacontene, CASRN 6971-40-0; 2,6,10-Trimethyldodecane, CASRN 3891-98-3; 2,6,10,14-Tetramethylhexadecane, CASRN 638-36-8; Cyclotriacontane, CASRN 297-35-8; Docosane, CASRN 629-97-0; Hexacosane, CASRN 630-01-3; Pentacosane, CASRN 629-99-2; Tetracosane, CASRN 646-31-1; and Tricosane, CASRN 638-67-5. Ten out of 11 of these chemicals appear to have physicochemical properties available (all except Cyclotriacontane) and none have selected toxicity data.

[b] Some chemicals are designated as "NOCAS_" which are DSSTox database-specific CAS-like identifiers assigned to a listed substance name.

[c] Chemicals are flagged as having selected toxicity data available if they have one or more oral reference values, oral slope factors, or qualitative cancer classifications available from the sources presented in Appendix G.

[d] Chemicals indicated as having a "CASRN" or "Name" change were changed from one or more of the original references cited in the table during the matching process.

Table H-5. Chemicals detected in produced water for which a specific, valid CASRN could not be identified.

These chemicals are either chemically ambiguous or too general for a definitive CASRN to be assigned (e.g., stereoisomerism not defined, groups of related compounds).

Chemical Name	Formation Type	Reference
1,1,3-Trimethylcyclopentane	S	18
1,2,4,5-Tetramethylbenzene	S	18
1,6,7-Trimethylnaphthalene	S,C	14, 18
1,7,11-Trimethyl-4-(1-methylethyl)-cyclotetradecane	S	18
1-Chloro-octadecane	S	18
1-Docosene	C	20
1-Methyl-3-propylbenzene	S	18
2,6,10-Trimethyl-dodecane	S	18
2,6-Dimethyloctane	S	18
2,6-Dimethylundecane	S	18
Decane	S	18
Eicosane	S	18
Heptadecane	S	18
Hexadecane	S	18
N-dodecyl-N,N-dimethylamine	S	22
N-tetradecyl-N,N-dimethylamine	S	22
Octacosane	S	18
Octadecane	S	18
Pentadecane	S,C	18, 21
Tetratetracontane	S	18
Trimethylbenzenes	S,C	20
Alkyl naphthalene	S	16
Alkyl propo-benzene	S	16
Trimethyl-piperdine	S,C	21
Ethyl-tetrahydronaphthalene	C	20
Alkyl benzene	S	16
Alkyl phosphates	C	21

Chemical Name	Formation Type	Reference
Alkyl phthalates	S,C	21
Bis(2-ethylhexyl)-hexanedioic acid	C	20
C11–C37 alkanes/alkenes	S,C	21
C12, C14, C16, C18 fatty acids	S,C	21
C23–C35 alkanes	C	21
C23–C36 alkanes	C	21
Dimethylphenanthrene	C	20
Dioctyldiphenylamine	C	20, 21
Ethyl-cyclodocosane	C	20
Methyl-(2,5-dimethoxyphenol)-methanoate	C	20
Octahydroanthracene	C	20
Octylphenyl ethoxylate	S,C	21
Phenanthrenone	C	20, 21
Tetramethylacenaphthylene	C	20
Tetramethylbenzenes	S	12
Tetramethylnaphthalene	C	20
Tetramethylphenanthrene	C	20, 21
Trimethylnaphthalene	S,C	12, 20
Trimethylphenanthrene	C	20
1,2-Benzenedicarboxylic acid, 1,2-didecyl ester	S	18
N-hexadecanoic acid	C	20, 21
Methyl-biphenyl	S,C	21
1,7,11-Trimethyl-cyclotetradecane	C	20
P-tert-butyl-phenol, p-tert-butyl-	C	21
2a,7a-(Epoxymethano)-2H-cyclobutyl	C	20
Di-tetra-butyl-4-hydroxbenzaldehyde	C	20
Phenanthrene derivative	C	20
Bisphenol F Isomer	S	22
Methoxynaphthalene derivative	C	20
Naphthalenone derivative	C	20

Chemical Name	Formation Type	Reference
Other alkyl phenols	C	20
Other aromatic compounds	C	20
Other benzenamines	C	20
Other benzene alkyl compounds	C	20
Other heterocyclics	C	20
Other indene derivatives	C	20
Other naphthalene alkyl compounds	C	20
Other phthalates	C	20
Other terpenoid compounds	C	20
Quinolo-furazan derivative	C	20
Benzisothiazole derivative	C	20
1-Methylphenanthrene	S	18
Poly(ethylene glycol) bis(carboxymethyl) ether	S	23
Squalene	C	20
Tetrahydro-dimethylnaphthalene	C	21
Trimethoxy-benzaldehyde	C	20
Methylpyrene	C	20
Quinindoline	S,C	21
Benzisothiazole	C	21
Ethyl phenylmethyl benzene	C	20
Dimethyl-ethylindene	C	20
Dihydrophenanthrene	C	20
9-Phenyl-tetrahydro-1H-benz[f]isoindol-1-one	C	20
Methylanthracene	C	20
Methoxyanthracene	C	20
Dimethyl-biphenyl	C	20
Methoxy-methylphenol	C	21
Methylphenanthrene	S,C	20, 21
Tert-butyl-phenol	S,C	21
Methyl-2-quinolinecarboxylic acid	C	20

Chemical Name	Formation Type	Reference
Methylethylnaphthalene	C	20
Tetrahydromethylnaphthalene	C	20
Tetrahydrophenanthrene	C	20
Dihydro-1-methylphenanthrene	C	20
1,7,11-Trimethylcyclotetradecane	C	20
Tetrahydro-trimethylnaphthalene	C	20
1-(2-Hydroxy-5-methylphenyl)-2-hexen-1-one	C	20
Ethyl dimethyl azulene	C	20
9-Methoxyfluorene	C	20
1,2-Di-but-2-enyl-cyclohexanone	C	20
9H-Fluoren-9-ol	C	20
1,4-[13C]-1,2,3,4-Tetrahydro-5-naphthaleneamine	C	20
Dihydro-(-)-neocloven-(II)	C	20
4-(4-Ethylcyclohexyl)-cyclohexene	C	20
Methyl-2-octylcyclopropene-1-octane	C	20
Decahydro-4,4,8,9,10-pentamethylnaphthalene	S,C	21
Hexahydro-1,3,5-trimethyl-1,3,5-triazine-2-thione (a biocide)	S,C	21
8-Isopropyl-2,5-dimethyl-terralin	C	20
9,10-Dimethoxy-2,3-dihydroanthracene	C	20
PEG-C-EO10[a]	S	22
PEG-C-EO2[a]	S	23
PEG-C-EO3[a]	S	23
PEG-C-EO4[a]	S	23
PEG-C-EO5[a]	S	23
PEG-C-EO6[a]	S	23
PEG-C-EO7[a]	S	23
PEG-C-EO8[a]	S	23
PEG-C-EO9[a]	S	23
PEG-EO10[b]	S	23
PEG-EO4[b]	S	23

Chemical Name	Formation Type	Reference
PEG-EO5[b]	S	23
PEG-EO6[b]	S	23
PEG-EO7[b]	S	23
PEG-EO8[b]	S	23
PEG-EO9[b]	S	23
PPG-PO10[c]	S	23
PPG-PO2[c]	S	23
PPG-PO3[c]	S	23
PPG-PO4[c]	S	23
PPG-PO5[c]	S	23
PPG-PO6[c]	S	23
PPG-PO7[c]	S	23
PPG-PO8[c]	S	23
PPG-PO9[c]	S	23

[a] Polyethylene glycol carboxylates containing between four to 10 ethylene oxide monomers.

[b] Polyethylene glycols containing between four to 10 ethylene oxide monomers.

[c] Polypropylene glycols containing between two and 10 proplyene oxide monomers.

This page is intentionally left blank.

Appendix I. Unit Conversions

This page is intentionally left blank.

Appendix I. Unit Conversions

LENGTH

1 in (inch)	=	2.54 cm (centimeters)
		25.4 mm (millimeters)
		25,400 μm (microns)
1 ft (foot)	=	0.3048 m (meters)
		30.48 cm
1 mi (mile)	=	5,280 ft
		1,609.344 m
		1.6093 km (kilometers)

AREA

1 ft^2 (square foot)	=	0.0929 m^2 (square meters)
1 acre	=	43,560 ft^2
	=	0.0016 mi^2 (square miles)
	=	0.4047 ha (hectares)
	=	4,046.825 m^2
1 mi^2	=	639.9974 ac
	=	258.9988 ha
	=	2.5899 km^2 (square kilometers)

MASS

1 lb (pound)	=	453.5924 g (grams)
	=	0.4536 kg (kilograms)
1 ton (short ton, U.S.)	=	2,000 lb
	=	907.185 kg
	=	0.9072 metric tons

VOLUME OR CAPACITY (LIQUID MEASURE)

1 bbl (barrel)	=	42 gal (gallons, U.S.)
	=	158.9873 L (liters)
1 gal	=	231 in^3 (cubic inches)
	=	0.1337 ft^3 (cubic feet)
	=	3.7854 L
	=	0.0039 m^3 (cubic meters)
	=	3.7854 × 10^{-9} Mm^3 (million cubic meters)
1 Mgal (million gallons)	=	1.3368 × 10^5 ft^3
1 ft^3	=	1,728 in^3
	=	7.4805 gal
	=	28.3169 L
	=	0.0283 m^3
1 mi^3 (cubic mile)	=	4.1682 km^3 (cubic kilometers)

CONCENTRATION

1 mg/L (milligram per liter)	=	1.0 × 10^{-6} kg/L (kilograms per liter)
	=	1.0 × 10^{-3} g/L (grams per liter)
	=	1,000 µg/L (micrograms per liter)
	=	1.001 ppm (parts per million)
	=	8.3454 × 10^{-6} lb/gal (pounds per gallon)
	=	6.2428 × 10^{-5} lb/ft^3 (pounds per cubic foot)

SPEED

1 mi/hr (mile per hour)	=	1.466$\overline{6}$ ft/s (feet per second)
	=	0.4470 m/s (meters per second)

DENSITY

1 g/mL	=	1,000 g/L
	=	1.0 × 10^6 mg/L

VOLUME PER UNIT TIME

1 ft³/s (cubic foot per second)	=	448.8312 gpm (gallons per minute)
	=	0.6163 Mgpd (million gallons per day)
	=	28.3169 L/s (liters per second)
	=	0.0283 m³/s (cubic meters per second)
1 ft³/day (cubic feet per day)	=	0.0052 gpm
	=	7.4805 gpd
	=	0.0283 m³/d (cubic meters per day)
1 bbl/day (barrel per day)	=	42 gpd
	=	158.9873 L/d (liters per day)

PRESSURE

1 psi (pound per square inch)	=	6,894.7573 Pa (pascals)
	=	0.068 atm (standard atmospheres)

RADIATION

Activity

1 Ci (curie)	=	3.7×10^{10} decays per second
1 Bq (becquerel)	≈	2.703×10^{-11} Ci
	≈	27.027 pCi (picocuries)
1 pCi	=	0.037 Bq
	=	0.037 decays per second
	=	2.22 decays per minute

Exposure

1 rem (röentgen equivalent in man)	=	0.01 Sv (sieverts)
1 Sv	=	1 J/kg (joule per kilogram)

ELECTRIC CONDUCTANCE

1 S (siemen)	=	1 Ω-1 (reciprocal of resistance)
	=	1 A/V (ampere per volt)
	=	1 kg-1 • m-2 • s³ • A² (second cubed- ampere squared per kilogram-square meter)
	=	1.0×10^{6} µS (microsiemens)

TEMPERATURE

[°F (degrees, Fahrenheit) - 32] × 5/9	=	°C (degrees, Celsius)

PERMEABILITY

1 cm^2	=	$1.0 \times 10\text{-}4$ m^2
	≈	1.0×108 D (darcys)
1 D	≈	$1.0 \times 10\text{-}12$ m^2
	=	1,000 mD (millidarcys)
	=	1.0×106 μD (microdarcys)

Appendix J. Glossary

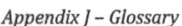

This page is intentionally left blank.

Appendix J. Glossary

J.1. Introduction

This glossary is intended to provide definitions for scientific and technical terms used in the rest of the document. For most terms, a citation is provided that indicates the reference from which a definition was reprinted or adapted. For terms without a citation, the definition was developed for the purposes of this assessment. In some cases, terms in this glossary may also have a legal or regulatory definition in addition to the definition provided; the definitions of these terms in the glossary are not intended to replace or modify any such legal or regulatory definitions. The terms in this glossary do not constitute terms of art for legal or regulatory purposes.

J.2. Glossary Terms and Definitions

Abandoned well: A well that is no longer being used, either because it is not economically producing or it cannot be used because of its poor condition.

Acid mine drainage: Flow of water from areas that have been mined for coal or other mineral ores. The water has a low pH because of its contact with sulfur-bearing material and is harmful to aquatic organisms (U.S. EPA, 2013d).

Additive: A single chemical or chemical mixture designed to serve a specific purpose in the hydraulic fracturing fluid.

Adsorption: Adhesion of molecules of gas, liquid, or dissolved solids to a surface (U.S. EPA, 2013d).

Advection: A mechanism for moving chemicals in flowing water, where a chemical moves along with the flow of the water itself.

Aeration: The process of mixing air with water or soil. It promotes biological degradation of organic matter in water. The process may be passive (as when waste is exposed to air) or active (as when a mixing or bubbling device introduces the air) (U.S. EPA, 2013d).

Aerobic mesophiles: Microorganisms that use oxygen for energy production and are tolerant of moderate temperatures.

Analyte: The element, ion, or compound that an analysis seeks to identify; the compound of interest (U.S. EPA, 2013d).

Annulus: Refers to either the space between the casing of a well and the wellbore or the space between any two strings of tubing or casing (U.S. EPA, 2013d).

API number: A unique identifying number for all oil and gas wells drilled in the United States. The system was developed by the American Petroleum Institute (Oil and Gas Mineral Services, 2010).

Aquifer: A water-bearing geologic formation, group of formations, or part of a formation. Groundwater is the water in an aquifer.

Base fluid: The fluid into which additives and proppants are mixed to formulate a hydraulic fracturing fluid.

Basin: A depression in the crust of the earth, caused by plate tectonic activity and subsidence, into which sediments accumulate. Sedimentary basins vary from bowl-shaped to elongated troughs. Basins can be bounded by faults. Rift basins are commonly symmetrical; basins along continental margins tend to be asymmetrical. If rich hydrocarbon source rocks occur in combination with appropriate depth and duration of burial, then a petroleum system can develop within the basin. Most basins contain some amount of shale, thus providing opportunities for shale gas exploration and production (Schlumberger, 2014).

Bedding plane: The surface that separates two layers of stratified rocks.

Biogenic: Methane that is produced in shallower formations by bacterial activity in anaerobic conditions. It is the ultimate dissimilation product of microbially mediated reactions of organic molecules.

Blowout preventer (BOP): Casinghead equipment that prevents the uncontrolled flow of oil, gas, and mud from the well by closing around the drill pipe or sealing the hole (Oil and Gas Mineral Services, 2010). BOPs are typically a temporary component of the well, in place only during drilling and perhaps through hydraulic fracturing operations

Brackish water: A general term used for water having a salinity content intermediate between fresh water and sea water, although it may also have a more specific definition, such as the 1,000 – 10,000 mg/L TDS value used in some USGS publications.

BTEX: An acronym for benzene, toluene, ethylbenzene, and xylenes. These chemicals are a group of single ringed aromatic hydrocarbon based on the benzene structure. These compounds are found in petroleum and are of specific importance because of their health effects.

British thermal unit (Btu): A measure of the heat (or energy) content of fuels.

Caliper log: A log that is used to check for any wellbore irregularities. It is run prior to primary cementing as a means of calculating the amount of cement needed. Also run in conjunction with other open hole logs for log corrections or run on cased holes to evaluate metal loss (NYSDEC, 2011).

Capillarity: The action by which the surface of a liquid in contact with a solid is elevated or lowered depending on the relative attraction of the molecules of the liquid for each other (cohesion) and for those of the solid (adhesion). Capillary forces arise from the differential attraction between immiscible fluids and solid surfaces; these are the forces responsible for capillary rise in small-diameter tubes and porous materials (adapted from Dake, 1978).

Casing: Steel pipe that is lowered into a wellbore. Casing extends from the bottom of the hole to the surface (Schlumberger, 2014).

Casing, fully cemented: Casing that had a continuous cement sheath from the bottom of the casing to at least the next larger and overlying casing (or the ground surface, if it is a surface casing).

Casing, partially cemented: Casing that had some portion of the casing that was cemented from the bottom of the casing to at least the next larger and overlying casing (or ground surface, if it is a surface casing), but is not fully cemented.

Casing, uncemented: Casing with no cement anywhere along the casing, from the bottom of the casing to at least the next larger and overlying casing (or ground surface, if it is a surface casing).

Casing inspection log: An in situ record of casing thickness and integrity, to determine whether and to what extent the casing has undergone corrosion. The term refers to an individual measurement, or a combination of measurements using acoustic, electrical, and mechanical techniques, to evaluate the casing thickness and other parameters. The log is usually presented with the basic measurements and an estimate of metal loss. Today the terms casing-evaluation log and pipe-inspection log are used synonymously (Schlumberger, 2014).

Casing string: An assembled length of steel pipe configured to suit a specific wellbore.

Chemical Abstract Service Registry Number (CASRN): A unique numeric identifier for only one substance, which serves as a link to information about a specific chemical substance. The CAS registry covers substances identified from the scientific literature from 1957 to the present, with additional substances going back to the early 1900s (CAS Registry Service, 2016). For simplicity, we refer to both pure chemicals and chemical substances that are mixtures, which have a single CASRN, as "chemicals."

Cation exchange capacity: The total amount of cations (positively charged ions) that a soil can hold.

Cement: Material used to support and seal the well casing to the rock formations exposed in the wellbore. Cement also protects the casing from corrosion and prevents movement of fluids up the borehole (U.S. EPA, 2013d).

Cement bond log: A representation of the integrity of the cement job, specifically whether the cement is adhering solidly to the outside of the casing (Schlumberger, 2014). Used to calculate a bond index, which varies between 0 and 1, with 1 representing the strongest bond and 0 representing the weakest bond.

Cement squeeze: A remedial cementing operation designed to force cement into leak paths in wellbore tubulars. The required squeeze pressure is achieved by carefully controlling pump pressure. Squeeze cementing operations may be performed to repair poor primary cement jobs, isolate perforations, or repair damaged casing or liner (Schlumberger, 2014).

Centralized waste treatment facility (CWT): Any facility that treats (for disposal, recycling or recovery of material) any hazardous or non-hazardous industrial wastes, hazardous or non-hazardous industrial wastewater, and/or used material received from off-site (U.S. EPA, 2012c).

Coalbed methane: Methane contained in coal seams. A coal seam is a layer or stratum of coal parallel to the rock stratification (U.S. EPA, 2013d).

Collapse pressure: The pressure at which a tube, or vessel, will catastrophically deform as a result of differential pressure acting from outside to inside of the vessel or tube (Schlumberger, 2014).

Collar: A threaded coupling used to join two lengths of pipe such as production tubing, casing, or liner. The type of thread and style of collar varies with the specifications and manufacturer of the tubing (Schlumberger, 2014).

Combination truck: A truck tractor or a truck tractor pulling any number of trailers (U.S. Department of Transportation, 2012).

Community water system: A public water system which serves at least 15 service connections used by year-round residents or regularly serves at least 25 year-round residents (U.S. EPA, 2013d).

Completion: A term used to describe the assembly of equipment at the bottom of the well that is needed to enable production from an oil or gas well. It can also refer to the activities and methods (including hydraulic fracturing) used to prepare a well for production following drilling.

Complexation: A reaction between two chemicals that form a new complex, either through covalent bonding or ionic forces. This often results in one chemical solubilizing the other.

Compressive strength: Measure of the ability of a substance to withstand compression (NYSDEC, 2011).

Conductor casing: This large diameter casing is usually the first string of casing in a well. It is set or driven into the unconsolidated material where the well will be drilled to keep the loose material from caving in (NYSDEC, 2011).

Confidential business information (CBI): Information that is claimed by the submitter to be entitled to confidential treatment, such as trade secrets, commercial or financial information, or other information that has been claimed as confidential. This information is generally not publicly available. The EPA may have special procedures for handling such information. Further discussion of information claimed to be CBI, including the EPA's process for determining the validity of such claims, is contained in 40 CFR, Part 2.

Contaminant: A substance that is either present in an environment where it does not belong or is present at levels that might cause harmful (adverse) health effects (U.S. EPA, 2013d).

Conventional rock formation: Permeable groups of rock with many large, well-connected pore spaces that allow fluids to move within the rock formation. See also conventional reservoir.

Crosslinked gel: A fluid with polymers that have been linked together through a chemical bond. The polymer chains link together to form larger chemical structures with higher viscosity. Increased viscosity allows the fracturing fluid to carry more proppant into the fractures.

Crude oil: A general term for unrefined petroleum or liquid petroleum (Schlumberger, 2014).

Cumulative effect: Combined changes in the environment that can take place as a result of multiple activities over time and/or space.

Cumulative water use/cumulative water: Refers to the amount of water used or consumed by all hydraulic fracturing wells in a given area per year.

Cyclical stress: Refers to stress caused by frequent or rapid changes in temperature or pressure.

Deviated well: Any non-horizontal well in which the well bottom is intentionally located at a distance (e.g., hundreds of feet) laterally from the wellhead.

Directional drilling: The practice of controlling the direction and deviation (angle) of a wellbore during drilling (SPE, 2016). This enables drilling the wellbore in a predetermined orientation to a targeted area in the subsurface. Directional drilling is required for drilling a deviated or horizontal well and is common in unconventional reservoirs.

Discharge: Any emission (other than natural seepage), intentional or unintentional. Includes, but is not limited to, spilling, leaking, pumping, pouring, emitting, emptying, or dumping (U.S. EPA, 2013d). Or, where groundwater flows to the surface at springs or through the bottoms of lakes and rivers.

Disclosure: With respect to the FracFocus Registry, all data submitted for a specific oil and gas production well for a specific fracture date.

Disinfection byproduct (DBP): A compound formed by the reaction of a disinfectant such as chlorine with organic material in the water supply (U.S. EPA, 2013d).

Domestic water use: Includes indoor and outdoor water uses at residences, and includes, but is not limited to, uses such as drinking, food preparation, bathing, washing clothes and dishes, flushing toilets, watering lawns and gardens, and maintaining pools (USGS, 2015).

Drill bit: The tool used to crush or cut rock during drilling of the well. Most bits work by scraping or crushing the rock as part of a rotational motion, while some bits work by pounding the rock vertically (Schlumberger, 2014).

Drill collar: A component of a drill string that provides weight on the drill bit for drilling the well. Drill collars are thick-walled tubular pieces machined from solid bars of steel, usually plain carbon steel but sometimes of nonmagnetic nickel-copper alloy or other nonmagnetic premium alloys (Schlumberger, 2014).

Drill cutting: The small pieces of broken and ground-up rock generated during the well drilling process.

Drill string: The combination of the drillpipe, the bottomhole assembly, and any other tools used to make the drill bit turn at the bottom of the wellbore (Schlumberger, 2014).

Drilling fluid: Any of a number of liquid and gaseous fluids and mixtures of fluids and solids used when drilling wellbores (adapted from Schlumberger, 2014).

Drinking water resource: Any groundwater or surface water that now serves, or in the future could serve, as a source of drinking water for public or private use (U.S. EPA, 2013d).

Dry gas: Refers to natural gas that occurs in the absence of liquid hydrocarbons (adapted from Schlumberger, 2014).

Effluent: Waste material being discharged into the environment, either treated or untreated (U.S. EPA, 2013d). For the purposes of this assessment, effluent refers to liquid waste material.

Facultative anaerobes: Microorganisms that can use oxygen for energy production if it is present in their environment, but can also use alternatives for energy production if no oxygen is present.

Factor: A feature of hydraulic fracturing operations or an environmental condition that affects the frequency or severity of impacts.

Fault: A fracture or fracture zone along which there has been displacement of the sides relative to each other (NYSDEC, 2011).

Field: Area of oil and gas production with at least one common reservoir for the entire area (Oil and Gas Mineral Services, 2010).

Flowback: The term is defined multiple ways in the literature. In general, it is either fluids predominantly containing hydraulic fracturing fluid that return from a well to the surface or a process used to prepare the well for production.

Fluid: A substance that flows when exposed to an external pressure; fluids include both liquids and gases.

Fluid formulation: The entire suite of chemicals, proppant, and base fluid injected into a well during hydraulic fracturing (U.S. EPA, 2013d).

Formation: A body of earth material with distinctive and characteristic properties and a degree of homogeneity in its physical properties (U.S. EPA, 2013d).

Formation packer: A specialized well casing part that has the same inner diameter as the casing but whose outer diameter expands to make contact with the formation and seal the annulus between the uncemented casing and formation, preventing migration of fluids.

Formation fluid: Fluid that occurs naturally within the pores of rock. These fluids consist primarily of water, with varying concentrations of total dissolved solids, but may also contain oil or gas. Sometimes referred to as native fluids, native brines, or reservoir fluids.

FracFocus Registry: A registry for oil and gas well operators to disclose information about hydraulic fracturing well locations, and water and chemical use during hydraulic fracturing operations. The registry was developed by the Ground Water Protection Council and the Interstate Oil and Gas Compact Commission.

Fracture: A crack or breakage surface within a rock.

Fracture complexity: The ratio of horizontal-to-vertical fracture volume distribution, as defined by Fisher and Warpinski (2012). Fracture complexity is higher in fractures with a larger horizontal component.

Fracture geometry: Refers to characteristics of the fracture such as height, aperture, orientation, and azimuth.

Fracture half-length: The radial distance from a wellbore to the outer tip of a fracture propagated from that well (Schlumberger, 2014).

Freeboard: The vertical distance between the level of the water in an impoundment and the overflow elevation (an outfall or the lowest part of the berm).

Fresh water: Qualitatively refers to water with relatively low TDS (total dissolved solids) that is most readily available for drinking water currently. We do not use the term to imply an exact TDS limit, except in Chapter 2 where it refers to water having TDS content up to 3,000 milligrams per liter.

Frequency: The number of impacts per a given unit (e.g., per geographic area, per unit time, per number of hydraulically fractured wells, per number of water bodies). Reflecting the scientific literature, the most common representation of frequency in this assessment is the number of impacts per hydraulically fractured well.

Gelation: The process in the setting of the cement where it begins to solidify and lose its ability to transmit pressure to the formation.

Gelled fluid: Fracturing fluids that are usually water-based with added gels to increase the fluid viscosity to aid in the transport of proppants (Spellman, 2012; Gupta and Valkó, 2007).

Groundwater: In the broadest sense, all subsurface water; more commonly that part of the subsurface water in the saturated zone (Solley et al., 1998).

Groundwater age: Refers to how long the water has been in the ground.

Groundwater availability: The amount of groundwater that is available regardless of legal or physical availability (TWDB, 2012).

Groundwater supply: The amount of groundwater that can be produced given current permits and existing infrastructure (TWDB, 2012).

Halite: A soft, soluble evaporate mineral commonly known as salt or rock salt. Can be critical in forming hydrocarbon traps and seals because it tends to flow rather than fracture during deformation, thus preventing hydrocarbons from leaking out of a trap even during and after some types of deformation (Schlumberger, 2014).

Hazard evaluation: A component of risk assessment that involves gathering and evaluating data on the types of health injuries or diseases (e.g., cancer) that may be produced by a chemical and on the conditions of exposure under which such health effects are produced.

Hazard identification: A process for determining if a chemical or a microbe can cause adverse health effects in humans and what those effects might be (U.S. EPA, 2013d).

Henry's law constant: Ratio of a chemical's vapor pressure in the atmosphere to its solubility in water. The higher the Henry's law constant, the more volatile the compound will be from water (NYSDEC, 2011).

Horizontal drilling: Drilling a portion of a well horizontally to expose more of the formation surface area to the wellbore (Oil and Gas Mineral Services, 2010). This is a type of directional drilling.

Horizontal well: A well that is drilled vertically downward up to a point known as the kickoff point, where the well turns toward the horizontal, extending into and parallel with the approximately horizontal targeted producing formation. Directional drilling is required to drill a horizontal well.

Hydraulic fracturing: A stimulation technique used to increase production of oil and gas. Hydraulic fracturing involves the injection of fluids under pressures great enough to fracture the oil- and gas-production formations (U.S. EPA, 2011a).

Hydraulic fracturing fluids: Engineered fluids, typically consisting of a base fluid, additives, and proppant that are pumped under high pressure into the well to create and hold open fractures in the formation.

Hydraulic fracturing wastewater: Produced water that is managed using practices that include, but are not limited to, reuse in subsequent hydraulic fracturing operations, treatment and discharge, and injection into disposal wells. The term is being used in this study as a general description of certain waters and is not intended to constitute a term of art for legal or regulatory purposes.[1]

Hydraulic fracturing water cycle: The cycle of water in the hydraulic fracturing process, encompassing the acquisition of water, chemical mixing of the fracturing fluid, injection of the fluid into the formation, the production and handling of produced water, and the ultimate treatment and disposal of hydraulic fracturing wastewaters.

---

[1] This general description does not, and is not intended to, provide that the production, recovery, or recycling of oil, including the production, recovery, or recycling of produced water, constitutes "wastewater treatment" for the purposes of the Oil Pollution Prevention regulation (with the exception of dry gas operations), which includes the Spill Prevention, Control, and Countermeasure rule and the Facility Response Plan rule, 40 CFR 112 et seq.

Hydraulic gradient: Slope of a water table or potentiometric surface. More specifically, change in the hydraulic head per unit of distance in the direction of the maximum rate of decrease (U.S. EPA, 2013d).

Hydrocarbon: An organic compound containing only hydrogen and carbon, often occurring in petroleum, natural gas, and coal (U.S. EPA, 2013d).

Hydrophilic: A chemical property that describes a tendency to dissolve in water. Literally, "water loving."

Hydrophobic: A chemical property that describes a tendency to be soluble in nonpolar solvents and sparingly soluble in water. Literally, "water fearing."

Hydrostatic pressure: The pressure exerted by a column of fluid at a given depth. In Chapter 6, it refers to the pressure exerted by a column of drilling mud or cement on the formation at a particular depth.

Imbibition: The displacement of a non-wet fluid (i.e., gas) by a wet fluid (typically water) (adapted from Dake, 1978).

Immiscible: The chemical property in which two or more liquids or phases are incapable of attaining homogeneity (U.S. EPA, 2013d).

Impact: Any change in the quality or quantity of drinking water resources, regardless of severity, that results from an activity in the hydraulic fracturing water cycle.

Induced fracture: A fracture created during hydraulic fracturing.

Integrated risk information system (IRIS): An electronic database that contains the EPA's latest descriptive and quantitative regulatory information about chemical constituents. Files on chemicals maintained in IRIS contain information related to both noncarcinogenic and carcinogenic health effects (U.S. EPA, 2013d).

Intermediate casing: Casing that seals off intermediate depths and geologic formations that may have considerably different reservoir pressures than deeper zones to be drilled (Devereux, 1998; Baker, 1979).

Karst: A type of topography that results from dissolution and collapse of carbonate rocks, such as limestone, dolomite, and gypsum, and that is characterized by closed depressions or sinkholes, caves, and underground drainage (Solley et al., 1998).

Kill fluid: A weighted fluid with a density that is sufficient to overcome the formation pressure and prevent fluids from flowing up the wellbore.

Large truck: A truck with a gross vehicle weight rating greater than 10,000 pounds (U.S. Department of Transportation, 2012).

Lateral: A horizontal section of a well.

Leakoff: The fraction of the injected fluid that infiltrates into the formation (e.g., through an existing natural fissure) and is not recovered during production (i.e., it does not return through the well to the surface) (Economides et al., 2007). Fluids that leak off and are not recovered are sometimes referred to as "lost" fluids.

Linear gel: A series of chemicals linked together so that they form a chain.

Liner: A casing string that does not extend to the top of the wellbore, but instead is anchored or suspended from inside the bottom of the previous casing string (Schlumberger, 2014).

Lost cement: Refers to a failure of the cement to be circulated back to the surface during construction of the well, indicating that the cement has escaped into the formation.

Lowest-observable-adverse effect level (LOAEL): The lowest exposure level at which there are biologically significant increases in the frequency or severity of adverse effects between the exposed population and its appropriate control group (U.S. EPA, 2011c).

Maximum allowable daily level (MADL): The maximum allowable daily level of a reproductive toxicant at which the chemical would have no observable adverse reproductive effect, assuming exposure at 1,000 times that level (OEHHA, 2012).

Maximum contaminant level (MCL): The highest level of a contaminant that is allowed in drinking water. MCLs are set as close to the MCLG as feasible using the best available analytical and treatment technologies and taking cost into consideration. MCLs are enforceable standards (U.S. EPA, 2012a).

Maximum contaminant level goal (MCLG): A non-enforceable health benchmark goal which is set at a level at which no known or anticipated adverse effect on the health of persons is expected to occur and which allows an adequate margin of safety (U.S. EPA, 2012a)

Mechanical integrity: The absence of significant leakage within the injection tubing, casing, or packer (known as internal mechanical integrity), or outside of the casing (known as external mechanical integrity) (U.S. EPA, 2013d).

Microaerophiles: Microorganisms that require small amounts of oxygen for energy production.

Microannuli: Very small openings that form between the cement and its surroundings and that may serve as pathways for fluid migration to drinking water resources.

Microseismic monitoring: A technique to track the propagation of a hydraulic fracture as it advances through a formation (Schlumberger, 2014).

Minimal risk level (MRL): An ATSDR estimate of daily human exposure to a hazardous substance at or below which the substance is unlikely to pose a measurable risk of harmful (adverse), noncancerous effects. MRLs are calculated for a route of exposure (inhalation or oral) over a specified time period (acute, intermediate, or chronic). MRLs should not be used as predictors of harmful (adverse) health effects (ATSDR, 2016).

Mobility: The ratio of effective permeability to phase viscosity. The overall mobility is a sum of the individual phase viscosities. Well productivity is directly proportional to the product of the mobility and the layer thickness product (Schlumberger, 2014).

National Pollution Discharge Elimination System (NPDES): A national program under Section 402 of the Clean Water Act for regulation of discharges of pollutants from point sources to waters of the United States. The Clean Water Act prohibits the discharge of pollutants from any point source into waters of the United States, except as authorized by the Act, which may include issuance of an NPDES permit.

National Secondary Drinking Water Regulations (NSDWR): Non-enforceable guidelines regulating contaminants that may cause cosmetic effects (such as skin or tooth discoloration) or aesthetic effects (such as taste, odor, or color) in drinking water (also referred to as secondary standards) (U.S. EPA, 2014c).

Natural gas: A naturally occurring mixture of hydrocarbon and nonhydrocarbon gases in porous formations beneath the earth's surface, often in association with petroleum. The principal constituent of natural gas is methane (Schlumberger, 2014).

Natural organic matter (NOM): Complex organic compounds that are formed from decomposing plant animal and microbial material in soil and water (U.S. EPA, 2013d).

Naturally Occurring Radioactive Materials (NORM): Radioactive materials found in nature that have not been moved or concentrated by human activities.

No-observed-adverse-effect level (NOAEL): NOAEL is defined as the highest exposure level at which there are no biologically significant increases in the frequency or severity of adverse effect between the exposed population and its appropriate control; some effects may be produced at this level, but they are not considered adverse or precursors of adverse effects (U.S. EPA, 2011c).

Non-community water system: Water systems that supply water to at least 25 of the same people or have 15 service connections at least six months per year, but not year-round (U.S. EPA, 2013c).

National Toxicology Program (NTP): The NTP describes the results of individual experiments on a chemical agent and notes the strength of the evidence for conclusions regarding each study. For more information, see Appendix G.

Octanol-water partition coefficient (K_{OW}): A coefficient representing the ratio of the solubility of a compound in octanol (a nonpolar solvent) to its solubility in water (a polar solvent). The higher the K_{OW}, the more nonpolar the compound. Log K_{OW} is generally used as a relative indicator of the tendency of an organic compound to adsorb to soil. Log K_{OW} values are generally inversely related to aqueous solubility and directly proportional to molecular weight (U.S. EPA, 2013d).

Offset well: An abandoned (i.e., plugged), inactive, or actively producing well near a well that is used for hydraulic fracturing.

Open hole completion: A well completion that has no casing or liner set across the reservoir formation, allowing the produced fluids to flow directly into the wellbore (Schlumberger, 2014).

Oral slope factor (OSF): An upper-bound, approximating a 95% confidence limit, on the increased cancer risk from a lifetime oral exposure to an agent. This estimate, usually expressed in units of proportion (of a population) affected per mg/kg day, is generally reserved for use in the low dose region of the dose response relationship, that is, for exposures corresponding to risks less than 1 in 100 (U.S. EPA, 2011c).

Soil adsorption coefficient (K_{oc}): A coefficient that provides a measure of the ability of a chemical to sorb (adhere) to the organic portion of soil, sediment, and sludge. The higher the K_{oc}, the more likely a compound is to adsorb to soils and sediments, and the less likely it is to migrate with water. Along with log K_{ow}, log K_{oc} is used as a relative indicator of the tendency of an organic compound to adsorb to soil.

Orphaned well: An inactive oil or gas well with no known (or financially solvent) owner.

Overburden: Material of any nature, consolidated or unconsolidated, that overlies a deposit of useful minerals or ores (U.S. EPA, 2013d).

Packer: A mechanical device that expands to selectively seal off certain sections of the wellbore to keep fluid from migrating within the annulus. Packers can be used to seal the space between the tubing and casing, between two casings, or between the production casing and the surrounding rock formation (Schlumberger, 2014).

Pad fluid: A mixture of base fluid, typically water and additives without solid, designed to create, elongate, and enlarge fractures along the natural channels of the formation when injected under high pressure at the start of the hydraulic fracturing process.

Partial cementing: Cementing a casing string of a well along only a portion of its length.

Passby flow: A prescribed, low-streamflow threshold below which withdrawals are not allowed (U.S. EPA, 2015d).

Peer review: A documented critical review of a specific major scientific and/or technical work product. Peer review is intended to uncover any technical problems or unresolved issues in a preliminary or draft work product through the use of independent experts. This information is then used to revise the draft so that the final work product will reflect sound technical information and analyses. The process of peer review enhances the scientific or technical work product so that the decision or position taken by the EPA, based on that product, has a sound and credible basis (U.S. EPA, 2013d).

Perforation: The communication tunnel created from the casing or liner into the reservoir formation through which injected fluids and oil or gas flows. Also refers to the process of creating communication channels, e.g., via the use of a jet perforating gun.

Permeability: The ability of fluids (including oil and gas) to flow through well-connected pores or small openings in the rock. Also referred to as intrinsic or absolute permeability.

Persistence: The length of time a compound stays in the environment, once introduced. A compound may persist for less than a second or indefinitely.

Physicochemical property: The inherent physical and chemical properties of a molecule such as boiling point, density, physical state, molecular weight, vapor pressure, etc. These properties define how a chemical interacts with its environment (U.S. EPA, 2013d).

Play: A set of oil or gas accumulations sharing similar geologic, geographic properties, such as source rock, hydrocarbon type, and migration pathways (Oil and Gas Mineral Services, 2010).

Poisson's ratio: A ratio of transverse-to-axial (or latitudinal-to-longitudinal) strain; characterizes how a material is deformed under pressure.

Polar molecule: A molecule with a slightly positive charge at one part of the molecule and a slightly negative charge on another. The water molecule, H_2O, is an example of a polar molecule, where the molecule is slightly positive around the hydrogen atoms and negative around the oxygen atom.

Porosity: A measure of empty space for a given volume of material, or the percentage of the material (e.g., rock or soil) volume that can be occupied by oil, gas, or water.

Principal aquifer: A regionally extensive aquifer or aquifer system that has the potential to be used as a source of potable water.

Private (non-public) water system: Water systems that serve fewer than 15 connections and fewer than 25 individuals (U.S. EPA, 1991).

Produced water: Water that flows from the subsurface through oil and gas wells to the surface as a by-product of oil and gas production.

Production casing: The deepest casing set in a well that serves primarily as the conduit for producing fluids, although when cemented to the wellbore, this casing can also serve to seal off other subsurface zones including groundwater resources (Devereux, 1998; Baker, 1979).

Production well: A well that is used to bring fluids (such as oil or gas) to the surface.

Production zone: Refers to the portion of a subsurface rock zone that contains oil or gas to be extracted (sometimes using hydraulic fracturing). The production zone is sometimes referred to as the target zone or targeted rock formation.

Proppant/propping agent: A granular substance (sand grains, aluminum pellets, or other material) that is carried in suspension by the fracturing fluid and that serves to keep the cracks open when fracturing fluid is withdrawn after a fracture treatment (U.S. EPA, 2013d).

Protected groundwater resource: All aquifers, or their portions, that the state or other regulatory agency requires to be protected from fluid migration through or along wellbores.

Public water system source: The source of the surface water or groundwater used by a public water system, including source wells, intakes, reservoirs, infiltration galleries, and springs.

Public water system: Water systems that provide water for human consumption from surface water or groundwater through pipes or other infrastructure to at least 15 service connections or serve an average of at least 25 people for at least 60 days a year (Safe Drinking Water Act, 2002).

Publicly owned treatment works (POTW): Any device or system used in the treatment (including recycling and reclamation) of municipal sewage or industrial wastes of a liquid nature that is owned by a state or municipality. This definition includes sewers, pipes, or other conveyances only if they convey wastewater to a POTW providing treatment (U.S. EPA, 2013d).

Quality assurance (QA): An integrated system of management activities involving planning, implementation, documentation, assessment, reporting, and quality improvement to ensure that a process, item, or service is of the type and quality needed and expected by the customer (U.S. EPA, 2013d).

Quality assurance project plan (QAPP): A formal document describing in comprehensive detail the necessary quality assurance procedures, quality control activities, and other technical activities that need to be implemented to ensure that the results of the work performed will satisfy the stated performance or acceptance criteria (U.S. EPA, 2013d).

Quality management plan: A document that describes a quality system in terms of the organizational structure, policy and procedures, functional responsibilities of management and staff, lines of authority, and required interfaces for those planning, implementing, documenting, and assessing all activities conducted (U.S. EPA, 2013d).

Radioactive tracer log: A record of the presence of radioactive tracer material placed in or around the wellbore to measure fluid movement in injection wells (Schlumberger, 2014).

Radionuclide: Radioactive particle, man-made or natural, with a distinct atomic weight number. Emits radiation in the form of alpha or beta particles, or as gamma rays. Can have a long life as soil or water pollutant. Prolonged exposure to radionuclides increases the risk of cancer (U.S. EPA, 2013d).

Reference dose (RfD): An estimate (with uncertainty spanning perhaps an order of magnitude) of a daily oral exposure to the human population (including sensitive subgroups) that is likely to be without an appreciable risk of deleterious effects during a lifetime. It can be derived from a NOAEL, LOAEL, or benchmark dose, with uncertainty factors generally applied to reflect limitations of the data used. Generally used in EPA's noncancer health assessments (U.S. EPA, 2011c).

Reference value (RfV): An estimate of an exposure or dose for a given duration to the human population (including susceptible subgroups) that is likely to be without an appreciable risk of adverse health effects over a lifetime. RfV is a generic term not specific to a given route of exposure (U.S. EPA, 2011c). In the context of this report, the term RfV refers to reference values for noncancer effects occurring via the oral route of exposure and for chronic durations, except where noted.

Relative permeability: A dimensionless property allowing for comparison of the different abilities of fluids to flow in multiphase settings. If a single fluid is present, its relative permeability is equal to 1, but the presence of multiple fluids generally inhibits flow and decreases the relative permeability.

Reservoir: A geologic formation where hydrocarbons collect under pressure over geological time.

> **Conventional reservoir:** A reservoir in which buoyant forces keep hydrocarbons in place below a sealing caprock. Reservoir and fluid characteristics of conventional reservoirs typically permit oil or natural gas to flow readily into wellbores. The term is used to make a distinction from shale and other unconventional reservoirs, in which gas might be distributed throughout the reservoir at the basin scale, and in which buoyant forces or the influence of a water column on the location of hydrocarbons within the reservoir are not significant (Schlumberger, 2014).

> **Unconventional reservoir:** A reservoir characterized by lower permeability than conventional reservoirs. It can be the same formation where hydrocarbons are formed and also serve as the source for hydrocarbons that migrate and accumulate in conventional reservoirs. Unconventional reservoirs can include methane-rich coalbeds and oil- and/or gas-bearing shales and tight sands.

Residuals: The solids generated or retained during the treatment of wastewater (U.S. EPA, 2013d).

Safe Drinking Water Act (SDWA): The act designed to protect the nation's drinking water supply by establishing national drinking water standards (maximum contaminant levels or specific treatment techniques) and by regulating underground injection control wells (U.S. EPA, 2013d).

Sandstone: A clastic sedimentary rock whose grains are predominantly sand sized. The term is commonly used to imply consolidated sand or a rock made of predominantly quartz sand, although sandstones often contain feldspar, rock fragments, mica, and numerous additional mineral grains held together with silica or another type of cement. The relatively high porosity and permeability of sandstones make them good reservoir rocks (Schlumberger, 2014).

Science Advisory Board (SAB): A federal advisory committee that provides a balanced, expert assessment of scientific matters relevant to the EPA. An important function of the Science Advisory Board is to review EPA's technical programs and research plans (U.S. EPA, 2013d).

Service company: A company that assists well operators by providing specialty services, including hydraulic fracturing (U.S. EPA, 2013d).

Severity: The magnitude of change in the quality or quantity of a drinking water resource as measured by a given metric (e.g., duration, spatial extent, contaminant concentration).

Shale: A fine-grained, fissile, detrital sedimentary rock formed by consolidation of clay- and silt-sized particles into thin, relatively impermeable layers (Schlumberger, 2014).

Shale gas: Natural gas generated and stored in shale.

Shale oil: Oil present in reservoirs that are made up of shale.

Shut in: The process of sealing off a well by either closing the valves at the wellhead, a downhole safety valve, or a blowout preventer.

Slickwater: A type of fracturing fluid designed to have a low viscosity to reduce friction loss when pumping the fracturing fluid downhole. The critical additive in a slickwater is friction reducer, which allows pumping at high rates (Barati and Liang, 2014).

Solubility: The amount of mass of a compound that will dissolve in a unit volume of solution (U.S. EPA, 2013d).

Sorption: The general term used to describe the partitioning of a chemical between soil and water and depends on the nature of the solids and the properties of the chemical.

Source water: Surface water or groundwater, or reused wastewater, acquired for use in hydraulic fracturing.

Spacer fluid: A fluid pumped into the well during construction before the cement to clean drilling mud out of the wellbore.

Spud (spud a well): To start the well drilling process by removing rock, dirt, and other sedimentary material with the drill bit (U.S. EPA, 2013d).

Spill: Any unintended release of fluids. Hydraulic fracturing-related spills are spills that occur at any phase within the hydraulic fracturing water cycle. These include chemicals, additives, hydraulic fracturing fluids (chemical mixing phase); flowback and produced water; wastewater.

Stages (frac stages): A single reservoir interval that is hydraulically stimulated in succession with other intervals.

Stimulation: Refers to (1) injecting fluids to clear the well or pore spaces near the well of drilling mud or other materials that create blockage and inhibit optimal production (i.e., matrix treatment) and (2) injecting fluid to fracture the rock to optimize the production of oil or gas.

Stray gas: Refers to the phenomenon of natural gas (primarily methane) migrating into shallow drinking water resources or to the surface.

Subsurface formation: a mappable body of rock of distinctive rock type(s) and characteristics (such as permeability and porosity) with a unique stratigraphic position.

Surface casing: The shallowest cemented casing, with the widest diameter. Cemented surface casing generally serves as an anchor for blowout protection equipment and to seal off drinking water resources (Baker, 1979).

Surface water: All water naturally open to the atmosphere (rivers, lakes, reservoirs, ponds, streams, impoundments, seas, estuaries, etc.) (U.S. EPA, 2013d).

Surfactant: Used during the hydraulic fracturing process to decrease liquid surface tension and improve fluid passage through the pipes (U.S. EPA, 2013d).

Sustained casing pressure: The pressure in any well annulus that is measurable at the wellhead and rebuilds after it is bled down, not caused solely by temperature fluctuations or imposed by the operator. If the pressure is relieved by venting natural gas from the annulus to the atmosphere, it will build up again once the annulus is closed (i.e., the pressure is sustained) (Skjerven et al., 2011). The return of pressure indicates that there is a small leak in a casing or through uncemented or poorly cemented intervals that exposes the annulus to a pressured source of gas. It is possible to have pressure in more than one of the annuli.

Targeted rock formation: The portion of a subsurface rock formation that contains oil or gas to be extracted (sometimes called the "target zone" or the "production zone").

Tolerable daily intake (TDI): An estimate of the intake of a substance, expressed on a body mass basis, to which an individual in a (sub) population may be exposed daily over its lifetime without appreciable health risk (WHO, 2015).

Technically recoverable resource: The volumes of oil and natural gas that could be produced with current technology, regardless of oil and natural gas prices and production costs (EIA, 2013).

Technologically Enhanced Naturally Occurring Radioactive Material (TENORM): defined by EPA as naturally occurring radioactive materials (NORM) that have been concentrated or exposed to the accessible environment as a result of human activities such as manufacturing, mineral extraction, or water processing.

Temperature log: A log of the temperature of the fluids in the wellbore; a differential temperature log records the rate of change in temperature with depth and is sensitive to very small changes (U.S. EPA, 2013d).

Tensile strength: The force per unit cross-sectional area required to pull a substance apart (Schlumberger, 2014).

Thermogenic: Methane that is produced by high temperatures and pressures in deep formations over geologic timescales. Thermogenic methane is formed by the thermal breakdown, or cracking, of organic material that occurs during deep burial of sediment.

Tight oil: Oil found in relatively impermeable reservoir rock (Schlumberger, 2014).

Total dissolved solids (TDS): The quantity of dissolved material in a given volume of water. Total dissolved solids can include salts (e.g., sodium chloride), dissolved metals, radionuclides, and dissolved organics (U.S. EPA, 2013d). Salinity and total dissolved solids are frequently interchangeable terms.

Total petroleum hydrocarbons (TPH): A large family of several hundred chemical compounds that originally come from crude oil. TPH is a mixture of chemicals, but they are all made mainly from hydrogen and carbon, called hydrocarbons. TPH are divided into groups of petroleum hydrocarbons that act alike in soil or water. These groups are called petroleum hydrocarbon fractions. Each hydrocarbon fraction contains many individual chemicals. Some chemicals that may be found in TPH are hexane, jet fuels, mineral oils, benzene, toluene, xylenes, naphthalene, and fluorene, as well as other petroleum products and gasoline components (ATSDR, 2011).

Toxicity: The degree to which a substance or mixture of substances can harm humans or animals. Acute toxicity involves harmful effects in an organism through a single or short-term exposure. Chronic toxicity is the ability of a substance or mixture of substances to cause harmful effects over an extended period, usually upon repeated or continuous exposure, sometimes lasting for the entire life of the exposed organism. Subchronic toxicity is the ability of the substance to cause effects for more than 1 year but less than the lifetime of the exposed organism (U.S. EPA, 2013d).

Tubing: The smallest, innermost steep pipe set within a completed well, either hung directly from the wellhead or secured at its bottom using a packer. Tubing is not typically cemented in the well.

Underground Injection Control (UIC): The program under the Safe Drinking Water Act that regulates the use of wells to emplace fluids into the ground (U.S. EPA, 2013d).

Underground Injection Control (UIC) Class II well: Refers to wells that inject fluids associated with oil and gas production, including for (1) disposal of fluids brought to the surface in connection with oil or natural gas production, (2) for enhanced recovery of oil or natural gas, and (3) for storage of hydrocarbons which are liquid at standard temperature and pressure. Adapted from § 144.6(b).

Underground Injection Control (UIC) Class IID well: Within the types of operations that can occur for UIC Class II wells (see above), refers to wells used for the disposal of fluids brought to the surface in connection with oil or natural gas production. Also known as wells for salt water disposal.

Underground source of drinking water (USDW): An aquifer or its portion that currently supplies a public water system; or which contains a sufficient quantity of groundwater to supply a public water system, and either now supplies water for human consumption, or contains fewer than 10,000 mg/L TDS and is not exempted. Defined in the federal regulations that implement the UIC program (20 CFR 144.3).

Unsaturated zone: The soil zone above the water table that is only partially filled by water; also referred to as the "vadose zone."

Vapor pressure: The force per unit area exerted by a vapor in an equilibrium state with its pure solid, liquid, or solution at a given temperature. Vapor pressure is a measure of a substance's propensity to evaporate. Vapor pressure increases exponentially with an increase in temperature (U.S. EPA, 2013d).

Vertical separation distance: Measured vertically from the shallowest point of hydraulic fracturing to the bottom of the drinking water resource. If measured along a wellbore from the shallowest point of hydraulic fracturing to the bottom of the drinking water resource, this is referred to as measured depth, which may be a straight vertical distance below ground or may follow a more complicated path if the wellbore is not straight and vertical.

Vertical well: A well in which the wellbore is vertical throughout its entire length, from the wellhead at the surface to the production zone.

Viscosity: A measure of the internal friction of a fluid that provides resistance to shear within the fluid, informally referred to as how "thick" a fluid is.

Volatile: Readily vaporizable at a relatively low temperature (U.S. EPA, 2013d).

Volatilization: The process in which a chemical leaves the liquid phase and enters the gas phase.

Wastewater: See hydraulic fracturing wastewater.

Wastewater treatment: Chemical, biological, and mechanical procedures applied to an industrial or municipal discharge or to any other sources of contaminated water in order to remove, reduce, or neutralize contaminants (U.S. EPA, 2013d).

Water availability: There is no standard definition for water availability, and it has not been assessed recently at the national scale (U.S. GAO, 2014). Instead, a number of water availability indicators have been suggested (e.g., Roy et al., 2005). Here, availability is most often used to qualitatively refer to the amount of a location's water that could, currently or in the future, serve as a source of drinking water (U.S. GAO, 2014), which is a function of water inputs to a hydrologic system (e.g., rain, snowmelt, groundwater recharge) and water outputs from that system occurring either naturally or through competing demands of users.

Water consumption: Water that is removed from the local hydrologic cycle following its use (e.g., via evaporation, transpiration, incorporation into products or crops, consumption by humans or livestock), and is therefore unavailable to other water users (Maupin et al., 2014).

Water intensity: The amount of water used per unit of energy obtained (Nicot et al., 2014; Laurenzi and Jersey, 2013)

Water reuse: Any hydraulic fracturing wastewater that is used to offset total fresh water withdrawals for hydraulic fracturing, regardless of the level of treatment required.

Water sensitivity: a formation's physicochemical properties are affected in the presence of water. An example of a water sensitive formation would be one where the soil particles swell when water is added, reducing the permeability of the formation.

Water table: The top, or uppermost surface, of groundwater. Below the water table, the ground is saturated with water.

Water use: Water withdrawn for a specific purpose, part or all of which may be returned to the local hydrologic cycle.

Water withdrawal: The volume of water removed from its source, either the groundwater or diverted from a surface water source, for use, regardless of how much of that volume is returned to the local hydrologic cycle or consumed without being returned to the hydrologic cycle (Nicot et al., 2014; Laurenzi and Jersey, 2013).

Weight-of-evidence (WOE) characterization for carcinogenicity: A system used for characterizing the extent to which the available data support the hypothesis that an agent causes cancer in humans. The U.S. EPA issued guidelines in 1986, 1996, 1999, and 2005. For more information, see Appendix G.

Well blowout: The uncontrolled flow of fluids out of a well.

Well communication: When activities in a well that is being stimulated affect abandoned or active (producing) offset wells or their fracture networks. Also referred to as a "frac hit".

Well logging: A continuous measurement of physical properties in or around the well with electrically powered instruments to infer formation properties. Measurements may include electrical properties (resistivity and conductivity), sonic properties, active and passive nuclear measurements, measurements of the wellbore, pressure measurement, formation fluid sampling, sidewall coring tools, and others. Measurements may be taken via a wireline, which is a wire or cable that is used to deploy tools and instruments downhole and that transmits data to the surface (adapted from Schlumberger, 2014).

Well operator: A company that controls and operates oil and gas wells (U.S. EPA, 2013d).

Well orientation: A well's inclination from verticality. Wells drilled straight downward are considered to be vertical, wells drilled directionally to end up parallel to the production zone's bedding plane are considered horizontal, and directionally drilled wells that are neither vertical nor horizontal are referred to as deviated. In industry usage, a well's orientation commonly refers both to its inclination from vertical and the azimuthal (compass) direction of a directionally drilled wellbores.

Well pad: A temporary drilling site, usually constructed of local materials such as sand and gravel. After the drilling operation is over, most of the pad is usually removed or plowed back into the ground (NYSDEC, 2011).

Wellbore: The drilled hole or borehole, including the open hole or uncased portion of the well.

Wet gas: Refers to natural gas that typically contains less than 85% methane along with ethane and more complex hydrocarbons.

Wettability: The ability of a liquid to maintain contact with a solid surface. When wettability is high, a liquid droplet will lie flat across a surface, maximizing the area of contact between the liquid and the solid. When wettability is low, a liquid droplet will approach a spherical shape, minimizing the area of contact between the liquid and solid.

Wetting/nonwetting: The preferential attraction of a fluid to the surface. In typical reservoirs, water preferentially wets the surface, and gas is nonwetting (adapted from Dake, 1978).

Workover: Refers to any maintenance activity performed on a well that involves ceasing operations and removing the wellhead.

Young's modulus: A ratio of stress to strain that is a measure of the rigidity of a material.

This page is intentionally left blank.

Appendix K. Appendix References

This page is intentionally left blank.

Appendix K. Appendix References

Hyperlinks to the reference citations throughout this document will take you to the ORD National Center for Environmental Assessment HERO database (Health and Environmental Research Online) at https://hero.epa.gov/hero. HERO is a database of scientific literature used by the U.S. EPA in the process of developing selected science assessments.

Abrams, R. (2013). Advanced oxidation frac water recycling system. Presented at 20th International Petroleum Environmental Conference, November 12-14, 2013, San Antonio, TX. http://ipec.utulsa.edu/Conf2013/Manuscripts_pdfs/FracCleansetechnology_Abrams.pdf

Acharya, HR; Henderson, C; Matis, H; Kommepalli, H; Moore, B; Wang, H. (2011). Cost effective recovery of low-TDS frac flowback water for reuse. (Department of Energy: DE-FE0000784). Niskayuna, NY: GE Global Research. http://www.netl.doe.gov/file%20library/Research/oil-gas/FE0000784_FinalReport.pdf

Afzal, W; Mohammadi, AH; Richon, D. (2009). Volumetric properties of mono-, di-, tri-, and polyethylene glycol aqueous solutions from (273.15 to 363.15) K: Experimental measurements and correlations. Journal of Chemical and Engineering Data 54: 1254-1261. http://dx.doi.org/10.1021/je800694a

Ahmann, D; Roberts, AL; Krumholz, LR; Morel, FM. (1994). Microbe grows by reducing arsenic [Letter]. Nature 371: 750. http://dx.doi.org/10.1038/371750a0

Akob, DM; Cozzarelli, IM; Dunlap, DS; Rowan, EL; Lorah, MM. (2015). Organic and inorganic composition and microbiology of produced waters from Pennsylvania shale gas wells. Appl Geochem 60: 116-125. http://dx.doi.org/10.1016/j.apgeochem.2015.04.011

Alain, K; Pignet, P; Zbinden, M; Quillevere, M; Duchiron, F; Donval, JP; Lesongeur, F; Raguenes, G; Crassous, P; Querellou, J; Cambon-Bonavita, MA. (2002). Caminicella sporogenes gen. nov., sp. nov., a novel thermophilic spore-forming bacterium isolated from an East-Pacific Rise hydrothermal vent. Int J Syst Evol Microbiol 52: 1621-1628.

Alfa Aesar. (2015). A16163: Formaldehyde, 37% w/w aq. soln., stab. with 7-8% methanol. https://www.alfa.com/en/catalog/A16163 (accessed May 4, 2015).

Ali, M; Taoutaou, S; Shafqat, AU; Salehapour, A; Noor, S. (2009). The use of self healing cement to ensure long term zonal isolation for HPHT wells subject to hydraulic fracturing operations in Pakistan. Presented at International Petroleum Technology Conference, December 7-9, 2009, Doha, Qatar.

ALL Consulting (ALL Consulting, LLC). (2013). Water treatment technology fact sheet: Electrodialysis [Fact Sheet]. Tulsa, OK. http://www.all-llc.com/publicdownloads/ED-EDRFactSheet.pdf

Alzahrani, S; Mohammad, AW; Hilal, N; Abdullah, P; Jaafar, O. (2013). Comparative study of NF and RO membranes in the treatment of produced water-Part I: Assessing water quality. Desalination 315: 18-26. http://dx.doi.org/10.1016/j.desal.2012.12.004

André, L; Rabemanana, V; Vuataz, FD. (2006). Influence of water-rock interactions on fracture permeability of the deep reservoir at Soultz-sous-Forêts, France. Geothermics 35: 507-531. http://dx.doi.org/10.1016/j.geothermics.2006.09.006

API (American Petroleum Institute). (1999). Recommended practice for care and use of casing and tubing [Standard] (18th ed.). (API RP 5C1). Washington, D.C.: API Publishing Services.

API (American Petroleum Institute). (2004). Recommended practice for centralizer placement and stop collar testing (1st ed.). (API RP 10D-2 (R2010)). Washington, D.C. http://www.techstreet.com/products/1173247

API (American Petroleum Institute). (2009a). Hydraulic fracturing operations - Well construction and integrity guidelines [Standard] (1st ed.). Washington, D.C.: API Publishing Services. http://www.shalegas.energy.gov/resources/HF1.pdf

API (American Petroleum Institute). (2009b). Packers and bridge plugs (2nd ed.). (API SPEC 11D1). Washington, D.C. http://www.techstreet.com/api/products/1634486

API (American Petroleum Institute). (2010a). Isolating potential flow zones during well construction [Standard] (1st ed.). (RP 65-2). Washington, D.C.: API Publishing Services. http://www.techstreet.com/products/preview/1695866

API (American Petroleum Institute). (2010b). Specification for cements and materials for well cementing [Standard] (24th ed.). (ANSI/API SPECIFICATION 10A). Washington, D.C.: API Publishing Services. http://www.techstreet.com/products/1757666

API (American Petroleum Institute). (2010c). Water management associated with hydraulic fracturing. Washington, D.C.: API Publishing Services. http://www.api.org/~/media/Files/Policy/Exploration/HF2_e1.pdf

API (American Petroleum Institute). (2011). Specification for casing and tubing [Standard] (9th ed.). (API SPEC 5CT). Washington, D.C.: API Publishing Services. http://www.techstreet.com/products/1802047

API (American Petroleum Institute). (2013). Recommended practice for testing well cements [Standard] (2nd ed.). (RP 10B-2). Washington, DC: API Publishing Services. http://www.techstreet.com/products/1855370

API (American Petroleum Institute). (2015). Hydraulic fracturing well integrity and fracture containment (1st ed.) (RP 100-1). Washington, DC: API Publishing Services. http://www.api.org/~/media/files/policy/exploration/100-1_e1.pdf

Arthur, JD. (2012). Understanding and assessing well integrity relative to wellbore stray gas intrusion issues. Presented at Ground Water Protection Council Stray Gas - Incidence & Response Forum, July 24-26, 2012, Cleveland, OH.

Arthur, JD; Bohm, B; Cornue, D. (2009). Environmental considerations of modern shale gas development. Presented at SPE Annual Technical Conference and Exhibition, October 4-7, 2009, New Orleans, LA. https://www.onepetro.org/conference-paper/SPE-122931-MS

Arthur, JD; Langhus, BG; Patel, C. (2005). Technical summary of oil and gas produced water treatment technologies. Tulsa, OK: ALL Consulting, LLC. http://www.all-llc.com/publicdownloads/ALLConsulting-WaterTreatmentOptionsReport.pdf

ATSDR (Agency for Toxic Substances and Disease Registry). (2009). Glossary of terms. http://www.atsdr.cdc.gov/glossary.html

ATSDR (Agency for Toxic Substances and Disease Registry). (2011). Total petroleum hydrocarbons (TPH). http://www.atsdr.cdc.gov/substances/toxsubstance.asp?toxid=75

ATSDR (Agency for Toxic Substances and Disease Registry). (2016). Minimal risk levels (MRLs). March 2016. Atlanta, GA: Agency for Toxic Substances and Disease Registry (ATSDR). http://www.atsdr.cdc.gov/mrls/index.asp

AWWA (American Water Works Association). (2010). Anomalous DBP speciation patterns: Examples and explanations. In 2010 Water quality and technology conference and exposition proceedings. Denver, CO.

AWWA (American Water Works Association). (2013). Water and hydraulic fracturing: A white paper from the American Water Works Association. Denver, CO. http://www.awwa.org/Portals/0/files/legreg/documents/AWWAFrackingReport.pdf

AWWA (American Water Works Association). (1999). Residential end uses of water. In PW Mayer; WB DeOreo (Eds.). Denver, CO: AWWA Research Foundation and American Water Works Association. http://www.waterrf.org/PublicReportLibrary/RFR90781_1999_241A.pdf

Bair, ES; Freeman, DC; Senko, JM. (2010). Subsurface gas invasion Bainbridge Township, Geauga County, Ohio. (Expert Panel Technical Report). Columbus, OH: Ohio Department of Natural Resources. https://oilandgas.ohiodnr.gov/portals/oilgas/pdf/bainbridge/DMRM%200%20Title%20P age,%20Preface,%20Acknowledgements.pdf

Baker, R. (1979). A primer of oilwell drilling (4th ed.). Austin, TX: Petroleum Extension Service (PETEX).

Banasiak, LJ; Schäfer, AI. (2009). Removal of boron, fluoride and nitrate by electrodialysis in the presence of organic matter. J Memb Sci 334: 101-109. http://dx.doi.org/10.1016/j.memsci.2009.02.020

Bank, T. (2011). Trace metal geochemistry and mobility in the Marcellus shale. In Proceedings of the Technical Workshops for the Hydraulic Fracturing Study: Chemical & Analytical Methods. http://www2.epa.gov/sites/production/files/documents/tracemetalgeochemistryandmobi lityinthemarcellusformation1.pdf

Bank, T; Fortson, LA; Malizia, TR; Benelli, P. (2012). Trace metal occurrences in the Marcellus Shale [Abstract]. Geological Society of America Abstracts with Programs 44: 313.

Baragi, JG; Maganur, S; Malode, V; Baragi, SJ. (2013). Excess molar volumes and refractive indices of binary liquid mixtures of acetyl acetone with n-nonane, n-decane and n-dodecane at (298.15, 303.15, and 308.15) K. Journal of Molecular Liquids 178: 175-177. http://dx.doi.org/10.1016/j.molliq.2012.11.022

Barati, R; Liang, JT. (2014). A review of fracturing fluid systems used for hydraulic fracturing of oil and gas wells. J Appl Polymer Sci Online pub. http://dx.doi.org/10.1002/app.40735

Barbot, E; Vidic, NS; Gregory, KB; Vidic, RD. (2013). Spatial and temporal correlation of water quality parameters of produced waters from Devonian-age shale following hydraulic fracturing. Environ Sci Technol 47: 2562-2569.

Barrett, ME. (2010). Evaluation of sand filter performance. (CRWR Online Report 10-7). Austin, TX: Center for Research in Water Resources, University of Texas at Austin. http://www.crwr.utexas.edu/reports/pdf/2010/rpt10-07.pdf

Benko, KL; Drewes, JE. (2008). Produced water in the Western United States: Geographical distribution, occurrence, and composition. Environ Eng Sci 25: 239-246.

Bennett, GM; Yuill, JL. (1935). The crystal form of anhydrous citric acid. J Chem Soc 1935: 130. http://dx.doi.org/10.1039/JR9350000130

Bethke, CM, : Yeakel, S. (2014). The geochemists workbench. Release 10.0. GWB essentials guide (Version Release 10.0). Champaign, Il: Aqueous Solutions, LLC. http://www.gwb.com/pdf/GWB10/GWBessentials.pdf

Biilmann, E. (1906). [Studien über organische Thiosäuren III]. Justus Liebigs Annalen der Chemie 348: 133-143. http://dx.doi.org/10.1002/jlac.19063480110

Biltz, W; Balz, G. (1928). [Über molekular- und atomvolumina. XVIII. Das volumen des ammoniaks in kristallisierten ammoniumsalzen]. Zeitschrift für Anorganische und Allgemeine Chemie 170: 327-341. http://dx.doi.org/10.1002/zaac.19281700141

Blanco, A; Garcia-Abuin, A; Gomez-Diaz, D; Navaza, JM; Villaverde, OL. (2013). Density, speed of sound, viscosity, surface tension, and excess volume of n-ethyl-2-pyrrolidone plus ethanolamine (or diethanolamine or triethanolamine) from T = (293.15 to 323.15) K. Journal of Chemical and Engineering Data 58: 653-659. http://dx.doi.org/10.1021/je301123j

Blauch, ME; Myers, RR; Moore, TR; Lipinski, BA. (2009). Marcellus shale post-frac flowback waters - Where is all the salt coming from and what are the implications? In Proceedings of the SPE Eastern Regional Meeting, 23-25 September, 2009, Charleston, WV: Society of Petroleum Engineers. http://dx.doi.org/10.2118/125740-MS

Blondes, MS; Gans, KD; Thordsen, JJ; Reidy, ME; Thomas, B; Engle, MA; Kharaka, YK; Rowan, EL. (2014). Data: U.S. Geological Survey National Produced Waters Geochemical Database v2.0 (Provisional) [Database]: U.S. Geological Survey. http://energy.usgs.gov/EnvironmentalAspects/EnvironmentalAspectsofEnergyProduction andUse/ProducedWaters.aspx#3822349-data

Bloomfield, C; Kelson, W; Pruden, G. (1976). Reactions between metals and humidified organic matter. Journal of Soil Science 27: 16-31. http://dx.doi.org/10.1111/j.1365-2389.1976.tb01971.x

Borrirukwisitsak, S; Keenan, HE; Gauchotte-Lindsay, C. (2012). Effects of salinity, pH and temperature on the octanol-water partition coefficient of bisphenol A. IJESD 3: 460-464. http://dx.doi.org/10.7763/IJESD.2012.V3.267

Boschee, P. (2012). Handling produced water from hydraulic fracturing. Oil and Gas Facilities 1: 23-26.

Boschee, P. (2014). Produced and flowback water recycling and reuse: Economics, limitations, and technology. Oil and Gas Facilities 3: 16-22.

Bottero, S; Picioreanu, C; Delft, TU; Enzien, M; van Loosdrecht, MCM; Bruining, H; Heimovaara, T. (2010). Formation damage and impact on gas flow caused by biofilms growing within proppant packing used in hydraulic fracturing. Presented at SPE International Symposium and Exhibiton on Formation Damage Control, February 10-12, 2010, Lafayette, Louisiana. https://www.onepetro.org/conference-paper/SPE-128066-MS

Brantley, SL; Yoxtheimer, D; Arjmand, S; Grieve, P; Vidic, R; Pollak, J; Llewellyn, GT; Abad, J; Simon, C. (2014). Water resource impacts during unconventional shale gas development: The Pennsylvania experience. Int J Coal Geol 126: 140-156. http://dx.doi.org/10.1016/j.coal.2013.12.017

Brufatto, C; Cochran, J; Conn, L; El-Zeghaty, SZAA; Fraboulet, B; Griffin, T; James, S; Munk, T; Justus, F; Levine, JR; Montgomery, C; Murphy, D; Pfeiffer, J; Pornpoch, T; Rishmani, L. (2003). From mud to cement - Building gas wells. Oilfield Rev 15: 62-76.

Bruff, M; Jikich, SA. (2011). Field demonstration of an integrated water treatment technology solution in Marcellus shale. Presented at SPE Eastern Regional Meeting, August 17-19, 2011, Columbus, OH. http://www.onepetro.org/mslib/servlet/onepetropreview?id=SPE-149466-MS&soc=SPE

Bukhari, AA. (2008). Investigation of the electro-coagulation treatment process for the removal of total suspended solids and turbidity from municipal wastewater. Bioresour Technol 99: 914-921. http://dx.doi.org/10.1016/j.biortech.2007.03.015

California Department of Water Resources. (2015). California state water project overview. http://www.water.ca.gov/swp/ (accessed February 20, 2015).

Camacho, LM, ar; Dumee, L; Zhang, J; Li, J; Duke, M; Gomez, J; Gray, S. (2013). Advances in membrane distillation for water desalination and purification applications. Water 5: 94-196. http://dx.doi.org/10.3390/w5010094

CAPP (Canadian Association of Petroleum Producers). (2013). CAPP hydraulic fracturing operating practice: Wellbore construction and quality assurance. (2012-0034). http://www.capp.ca/getdoc.aspx?DocId=218137&DT=NTV

Carpenter, EL; Davis, HS. (1957). Acrylamide. Its preparation and properties. Journal of Applied Chemistry 7: 671-676. http://dx.doi.org/10.1002/jctb.5010071206

CAS Registry Service (Chemical Abstracts Service). (2016). CAS registry and CAS registry number FAQs. http://www.cas.org/content/chemical-substances/faqs

Casanova, C; Wilhelm, E; Grolier, JPE; Kehiaian, HV. (1981). Excess volumes and excess heat-capacities of (water + alkanoic acid). The Journal of Chemical Thermodynamics 13: 241-248. http://dx.doi.org/10.1016/0021-9614(81)90123-3

Cavanagh, PH; Johnson, CR; Le Roy-Delage, S; DeBruijn, GG; Cooper, I; Guillot, DJ; Bulte, H; Bargaud, B. (2007). Self-healing cement - Novel technology to achieve leak-free wells. In SPE/IADC drilling conference 2007 (Proceedings): Reaching out to discover and recover. Richardson, TX: Society of Petroleum Engineers. http://dx.doi.org/10.2118/105781-MS

Cayol, JL; Ollivier, B; Lawson anani soh, A; Fardeau, ML; Ageron, E; Grimont, PAD; Prensier, G; Guezennec, J; Magot, M; Garcia, JL. (1994). Haloincola saccharolytica subsp. senegalensis subsp. nov., isolated from the sediments of a hypersaline lake, and emended description of Haloincola saccharolytica. International Journal of Systematic Bacteriology 44: 805-811. http://dx.doi.org/10.1099/00207713-44-4-805

CCG (Chemical Computing Group). (2011). Molecular operating environment (MOE) linux (Version 2011.10) [Computer Program]. Montreal, Quebec. http://www.chemcomp.com/software.htm

CCST (California Council on Science and Technology). (2014). Advanced well stimulation technologies in California: An independent review of scientific and technical information. Sacramento, CA. http://ccst.us/publications/2014/2014wst.pdf

CCST (California Council on Science and Technology). (2015a). An independent scientific assessment of well stimulation in California Volume II: Potential environmental impacts of hydraulic fracturing and acid stimulations. Sacramento, CA. https://ccst.us/publications/2015/2015SB4-v2.pdf

CCST (California Council on Science and Technology). (2015b). An independent scientific assessment of well stimulation in California, Volume 1: Well stimulation technologies and their past, present, and potential future use in California. Sacramento, CA. http://www.ccst.us/publications/2015/2015SB4-v1.pdf

Chapman, EC; Capo, RC; Stewart, BW; Kirby, CS; Hammack, RW; Schroeder, KT; Edenborn, HM. (2012). Geochemical and strontium isotope characterization of produced waters from Marcellus Shale natural gas extraction. Environ Sci Technol 46: 3545-3553.

Chasib, KF. (2013). Extraction of phenolic pollutants (phenol and p-chlorophenol) from industrial wastewater. Journal of Chemical and Engineering Data 58: 1549-1564. http://dx.doi.org/10.1021/je4001284

ChemicalBook (ChemicalBook Inc.). (2010). Sorbitan trioleate. Available online at http://www.chemicalbook.com/chemicalproductproperty_en_cb4677178.htm (accessed April 6, 2015).

Cheremisinoff, NP; Davletshin, A. (2015). Well construction and integrity. In M Dayal (Ed.), Hydraulic fracturing operations: Handbook of environmental management practices (pp. 437-476). Salem, MA: Scrivener Publishing, LLC.

Chermak, JA; Schreiber, ME. (2014). Mineralogy and trace element geochemistry of gas shales in the United States: Environmental implications. Int J Coal Geol 126: 32-44. http://dx.doi.org/10.1016/j.coal.2013.12.005

Cheung, K; Klassen, P; Mayer, B; Goodarzi, F; Aravena, R. (2010). Major ion and isotope geochemistry of fluids and gases from coalbed methane and shallow groundwater wells in Alberta, Canada. Appl Geochem 25: 1307-1329. http://dx.doi.org/10.1016/j.apgeochem.2010.06.002

Choppin, GR. (2006). Actinide speciation in aquatic systems. Mar Chem 99: 83-92. http://dx.doi.org/10.1016/j.marchem.2005.003.011

Choppin, GR. (2007). Actinide speciation in the environment. Journal of Radioanal Chem 273: 695-703. http://dx.doi.org/10.1007/s10967-007-0933-3

Clark, CE; Veil, JA. (2009). Produced water volumes and management practices in the United States. (ANL/EVS/R-09/1). Argonne, IL: Argonne National Laboratory. http://www.ipd.anl.gov/anlpubs/2009/07/64622.pdf

Cluff, M; Hartsock, A; Macrae, J; Carter, K; Mouser, PJ. (2014). Temporal changes in microbial ecology and geochemistry in produced water from hydraulically fractured Marcellus Shale gas wells. Environ Sci Technol 48: 6508-6517. http://dx.doi.org/10.1021/es501173p

COGCC (Colorado Oil and Gas Conservation Commission). (2016). COGIS - Facility inquiry [Database]. Denver, CO. Retrieved from http://cogcc.state.co.us/cogis/FacilitySearch.asp

Colborn, T; Kwiatkowski, C; Schultz, K; Bachran, M. (2011). Natural gas operations from a public health perspective. Hum Ecol Risk Assess 17: 1039-1056. http://dx.doi.org/10.1080/10807039.2011.605662

Collado, L; Cleenwerck, I; Van Trappen, S; De Vos, P; Figueras, MJ. (2009). Arcobacter mytili sp. nov., an indoxyl acetate-hydrolysis-negative bacterium isolated from mussels. Int J Syst Evol Microbiol 59: 1391-1396. http://dx.doi.org/10.1099/ijs.0.003749-0

Craft, R. (2004). Crashes involving trucks carrying hazardous materials. (FMCSA-RI-04-024). Washington, D.C.: U.S. Department of Transportation. http://ntl.bts.gov/lib/51000/51300/51302/fmcsa-ri-04-024.pdf

Craig, MS; Wendte, SS; Buchwalter, JL. (2012). Barnett shale horizontal restimulations: A case study of 13 wells. SPE Americas unconventional resources conference, June 5-7, 2012, Pittsburgh, PA.

Cramer, DD. (2008). Stimulating unconventional reservoirs: Lessons learned, successful practices, areas for improvement. SPE Unconventional Reservoirs Conference, February 10-12, 2008, Keystone, CO.

Cramer, GM; Ford, RA; Hall, RL. (1978). Estimation of toxic hazard: A decision tree approach [Review]. Food Cosmet Toxicol 16: 255-276.

Crescent (Crescent Consulting, LLC). (2011). East Mamm creek project drilling and cementing study. Oklahoma City, OK. http://cogcc.state.co.us/Library/PiceanceBasin/EastMammCreek/ReportFinal.pdf

Criquet, J; Allard, S; Salhi, E; Joll, CA; Heitz, A; von Gunten, U. (2012). Iodate and iodo-trihalomethane formation during chlorination of iodide-containing waters: Role of bromide. Environ Sci Technol 46: 7350-7357. http://dx.doi.org/10.1021/es301301g

Crook, R. (2008). Cementing: Cementing horizontal wells. Halliburton.

Curtis, JB. (2002). Fractured shale-gas systems. AAPG Bulletin 86: 1921-1938. http://dx.doi.org/10.1306/61EEDDBE-173E-11D7-8645000102C1865D

Dahm, K; Chapman, M. (2014). Produced water treatment primer: Case studies of treatment applications. (S&T Research Project #1617). Denver CO: U.S. Department of the Interior. http://www.usbr.gov/research/projects/download_product.cfm?id=1214.

Dahm, KG; Guerra, KL; Xu, P; Drewes, JE. (2011). Composite geochemical database for coalbed methane produced water quality in the Rocky Mountain region. Environ Sci Technol 45: 7655-7663. http://dx.doi.org/10.1021/es201021n

Dake, LP. (1978). Fundamentals of reservoir engineering. Boston, MA: Elsevier. http://www.ing.unp.edu.ar/asignaturas/reservorios/Fundamentals%20of%20Reservoir%20Engineering%20%28L.P.%20Dake%29.pdf

Dao, TD; Mericq, JP; Laborie, S; Cabassud, C. (2013). A new method for permeability measurement of hydrophobic membranes in Vacuum Membrane Distillation process. Water Res 47: 20962104. http://dx.doi.org/10.1016/j.watres.2013.01.040

Davis, JP; Struchtemeyer, CG; Elshahed, MS. (2012). Bacterial communities associated with production facilities of two newly drilled thermogenic natural gas wells in the Barnett Shale (Texas, USA). Microb Ecol 64: 942-954. http://dx.doi.org/10.1007/s00248-012-0073-3

De Andrade, J; Sangesland, S; Todorovic, J; Vrålstad, T. (2015). Cement sheath integrity during thermal cycling: A novel approach for experimental tests of cement systems. SPE Bergen One Day Seminar, April 22, 2015, Bergen, Norway.

de Oliveira, LH; da Silva, JL, Jr; Aznar, M. (2011). Apparent and partial molar volumes at infinite dilution and solid-liquid equilibria of dibenzothiophene plus alkane systems. Journal of Chemical and Engineering Data 56: 3955-3962. http://dx.doi.org/10.1021/je200327s

DeArmond, PD; DiGoregorio, AL. (2013a). Characterization of liquid chromatography-tandem mass spectrometry method for the determination of acrylamide in complex environmental samples. Anal Bioanal Chem 405: 4159-4166. http://dx.doi.org/10.1007/s00216-013-6822-4

DeArmond, PD; DiGoregorio, AL. (2013b). Rapid liquid chromatography-tandem mass spectrometry-based method for the analysis of alcohol ethoxylates and alkylphenol ethoxylates in environmental samples. J Chromatogr A 1305: 154-163. http://dx.doi.org/10.1016/j.chroma.2013.07.017

Dejoye Tanzi, C; Abert Vian, M; Ginies, C; Elmaataoui, M; Chemat, F. (2012). Terpenes as green solvents for extraction of oil from microalgae. Molecules 17: 8196-8205. http://dx.doi.org/10.3390/molecules17078196

Devereux, S. (1998). Practical well planning and drilling manual. Tulsa, OK: PennWell Publishing Company. http://www.pennwellbooks.com/practical-well-planning-and-drilling-manual/

Dhondge, SS; Pandhurnekar, CP; Parwate, DV. (2010). Density, speed of sound, and refractive index of aqueous binary mixtures of some glycol ethers at T=298.15 K. Journal of Chemical and Engineering Data 55: 3962-3968. http://dx.doi.org/10.1021/je901072c

Diehl, SF; Goldhaber, MB; Hatch, JR. (2004). Modes of occurrence of mercury and other trace elements in coals from the warrior field, Black Warrior Basin, Northwestern Alabama. Int J Coal Geol 59: 193-208. http://dx.doi.org/10.1016/j.coal.2004.02.003

Diehl, TH; Harris, MA. (2014). Withdrawal and consumption of water by thermoelectric power plants in the United States, 2010. (Scientific Investigations Report 20145184). Reston, VA: U.S. Geological Survey. http://dx.doi.org/10.3133/sir20145184

Digiulio, DC; Jackson, RB. (2016). Impact to underground sources of drinking water and domestic wells from production well stimulation and completion practices in the Pavillion, Wyoming, Field. Environ Sci Technol 50: 4524-4536. http://dx.doi.org/10.1021/acs.est.5b04970

DOE (U.S. Department of Energy). (2006). A guide to practical management of produced water from onshore oil and gas operations in the United States. Washington, DC: U.S. Department of Energy, National Petroleum Technology Office. http://fracfocus.org/sites/default/files/publications/a_guide_to_practical_management_of_produced_water_from_onshore_oil_and_gas_operations_in_the_united_states.pdf

DOE (U.S. Department of Energy). (2011). A comparative study of the Mississippian Barnett shale, Fort Worth basin, and Devonian Marcellus shale, Appalachian basin. (DOE/NETL-2011/1478). http://www.netl.doe.gov/technologies/oil-gas/publications/brochures/DOE-NETL-2011-1478%20Marcellus-Barnett.pdf

DOE (U.S. Department of Energy). (2014). Water management strategies for improved coalbed methane production in the Black Warrior Basin. http://www.netl.doe.gov/research/oil-and-gas/project-summaries/natural-gas-resources/de-fe0000888

Dresel, PE; Rose, AW. (2010). Chemistry and origin of oil and gas well brines in western Pennsylvania (pp. 48). (Open-File Report OFOG 1001.0). Harrisburg, PA: Pennsylvania Geological Survey, 4th ser. http://www.marcellus.psu.edu/resources/PDFs/brines.pdf

Drewes, J; Cath, T; Debroux, J; Veil, J. (2009). An integrated framework for treatment and management of produced water - Technical assessment of produced water treatment technologies (1st ed.). (RPSEA Project 07122-12). Golden, CO: Colorado School of Mines. http://aqwatec.mines.edu/research/projects/Tech_Assessment_PW_Treatment_Tech.pdf

DrillingInfo, Inc.. (2012). DI Desktop August 2012 download [Database]. Austin, TX. http://info.drillinginfo.com/

Dubey, GP; Kumar, K. (2011). Thermodynamic properties of binary liquid mixtures of diethylenetriamine with alcohols at different temperatures. Thermochim Acta 524: 7-17. http://dx.doi.org/10.1016/j.tca.2011.06.003

Dubey, GP; Kumar, K. (2013). Studies of thermodynamic, thermophysical and partial molar properties of liquid mixtures of diethylenetriamine with alcohols at 293.15 to 313.15 K. Journal of Molecular Liquids 180: 164-171. http://dx.doi.org/10.1016/j.molliq.2013.01.011

Duhon, H. (2012). Produced water treatment: Yesterday, today, and tomorrow. Oil and Gas Facilities 3: 29-31.

Dunkel, M. (2013). Reducing fresh water use in upstream oil and gas hydraulic fracturing. In Summary of the technical workshop on wastewater treatment and related modeling (pp. A37-A43). Irving, TX: Pioneer Natural Resources USA, Inc. http://www2.epa.gov/hfstudy/summary-technical-workshop-wastewater-treatment-and-related-modeling

Duraisamy, RT; Beni, AH; Henni, A. (2013). State of the art treatment of produced water. In W Elshorbagy; RK Chowdhury (Eds.), Water treatment (pp. 199-222). Rijeka, Croatia: InTech. http://dx.doi.org/10.5772/53478

Dusseault, MB; Gray, MN; Nawrocki, PA. (2000). Why oilwells leak: Cement behavior and long-term consequences. Paper presented at SPE International Oil and Gas Conference and Exhibition in China, November 7-10, 2000, Beijing, China. https://www.onepetro.org/conference-paper/SPE-64733-MS

Dyshin, AA; Eliseeva, OV; Kiselev, MG; Al'per, GA. (2008). The volume characteristics of solution of naphthalene in heptane-ethanol mixtures at 298.15 K. Russian Journal of Physical Chemistry A, Focus on Chemistry 82: 1258-1261. http://dx.doi.org/10.1134/S0036024408080037

Easton, J. (2014). Optimizing fracking wastewater management. Pollution Engineering January 13.

Economides, MJ; Mikhailov, DN; Nikolaevskiy, VN. (2007). On the problem of fluid leakoff during hydraulic fracturing. Transport in Porous Media 67: 487-499. http://dx.doi.org/10.1007/s11242-006-9038-7

Egorov, GI; Makarov, DM; Kolker, AM. (2013). Volume properties of liquid mixture of water plus glycerol over the temperature range from 278.15 to 348.15 K at atmospheric pressure. Thermochim Acta 570: 16-26. http://dx.doi.org/10.1016/j.tca.2013.07.012

EIA (U.S. Energy Information Administration). (2013). Technically recoverable shale oil and shale gas resources: an assessment of 137 shale formations in 41 countries outside the United States (pp. 730). Washington, D.C.: Energy Information Administration, U.S. Department of Energy. http://www.eia.gov/analysis/studies/worldshalegas/

EIA (U.S. Energy Information Administration). (2015). Lower 48 states shale plays. Washington, D.C.: Energy Information Administration, U.S. Department of Energy. http://www.eia.gov/oil_gas/rpd/shale_gas.pdf

Ely, JW; Horn, A; Cathey, R; Fraim, M; Jakhete, S. (2011). Game changing technology for treating and recycling frac water. Paper presented at SPE Annual Technical Conference and Exhibition, October 30 - November 2, 2011, Denver, CO.

Enform. (2013). Interim industry recommended practice 24: Fracture stimulation: Interwellbore communication 3/27/2013 (1st ed.). (IRP 24). Calgary, Alberta: Enform Canada.

Engle, MA; Rowan, EL. (2014). Geochemical evolution of produced waters from hydraulic fracturing of the Marcellus Shale, northern Appalachian Basin: A multivariate compositional data analysis approach. Int J Coal Geol 126: 45-56. http://dx.doi.org/10.1016/j.coal.2013.11.010

ER (Eureka Resources, LLC). (2014). Crystallization technology. http://www.eureka-resources.com/wp-content/uploads/2013/07/EURE-022_Crystallization_53013.pdf (accessed March 4, 2015).

Ertel, D; McManus, K; Bogdan, J. (2013). Marcellus wastewater treatment: Case study. In Summary of the technical workshop on wastewater treatment and related modeling (pp. A56-A66). Williamsport, PA: Eureka Resources, LLC. http://www2.epa.gov/hfstudy/summary-technical-workshop-wastewater-treatment-and-related-modeling

Fadeeva, YA; Shmukler, LE; Safonova, LP. (2004). Physicochemical properties of the H3PO4-dimethylformamide system. Russian Journal of General Chemistry 74: 174-178. http://dx.doi.org/10.1023/B:RUGC.0000025496.07304.66

Fakhru'l-Razi, A; Pendashteh, A; Abdullah, LC; Biak, DR; Madaeni, SS; Abidin, ZZ. (2009). Review of technologies for oil and gas produced water treatment [Review]. J Hazard Mater 170: 530-551.

Faria, MAF; Martins, RJ; Cardoso, MJE, M; Barcia, OE. (2013). Density and viscosity of the binary systems ethanol + butan-1-ol, + pentan-1-ol, + heptan-1-ol, + octan-1-ol, nonan-1-ol, + decan-1-ol at 0.1 mpa and temperatures from 283.15 K to 313.15 K. Journal of Chemical and Engineering Data 58: 3405-3419. http://dx.doi.org/10.1021/je400630f

Fels, G. (1900). Ueber die Frage der isomorphen vertretung von halogen und hydroxyl. In Zeitschrift fur Kristallographie, Kristallgeometrie, Kristallphysik, Kristallchemie. Frankfurt: Leipzig. http://babel.hathitrust.org/cgi/pt?id=uc1.b3327977;view=1up;seq=5

Ferrar, KJ; Michanowicz, DR; Christen, CL; Mulcahy, N; Malone, SL; Sharma, RK. (2013). Assessment of effluent contaminants from three facilities discharging Marcellus Shale wastewater to surface waters in Pennsylvania. Environ Sci Technol 47: 3472-3481.

Fertl, WH; Chilingar, GV. (1988). Total organic carbon content determined from well logs. SPE Formation Evaluation 3: 407-419. http://dx.doi.org/10.2118/15612-PA

Fichter, J; Moore, R; Braman, S; Wunch, K; Summer, E; Holmes, P. (2012). How hot is too hot for bacteria? A technical study assessing bacterial establishment in downhole drilling, fracturing, and stimulation operations. Presented at NACE International Corrosion Conference & Expo, March 11-15, 2012, Salt Lake City, UT. https://www.onepetro.org/conference-paper/NACE-2012-1310

Filgueiras, AV; Lavilla, I; Bendicho, C. (2002). Chemical sequential extraction for metal partitioning in environmental solid samples. J Environ Monit 4: 823-857. http://dx.doi.org/10.1039/b207574c

finemech (finemech Precision Mechanical Components). (2012). Technical resources: Liquid nitrogen, LN2. http://www.finemech.com/liquid_nitrogen.html

Fisher, JB; Sublette, KL. (2005). Environmental releases from exploration and production operations in Oklahoma: Type, volume, causes, and prevention. Environmental Geosciences 12: 89-99. http://dx.doi.org/10.1306/eg.11160404039

Fisher, JG; Santamaria, A. (2002). Dissolved organic constituents in coal-associated waters and implications for human and ecosystem health. Paper presented at 9th Annual International Petroleum Environmental Conference, October 22-25, 2002, Albuquerque, NM. http://ipec.utulsa.edu/Conf2002/fisher_santamaria_120.pdf

Fisher, M; Warpinski, N. (2012). Hydraulic fracture height growth: Real data. S P E Prod Oper 27: 8-19. http://dx.doi.org/10.2118/145949-PA

Fisher, RS. (1998). Geologic and geochemical controls on naturally occurring radioactive materials (NORM) in produced water from oil, gas, and geothermal operations. Environmental Geosciences 5: 139-150.

Fleckenstein, WW; Eustes, AW; Stone, CH; Howell, PK. (2015). An assessment of risk of migration of hydrocarbons or fracturing fluids to fresh water aquifers: Wattenberg Field, CO. Richardson, TX: Society of Petroleum Engineers. http://dx.doi.org/10.2118/175401-MS

Francis, AJ. (2007). Microbial mobilization and immobilization of plutonium. J Alloy Comp 444: 500-505. http://dx.doi.org/10.1016/j.jallcom.2007.01.132

Francis, RA; Small, MJ; Vanbriesen, JM. (2009). Multivariate distributions of disinfection by-products in chlorinated drinking water. Water Res 43: 3453-3468. http://dx.doi.org/10.1016/j.watres.2009.05.008

Fuess, H; Bats, JW; Dannohl, H; Meyer, H; Schweig, A. (1982). Comparison of observed and calculated densities. XII. Deformation density in complex anions. II. Experimental and theoretical densities in sodium formate. Acta Crystallogr B 38: 736-743. http://dx.doi.org/10.1107/S0567740882003999

Fujino, S; Hwang, C; Morinaga, K. (2004). Density, surface tension, and viscosity of PbO-B2O3-SiO2 glass melts. Journal of the American Ceramic Society 87: 10-16. http://dx.doi.org/10.1111/j.1151-2916.2004.tb19937.x

Gadd, GM. (2004). Microbial influence on metal mobility and application for bioremediation. Geoderma 122: 109-119. http://dx.doi.org/10.1016/j.geoderma.2004.01.002

Gallegos, TJ; Varela, BA; Haines, SS; Engle, MA. (2015). Hydraulic fracturing water use variability in the United States and potential environmental implications. Water Resour Res 51: 5839-5845. http://dx.doi.org/10.1002/2015WR017278

García, MT; Mellado, E; Ostos, JC; Ventosa, A. (2004). Halomonas organivorans sp. nov., a moderate halophile able to degrade aromatic compounds. Int J Syst Evol Microbiol 54: 1723-1728. http://dx.doi.org/10.1099/ijs.0.63114-0

Gauthier, MJ; Lafay, B; Christen, R; Fernandez, L; Acquaviva, M; Bonin, P; Bertrand, JC. (1992). Marinobacter hydrocarbonoclasticus gen. nov., sp. nov., a new, extremely halotolerant, hydrocarbon-degrading Marine Bacterium. International Journal of Systematic Bacteriology 42: 568-576. http://dx.doi.org/10.1099/00207713-42-4-568

Geological Survey of Alabama. (2014). Water management strategies for improved coalbed methane production in the Black Warrior Basin. (DE-FE0000888). Washington, DC: U.S. Department of Energy, National Energy Technology Library. https://www.netl.doe.gov/research/oil-and-gas/natural-gas-resources/00888-geosurveyalabama

Gilmore, K; Hupp, R; Glathar, J. (2013). Transport of hydraulic fracturing water and wastes in the Susquehanna River basin, Pennsylvania. J Environ Eng 140: B4013002. http://dx.doi.org/10.1061/(ASCE)EE.1943-7870.0000810

Glorius, M; Moll, H; Geipel, G; Bernhard, G. (2008). Complexation of uranium(VI) with aromatic acids such as hydroxamic and benzoic acid investigated by TRLFS. Journal of Radioanal Chem 277: 371-377. http://dx.doi.org/10.1007/s10967-007-7082-6

Gomes, J; Cocke, D; Das, K; Guttula, M; Tran, D; Beckman; J. (2009). Treatment of produced water by electrocoagulation. Shiner, TX: KASELCO, LLC. http://www.kaselco.com/index.php/library/industry-white-papers

Goodwin, S; Carlson, K; Knox, K; Douglas, C; Rein, L. (2014). Water intensity assessment of shale gas resources in the Wattenberg field in northeastern Colorado. Environ Sci Technol 48: 5991-5995. http://dx.doi.org/10.1021/es404675h

Grabowski, A; Nercessian, O; Fayolle, F; Blanchet, D; Jeanthon, C. (2005). Microbial diversity in production waters of a low-temperature biodegraded oil reservoir. FEMS Microbiol Ecol 54: 427-443. http://dx.doi.org/10.1016/j.femsec.2005.05.007

Gradient. (2013). National human health risk evaluation for hydraulic fracturing fluid additives. Gradient. http://www.energy.senate.gov/public/index.cfm/files/serve?File_id=53a41a78-c06c-4695-a7be-84225aa7230f

Gross, SA; Avens, HJ; Banducci, AM; Sahmel, J; Panko, JM; Tvermoes, BE. (2013). Analysis of BTEX groundwater concentrations from surface spills associated with hydraulic fracturing operations. J Air Waste Manag Assoc 63: 424-432. http://dx.doi.org/10.1080/10962247.2012.759166

GTI (Gas Technology Institute). (2012). Barnett and Appalachian shale water management and resuse technologies. (Report no. 08122-05.FINAL.1). Sugar Land, TX: Research Partnership to Secure Energy for America, RPSEA. https://www.netl.doe.gov/file%20library/research/oil-gas/Natural%20Gas/shale%20gas/08122-05-final-report.pdf

Guerra, K; Dahm, K; Dundorf, S. (2011). Oil and gas produced water management and beneficial use in the western United States. (Science and Technology Program Report No. 157). Denver, CO: U.S. Department of the Interior Bureau of Reclamation.

Guolin, J; Xiaoyu, W; Chunjie, H. (2008). The effect of oilfield polymer-flooding wastewater on anion exchange membrane performance. Desalination 220: 386-393.

Gupta, DVS; Valkó, P. (2007). Fracturing fluids and formation damage. In M Economides; T Martin (Eds.), Modern fracturing: enhancing natural gas production (pp. 227-279). Houston, TX: Energy Tribune Publishing Inc.

Gurdak, JJ; McMahon, PB; Dennehy, K; Qi, SL. (2009). Water quality in the high plains aquifer, Colorado, Kansas, Nebraska, New Mexico, Oklahoma, South Dakota, Texas, and Wyoming, 1999 - 2004. (GSC 1337). Reston, VA: U.S. Geological Survey. http://pubs.usgs.gov/circ/1337/

GWPC (Groundwater Protection Council). (2014). State oil and natural gas regulations designed to protect water resources. Morgantown, WV: U.S. Department of Energy, National Energy Technology Laboratory. http://www.gwpc.org/sites/default/files/files/Oil%20and%20Gas%20Regulation%20Report%20Hyperlinked%20Version%20Final-rfs.pdf

GWPC and ALL Consulting (Ground Water Protection Council and ALL Consulting). (2009). Modern shale gas development in the United States: A primer. (DE-FG26-04NT15455). Washington, DC: U.S. Department of Energy, Office of Fossil Energy and National Energy Technology Laboratory. http://www.gwpc.org/sites/default/files/Shale%20Gas%20Primer%202009.pdf

Habuda-Stanic, M; Ravancic, ME; Flanagan, A. (2014). A Review on adsorption of fluoride from aqueous solution. Materials 7: 6317-6366. http://dx.doi.org/10.3390/ma7096317

Hagen, R; Kaatze, U. (2004). Conformational kinetics of disaccharides in aqueous solutions. J Chem Phys 120: 9656-9664. http://dx.doi.org/10.1063/1.1701835

Halldorson, B. (2013). Successful oilfield water management: Five unique case studies. Presented at EPA Technical Workshop - Wastewater Treatment and Related Modeling Research, April 18, 2013, Research Triangle Park, NC. http://www2.epa.gov/sites/production/files/documents/halldorson.pdf

Halliburton. (2014). Hydraulic fracturing 101.

Haluszczak, LO; Rose, AW; Kump, LR. (2013). Geochemical evaluation of flowback brine from Marcellus gas wells in Pennsylvania, USA. Appl Geochem 28: 55-61. http://dx.doi.org/10.1016/j.apgeochem.2012.10.002

Hamieh, BM; Beckman, JR. (2006). Seawater desalination using Dewvaporation technique: theoretical development and design evolution. Desalination 195: 1-13. http://dx.doi.org/10.1016/j.desal.2005.09.034

Hammer, R; VanBriesen, J. (2012). In frackings wake: New rules are needed to protect our health and environment from contaminated wastewater. New York, NY: Natural Resources Defense Council. http://www.nrdc.org/energy/files/fracking-wastewater-fullreport.pdf

Hansen, E; Mulvaney, D; Betcher, M. (2013). Water resource reporting and water footprint from Marcellus Shale development in West Virginia and Pennsylvania. Durango, CO: Earthworks Oil & Gas Accountability Project. http://www.downstreamstrategies.com/documents/reports_publication/marcellus_wv_pa. pdf

Harkness, JS; Dwyer, GS; Warner, NR; Parker, KM; Mitch, WA; Vengosh, A. (2015). Iodide, bromide, and ammonium in hydraulic fracturing and oil and gas wastewaters: Environmental implications. Environ Sci Technol 49: 1955-1963. http://dx.doi.org/10.1021/es504654n

Harlow, A; Wiegand, G; Franck, EU. (1997). The density of ammonia at high pressures to 723 K and 950 MPa. Berichte der Bunsengesellschaft für physikalische Chemie 101: 1461-1465. http://dx.doi.org/10.1002/bbpc.199700007

Harwood, DW; Viner, JG; Russell, ER. (1993). Procedure for developing truck accident and release rates for hazmat routing. Journal of Transportation Engineering 119: 189-199. http://dx.doi.org/10.1061/(ASCE)0733-947X(1993)119:2(189)

Hayes, T. (2009). Sampling and analysis of water streams associated with the development of Marcellus shale gas. Des Plaines, IL: Marcellus Shale Coalition. http://energyindepth.org/wp-content/uploads/marcellus/2012/11/MSCommission-Report.pdf

Hayes, T; Severin, B. (2012a). Characterization of flowback water from the the Marcellus and the Barnett shale regions. Barnett and Appalachian shale water management and reuse technologies. (08122-05.09; Contract 08122-05). http://www.rpsea.org/media/files/project/2146b3a0/08122-05-RT-Characterization_Flowback_Waters_Marcellus_Barnett_Shale_Regions-03-20-12.pdf

Hayes, T; Severin, BF. (2012b). Evaluation of the aqua-pure mechanical vapor recompression system in the treatment of shale gas flowback water - Barnett and Appalachian shale water management and reuse technologies. (08122-05.11). http://barnettshalewater.org/documents/08122-05.11-EvaluationofMVR-3-12-2012.pdf

Hayes, TD; Arthur, D. (2004). Overview of emerging produced water treatment technologies. Paper presented at 11th Annual International Petroleum Environmental Conference, October 12-15, 2004, Albuquerque, NM. http://ipec.utulsa.edu/Conf2004/Papers/hayes_arthur.pdf

Hayes, TD; Halldorson, B; Horner, P; Ewing, J; Werline, JR; Severin, BF. (2014). Mechanical vapor recompression for the treatment of shale-gas flowback water. Oil and Gas Facilities 3: 54-62.

Haynes, WM. (2014). CRC handbook of chemistry and physics. In WM Haynes (95th ed.). Boca Raton, FL: CRC Press. http://www.hbcponline.com/

He, YM; Jiang, RF; Zhu, F; Luan, TG; Huang, ZQ; Ouyang, GF. (2008). Excess molar volumes and surface tensions of 1,2,4-trimethylbenzene and 1,3,5-trimethylbenzene with isopropyl acetate and isobutyl acetate at (298.15, 308.15, and 313.15)K. Journal of Chemical and Engineering Data 53: 1186-1191. http://dx.doi.org/10.1021/je800046k

Hedlund, BP; Geiselbrecht, AD; Staley, JT. (2001). Marinobacter strain NCE312 has a pseudomonas-like naphthalene dioxygenase. FEMS Microbiol Lett 201: 47-51.

Hernlem, BJ; Vane, LM; Sayles, GD. (1999). The application of siderophores for metal recovery and waste remediation: Examination of correlations for prediction of metal affinities. Water Res 33: 951-960.

Horsey, CA. (1981). Depositional environments of the Pennsylvanian Pottsville Formation in the Black Warrior Basin of Alabama. Journal of Sedimentary Research 51: 799-806. http://dx.doi.org/10.1306/212F7DB5-2B24-11D7-8648000102C1865D

House of Representatives (U.S. House of Representatives). (2011). Chemicals used in hydraulic fracturing. Washington, D.C.: U.S. House of Representatives, Committee on Energy and Commerce, Minority Staff. http://www.conservation.ca.gov/dog/general_information/Documents/Hydraulic%20Fracturing%20Report%204%2018%2011.pdf

Hua, GH; Reckhow, DA; Kim, J. (2006). Effect of bromide and iodide ions on the formation and speciation of disinfection byproducts during chlorination. Environ Sci Technol 40: 3050-3056. http://dx.doi.org/10.1021/es0519278

Huffman, HM; Fox, SW. (1938). Thermal data. X. The heats of combustion and free energies, at 25, of some organic compounds concerned in carbohydrate metabolism. J Am Chem Soc 60: 1400-1403. http://dx.doi.org/10.1021/ja01273a036

Hyne, NJ. (2012). Nontechnical guide to petroleum geology, exploration, drilling and production. (3rd ed.). Tulsa, OK: PennWell Corporation.

IARC (International Agency for Research on Cancer). (2015). IARC monographs - Classifications. http://monographs.iarc.fr/ENG/Classification/index.php

Igunnu, ET; Chen, GZ. (2014). Produced water treatment technologies. International Journal of Low-Carbon Technologies 9: 157-177. http://dx.doi.org/10.1093/ijlct/cts049

IUPAC (International Union of Pure and Applied Chemistry). (2014). Global availability of information on agrochemicals: Triisopropanolamine. http://sitem.herts.ac.uk/aeru/iupac/Reports/1338.htm

Jackson, G; Flores, C; Abolo, N; Lawal, H. (2013a). A novel approach to modeling and forecasting frac hits in shale gas wells. Presented at EAGE Annual Conference & Exhibition incorporating SPE Europec, June 10-13, 2013, London, UK. https://www.onepetro.org/conference-paper/SPE-164898-MS

Jackson, RE; Gorody, AW; Mayer, B; Roy, JW; Ryan, MC; Van Stempvoort, DR. (2013b). Groundwater protection and unconventional gas extraction: The critical need for field-based hydrogeological research. Ground Water 51: 488-510. http://dx.doi.org/10.1111/gwat.12074

Jackson, RE; Dussealt, MB. (2014). Gas release mechanisms from energy wellbores. Presented at 48th US Rock Mechanics/Geomechanics Symposium, June 1-4, 2014, Minneapolis, Minnesota. https://www.onepetro.org/conference-paper/ARMA-2014-7753

Jiang, L; Guillot, D; Meraji, M; Kumari, P; Vidick, B; Duncan, B; Gaafar, GR; Sansudin, SB. (2012). Measuring isolation integrity in depleted reservoirs. SPWLA 53rd Annual Logging Symposium, June 16 - 20, 2012, Cartagena, Colombia. http://www.onepetro.org/mslib/app/Preview.do?paperNumber=SPWLA-2012-078&societyCode=SPWLA

Jones, DB; Saglam, A; Song, H; Karanfil, T. (2012). The impact of bromide/iodide concentration and ratio on iodinated trihalomethane formation and speciation. Water Res 46: 11-20. http://dx.doi.org/10.1016/j.watres.2011.10.005

Judson, RS; Kavlock, RJ; Setzer, RW; Hubal, EA; Martin, MT; Knudsen, TB; Houck, KA; Thomas, RS; Wetmore, BA; Dix, DJ. (2011). Estimating toxicity-related biological pathway altering doses for high-throughput chemical risk assessment. Chem Res Toxicol 24: 451-462. http://dx.doi.org/10.1021/tx100428e

Julian, JY; King, GE; Johns, JE; Sack, JK; Robertson, DB. (2007). Detecting ultrasmall leaks with ultrasonic leak detection, case histories from the North Slope, Alaska. Presented at International Oil Conference and Exhibition in Mexico, June 27-30, 2007, Veracruz, Mexico. http://www.onepetro.org/mslib/app/Preview.do?paperNumber=SPE-108906-MS&societyCode=SPE

Kahrilas, GA; Blotevogel, J; Corrin, ER; Borch, T. (2016) Downhole transformation of the hydraulic fracturing fluid biocide glutaraldehyde: Implications for flowback and produced water quality. Environ Sci Technol 50 (20): 11414-11423. http://dx.doi.org/10.1021/acs.est.6b02881

Kansas Water Office. (2014). How is water used in oil and gas exploration in Kansas? Topeka, KA. http://www.kwo.org/about_us/BACs/KWIF/rpt_Hydraulic%20Fracturing_KS_Water_FAQ_03082012_final_ki.pdf

Kashem, MA; Singh, BR; Kondo, T; Huq, SMI; Kawai, S. (2007). Comparison of extractability of Cd, Cu, Pb and Zn with sequential extraction in contaminated and non-contaminated soils. Int J Environ Sci Tech 4: 169-176. http://dx.doi.org/10.1007/BF03326270

Kekacs, D; Drollette, BD; Brooker, M; Plata, DL; Mouser, PJ. (2015). Aerobic biodegradation of organic compounds in hydraulic fracturing fluids. Biodegradation 26: 271-287. http://dx.doi.org/10.1007/s10532-015-9733-6

Kennedy/Jenks Consultants. (2002). Evaluation of technical and economic feasibility of treating oilfield produced water to create a new water resource. http://www.gwpc.org/sites/default/files/event-sessions/Roger_Funston_PWC2002_0.pdf

Khan, NA; Engle, M; Dungan, B; Holguin, FO; Xu, P; Carroll, KC. (2016). Volatile-organic molecular characterization of shale-oil produced water from the Permian Basin. Chemosphere 148: 126-136. http://dx.doi.org/10.1016/j.chemosphere.2015.12.116

Kim, HM; Hwang, CY; Cho, BC. (2010). Arcobacter marinus sp. nov. Int J Syst Evol Microbiol 60: 531-536. http://dx.doi.org/10.1099/ijs.0.007740-0

Kim, J; Moridis, GJ. (2013). Development of the T+M coupled flowgeomechanical simulator to describe fracture propagation and coupled flowthermalgeomechanical processes in tight/shale gas systems. Computers and Geosciences 60: 184-198. http://dx.doi.org/10.1016/j.cageo.2013.04.023

Kim, J; Moridis, GJ. (2015). Numerical analysis of fracture propagation during hydraulic fracturing operations in shale gas systems. International Journal of Rock Mechanics and Mining Sciences 76: 127-137.

Kim, J; Moridis, GJ; Martinez, ER. (2016). Investigation of possible wellbore cement failures during hydraulic fracturing operations. Journal of Petroleum Science and Engineering 139: 254-263. http://dx.doi.org/10.1016/j.petrol.2016.01.035

Kim, J; Um, ES; Moridis, GJ. (2014). Fracture propagation, fluid flow, and geomechanics of water-based hydraulic fracturing in shale gas systems and electromagnetic geophysical monitoring of fluid migration. SPE Hydraulic Fracturing Technology Conference, February 4-6, 2014, The Woodlands, Texas. http://dx.doi.org/10.2118/168578-MS

Kimball, B. (2010). Water treatment technologies for global unconventional gas plays. Presented at US - China Industry Oil and Gas Forum, September 16, 2010, Fort Worth, TX. http://www.uschinaogf.org/Forum10/pdfs/5%20-%20CDM%20-%20Kimball%20-%20EN.pdf

King, GE. (2012). Hydraulic fracturing 101: What every representative, environmentalist, regulator, reporter, investor, university researcher, neighbor and engineer should know about estimating frac risk and improving frac performance in unconventional gas and oil wells. SPE Hydraulic Fracturing Technology Conference, February 6-8, 2012, The Woodlands, TX. http://fracfocus.org/sites/default/files/publications/hydraulic_fracturing_101.pdf

King, GE; Valencia, RL. (2016). Well integrity for fracturing and re-fracturing: What is needed and why? SPE Hydraulic Fracturing Technology Conference, February 9-11, 2016, The Woodlands, Texas, USA. https://www.onepetro.org/conference-paper/SPE-179120-MS

Kirksey, J. (2013). Optimizing wellbore integrity in well construction. Presented at North American Wellbore Integrity Workshop, October 16-17, 2013, Denver, CO. http://ptrc.ca/+pub/document/Kirksey%20-%20Optimizing%20Wellbore%20Integrity.pdf

Kiselev, VD; Kashaeva, HA; Shakirova, II; Potapova, LN; Konovalov, AI. (2012). Solvent effect on the enthalpy of solution and partial molar volume of the ionic liquid 1-butyl-3-methylimidazolium tetrafluoroborate. Journal of Solution Chemistry 41: 1375-1387. http://dx.doi.org/10.1007/s10953-012-9881-9

Konschnik, K; Dayalu, A. (2016). Hydraulic fracturing chemicals reporting: Analysis of available data and recommendations for policymakers. Energy Policy 88: 504-514. http://dx.doi.org/10.1016/j.enpol.2015.11.002

Kose, B; Ozgun, H; Ersahin, ME; Dizge, N; KoseogluImer, DY; Atay, B; Kaya, R; Altinbas, M; Sayili, S; Hoshan, P; Atay, D; Eren, E; Kinaci, C; Koyuncu, I. (2012). Performance evaluation of a submerged membrane bioreactor for the treatment of brackish oil and natural gas field produced water. Desalination 285: 295-300.

Kraemer, TF; Reid, DF. (1984). The occurrence and behavior of radium in saline formation water of the U.S. Gulf Coast region. Isotope Geoscience 2: 153-174.

Krakowiak, J; Bobicz, D; Grzybkowski, W. (2001). Limiting partial molar volumes of tetra-n-alkylammonium perchlorates in N,N-dimethylacetamide, triethylphosphate and dimethyl sulfoxide at T=298.15 K. The Journal of Chemical Thermodynamics 33: 121-133. http://dx.doi.org/10.1006/jcht.2000.0725

Krasner, SW. (2009). The formation and control of emerging disinfection by-products of health concern [Review]. Philos Transact A Math Phys Eng Sci 367: 4077-4095. http://dx.doi.org/10.1098/rsta.2009.0108

Kroes, R; Kleiner, J; Renwick, A. (2005). The threshold of toxicological concern concept in risk assessment. Toxicol Sci 86: 226-230. http://dx.doi.org/10.1093/toxsci/kfi169

Kroes, R; Renwick, AG; Cheeseman, M; Kleiner, J; Mangelsdorf, I; Piersma, A; Schilter, B; Schlatter, J; van Schothorst, F; Vos, JG; Würtzen, G. (2004). Structure-based thresholds of toxicological concern (TTC): Guidance for application to substances present at low levels in the diet [Review]. Food Chem Toxicol 42: 65-83.

Kuthnert, N; Werline, R; Nichols, K. (2012). Water reuse and recycling in the oil and gas industry: Devons water management success. Presentation presented at 2nd Annual Texas Water Reuse Conference, July 20, 2012, Forth Worth, TX. http://www.weat.org/Presentations/A_22_NICHOLS.pdf

LA Ground Water Resources Commission (Louisiana Ground Water Resources Commission). (2012). Managing Louisiana's groundwater resources: An interim report to the Louisiana Legislature. Baton Rouge, LA: Louisiana Department of Natural Resources. http://dnr.louisiana.gov/index.cfm?md=pagebuilder&tmp=home&pid=907

Laavi, H; Pokki, JP; Uusi-Kyyny, P; Massimi, A; Kim, Y; Sapei, E; Alopaeus, V. (2013). Vapor-liquid equilibrium at 350 K, excess molar enthalpies at 298 K, and excess molar volumes at 298 K of binary mixtures containing ethyl acetate, butyl acetate, and 2-butanol. Journal of Chemical and Engineering Data 58: 1011-1019. http://dx.doi.org/10.1021/je400036b

Laavi, H; Zaitseva, A; Pokki, JP; Uusi-Kyyny, P; Kim, Y; Alopaeus, V. (2012). Vapor-liquid equilibrium, excess molar enthalpies, and excess molar volumes of binary mixtures containing methyl isobutyl ketone (MIBK) and 2-butanol, tert-pentanol, or 2-ethyl-1-hexanol. Journal of Chemical and Engineering Data 57: 3092-3101. http://dx.doi.org/10.1021/je300678r

Lalucat, J; Bennasar, A; Bosch, R; Garcia-Valdes, E; Palleroni, NJ. (2006). Biology of Pseudomonas stutzeri [Review]. Microbiol Mol Biol Rev 70: 510-547. http://dx.doi.org/10.1128/MMBR.00047-05

Landry, G; Welty, RD; Thomas, M; Vaughan, ML; Tatum, D. (2015). Bridging the gap: An integrated approach to solving sustained casing pressure in the Cana Woodford Shale. SPE Well Integrity Symposium, June 2-3, 2015, Galveston, Texas, USA. https://www.onepetro.org/conference-paper/SPE-174525-MS

Langmuir, D; Herman, JS. (1980). The mobility of thorium in natural waters at low temperatures. Geochim Cosmo Act 44: 1753-1766. http://dx.doi.org/10.1016/0016-7037(80)90226-4

Langmuir, D; Riese, AC. (1985). The thermodynamic properties of radium. Geochim Cosmo Act 49: 1593-1601.

Lapenna, S; Worth, A. (2011). Analysis of the Cramer classification scheme for oral systemic toxicity - implications for its implementation in Toxtree. (EUR 24898 EN - 2011). Luxembourg: Publications Office of the European Union. http://dx.doi.org/10.2788/39716

Laurenzi, IJ; Jersey, GR. (2013). Life cycle greenhouse gas emissions and freshwater consumption of Marcellus shale gas. Environ Sci Technol 47: 4896-4903. http://dx.doi.org/10.1021/es305162w

Leadscope, Inc. (2012). Leadscope [Computer Program]. Columbus, Ohio. Retrieved from http://www.leadscope.com

LEau LLC. (2008). Dew vaporation desalination 5,000-gallon-per-day pilot plant. (Desalination and Water Purification Research and Development Program Report No. 120). Denver, CO: Bureau of Reclamation, U.S. Department of the Interior. http://www.usbr.gov/research/AWT/reportpdfs/report120.pdf

Lee, K; Neff, J. (2011). Produced water: Environmental risks and advances in mitigation technologies. New York, NY: Springer. http://dx.doi.org/10.1007/978-1-4614-0046-2

Lester, Y; Ferrer, I; Thurman, EM; Sitterley, KA; Korak, JA; Aiken, G; Linden, KG. (2015). Characterization of hydraulic fracturing flowback water in Colorado: Implications for water treatment. Sci Total Environ 512-513: 637-644. http://dx.doi.org/10.1016/j.scitotenv.2015.01.043

Lovley, DR; Chapelle, FH. (1995). Deep subsurface microbial processes. Rev Geophys 33: 365-381. http://dx.doi.org/10.1029/95RG01305

Lovley, DR; Phillips, EJ. (1986). Organic matter mineralization with reduction of ferric iron in anaerobic sediments. Appl Environ Microbiol 51: 683-689.

Ludzack, FJ; Noran, DK. (1965). Tolerance of high salinities by conventional wastewater treatment processes. J Water Pollut Control Fed 37: 1404-1416.

Luh, J; Mariñas, BJ. (2012). Bromide ion effect on N-nitrosodimethylamine formation by monochloramine. Environ Sci Technol 46: 5085-5092. http://dx.doi.org/10.1021/es300077x

Lutz, BD; Lewis, AN; Doyle, MW. (2013). Generation, transport, and disposal of wastewater associated with Marcellus Shale gas development. Water Resour Res 49: 647-656.

Lyons, WC; Pligsa, GJ. (2004). Standard handbook of petroleum and natural gas engineering (2nd ed.). Houston, TX: Gulf Professional Publishing. http://www.elsevier.com/books/standard-handbook-of-petroleum-and-natural-gas-engineering/lyons-phd-pe/978-0-7506-7785-1

Ma, G; Geza, M; Xu, P. (2014). Review of flowback and produced water management, treatment, and beneficial use for major shale gas development basins. Shale Energy Engineering Conference 2014, Pittsburgh, Pennsylvania.

Maguire-Boyle, SJ; Barron, AR. (2014). Organic compounds in produced waters from shale gas wells. Environ Sci Process Impacts 16: 2237-2248. http://dx.doi.org/10.1039/c4em00376d

Mak, TCW. (1965). Hexamethylenetetramine hexahydrate: A new type of clathrate hydrate. J Chem Phys 43: 2799-2805. http://dx.doi.org/10.1063/1.1697212

Maloney, KO; Yoxtheimer, DA. (2012). Production and disposal of waste materials from gas and oil extraction from the Marcellus shale play in Pennsylvania. Environmental Practice 14: 278-287. http://dx.doi.org/10.1017/S146604661200035X

Manios, T; Stentiford, EI; Millner, P. (2003). Removal of total suspended solids from wastewater in constructed horizontal flow subsurface wetlands. J Environ Sci Health A Tox Hazard Subst Environ Eng 38: 1073-1085. http://dx.doi.org/10.1081/ESE-120019865

Mantell, ME. (2013). Recycling and reuse of produced water to reduce freshwater use in hydraulic fracturing operations. In Summary of the Technical Workshop on Water Acquisition Modeling: Assessing Impacts through Modeling and Other Means (pp. A20-A27). Washington, D.C.: U.S. Environmental Protection Agency. http://www2.epa.gov/hfstudy/summary-technical-workshop-water-acquisition-modeling-assessing-impacts-through-modeling-and

Martinez-Reina, M; Amado-Gonzalez, E; Mauricio Munoz-Munoz, Y. (2012). Study of liquid-liquid equilibria of toluene plus (hexane, heptane, or cyclohexane) with 1-ethyl-3-methylimidazolium ethylsulfate at 308.15 K. Bull Chem Soc Jpn 85: 1138-1144. http://dx.doi.org/10.1246/bcsj.20120112

Martinez, CE; McBride, MB. (2001). Cd, Cu, Pb, and Zn coprecipitates in Fe oxide formed at different pH: Aging effects on metal solubility and extractability by citrate. Environ Toxicol Chem 20: 122-126. http://dx.doi.org/10.1002/etc.5620200112

Masood, AKM; Pethrick, RA; Swinton, FL. (1976). Physicochemical studies of super-cooled liquids - cyclic carbonates and alpha,beta-unsaturated aldehydes. Faraday Trans 1 72: 20-28. http://dx.doi.org/10.1039/f19767200020

Mata, JA; Martínez-Cánovas, J; Quesada, E; Béjar, V. (2002). A detailed phenotypic characterisation of the type strains of Halomonas species. Syst Appl Microbiol 25: 360-375. http://dx.doi.org/10.1078/0723-2020-00122

Matamoros, V; Mujeriego, R; Bayona, JM. (2007). Trihalomethane occurrence in chlorinated reclaimed water at full-scale wastewater treatment plants in NE Spain. Water Res 41: 3337-3344. http://dx.doi.org/10.1016/j.watres.2007.04.021

Material Safety Data Sheets. (a) Encana/Halliburton Energy Services, Inc.: Duncan, Oklahoma. Provided by Halliburton Energy Services during an onsite visit by the EPA on May 10, 2010; (b) Encana Oil and Gas (USA), Inc.: Denver, Colorado. Provided to US EPA Region 8.

Maupin, MA; Kenny, JF; Hutson, SS; Lovelace, JK; Barber, NL; Linsey, KS. (2014). Estimated use of water in the United States in 2010. (USGS Circular 1405). Reston, VA: U.S. Geological Survey. http://dx.doi.org/10.3133/cir1405

Maxwell, SC. (2011). Hydraulic fracture height growth. Recorder 36: 18-22.

McDaniel, BW; Rispler, KA. (2009). Horizontal wells with multistage fracs prove to be best economic completion for many low permeability reservoirs. Presented at SPE Eastern Regional Meeting, September 23-15, 2009, Charleston, WV. https://www.onepetro.org/conference-paper/SPE-125903-MS

McDaniel, J; Watters, L; Shadravan, A. (2014). Cement sheath durability: Increasing cement sheath integrity to reduce gas migration in the Marcellus Shale Play. In SPE hydraulic fracturing technology conference proceedings, 4-6 February, 2014, The Woodlands, TX: Society of Petroleum Engineers. http://dx.doi.org/10.2118/168650-MS

McGowan, L; Herbert, R; Muyzer, G. (2004). A comparative study of hydrocarbon degradation by Marinobacter sp., Rhodococcus sp. and Corynebacterium sp. isolated from different mat systems. Ophelia 58: 271-281. http://dx.doi.org/10.1080/00785236.2004.10410235

McGuire, MJ; Karanfil, T; Krasner, SW; Reckhow, DA; Roberson, JA; Summers, RS; Westerhoff, P; Xie, Y. (2014). Not your granddad's disinfection by-product problems and solutions. JAWWA 106: 54-73. http://dx.doi.org/10.5942/jawwa.2014.106.0128

McLin, K; Brinton, D; Moore, J. (2011). Geochemical modeling of water-rock-proppant interactions. Thirty-Sixth Workshop on Geothermal Reservoir Engineering, January 31 - February 2, 2011, Stanford University, Stanford, California. https://pangea.stanford.edu/ERE/db/IGAstandard/record_detail.php?id=7234

Miller, P. (2011). Future of hydraulic fracturing depends on effective water treatment. Hydrocarbon Process 90: 13-13.

Minnich, K. (2011). A water chemistry perspective on flowback reuse with several case studies. In Proceedings of the Technical Workshops for the Hydraulic Fracturing Study: Water Resources Management. http://www2.epa.gov/sites/production/files/documents/10_Minnich_-_Chemistry_508.pdf

Mitchell, J; Pabon, P; Collier, ZA; Egeghy, PP; Cohen-Hubal, E; Linkov, I; Vallero, DA. (2013). A decision analytic approach to exposure-based chemical prioritization. PLoS ONE 8: e70911. http://dx.doi.org/10.1371/journal.pone.0070911

Mohan, AM; Gregory, KB; Vidic, RD; Miller, P; Hammack, RW. (2011). Characterization of microbial diversity in treated and untreated flowback water impoundments from gas fracturing operations. Presented at SPE Annual Technical Conference and Exhibition, October 30 - November 2, 2011, Denver, CO. http://www.onepetro.org/mslib/servlet/onepetropreview?id=SPE-147414-MS

Montgomery, C. (2013). Fracturing fluid components. In A Bunder; J McLennon; R Jeffrey (Eds.), Effective and Sustainable Hydraulic Fracturing. Croatia: InTech. http://dx.doi.org/10.5772/56422

Moosavi, M; Motahari, A; Omrani, A; Rostami, AA. (2013). Thermodynamic study on some alkanediol solutions: Measurement and modeling. Thermochim Acta 561: 1-13. http://dx.doi.org/10.1016/j.tca.2013.03.010

MSC (Marcellus Shale Coalition). (2013). Recommended practices: Drilling and completions. (MSC RP 2013-3). Pittsburgh, Pennsylvania.

Munro, IC; Ford, RA; Kennepohl, E; Sprenger, JG. (1996). Correlation of structural class with no-observed-effect levels: a proposal for establishing a threshold of concern. Food Chem Toxicol 34: 829-867.

Munter, R. (2000). Industrial wastewater treatment. In LC Lundin (Ed.), Sustainable water management in the Baltic Sea Basin book II: Water use and management (pp. 195-210). Sida, Sweden: Baltic University Programme Publication. http://www.balticuniv.uu.se/index.php/boll-online-library/831-swm-2-water-use-and-management

Murali Mohan, A; Hartsock, A; Bibby, KJ; Hammack, RW; Vidic, RD; Gregory, KB. (2013a). Microbial community changes in hydraulic fracturing fluids and produced water from shale gas extraction. Environ Sci Technol 47: 13141-13150. http://dx.doi.org/10.1021/es402928b

Murali Mohan, A; Hartsock, A; Hammack, RW; Vidic, RD; Gregory, KB. (2013b). Microbial communities in flowback water impoundments from hydraulic fracturing for recovery of shale gas. FEMS Microbiol Ecol. http://dx.doi.org/10.1111/1574-6941.12183

Murray, KE. (2013). State-scale perspective on water use and production associated with oil and gas operations, Oklahoma, U.S. Environ Sci Technol 47: 4918-4925. http://dx.doi.org/10.1021/es4000593

Myers, CR; Nealson, KH. (1988). Bacterial manganese reduction and growth with manganese oxide as the sole electron acceptor. Science 240: 1319-1321. http://dx.doi.org/10.1126/science.240.4857.1319

National Drought Mitigation Center. (2015). U.S. drought monitor. http://droughtmonitor.unl.edu/Home.aspx (accessed February 27, 2015).

Newman, DK. (2001). Microbiology - How bacteria respire minerals. Science 292: 1312-1313. http://dx.doi.org/10.1126/science.1060572

Nicot, JP; Reedy, RC; Costley, RA; Huang, Y. (2012). Oil & gas water use in Texas: Update to the 2011 mining water use report. Austin, TX: Bureau of Economic Geology, University of Texas at Austin. http://www.twdb.texas.gov/publications/reports/contracted_reports/doc/0904830939_2012Update_MiningWaterUse.pdf

Nicot, JP; Scanlon, BR. (2012). Water use for shale-gas production in Texas. U.S. Environ Sci Technol 46: 3580-3586. http://dx.doi.org/10.1021/es204602t

Nicot, JP; Scanlon, BR; Reedy, RC; Costley, RA. (2014). Source and fate of hydraulic fracturing water in the Barnett Shale: A historical perspective. Environ Sci Technol 48: 2464-2471. http://dx.doi.org/10.1021/es404050r

NLM (National Institutes of Health, National Library of Medicine). (2014). ChemID plus advanced. http://chem.sis.nlm.nih.gov/chemidplus/

NM OSE (New Mexico Office of the State Engineer). (2013). New Mexico water use by categories 2010. (Technical Report 54). Santa Fe, NM: New Mexico Office of the State Engineer, Water Use and Conservation Bureau. http://www.ose.state.nm.us/Pub/TechnicalReports/TechReport%2054NM%20Water%20Use%20by%20Categories%20.pdf

NMSU DACC WUTAP (New Mexico State University, Doña Ana Community College, Water Utilities Technical Assistance Program). (2007). New Mexico wastewater systems operator certification study manual - Version 1.1. Santa Fe, NM: New Mexico Environment Department. http://www.nmrwa.org/sites/nmrwa.org/files/WastewaterOperatorStudyManual.pdf

North Dakota Department of Health. (2015). Oil field environmental incident summary, incident 20150107160242. http://www.ndhealth.gov/EHS/FOIA/Spills/Summary_Reports/20150107160242_Summary_Report.pdf

North Dakota Department of Mineral Resources. (2016). Bakken horizontal wells by producing zone. https://www.dmr.nd.gov/oilgas/bakkenwells.asp

North Dakota State Water Commission. (2014). Facts about North Dakota fracking and water use. Bismarck, ND. http://www.swc.nd.gov/pdfs/fracking_water_use.pdf

NPC (National Petroleum Council). (2011). Management of produced water from oil and gas wells. (Paper #2-17). Washington, D.C. http://www.npc.org/Prudent_Development-Topic_Papers/2-17_Management_of_Produced_Water_Paper.pdf

NTP (National Toxicology Program). (2014a). Definition of carcinogenicity results. http://ntp.niehs.nih.gov/results/pubs/longterm/defs/index.html

NTP (National Toxicology Program). (2014b). Report on carcinogens. Thirteenth edition. Research Triangle Park, NC: U.S. Department of Health and Human Services, Public Health Service. http://ntp.niehs.nih.gov/pubhealth/roc/roc13/index.html

NYSDEC (New York State Department of Environmental Conservation). (2011). Revised draft supplemental generic environmental impact statement (SGEIS) on the oil, gas and solution mining regulatory program: Well permit issuance for horizontal drilling and high-volume hydraulic fracturing to develop the Marcellus shale and other low-permeability gas reservoirs. Albany, NY. http://www.dec.ny.gov/energy/75370.html

Obolensky, A; Singer, PC. (2008). Development and interpretation of disinfection byproduct formation models using the Information Collection Rule database. Environ Sci Technol 42: 5654-5660. http://dx.doi.org/10.1021/es702974f

OECD (Organisation for Economic Co-operation and Development). (2016). The OECD QSAR toolbox. http://www.oecd.org/chemicalsafety/risk-assessment/theoecdqsartoolbox.htm

OEHHA (Office of Environmental Health Hazard Assessment) (2012). Title 27, California Code of Regulations Article 8. No Observable Effect Levels. http://oehha.ca.gov/media/downloads/crnr/regulation022610.pdf

Oil and Gas Mineral Services. (2010). MineralWise: Oil and gas terminology. http://www.mineralweb.com/library/oil-and-gas-terms/

Oka, S. (1962). Studies on lactone formation in vapor phase. III. Mechanism of lactone formation from diols. Bull Chem Soc Jpn 35: 986-989. http://dx.doi.org/10.1246/bcsj.35.986

Olsson, O; Weichgrebe, D; Rosenwinkel, KH. (2013). Hydraulic fracturing wastewater in Germany: Composition, treatment, concerns. Environ Earth Sci 70: 3895-3906. http://dx.doi.org/10.1007/s12665-013-2535-4

ONG Services. (2015). ONGList: Reserved Environmental Services. http://www.onglist.com/Home/Search?SearchString=Reserved+environmental+services&Distance=&searchAddress=&CategoryTypeID=1&SubCategoryID

Orem, W; Tatu, C; Varonka, M; Lerch, H; Bates, A; Engle, M; Crosby, L; McIntosh, J. (2014). Organic substances in produced and formation water from unconventional natural gas extraction in coal and shale. Int J Coal Geol 126: 20-31. http://dx.doi.org/10.1016/j.coal.2014.01.003

Orem, WH; Tatu, CA; Lerch, HE; Rice, CA; Bartos, TT; Bates, AL; Tewalt, S; Corum, MD. (2007). Organic compounds in produced waters from coalbed natural gas wells in the Powder River Basin, Wyoming, USA. Appl Geochem 22: 2240-2256. http://dx.doi.org/10.1016/j.apgeochem.2007.04.010

OSHA (Occupational Safety and Health Administration) (2013). Title 29 - Department of Labor. Subpart z Toxic and hazardous substances, hazard communication. http://www.gpo.gov/fdsys/pkg/CFR-2013-title29-vol6/xml/CFR-2013-title29-vol6-sec1910-1200.xml

OWRB (Oklahoma Water Resources Board). (2014). The Oklahoma comprehensive water plan. http://www.owrb.ok.gov/supply/ocwp/ocwp.php

Oyarhossein, M; Dusseault, MB. (2015). Wellbore stress changes and microannulus development because of cement shrinkage. 49th US Rock Mechanics/Geomechanics Symposium, June 28 - July 1, 2015, San Francisco, CA. https://www.onepetro.org/conference-paper/ARMA-2015-118

PA DEP (Pennsylvania Department of Environmental Protection). (2010). Chemicals used by hydraulic fracturing companies in Pennsylvania for surface and hydraulic fracturing activities. Harrisburg, PA: Pennsylvania Department of Environmental Protection (PADEP). http://files.dep.state.pa.us/oilgas/bogm/bogmportalfiles/MarcellusShale/Frac%20list%20 6-30-2010.pdf

PA DEP (Pennsylvania Department of Environmental Protection). (2015). Technologically enhanced naturally occurring radioactive materials (TENORM) study report. Harrisburg, PA: Pennsylvania Department of Environmental Protection (PADEP).

PA DEP (Pennsylvania Department of Environmental Protection). (2016). Oil and gas compliance - Report viewer. Harrisburg, PA: Pennsylvania Department of Environmental Protection (PADEP). http://www.depreportingservices.state.pa.us/ReportServer/Pages/ReportViewer.aspx?/Oi l_Gas/OG_Compliance

Pal, A; Kumar, H; Maan, R; Sharma, HK. (2013). Densities and speeds of sound of binary liquid mixtures of some n-alkoxypropanols with methyl acetate, ethyl acetate, and n-butyl acetate at T = (288.15, 293.15, 298.15, 303.15, and 308.15) K. Journal of Chemical and Engineering Data 58: 225-239. http://dx.doi.org/10.1021/je300789a

Parker, KM; Zeng, T; Harkness, J; Vengosh, A; Mitch, WA. (2014). Enhanced formation of disinfection byproducts in shale gas wastewater-impacted drinking water supplies. Environ Sci Technol 48: 11161-11169. http://dx.doi.org/10.1021/es5028184

Pashin, JC; Mcintyre-Redden, MR; Mann, SD; Kopaska-Merkel, DC; Varonka, M; Orem, W. (2014). Relationships between water and gas chemistry in mature coalbed methane reservoirs of the Black Warrior Basin. Int J Coal Geol 126: 92-105. http://dx.doi.org/10.1016/j.coal.2013.10.002

Pijper, WP. (1971). Molecular and crystal structure of glycollic acid. Acta Crystallogr B B27: 344-348. http://dx.doi.org/10.1107/S056774087100219X

Pope, PG; Martin-Doole, M; Speitel, GE; Collins, MR. (2007). Relative significance of factors influencing DXAA formation during chloramination. JAWWA 99: 144-156.

Radwan, MHS; Hanna, AA. (1976). Binary azeotropes containing butyric acids. Journal of Chemical and Engineering Data 21: 285-289. http://dx.doi.org/10.1021/je60070a032

Rahm, BG; Bates, JT; Bertoia, LR; Galford, AE; Yoxtheimer, DA; Riha, SJ. (2013). Wastewater management and Marcellus Shale gas development: Trends, drivers, and planning implications. J Environ Manage 120: 105-113. http://dx.doi.org/10.1016/j.jenvman.2013.02.029

Rahm, BG; Riha, SJ. (2012). Toward strategic management of shale gas development: Regional, collective impacts on water resources. Environ Sci Pol 17: 12-23. http://dx.doi.org/10.1016/j.envsci.2011.12.004

Rahm, BG; Vedachalam, S; Bertoia, LR; Mehta, D; Vanka, VS; Riha, SJ. (2015). Shale gas operator violations in the Marcellus and what they tell us about water resource risks. Energy Policy 82: 1-11. http://dx.doi.org/10.1016/j.enpol.2015.02.033

Rani, M; Maken, S. (2013). Excess molar enthalpies and excess molar volumes of formamide+1-propanol or 2-propanol and thermodynamic modeling by Prigogine-Flory-Patterson theory and Treszczanowicz-Benson association model. Thermochim Acta 559: 98-106. http://dx.doi.org/10.1016/j.tca.2013.02.010

Ravot, G; Magot, M; Ollivier, B; Patel, BK; Ageron, E; Grimont, PA; Thomas, P; Garcia, JL. (1997). Haloanaerobium congolense sp. nov., an anaerobic, moderately halophilic, thiosulfate- and sulfur-reducing bacterium from an African oil field. FEMS Microbiol Lett 147: 81-88.

Rawat, BS; Gulati, IB; Mallik, KL. (1976). Study of some sulphur-group solvents for aromatics extraction by gas chromatography. Journal of Applied Chemistry and Biotechnology 26: 247-252. http://dx.doi.org/10.1002/jctb.5020260504

Reagan, MT; Moridis, GJ; Johnson, JN; Keen, ND. (2015). Numerical simulation of the environmental impact of hydraulic fracturing of tight/shale gas reservoirs on near-surface groundwater: Background, base cases, shallow reservoirs, short-term gas and water transport. Water Resour Res 51: 1-31. http://dx.doi.org/10.1002/2014WR016086

Renpu, W. (2011). Advanced well completion engineering (Third ed.). Houston, TX: Gulf Professional Publishing.

Rice, CA; Flores, RM; Stricker, GD; Ellis, MS. (2008). Chemical and stable isotopic evidence for water/rock interaction and biogenic origin of coalbed methane, Fort Union Formation, Powder River Basin, Wyoming and Montana USA. Int J Coal Geol 76: 76-85. http://dx.doi.org/10.1016/j.coal.2008.05.002

Richardson, SD; Plewa, MJ; Wagner, ED; Schoeny, R; Demarini, DM. (2007). Occurrence, genotoxicity, and carcinogenicity of regulated and emerging disinfection by-products in drinking water: A review and roadmap for research [Review]. Mutat Res 636: 178-242. http://dx.doi.org/10.1016/j.mrrev.2007.09.001

Rodnikova, MN; Solonina, IA; Egorov, GI; Makarov, DM; Gunina, MA. (2012). The bulk properties of dioxane solutions in ethylene glycol at 2575C. Russian Journal of Physical Chemistry A, Focus on Chemistry 86: 330-332. http://dx.doi.org/10.1134/S0036024412020239

Ross, D; King, G. (2007). Well completions. In MJ Economides; T Martin (Eds.), Modern fracturing: Enhancing natural gas production (1st ed., pp. 169-198). Houston, Texas: ET Publishing.

Rotroff, DM; Wetmore, BA; Dix, DJ; Ferguson, SS; Clewell, HJ; Houck, KA; Lecluyse, EL; Andersen, ME; Judson, RS; Smith, CM; Sochaski, MA; Kavlock, RJ; Boellmann, F; Martin, MT; Reif, DM; Wambaugh, JF; Thomas, RS. (2010). Incorporating human dosimetry and exposure into high-throughput in vitro toxicity screening. Toxicol Sci 117: 348-358. http://dx.doi.org/10.1093/toxsci/kfq220

Rowan, EL; Engle, MA; Kirby, CS; Kraemer, TF. (2011). Radium content of oil- and gas-field produced waters in the northern Appalachian Basin (USA): Summary and discussion of data. (Scientific Investigations Report 20115135). Reston, VA: U.S. Geological Survey. http://pubs.usgs.gov/sir/2011/5135/

Roy, SB; Ricci, PF; Summers, KV; Chung, CF; Goldstein, RA. (2005). Evaluation of the sustainability of water withdrawals in the United States, 1995 to 2025. J Am Water Resour Assoc 41: 1091-1108. http://dx.doi.org/10.1111/j.1752-1688.2005.tb03787.x

Rozell, DJ; Reaven, SJ. (2012). Water pollution risk associated with natural gas extraction from the Marcellus Shale. Risk Anal 32: 13821393. http://dx.doi.org/10.1111/j.1539-6924.2011.01757.x

Rushton, L; Castaneda, C. (2014). Drilling into hydraulic fracturing and the associated wastewater management issues. Washington, D.C.: Paul Hastings, LLP. http://www.paulhastings.com/docs/default-source/PDFs/stay-current-hydraulic-fracturing-wastewater-management.pdf

Rutqvist, J; Rinaldi, AP; Cappa, F; Moridis, GJ. (2013). Modeling of fault reactivation and induced seismicity during hydraulic fracturing of shale-gas reservoirs. Journal of Petroleum Science and Engineering 107: 31-44. http://dx.doi.org/10.1016/j.petrol.2013.04.023

Rutqvist, J; Rinaldi, AP; Cappa, F; Moridis, GJ. (2015). Modeling of fault activation and seismicity by injection directly into a fault zone associated with hydraulic fracturing of shale-gas reservoirs. Journal of Petroleum Science and Engineering 127: 377-386. http://dx.doi.org/10.1016/j.petrol.2015.01.019

Sabins, F. (1990). Problems in cementing horizontal wells. J Pet Tech 42: 398-400. http://dx.doi.org/10.2118/20005-PA

SDWA (Safe Drinking Water Act). (2002). Title XIV of the Public Health Service Act Safety of Public Water Systems as amended through P.L. 107-377. http://www.epw.senate.gov/sdwa.pdf

Santa Cruz Biotechnology. (2015). Sorbitane trioleate (CAS 26266-58-0). http://www.scbt.com/datasheet-281154-Sorbitane-Trioleate.html (accessed April 6, 2015).

Sarkar, BK; Choudhury, A; Sinha, B. (2012). Excess molar volumes, excess viscosities and ultrasonic speeds of sound of binary mixtures of 1,2-dimethoxyethane with some aromatic liquids at 298.15 K. Journal of Solution Chemistry 41: 53-74. http://dx.doi.org/10.1007/s10953-011-9780-5

Sarkar, L; Roy, MN. (2009). Density, viscosity, refractive index, and ultrasonic speed of binary mixtures of 1,3-dioxolane with 2-methoxyethanol, 2-ethoxyethanol, 2-butoxyethanol, 2-propylamine, and cyclohexylamine. Journal of Chemical and Engineering Data 54: 3307-3312. http://dx.doi.org/10.1021/je900240s

Scanlon, BR; Reedy, RC; Nicot, JP. (2014). Comparison of water use for hydraulic fracturing for unconventional oil and gas versus conventional oil. Environ Sci Technol 48: 12386-12393. http://dx.doi.org/10.1021/es502506v

Schlumberger (Schlumberger Limited). (2014). Schlumberger oilfield glossary. http://www.glossary.oilfield.slb.com/

Schrodinger. (2012). Qikprop [Computer Program]. New York, New York: Schrodinger, LLC. http://www.schrodinger.com/products/14/17

Schwarzenbach, RP; Gschwend, PM; Imboden, DM. (2002). Environmental Organic Chemistry. In Environmental organic chemistry (2nd ed.). Hoboken, NJ: John Wiley & Sons, Inc.

Senters, CW; Snyder, DJ; Warren, MN; Leonard, RS; Woodroof, RA. (2016). Determining the effectiveness of isolation techniques using completion diagnostics and production analysis. SPE Hydraulic Fracturing Technology Conference, February 9-11, 2016, The Woodlands, Texas, USA. https://www.onepetro.org/conference-paper/SPE-179175-MS

Shadravan, A; Amani, M. (2015). A decade of self-sealing cement technology application to ensure long-term well integrity. SPE Kuwait Oil and Gas Show and Conference, October 11-14, 2015, Mishref, Kuwait. https://www.onepetro.org/conference-paper/SPE-175237-MS

Shafer, L. (2011). Water recycling and purification in the Pinedale anticline field: results from the anticline disposal project. In 2011 SPE America's E&P Health, Safety, Security & Environmental conference. Richardson, TX: Society of Petroleum Engineers. http://dx.doi.org/10.2118/141448-MS

Shaffer, DL; Arias Chavez, LH; Ben-Sasson, M; Romero-Vargas Castrillón, S; Yip, NY; Elimelech, M. (2013). Desalination and reuse of high-salinity shale gas produced water: Drivers, technologies, and future directions. Environ Sci Technol 47: 9569-9583.

Shammas, NK. (2010). Wastewater renovation by flotation. In LK Wang; NK Shammas; WA Selke; DB Aulenbach (Eds.), Flotation technology (pp. 327-345). New York, NY: Humana Press. http://dx.doi.org/10.1007/978-1-60327-133-2_9

Shanley, P; Collin, RL. (1961). The crystal structure of the high temperature form of choline chloride. Acta Cryst 14: 79-80. http://dx.doi.org/10.1107/S0365110X61000292

Sigma-Aldrich. (2007). Material safety data sheet: Tert-butyl hydroperoxide (70% solution in water). http://www.orcbs.msu.edu/msds/111607_DLI_027_TERT-BUTYL.PDF

Sigma-Aldrich. (2010). Product information: Sodium chloride. https://www.sigmaaldrich.com/content/dam/sigma-aldrich/docs/Sigma-Aldrich/Product_Information_Sheet/s7653pis.pdf

Sigma-Aldrich. (2014a). Material safety data sheet: Phosphorus acid. http://www.sigmaaldrich.com/catalog/product/sial/215112?lang=en&region=US

Sigma-Aldrich. (2014b). Material safety data sheet: Potassium carbonate. http://www.sigmaaldrich.com/catalog/product/aldrich/367877?lang=en&region=US

Sigma-Aldrich. (2015a). Material safety data sheet: Aluminum chloride [Fact Sheet]. St. Louis, MO. http://www.sigmaaldrich.com/catalog/product/aldrich/563919?lang=en&region=US

Sigma-Aldrich. (2015b). Material safety data sheet: Peracetic acid solution. http://www.sigmaaldrich.com/catalog/product/sial/269336?lang=en&region=US

Sigma-Aldrich. (2015c). Material safety data sheet: Sulfur dioxide. http://www.sigmaaldrich.com/catalog/product/aldrich/295698?lang=en&region=US

Sigma-Aldrich. (2015d). Material safety data sheet: Sulfuric acid. http://www.sigmaaldrich.com/catalog/product/aldrich/339741?lang=en&region=US

Sigma-Aldrich. (2015e). Material safety data sheet: Trimethyl borate. http://www.sigmaaldrich.com/catalog/product/aldrich/447218?lang=en&region=US

Šimunek, J; Šejna, M; van Genuchten, MT. (1998). The HYDRUS-1D software package for simulating the one-dimensional movement of water, heat, and multiple solutes in variably-saturated media, Version 2.0, IGWMC-TPS-70.

Sirivedhin, T; Dallbauman, L. (2004). Organic matrix in produced water from the Osage-Skiatook petroleum environmental research site, Osage county, Oklahoma. Chemosphere 57: 463-469.

Skjerven, T; Lunde, Ø; Perander, M; Williams, B; Farquhar, R; Sinet, J; Sæby, J; Haga, HB; Finnseth, Ø; Johnsen, S. (2011). Norwegian Oil and Gas Association recommended guidelines for well integrity. (117, Revision 4). Norway: Norwegian Oil and Gas Association.

Slutz, J; Anderson, J; Broderick, R; Horner, P. (2012). Key shale gas water management strategies: An economic assessment tool. Presented at International Conference on Health, Safety and Environment in Oil and Gas Exploration and Production, September 11-13, 2012, Perth, Australia. http://www.onepetro.org/mslib/app/Preview.do?paperNumber=SPE-157532-MS&societyCode=SPE

Smirnov, VI; Badelin, VG. (2013). Enthalpy characteristics of dissolution of L-tryptophan in water plus formamides binary solvents at 298.15 K. Russian Journal of Physical Chemistry A, Focus on Chemistry 87: 1165-1169. http://dx.doi.org/10.1134/S0036024413070285

Solley, WB; Pierce, RR; Perlman, HA. (1998). Estimated use of water in the United States in 1995. (USGS Circular: 1200). U.S. Geological Survey. http://pubs.er.usgs.gov/publication/cir1200

Spaulding, R. (2015) A quantitative assessment of atmospherically generated foam cements: Insights, impacts, and implications of wellbore integrity and stability. (Master's Thesis). University of Pittsburgh, Pittsburgh, PA. http://d-scholarship.pitt.edu/26018/

SPE (Society of Petroleum Engineers). (2016). Directional drilling. http://petrowiki.org/Directional_drilling

Spellman, FR. (2012). Chapter 7. Chemicals Used in Hydraulic Fracturing. In Environmental impacts of hydraulic fracturing. Boca Raton, Florida: CRC Press.

SRBC (Susquehanna River Basin Commission). (2016). Water use associated with natural gas development: An assessment of activities managed by the Susquehanna river basin commission July 2008 - December 2013. (Publication No. 299). Harrisburg, PA. http://www.srbc.net/pubinfo/techdocs/NaturalGasReport/docs/SRBC_Full_Gas_Report_fs_306397v1_20160408.pdf

States, S; Cyprych, G; Stoner, M; Wydra, F; Kuchta, J; Monnell, J; Casson, L. (2013). Marcellus Shale drilling and brominated THMs in Pittsburgh, Pa., drinking water. J Am Water Works Assoc 105: E432-E448. http://dx.doi.org/10.5942/jawwa.2013.105.0093

Stein, D; Griffin Jr, TJ; Dusterhoft, D. (2003). Cement pulsation reduces remedial cementing costs. GasTIPS 9: 22-24.

Steinhauser, O; Boresch, S; Bertagnolli, H. (1990). The effect of density variation on the structure of liquid hydrogen chloride. A Monte Carlo study. J Chem Phys 93: 2357-2363. http://dx.doi.org/10.1063/1.459015

Stepan, DJ; Shockey, RE; Kurz, BA; Kalenze, NS; Cowan, RM; Ziman, JJ; Harju, JA. (2010). Bakken water opportunities assessment: Phase I. (2010-EERC-04-03). Bismarck, ND: North Dakota Industrial Commission. http://www.nd.gov/ndic/ogrp/info/g-018-036-fi.pdf

Stewart, DR. (2013). Treatment for beneficial use of produced water and hydraulic fracturing flowback water. Presented at US EPA Technical Workshop on Wastewater Treatment and Related Modeling For Hydraulic Fracturing, April 18, 2013, Research Triangle Park, NC. http://www2.epa.gov/sites/production/files/documents/stewart_0.pdf

Struchtemeyer, CG; Elshahed, MS. (2012). Bacterial communities associated with hydraulic fracturing fluids in thermogenic natural gas wells in North Central Texas, USA. FEMS Microbiol Ecol 81: 13-25. http://dx.doi.org/10.1111/j.1574-6941.2011.01196.x

Stumm, W; Morgan, JJ. (1981). Aquatic chemistry: An introduction emphasizing chemical equilibria in natural waters (2nd ed.). New York, NY: Wiley.

Sturchio, NC; Banner, JL; Binz, CM; Heraty, LB; Musgrove, M. (2001). Radium geochemistry of ground waters in Paleozoic carbonate aquifers, midcontinent, USA. Appl Geochem 16: 109-122.

Swanson, VE. (1955). Uranium in marine black shales of the United States. In Contributions to the geology of uranium and thorium by the United States Geological Survey and Atomic Energy Commission for the United Nations International Conference on Peaceful Uses of Atomic Energy, Geneva, Switzerland, 1955 (pp. 451-456). Reston, VA: U.S. Geological Survey. http://pubs.usgs.gov/pp/0300/report.pdf

Syed, T; Cutler, T. (2010). Well integrity technical and regulatory considerations for CO2 injection wells. In 2010 SPE international conference on health, safety & environment in oil and gas exploration and production. Richardson, TX: Society of Petroleum Engineers. http://dx.doi.org/10.2118/125839-MS

Taylor, A. (2012). Watering the boom in Oklahoma: supplies, demands, and neighbors. Presented at 2012 Kansas Water Issues Forums, February 29-March 1, 2012, Wichita and Hays, Kansas.

Tchobanoglous, G; Burton, FL; Stensel, HD. (2013). Wastewater engineering: Treatment and reuse (4th ed.), (978-0070418783). Boston, MA: McGraw-Hill.

Thacker, JB; Carlton, DD, Jr; Hildenbrand, ZL; Kadjo, AF; Schug, KA. (2015). Chemical analysis of wastewater from unconventional drilling operations. Water 7: 1568-1579. http://dx.doi.org/10.3390/w7041568

Thalladi, VR; Nusse, M; Boese, R. (2000). The melting point alternation in alpha,omega-alkanedicarboxylic acids. J Am Chem Soc 122: 9227-9236. http://dx.doi.org/10.1021/ja0011459

Thurman, EM; Ferrer, I; Blotevogel, J; Borch, T. (2014). Analysis of hydraulic fracturing flowback and produced waters using accurate mass: Identification of ethoxylated surfactants. Anal Chem 86: 9653-9661. http://dx.doi.org/10.1021/ac502163k

Tian, Z; Shi, L; Qiao, L. (2015). Problems in the wellbore integrity of a shale gas horizontal well and corresponding countermeasures. Nat Gas Industry B 2: 522-529. http://dx.doi.org/10.1016/j.ngib.2015.12.006

Tiedeman, K; Yeh, S; Scanlon, BR; Teter, J; Mishra, GS. (2016). Recent trends in water use and production for California oil production. Environ Sci Technol 50: 7904-7912. http://dx.doi.org/10.1021/acs.est.6b01240

Tiemann, M; Folger, P; Carter, NT. (2014). Shale energy technology assessment: Current and emerging water practices. Washington, D.C.: Congressional Research Service. http://nationalaglawcenter.org/wp-content/uploads//assets/crs/R43635.pdf

Timmis, KN. (2010). Handbook of hydrocarbon and lipid microbiology. Berlin, Germany: Springer-Verlag. http://www.springer.com/life+sciences/microbiology/book/978-3-540-77584-3

Tourtelot, HA. (1979). Black shale - Its deposition and diagenesis. Clays and Clay Minerals 27: 313-321. http://dx.doi.org/10.1346/CCMN.1979.0270501

TWDB (Texas Water Development Board). (2012). Water for Texas 2012 state water plan. Austin, TX. http://www.twdb.state.tx.us/waterplanning/swp/2012/index.asp

U.S. Department of Transportation. (2012). Large truck and bus crash facts 2012. Washngton, D.C.: Federal Motor Carrier Safety Administration, U.S. Department of Transportation. http://ai.fmcsa.dot.gov/CarrierResearchResults/PDFs/LargeTruckandBusCrashFacts2012.pdf

U.S. DOI, Bureau of Reclamation (U.S. Department of the Interior, Bureau of Reclamation). (2016). Brackish desalination landing page. http://www.usbr.gov/research/AWT/brackish.html

U.S. EPA (U.S. Environmental Protection Agency). (1991). Manual of individual and non-public water supply systems [EPA Report]. (EPA 570/9-91-004). Washington, D.C.: U.S. Environmental Protection Agency, Office of Water. https://nepis.epa.gov/Exe/ZyPURL.cgi?Dockey=2000U9HN.txt

U.S. EPA (U.S. Environmental Protection Agency). (1996). Proposed guidelines for carcinogen risk assessment [EPA Report]. (EPA/600/P-92/003C). Washington, D.C.: U.S. Environmental Protection Agency, Risk Assessment Forum.

U.S. EPA (U.S. Environmental Protection Agency). (1999a). Guidelines for carcinogen risk assessment [review draft] [EPA Report]. (NCEA-F-0644). Washington, D.C.: U.S. Environmental Protection Agency, Office of the Science Advisor. https://ofmpub.epa.gov/eims/eimscomm.getfile?p_download_id=437005

U.S. EPA (U.S. Environmental Protection Agency). (1999b). Understanding oil spills and oil spill response [EPA Report]. (EPA 540-K-99-007). Washington, D.C.: U.S. Environmental Protection Agency, Office of Emergency and Remedial Response. http://www4.nau.edu/itep/waste/hazsubmap/docs/OilSpill/EPAUnderstandingOilSpillsAndOilSpillResponse1999.pdf

U.S. EPA (U.S. Environmental Protection Agency). (2004). Evaluation of impacts to underground sources of drinking water by hydraulic fracturing of coalbed methane reservoirs [EPA Report]. (EPA/816/R-04/003). Washington, D.C.: U.S. Environmental Protection Agency, Office of Solid Waste. https://nepis.epa.gov/Exe/ZyPURL.cgi?Dockey=P100A99N.txt

U.S. EPA (U.S. Environmental Protection Agency). (2005). Membrane filtration guidance manual. (EPA 815-R-06-009). Washington, D.C.: U.S. Environmental Protection Agency, Office of Water. https://nepis.epa.gov/Exe/ZyPURL.cgi?Dockey=901V0500.txt

U.S. EPA (U.S. Environmental Protection Agency). (2006). National Primary Drinking Water Regulations: Stage 2 Disinfectants and Disinfection Byproducts Rule. Washington, D.C.: U.S. Environmental Protection Agency, Office of Water. http://water.epa.gov/lawsregs/rulesregs/sdwa/stage2/

U.S. EPA (U.S. Environmental Protection Agency). (2007). Monitored natural attenuation of inorganic contaminants in ground water, Volume 1, Technical basis for assessment [EPA Report]. (EPA/600/R-07/139). Washington, D.C.: U.S. Environmental Protection Agency, Office of Research and Development. http://nepis.epa.gov/Adobe/PDF/60000N4K.pdf

U.S. EPA (U.S. Environmental Protection Agency). (2011a). Plan to study the potential impacts of hydraulic fracturing on drinking water resources [EPA Report]. (EPA/600/R-11/122). Washington, D.C.: U.S. Environmental Protection Agency, Office of Research and Development. http://www2.epa.gov/hfstudy/plan-study-potential-impacts-hydraulic-fracturing-drinking-water-resources-epa600r-11122

U.S. EPA (U.S. Environmental Protection Agency). (2011b). Sampling data for flowback and produced water provided to EPA by nine oil and gas well operators (non-confidential business information). http://www.regulations.gov/#!docketDetail;rpp=100;so=DESC;sb=docId;po=0;D=EPA-HQ-ORD-2010-0674

U.S. EPA (U.S. Environmental Protection Agency). (2011c). Terminology services (TS): Vocabulary catalog - IRIS glossary. http://ofmpub.epa.gov/sor_internet/registry/termreg/searchandretrieve/glossariesandkeywordlists/search.do?details=&glossaryName=IRIS%20Glossary (accessed May 21, 2015).

U.S. EPA (U.S. Environmental Protection Agency). (2012a). 2012 Edition of the drinking water standards and health advisories [EPA Report]. (EPA/822/S-12/001). Washington, D.C.: U.S. Environmental Protection Agency, Office of Water. http://www.epa.gov/sites/production/files/2015-09/documents/dwstandards2012.pdf

U.S. EPA (U.S. Environmental Protection Agency). (2012b). Estimation Programs Interface Suite for Microsoft Windows (EPI Suite) [Computer Program]. Washington D.C.: U.S. Environmental Protection Agency. https://www.epa.gov/tsca-screening-tools/epi-suitetm-estimation-program-interface

U.S. EPA (U.S. Environmental Protection Agency). (2012c). General definitions, 40 CFR. http://www.gpo.gov/fdsys/pkg/CFR-2012-title40-vol31/pdf/CFR-2012-title40-vol31-sec437-2.pdf

U.S. EPA (U.S. Environmental Protection Agency). (2012d). Quality Assurance Project Plan: Hydraulic Fracturing Retrospective Case Study, Bradford-Susquehanna Counties, Pennsylvania.

U.S. EPA (U.S. Environmental Protection Agency). (2012e). Study of the potential impacts of hydraulic fracturing on drinking water resources: Progress report. (EPA/601/R-12/011). Washington, D.C.: U.S. Environmental Protection Agency, Office of Research and Development. http://nepis.epa.gov/exe/ZyPURL.cgi?Dockey=P100FH8M.txt

U.S. EPA (U.S. Environmental Protection Agency). (2013a). Data received from oil and gas exploration and production companies, including hydraulic fracturing service companies 2011 to 2013. Non-confidential business information source documents are located in Federal Docket ID: EPA-HQ-ORD2010-0674. http://www.regulations.gov.

U.S. EPA (U.S. Environmental Protection Agency). (2013b). Distributed structure-searchable toxicity (DSSTOX) database network. http://www.epa.gov/ncct/dsstox/index.html

U.S. EPA (U.S. Environmental Protection Agency). (2013c). Fiscal year 2011: Drinking water and ground water statistics [EPA Report]. (EPA 816-R-13-003). Washington, D.C.: U.S. Environmental Protection Agency, Office of Water.

U.S. EPA (U.S. Environmental Protection Agency). (2013d). Terminology services (TS): Terms and acronyms. http://iaspub.epa.gov/sor_internet/registry/termreg/searchandretrieve/termsandacronyms/search.do

U.S. EPA (U.S. Environmental Protection Agency). (2014a). Announcement of preliminary regulatory determinations for contaminants on the third drinking water contaminant candidate list. U.S. Environmental Protection Agency, Office of Water. https://www.federalregister.gov/articles/2014/10/20/2014-24582/announcement-of-preliminary-regulatory-determinations-for-contaminants-on-the-third-drinking-water#page-62715

U.S. EPA (U.S. Environmental Protection Agency). (2014b). Development of rapid radiochemical method for gross alpha and gross beta activity concentration in flowback and produced waters from hydraulic fracturing operations [EPA Report]. (EPA/600/R-14/107). Washington, D.C.: U.S. Environmental Protection Agency. http://www2.epa.gov/hfstudy/development-rapid-radiochemical-method-gross-alpha-and-gross-beta-activity-concentration

U.S. EPA (U.S. Environmental Protection Agency). (2014c). Drinking water contaminants. http://water.epa.gov/drink/contaminants/

U.S. EPA (U.S. Environmental Protection Agency). (2014d). Retrospective case study in northeastern Pennsylvania: Study of the potential impacts of hydraulic fracturing on drinking water resources [EPA Report]. (EPA 600/R-14/088). Washington, D.C.: U.S. Environmental Protection Agency. http://www2.epa.gov/hfstudy/retrospective-case-study-northeastern-pennsylvania

U.S. EPA (U.S. Environmental Protection Agency). (2014e). Substance registry services [Database]. Washington, D.C.: U.S. Environmental Protection Agency. https://ofmpub.epa.gov/sor_internet/registry/substreg/searchandretrieve/substancesearch/search.do

U.S. EPA (U.S. Environmental Protection Agency). (2014f). The verification of a method for detecting and quantifying diethylene glycol, triethylene glycol, tetraethylene glycol, 2-butoxyethanol and 2-methoxyethanol in ground and surface waters [EPA Report]. (EPA/600/R-14/008). Washington, D.C.: U.S. Environmental Protection Agency, Office of Research and Development. http://www2.epa.gov/hfstudy/verification-method-detecting-and-quantifying-diethylene-glycol-triethylene-glycol

U.S. EPA (U.S. Environmental Protection Agency). (2015a). Analysis of hydraulic fracturing fluid data from the FracFocus chemical disclosure registry 1.0 [EPA Report]. (EPA/601/R-14/003). Washington, D.C.: U.S. Environmental Protection Agency, Office of Research and Development. http://www2.epa.gov/hfstudy/analysis-hydraulic-fracturing-fluid-data-fracfocus-chemical-disclosure-registry-1-pdf

U.S. EPA (U.S. Environmental Protection Agency). (2015b). Analysis of hydraulic fracturing fluid data from the FracFocus chemical disclosure registry 1.0: Data management and quality assessment report [EPA Report]. (EPA/601/R-14/006). Washington, D.C.: U.S. Environmental Protection Agency, Office of Research and Development. http://www2.epa.gov/sites/production/files/2015-03/documents/fracfocus_data_management_report_final_032015_508.pdf

U.S. EPA (U.S. Environmental Protection Agency). (2015c). Analysis of hydraulic fracturing fluid data from the FracFocus chemical disclosure registry 1.0: Project database [EPA Report]. (EPA/601/R-14/003). Washington, D.C.: U.S. Environmental Protection Agency, Office of Research and Development. http://www2.epa.gov/hfstudy/epa-project-database-developed-fracfocus-1-disclosures

U.S. EPA (U.S. Environmental Protection Agency). (2015d). Case study analysis of the impacts of water acquisition for hydraulic fracturing on local water availability [EPA Report]. (EPA/600/R-14/179). Washington, D.C.: U.S. Environmental Protection Agency, Office of Research and Development. https://www.epa.gov/sites/production/files/2015-07/documents/hf_water_acquisition_report_final_6-3-15_508_km.pdf

U.S. EPA (U.S. Environmental Protection Agency). (2015e). EPA Enforcement and Compliance History. Online: Effluent Charts: SEECO-Judsonia Water Reuse Recycling Facility. http://echo.epa.gov/effluent-charts#AR0052051

U.S. EPA (U.S. Environmental Protection Agency). (2015f). Retrospective case study in Killdeer, North Dakota: Study of the potential impacts of hydraulic fracturing on drinking water resources [EPA Report]. (EPA 600/R-14/103). Washington, D.C.: U.S. Environmental Protection Agency. http://www2.epa.gov/hfstudy/retrospective-case-study-killdeer-north-dakota

U.S. EPA (U.S. Environmental Protection Agency). (2015g). Retrospective case study in southwestern Pennsylvania: Study of the potential impacts of hydraulic fracturing on drinking water resources [EPA Report]. (EPA 600/R-14/084). Washington, D.C.: U.S. Environmental Protection Agency. http://www2.epa.gov/hfstudy/retrospective-case-study-southwestern-pennsylvania

U.S. EPA (U.S. Environmental Protection Agency). (2015h). Retrospective case study in the Raton Basin, Colorado: Study of the potential impacts of hydraulic fracturing on drinking water resources [EPA Report]. (EPA 600/R-14/091). Washington, D.C.: U.S. Environmental Protection Agency. http://www2.epa.gov/hfstudy/retrospective-case-study-raton-basin-colorado

U.S. EPA (U.S. Environmental Protection Agency). (2015i). Retrospective case study in Wise County, Texas: Study of the potential impacts of hydraulic fracturing on drinking water resources [EPA Report]. (EPA 600/R-14/090). Washington, D.C.: U.S. Environmental Protection Agency. http://www2.epa.gov/hfstudy/retrospective-case-study-wise-county-texas

U.S. EPA (U.S. Environmental Protection Agency). (2015j). Review of state and industry spill data: Characterization of hydraulic fracturing-related spills [EPA Report]. (EPA/601/R-14/001). Washington, D.C.: U.S. Environmental Protection Agenyc: Office of Research and Development. http://www2.epa.gov/hfstudy/review-state-and-industry-spill-data-characterization-hydraulic-fracturing-related-spills-1

U.S. EPA (U.S. Environmental Protection Agency). (2015k). Review of well operator files for hydraulically fractured oil and gas production wells: Well design and construction [EPA Report]. (EPA/601/R-14/002). Washington, D.C.: U.S. Environmental Protection Agency: Office of Research and Development. http://www2.epa.gov/hfstudy/review-well-operator-files-hydraulically-fractured-oil-and-gas-production-wells-well-design

U.S. EPA (U.S. Environmental Protection Agency). (2015l). Sources contributing bromide and inorganic species to drinking water intakes on the Allegheny River in western Pennsylvania [EPA Report]. (EPA/600/R-14/430). Washington, D.C.: U.S. Environmental Protection Agency. https://www.epa.gov/hfstudy/sources-contributing-inorganic-species-drinking-water-intakes-during-low-flow-conditions

U.S. EPA (U.S. Environmental Protection Agency). (2015m). Technical development document for proposed effluent limitation guidelines and standards for oil and gas extraction. (EPA-821-R-15-003). Washington, D.C.: U.S. Environmental Protection Agency. http://water.epa.gov/scitech/wastetech/guide/oilandgas/unconv.cfm

U.S. EPA (U.S. Environmental Protection Agency). (2016a). Review of well operator files for hydraulically fractured oil and gas production wells: Hydraulic fracturing operations [EPA Report]. (EPA/601/R-14/004). Washington, D.C.: U.S. Environmental Protection Agency, Office of Research and Development. https://www.epa.gov/sites/production/files/2016-07/documents/wfr2_final_07-28-16_508.pdf

U.S. EPA (U.S. Environmental Protection Agency). (2016b). Technical development document for the effluent limitations guidelines and standards for the oil and gas extraction point source category [EPA Report]. (EPA-820-R-16-003). Washington , DC: U.S. Environmental Protection Agency, Office of Water.

U.S. GAO (U.S. Government Accountability Office). (2014). Freshwater: Supply concerns continue, and uncertainties complicate planning. Report to Congressional requesters. (GAO-14-430). Washington, DC: U.S. Government Accountability Office (GAO). http://www.gao.gov/assets/670/663343.pdf

USGS (U.S. Geological Survey). (1961). Geology and geochemistry of uranium in marine black shales: A review. (U.S. Geological Survey Professional Paper 356-C). Reston, VA. http://pubs.usgs.gov/pp/0356c/report.pdf

USGS (U.S. Geological Survey). (1997). Radioactive elements in coal and fly ash: Abundance, forms, and environmental significance [Fact Sheet]. (U.S. Geological Survey Fact Sheet FS-163-97). http://pubs.usgs.gov/fs/1997/fs163-97/FS-163-97.pdf

USGS (U.S. Geological Survey). (2002). Water quality and environmental isotopic analyses of ground-water samples collected from the Wasatch and Fort Union formations in areas of coalbed methane developmentimplications to recharge and groundwater flow, eastern Powder River basin, Wyoming. (Report 02-4045). Reston, VA. http://pubs.usgs.gov/wri/wri024045/

USGS (U.S. Geological Survey). (2014). USGS investigations of water produced during hydrocarbon reservoir development [Fact Sheet]. Reston, VA. http://dx.doi.org/10.3133/fs20143104

USGS (U.S. Geological Survey). (2015). Water use in the United States. http://water.usgs.gov/watuse/

Vaidyanathan, G. (2014). Email communications between Gayathri Vaidyanathan and Ken Klewicki regarding the New Mexico Oil Conservation Division District 3 well communication data.

Van Voast, WA. (2003). Geochemical signature of formation waters associated with coalbed methane. AAPG Bulletin 87: 667-676.

Vanengelen, MR; Peyton, BM; Mormile, MR; Pinkart, HC. (2008). Fe(III), Cr(VI), and Fe(III) mediated Cr(VI) reduction in alkaline media using a Halomonas isolate from Soap Lake, Washington. Biodegradation 19: 841-850. http://dx.doi.org/10.1007/s10532-008-9187-1

Veil, JA; Puder, MG; Elock, D; Redweik, RJ. (2004). A white paper describing produced water from production of crude oil, natural gas, and coal bed methane. Morgantown, WV: Department of Energy (DOE), National Energy Technology Laboratory (NETL). http://www.ipd.anl.gov/anlpubs/2004/02/49109.pdf

Veil, JA. (2010). Water management technologies used by Marcellus shale gas producers - Final Report. (DOE Award No.: FWP 49462). http://fracfocus.org/sites/default/files/publications/water_management_in_the_marcellus.pdf

Vengosh, A; Jackson, RB; Warner, N; Darrah, TH; Kondash, A. (2014). A critical review of the risks to water resources from unconventional shale gas development and hydraulic fracturing in the United States. Environ Sci Technol 48: 36-52. http://dx.doi.org/10.1021/es405118y

Verdegem, MCJ; Bosma, RH. (2009). Water withdrawal for brackish and inland aquaculture, and options to produce more fish in ponds with present water use. Water Policy 11: 52-68. http://dx.doi.org/10.2166/wp.2009.003

Vidic, RD; Brantley, SL; Vandenbossche, JM; Yoxtheimer, D; Abad, JD. (2013). Impact of shale gas development on regional water quality [Review]. Science 340: 1235009. http://dx.doi.org/10.1126/science.1235009

Vijaya Kumar, R; Anand Rao, M; Venkateshwara Rao, M; Ravi Kumar, YVL; Prasad, DHL. (1996). Bubble temperature measurements on 2-propyn-1-ol with 1,2-dichloroethane, 1,1,1-trichloroethane, and 1,1,2,2-tetrachloroethane. Journal of Chemical and Engineering Data 41: 1020-1023. http://dx.doi.org/10.1021/je9600156

Vine, JD. (1956). Uranium-bearing coal in the United States. In Contributions to the geology of uranium and thorium by the United States Geological Survey and Atomic Energy Commission for the United Nations International Conference on Peaceful Uses of Atomic Energy, Geneva, Switzerland, 1955. Reston, VA: U.S. Geological Survey. http://pubs.er.usgs.gov/publication/pp300

Vine, JD; Tourtelot, EB. (1970). Geochemistry of black shale deposits: A summary report. Econ Geol 65: 253-272. http://dx.doi.org/10.2113/gsecongeo.65.3.253

Vinson, DS; Vengosh, A; Hirschfeld, D; Dwyer, GS. (2009). Relationships between radium and radon occurrence and hydrochemistry in fresh groundwater from fractured crystalline rocks, North Carolina (USA). Chem Geol 260: 159-171. http://dx.doi.org/10.1016/j.chemgeo.2008.10.022

Walsh, JM. (2013). Water management for hydraulic fracturing in unconventional resources: Part 1. Oil and Gas Facilities 2. https://www.spe.org/en/ogf/ogf-article-detail/?art=422

Wambaugh, JF; Setzer, RW; Reif, DM; Gangwal, S; Mitchell-Blackwood, J; Arnot, JA; Joliet, O; Frame, A; Rabinowitz, J; Knudsen, TB; Judson, RS; Egeghy, P; Vallero, D; Cohen Hubal, EA. (2013). High-throughput models for exposure-based chemical prioritization in the ExpoCast project. Environ Sci Technol 47: 8479-8488. http://dx.doi.org/10.1021/es400482g

Warner, NR; Christie, CA; Jackson, RB; Vengosh, A. (2013a). Impacts of shale gas wastewater disposal on water quality in western Pennsylvania. Environ Sci Technol 47: 11849-11857. http://dx.doi.org/10.1021/es402165b

Warner, NR; Jackson, RB; Darrah, TH; Osborn, SG; Down, A; Zhao, K; White, A; Vengosh, A. (2012). Reply to Engelder: Potential for fluid migration from the Marcellus formation remains possible. Proc Natl Acad Sci USA 109: E3626-E3626. http://dx.doi.org/10.1073/pnas.1217974110

Warner, NR; Kresse, TM; Hays, PD; Down, A; Karr, JD; Jackson, RB; Vengosh, A. (2013b). Geochemical and isotopic variations in shallow groundwater in areas of the Fayetteville Shale development, north-central Arkansas. Appl Geochem 35: 207-220.

Wasylishen, R; Fulton, S. (2012). Reuse of flowback & produced water for hydraulic fracturing in tight oil. Calgary, Alberta, Canada: The Petroleum Technology Alliance Canada (PTAC). http://www.ptac.org/projects/151

Watson, TL; Bachu, S. (2009). Evaluation of the potential for gas and CO2 leakage along wellbores. S P E Drilling & Completion 24: 115-126. http://dx.doi.org/10.2118/106817-PA

Weaver, JW; Xu, J; Mravik, SC. (2016). Scenario analysis of the impact on drinking water intakes from bromide in the discharge of treated oil and gas wastewater. J Environ Eng 142: 04015050. http://dx.doi.org/10.1061/(ASCE)EE.1943-7870.0000968

Weaver, TR; Frape, SK; Cherry, JA. (1995). Recent cross-formational fluid flow and mixing in the shallow Michigan Basin. Geol Soc Am Bulletin 107: 697-707. http://dx.doi.org/10.1130/0016-7606(1995)107<0697:RCFFFA>2.3.CO;2

Webb, CH; Nagghappan, L; Smart, G; Hoblitzell, J; Franks, R. (2009). Desalination of oilfield-produced water at the San Ardo water reclamation facility, CA. Presented at SPE Western regional meeting, 24-26 March, 2009, San Jose, CA: Society of Petroleum Engineers. http://dx.doi.org/10.2118/121520-MS

Webster, IT; Hancock, GJ; Murray, AS. (1995). Modelling the effect of salinity on radium desorption from sediments. Geochim Cosmo Act 59: 2469-2476. http://dx.doi.org/10.1016/0016-7037(95)00141-7

Wetmore, BA; Wambaugh, JF; Allen, B; Ferguson, SS; Sochaski, MA; Setzer, RW; Houck, KA; Strope, CL; Cantwell, K; Judson, RS; Lecluyse, E; Clewell, HJ; Thomas, RS; Andersen, ME. (2015). Incorporating high-throughput exposure predictions with dosimetry-adjusted in vitro bioactivity to inform chemical toxicity testing. Toxicol Sci 148: 121-136. http://dx.doi.org/10.1093/toxsci/kfv171

Wetmore, BA; Wambaugh, JF; Ferguson, SS; Sochaski, MA; Rotroff, DM; Freeman, K; Clewell, HJ; Dix, DJ; Andersen, ME; Houck, KA; Allen, B; Judson, RS; Singh, R; Kavlock, RJ; Richard, AM; Thomas, RS. (2012). Integration of dosimetry, exposure, and high-throughput screening data in chemical toxicity assessment. Toxicol Sci 125: 157-174. http://dx.doi.org/10.1093/toxsci/kfr254

White, GJ. (1992). Naturally occurring radioactive materials (NORM) in oil and gas industry equipment and wastes: A literature review. (DOE/ID/01570-T158). Bartlesville, OK: U.S. Department of Energy.

WHO (World Health Organization). (2015). Concise international chemical assessment documents. http://www.who.int/ipcs/publications/cicad/en/

Wignall, PG; Myers, KJ. (1988). Interpreting benthic oxygen levels in mudrocks: A new approach. Geology 16: 452-455. http://dx.doi.org/10.1130/0091-7613(1988)016<0452:IBOLIM>2.3.CO;2

Wilt, JW. (1956). Notes - the halodecarboxylation of cyanoacetic acid. J Org Chem 21: 920-921. http://dx.doi.org/10.1021/jo01114a607

Wojtanowicz, AK. (2008). Environmental control of well integrity. In ST Orszulik (Ed.), Environmental technology in the oil industry (pp. 53-75). Houten, Netherlands: Springer Netherlands.

Woolard, CR; Irvine, RL. (1995). Treatment of of hypersaline wastewater in the sequencing batch reactor. Water Res 29: 1159-1168.

Wuchter, C; Banning, E; Mincer, TJ; Drenzek, NJ; Coolen, MJ. (2013). Microbial diversity and methanogenic activity of Antrim Shale formation waters from recently fractured wells. FMICB 4: 1-14. http://dx.doi.org/10.3389/fmicb.2013.00367

WVWRI (West Virginia Water Research Institute, West Virginia University). (2012). Zero discharge water management for horizontal shale gas well development. (DE-FE0001466). https://www.netl.doe.gov/File Library/Research/Oil-Gas/Natural Gas/shale gas/fe0001466-final-report.pdf

WYOGCC (Wyoming Oil and Gas Conservation Commission). (2014). Pavillion Field Well Integrity Review. Casper, Wyoming. http://wogcc.state.wy.us/pavillionworkinggrp/PAVILLION REPORT 1082014 Final Report.pdf

Xiao, LN; Xu, JN; Hu, YY; Wang, LM; Wang, Y; Ding, H; Cui, XB; Xu, JQ. (2013). Synthesis and characterizations of the first [V16O39Cl]6- (V16O39) polyanion. Dalton Transactions (Online) 42: 5247-5251. http://dx.doi.org/10.1039/c3dt33081h

Yakimov, MM; Denaro, R; Genovese, M; Cappello, S; D'Auria, G; Chernikova, TN; Timmis, KN; Golyshin, PN; Giuliano, L. (2005). Natural microbial diversity in superficial sediments of Milazzo Harbor (Sicily) and community successions during microcosm enrichment with various hydrocarbons. Environ Microbiol 7: 1426-1441. http://dx.doi.org/10.1111/j.1462-5822.2005.00829.x

Yang, JS; Lee, JY; Baek, K; Kwon, TS; Choi, J. (2009). Extraction behavior of As, Pb, and Zn from mine tailings with acid and base solutions. J Hazard Mater 171: 1-3. http://dx.doi.org/10.1016/j.jhazmat.2009.06.021

Yang, X; Shang, C. (2004). Chlorination byproduct formation in the presence of humic acid, model nitrogenous organic compounds, ammonia, and bromide. Environ Sci Technol 38: 4995-5001. http://dx.doi.org/10.1021/es049580g

Yoshizawa, S; Wada, M; Kita-Tsukamoto, K; Ikemoto, E; Yokota, A; Kogure, K. (2009). Vibrio azureus sp. nov., a luminous marine bacterium isolated from seawater. Int J Syst Evol Microbiol 59: 1645-1649. http://dx.doi.org/10.1099/ijs.0.004283-0

Yoshizawa, S; Wada, M; Yokota, A; Kogure, K. (2010). Vibrio sagamiensis sp. nov., luminous marine bacteria isolated from sea water. J Gen Appl Microbiol 56: 499-507.

Yost, EE; Stanek, J; Burgoon, LD. (In Press) A decision analysis framework for estimating the potential hazards for drinking water resources of chemicals used in hydraulic fracturing fluids. Sci Total Environ. http://dx.doi.org/10.1016/j.scitotenv.2016.08.167

Yost, EE; Stanek, J; Dewoskin, RS; Burgoon, LD. (2016a). Estimating the potential toxicity of chemicals associated with hydraulic fracturing operations using quantitative structure-activity relationship modeling. Environ Sci Technol 50: 7732-7742. http://dx.doi.org/10.1021/acs.est.5b05327

Yost, EE; Stanek, J; Dewoskin, RS; Burgoon, LD. (2016b). Overview of chronic oral toxicity values for chemicals present in hydraulic fracturing fluids, flowback, and produced waters. Environ Sci Technol 50: 4788-4797. http://dx.doi.org/10.1021/acs.est.5b04645

Younos, T; Tulou, KE. (2005). Overview of desalination techniques. Journal of Contemporary Water Research & Education 132: 3-10. http://dx.doi.org/10.1111/j.1936-704X.2005.mp132001002.x

Zapecza, OS; Szabo, Z. (1988). Natural radioactivity in ground watera review. In National Water Summary 1986 Hydrologic Events and Ground-Water Quality, Water-Supply Paper 2325. Reston, VA: U.S. Geological Survey. http://pubs.er.usgs.gov/publication/wsp2325

Zeikus, JG; Hegge, PW; Thompson, TE; Phelps, TJ; Langworthy, TA. (1983). Isolation and description of Haloanaerobium praevalens gen. nov. and sp. nov., an obligately anaerobic halophile common to Great Salt Lake sediments. Curr Microbiol 9: 225-233. http://dx.doi.org/10.1007/BF01567586

Zhang, L; Guo, Y; Xiao, J; Gong, X; Fang, W. (2011). Density, refractive index, viscosity, and surface tension of binary mixtures of exo-tetrahydrodicyclopentadiene with some n-alkanes from (293.15 to 313.15) K. Journal of Chemical and Engineering Data 56: 4268-4273. http://dx.doi.org/10.1021/je200757a

Zhang, T; Gregory, K; Hammack, RW; Vidic, RD. (2014). Co-precipitation of radium with barium and strontium sulfate and its impact on the fate of radium during treatment of produced water from unconventional gas extraction. Environ Sci Technol 48: 4596-4603. http://dx.doi.org/10.1021/es405168b

Zhang, Z; Yang, L; Xing, Y; Li, W. (2013). Vapor-liquid equilibrium for ternary and binary mixtures of 2-isopropoxypropane, 2-propanol, and n,n-dimethylacetamide at 101.3 kPa. Journal of Chemical and Engineering Data 58: 357-363. http://dx.doi.org/10.1021/je300994y

Ziemkiewicz, P; Quaranta, JD; Mccawley, M. (2014). Practical measures for reducing the risk of environmental contamination in shale energy production. Environ Sci Process Impacts 16: 1692-1699. http://dx.doi.org/10.1039/c3em00510k

Ziemkiewicz, PF; He, YT. (2015). Evolution of water chemistry during Marcellus Shale gas development: A case study in West Virginia. Chemosphere 134: 224-231. http://dx.doi.org/10.1016/j.chemosphere.2015.04.040

Front cover (top): Illustrations of activities in the hydraulic fracturing water cycle.
From left to right: Water Acquisition, Chemical Mixing, Well Injection, Produced Water Handling, and Wastewater Disposal and Reuse.

Front cover (bottom): Aerial photographs of hydraulic fracturing activities.
Left: Near Williston, North Dakota. Image ©J Henry Fair / Flights provided by LightHawk.
Right: Springville Township, Pennsylvania. Image ©J Henry Fair / Flights provided by LightHawk.

Back cover: Top left: DOE/NETL. All other images courtesy of the U.S. EPA.

www.ingramcontent.com/pod-product-compliance
Lightning Source LLC
Chambersburg PA
CBHW081253170526
45165CB00011B/3301